AF527780

POLLUTION ABATEMENT AND CONTROL

S.N. Kaul
D.R. Saini
B.N.Rai
Prateek Kaul

2024

Daya Publishing House®
A Division of
Astral International Pvt. Ltd.
New Delhi – 110 002

Published by Daya Publishing House

Website: www.astralint.com

Printed at Replika Press Pvt. Ltd.

Preface

The past few years we have seen the emergence of a growing, widespread desire in this country and indeed everywhere, that positive action to be taken to restore the quality of our environment and to protect it from the degrading effects of all forms of pollution-air, water, noise, solid waste (Electronic, Nuclear etc). Since the seemingly idealistic demand for "zero discharge" can be construed as a demand for zero waste. However, as long as, there is waste, we can only attempt to abate the consequent pollution by converting it to a less noxious form. In those instances, in which a particular type of pollution has been recognized, three major questions usually arise; how serious is the pollution? Is the technology to abate it available? and Do the costs of abatement justify the degree of abatement achieved ? The principal objectives of this book on Pollution Abatement and Control are to help in formulating the useful answers to these questions and to examine the benefits that could accrue to the society form abatement of pollution; otherwise nature has its own plans.

Unplanned and haphazard waste disposal is a crime against humanity worldwide. Unplanned and uncontrolled urbanization has led to generation of large amount of solid and municipal waste. Uncontrolled dumping and disposal of wastes generated has led to pollution of air, land and water sources and also has significant impact on eco-systems and environment. The most important crisis that we are facing today is inequality and poverty in our immediate midst and growing climate insecurity in our world. Pollution due to urbanization and industrialization has global impact which can be seen in the form of irregular seasonal patterns. The problem of pollution and its impact gets aggravated due to lack of awareness about appropriate waste management technologies. Also these technologies should be low cost, sustainable and eco-friendly.

Pollution Abatement and Control focuses on issues relating to Environment in our quest for sustainable development. This book emphasizes on the following themes:

- ☆ Design and analysis of reactor system utilized in waste handling
- ☆ Bio-techniques for waste management
- ☆ Waste reuse and recycling
- ☆ Air and Water Environment related issues [Rising sea water level, Climate change, Groundwater, Surface Water, Rain Water harvesting, etc.]
- ☆ Land Environment related issues [Forest, Solid Waste, Nuclear Waste, e-waste, Bio-medical waste, etc.]
- ☆ Quality assessment and Control of waste
- ☆ Environmental policies and legal issues

The editors are pleased to acknowledge the efforts and time given by all contributors for their contributions in this book and trust that this book will be useful to researchers, practicing engineers, professionals students, teachers, NGOs, etc. working in the field of Environmental Science and Engineering. The editors would like to thank profusely to Mrs. Sushma Mutgi for her assistance and cooperation in the preparation of this book. The trend reposed by Daya Publishing House, a Division of the Astral International Pvt. Ltd., New Delhi is thankfully acknowledged by the editors.

Dr. S.N. Kaul

Dr. D.R. Saini

Dr. B.N. Rai

Er. Prateek Kaul

Contents

Chapter 1

Wastewater Treatment Plants: Application of Dimensional Analysis for Design Purposes

***S.N. Kaul*[1], *D.R. Saini*[2] *and P. Kaul*[3]**

[1]*MIT College of Engineering, Pune*

[2]*CSIR-NCL, Pune*

[3]*Symantec Corporation (USA), Pune*

ABSTRACT

Scale up equations have been developed for mechanical surface aerators [radial and pitch blade] and anaerobic fixed film fixed bed reactors treating tannery wastewater using the Buckingham π theory.

Keywords: *π Buckingham theory, Scale up, Aeration, Aerators, Fixed film, Fixed bed, Anaerobic.*

Introduction

Non-dimensionalization of an equation by inspectional analysis is useful only when one knows the equation to begin with. However, in many cases in real life engineering the equations are either not known or too difficult to solve; oftentimes experimentation is the only method of obtaining reliable information. In most experiments, to save time and money, tests are performed on a geometrically scaled model, rather than on the full scale prototype. In such cases, care must be taken to properly scale the results. We introduce here a powerful technique called dimensional analysis. While typically taught to fluid mechanics, dimensional analysis is useful in all disciplines, especially when it is necessary to design and conduct experiments.

You are encouraged to use this powerful tool in other subjects as well, not just in fluid mechanics. The three primary purposes of dimensional analysis are:

- ☆ To generate non-dimensional parameters that help in the design of experiments [physical and/or numerical] and in the reporting experimental results.
- ☆ To obtain scaling laws so that prototype performance can be predicted from model performance
- ☆ To [sometimes] predict trends in the relationship between parameters.

The technique of dimensional analysis, involves the underlying concept of dimensional analysis–the principle of similarity. There are three necessary conditions for complete similarity between a model and a prototype. The first condition is geometric similarity–the model must be the same shape as the prototype, but may be scaled by some constant scale factor. The second condition is kinematic similarity, which means that the velocity at any point in the model flow must be proportional.

The third and most restrictive similarity condition is that of the dynamic similarity. Dynamic similarity is achieved when all forces in the model flow scale by a constant factor to corresponding forces in the prototype flow.(force-scale equivalence).As with geometric and kinematic similarity, the scale factor for forces can be less than, equal to, or greater than one. Kinematic similarity is a necessary but insufficient condition for dynamic similarity. It is thus possible for a model flow and a prototype flow to achieve both geometric and kinematic similarity. All three similarity conditions must exist for complete similarity to be ensured.

In a general flow field, complete similarity between model and a prototype is achieved only when there is geometric, kinematic, and dynamic similarity.

Buckingham π theorem states that 'If' there are 'n' variables involved and if these variables contain 'm' primary dimensions [M, L, T], then the variables can be expressed in terms of $(n–m)$ dimensionless groups or parameters. This principle has been applied to determine the scale up equations for mechanical surface aerators [radial and pitch], and fixed film fixed bed anaerobic reactors for tannery wastewater.

1. Scale Up Formulations for Radial and Pitch Blade Surface Aerators

1.1. Experimental and System Analysis

Tables 1.1 and 1.2 describes the details of aerator size, blade size, number of blades, aeration basin size, and blade angles. The aerators were mounted on a platform in the centre of the aeration basin and driven by a 0.373 k W induction motor at speeds which were varied by a stepless speed controller in the range of 0.5-5.8 Hz.

Before any test runs were carried out, the dissolved oxygen in the test basin was reduced to virtually zero by addition of sodium sulphite using a cobalt salt as a catalyst. During re-aeration tests, samples of water were taken from various points in the basin, located so as to avoid excessive rates of increase in the dissolved oxygen, such as, for example, near to the points at which fluid flowed directly from the aerator and also from points of high and low turbulence. In the pilot-scale tests, samples were taken from points at the mid-water depth and the centre of the half diagonal between the centre of the shaft and the basin corner. The dissolved oxygen probe, incorporating a silver-lead electrode, was calibrated using Winkler's method and in the case of samples, the dissolved oxygen was fixed by adding $MnSO_4$ and alkaline amide solutions permitting analysis by a calorimetric technique at a later stage.

In carrying out the tests, the aeration basin was thoroughly cleansed and filled with tap water. After de-aeration had been completed, the dissolved oxygen was measured at various time intervals and the transfer coefficient $k_L\alpha$.

Table 1.1: Details of Model Surface Aerators Tested.

Aerator Diameter (D) (m)	*Blade Size*	*Number of Blades (n)*	*Blade Angle (Ø)*	*Aerator Basin Size (mxmxm)*
[A]	**Radial Surface Aerator**		Perpendicular	
0.102	D/4 x D/4), (D/4 x D/6), (D/6 x D/6)	4,6,8,12 and 16	to top horizontal plates.	[0.46x0.46x1.07]; [0.76x0.76x1.07];
0.152	(D/6xD/4),(D/8 x D/8), (D/6 x D/12)	–do–		[1.52x1.52x 1.22] For all aerator
0.229	(D/12 x D/6) for all aerator	–do–		diameters
0.305	(D/12 x D/6) for all aerator	–do–		
[B]	**Pitch Blade Surface Aerators**			
0.1524	D/3, D/4, D/6, D/9, D/12	2,3,6,8,12 and 16	10,20,30,45, 60 and 90	[0.45x0.45x0.91]; [0.91x0.91x0.91];
0.2286	–do–	–do–	–do–	[1.67x1.67x0.91]
0.3048	–do–	–do–	–do–	For all aerator
0.3810	–do–	–do–	–do–	diameters

Table 1.2: Basic Variables for Oxygen Transfer through Aerators.

Variables	*Symbols*	*Units [FLT Systems]*
Dia of Surface aerators	D_i	L
Size of aeration basin	T	L
Water depth	H	L
Submergence level	h	L
Kinematic Viscosity	ν	$L^2 T^{-1}$
Gravity constant	g	$L T^{-2}$
Molecular diffusivity	D_m	$L^2 T^{-2}$
Oxygen solubility	C_S	$F T^2 L^{-4}$
Oxygenation capacity	OC	$FT L^{-1}$
No. of blades	n	–
Rotational speed	N	T^{-1}
Water Density	ρ	$F T^2 L^{-4}$
Blade Angle	Ø	–

K_La was determined as slope of a plot of the oxygen deficit $[C_S-C]$ and time. Only points between 10 and 80 per cent of saturation were considered.

At a particular operating temperature, T, the transfer coefficient, is given by

$$K_La(T)=[\ln(C_S-C_1)-\ln(C_S-C_2)]/(t_2-t_1) \quad (1)$$

Where C_1 and C_2 are the oxygen concentrations at times t_1 and t_2.

A correction to 20° C is given by

$$K_La\ (20) = K_La\ (T)/1.024^{(T-20)} \quad (2)$$

The mass of oxygen dissolving in unit time under standard conditions of tap water at 20 degree centigrade with zero dissolved oxygen and at 1 atm is known as oxygenation capacity, given by

$$OC = K_La\ (20)\ CsV,\ kgh^{-1} \quad (3)$$

And oxygen efficiency by

$$OE = OC/P,\ kg\ kWh^{-1} \quad (4)$$

Where, P (kW) is the power consumed by the aerator. In processing test data, C_S at 20°C was taken as 9.1 mgL^{-1} and OC and OE were based on averaged K_La values in the tank. By plotting Reynolds number against speed of rotation, it was possible to obtain operating conditions equivalent to a constant Reynolds number and also a constant Froudes number.

1.2 Radial Blade Surface Aerator

As may be expected, the oxygenation capacity increased with higher rotational speeds (N) and also with the diameter (D) of the aerator. As the submergence of the aerator was increased, oxygenation capacity at first increased from effectively zero to some 13 kgh^{-1} at $h/H = 0.2$ but then fell to comparatively low levels. Up to the peak value, the oxygenation capacity was directly proportional to h with all three pilot scale tanks. A similar effect was observed with number of blades and, using eight as the standard, a value used in previous works, the oxygenation capacity was found to be proportional to $(n/n_8)^{0.2}$ the actual value depending on the speed of rotation. The oxygenation capacity was found to depend on the tank size and to be proportional to $(T/D)^{0.05.}$ When the depth of water was in excess of 2D, the oxygenation capacity decreased and most of the tests were carried out at H = 2D. The effects of both N and D on oxygenation were evaluated in terms of the Reynolds and Froude numbers and it was concluded that variations in the oxygenation capacity due to these groups was the result of the entrainment of the oxygen in the hydraulic jump, oxygen absorption from air bubble sand surface turbulences. The oxygenation capacity was found to be proportional to $Re^{1.9}$ and, at constant Reynolds number, oxygenation increased with increase in Froude number, presumably because the hydraulic jump formation creates small air bubbles and also increases the air water interfacial area. All the experimental data were correlated by the equation

$$OC=5.0(DCsDm)Re^{1.9}\ Fr^{0.15}x\ (D/H)^{0.02}x(h/H)\ (n/n_8)^{0.2}x(T/D)^{0.05}\ x(10^3);(kgO_2+/hr.) \quad (5)$$

The group OC/((DCs Dm) may regarded as an oxygenation capacity number which represents the ratio of mass transfer by convection + diffusion to that by diffusion alone and is analogous to Sherwood number, $K_L D/Dm$. With full scale trials the oxygenation capacity predicted by equation (5) was somewhat (10 per cent) higher than the experimental data. The full scale data were correlated using an equation proposed by Emde [2] modified as follows:

$$OC = 0.1\ [1+0.84h]\ D^{2.6}\ V_p^{\ 2.6} \quad (6)$$

Where V_p is peripheral speed (m/s). This may be compared with the data on typical commercial aerators, and correlated at optimal conditions by:

$$OC = 0.7\ D^6;\ lbh^{-1} \quad (7)$$

$$OC = 56\ Q^{1.43} \quad (8)$$

Where D(ft) and Q (m^3/min.) are the impeller diameter and volumetric flow rate respectively.

The relationship between diameter and rotational speed under optimal conditions based on oxygenation efficiency data is best expressed by,:

$$DN^{1.5} = 555 \quad (9)$$

In order to determine the oxygen efficiencies the power consumed by the aerators was required and in full scale trials, power values were obtained from Watt hour meter and using performance curves obtained as described by Kaul and Raman allowing for resistance loses and mechanical losses in the reduction gears and belts. In the pilot scale work the net power was determined by using Cleasby and Beaumann method. The power requirements were determined as functions of submergence, speed of rotation and blade size and data were correlated by equation:

$$(p/\rho N^3 D^5) = 6.2\ Fr^{-x}\ (n/n_8)^{\ 0.45}\ (T/D)^{\ 0.27}(h/H)^y \quad (10)$$

Where x = 0.43 or 0.50 depending on the speed of rotation and y = 1.0 when, H/h > 12 and 1.38 when H/h < 12.

The oxygenation efficiency increased with rotational speed to a peak at N = 150 rpm and then decrease and similar trend was observed on full scale trials. The optimum submergence was indicated, usually at H/h = 10 and oxygen efficiency increased with number of blades until (n/n_8) =1 and then remained more or less constant. When H>2D, the oxygenation efficiency decreased, presumably because air bubble do not reach the bottom of the tank, and released to atmosphere. With increase in tank size T the oxygenation efficiency decreased linearly with increase in T/H, and optimum conditions were found to be T/D = 5.0, H = 2D with bottom, velocity of 0.15 m/s. The oxygenation efficiency generally increased with blade size up to [D/6 x D/6], increased from 1.7 to 3.5 kg/O_2/kW h with specific power input to 67 W/m^3 when, H = 2D. The specific power should be > 15 Watts/m^3, in order that the bottom velocity of 0.15 m/s are attained, the value at which sludge solids are kept in suspension.

Floor velocities were found to increase, as may be expected, with and the submergence level all with resulting increases in specific power levels in aeration vessels. Some tests were carried out to determine the velocities in presence of 1 to 3 mg/L of surface active agent. In general this reduced the velocity at a given point from 40 to30 cm/s, presumably due to greater number of smaller bubbles being produced resulting in reduced surface tension causing density currents. These results appear to agree with the data of Kalbskopff, who worked with BSK type aerators.

Also,

$$P= 0.417\, D^{2.6}\, V_p^{\,2.4;}(kW) \tag{11}$$

and the data are found to compare favourably with that obtained from commercial units at optimal conditions which are correlated by:

$$P = 0.24\, D^{2.6}; (hp) \tag{12}$$

In this equation, D(ft) is the impeller diameter.

1.3 Pitch Blade Surface Aerator

The oxygenation capacity increases with increase in the rotational speed (N), Diameter of the aerator (D), and aeration basin size. As the submergence level (h/H) of the aerator was increased the oxygenation capacity first increased up to h/H = 0.18 and after that began to decline with increase in submergence load. The oxygenation varies as $(h/H)^{0.90}$. This has been found true with all three pilot scale tanks. The similar effect was observed with number of blades and, using 4 as the reference value. The oxygenation capacity was determined to be proportional to $(n/n_4)^{0.45}$. With increase in blade angle (θ) the oxygenation capacity was found to increase and was determined to be proportional to $(\text{Sin}\ \theta)^{0.45}$.

The increases in oxygenation capacity by increasing any of these variables are perhaps due to following three mechanisms. But it may be that all of them are operating simultaneously;

- ✰ Entertainment of oxygen in hydraulic jump
- ✰ Oxygen absorption by air bubble
- ✰ Oxygen absorption due to surface turbulence.

The oxygen capacity was found to depend on the tank size and proportional to $(T/D)^{0.04}$. Increasing the depth of the water beyond 1.8 D, oxygenation capacity decreased and most of the tests were carried out at H = 1.8 D.

The effects both N and D on oxygenation capacity were evaluated in terms of Reynolds and Froude numbers. The oxygenation capacity was found to be proportional to $(Re)^x$; x = 1.7 for speed > optimum speed and 1.8 for speed < optimum speed and at constant Reynolds number, oxygenation capacity increased with increase in Froude number, presumably because hydraulic jump formation creates small air bubble and also increases air water interfacial area. Oxygenation capacity was found be proportional to $(Fr)^{0.10}$. All the experimental data were correlated by:

$$OC=13.5(DC_sDm)Re \times Fr^{0.17}(h/H)^{0.9}(n/n_4)^{0.4} \times (T/D)^{0.04}(D/H)^{0.19} (\sin\theta)^{0.45}(10)^{-3}[kgO_2/hr] \tag{12}$$

The group (OC/DC_sD_m) may be regarded as an oxygenation capacity number which represents convection plus diffusion, rather in comparison and is analogous to the Sherwood number (k_LD/D_m). The oxygenation capacity was found to vary with the blade size. Only the proportionality constant was found to vary. The rate of increase of the proportionality constant was higher for blade size D/12 to D/6 and after that there was only marginal increase. The relationship between surface aerator diameter and speed has been determined under optimum conditions based on oxygenation efficiency and is given by:

$$DN^{1.49} = 601 \tag{13}$$

The oxygenation capacity increased with increase in water discharge rate of aerator. Under optimum conditions the relationship between oxygenation capacity and volumetric flow rate (Q) is given by:

$$OC = 62Q^{1.44} \tag{14}$$

Net power was estimated using the method described by Cleasby, Beaumann[4] and Kaul [3]. The power requirements were determined as functions of submergence, speed of rotation, blade number and other parameters and the data were correlated by:

$$P/\rho N^3D^5 = K_p(F_r)^{-x}(h/H)^y \times (n/n_4)^z \times (T/D)^{0.176} \times (\sin\theta)^m \tag{15}$$

Where K_p = 0.90 or 3.8 and x = 0.475 or 0.635, depending on the speed of rotation; y = 0.84 or 0.90, depending on submergence; z = 0.45 or 0.27, depending on the number of blades and m = 0.45 or 0.40, depending on the blade angle. The efficiency (OE) of the model surface aerator was found to increase with increase in rotational speed until it reached peak value and after that it remained constant. Similar behaviour was noticed for number of blades n/n_4, blade angle (θ) and submergence level (h/H). The values of these parameters under optimum conditions based on oxygenation efficiency are given as:

$$DN^{1.49} = 601 \tag{16}$$

$$\eta = 4$$

$$\Theta = 45^\circ$$

$$h/H = 0.1$$

Where H> 1.8D, the oxygenation efficiency decreased, presumably because air bubbles do not reach the bottom of the tank and are released to the atmosphere without any significant addition of oxygen. With increase in the tank size, the oxygenation efficiency decreased linearly with increase in T/H and optimum conditions were found to be T/D = 6.0, H = 1.8D with a velocity of 0.15 m/s. The specific power was > 15W/m^3 resulting in bottom velocities of greater than 0.15 m/s at which biological sludge solids are kept in suspension.

2. Anaerobic Fixed Film Fixed Bed Reactor

The basic variables involved in the determination of scale-up equations for an anaerobic fixed film fixed bed reactor are given in Table 1.3. The generalized scale-up equations involve values of various parameters such as diffusivity across the bio-film layer, kinematic viscosity of wastewater, half velocity saturation constant, growth rate constant, density, etc.

Table 1.3: Basic Variables for Development of Scale-up Equations.

Sl.No.	*Basic Variables*	*Symbol*	*Units inMLT*
1.	Influent subtrate concentration	S_i	ML^{-3}
2.	Attached biomass concentration	X	ML^{-3}
3.	Substrate reaction rate constant	K	T^{-1}
4.	Hydraulic detention period	θ	T
5.	Yield coefficient	Y	MM^{-1}
6.	Media particle size	d_p	L
7.	Diffusivity	D	L^2T^{-1}
8.	Kinematic viscosity	ν	L^2T^{-1}
9.	Maximum specific growth rate	μmax	T^{-1}
10.	Wastewater density	e	ML^{-3}
11.	Half substrate velocity constant	Ks	ML^{-3}
12.	Solids retention time	$Ø_c$	T^{-1}
13.	Diameter of reactor	D_i	L
14.	Media height of reactor	h	L
15.	Biomass decay co-efficient	K_d	T^{-1}

The Three generalized scale-up equations were developed each accounting for one important biological aspect, *viz.* substrate utilization, micro-organism growth and biogas generation. Depending upon the shape of the reactor rectangular square or circular, the terms relating to physical configuration can be changed to involve length, breadth, media height and diameter of reactor. The important dimensionless groups developed are: Reynolds's number, Thele number, peclet number, etc. The three general scale-up equations are presented in Table 1.4. These three equations are coupled. Each equation consists of four components, *viz.* hydraulics, micro-organism growth, substrate utilization and physical configuration of reactor. Close observation of biomass and biogas generation equation reveal that specific biomass yield M/ $Q(S_{i–}S_e)$, specific methane yield M /$Q(S_i–S_e)$ are by and large constant and do not vary much under normal conditions of optimum reactor performance. From the past experience also it has been observed that the values for specific biomass and biogas yields are small and, therefore, for all practical purposes, these are considered to be constant for specific type of substrate and reactor. This information leads to simplification of the procedure of scaling up the anaerobic fixed film fixed bed reactors and these scale-up equations are finally reduced to only one equation relating to substrate conversion.

It is not necessary to develop scale up equations for all types of reactors, particularly those processes for which enough experience and expertise are available. Laboratory treat ability studies, coupled with determination of kinetic constant, would

Table 1.4: Generalized Scale-up Equation for Anaerobic Fixed Film Fixed Bed Reactor.

SUBSTRATE CONVERSION

$$\frac{Si - S}{K\theta\ Si} = K_1 \underbrace{N_P^{a}\ N_R^{b}}_{I}\ \underbrace{N_T c\ (X/Ks)^{d} (\mu_{max}\ dp/V)^{e} (\theta_c\ V/dp)^{f} (K_d\ dp/V)^{g} (Kd/\mu_{max})^{h}}_{II}\ \underbrace{(K\ dp/V)^{i}}_{III}\ \underbrace{(Di/dp)^{j} (h/dp)^{k}}_{IV}$$

BIOMASS PRODUCTION

$$\frac{M}{Q\ (Si\text{-}S)} = K_2 \underbrace{N_P^{a'} N_R^{b'}}_{I}\ \underbrace{N_T^{c'} (X/Ks)^{d'} (\mu_{max}\ dp/V)^{e'} (\theta_c V/dp)^{f_1'} (K_d\ dp/V)^{g'} (K_d/\mu_{max})^{h'}}_{II}\ \underbrace{(K\ dp/V)^{i'}}_{III}\ \underbrace{(Di/dp)^{j'} (h/dp)^{k'}}_{IV}$$

BIOGAS GENERATION

$$\frac{M'}{Q\ (Si\text{-}S)} = K_3 \underbrace{N_P^{a''} N_R^{b''}}_{I}\ \underbrace{N_T^{c''} (X/Ks)^{d''} (\mu_{max}\ dp/V)^{e''} (\theta_c V/dp)^{f''} (K_d\ dp/V)^{g''} (K_d/\mu_{max})^{h''}}_{II}\ \underbrace{(K\ dp/V)^{i''}}_{III}\ \underbrace{(Di/dp)^{j''} (h/dp)^{k''}}_{IV}$$

Peclet Number, $N_P = V\ dp/D$; Reynold Number, $N_R = V\ dp/\nu$; Thiele Number, $N_T = dp\ \sqrt{\mu_{max}/Dç\ \ Ks}$

Constants - K_1, K_2, K_3; Exponents - a,b,c,.....,k ; a',b',c',....,k' ; a",b",c",....,k"

I - Hydraulic properties II - Microorganisms properties III - Substrate properties IV - Reactor configuration

M and M' refer to rate of biomass yield and biogas generation respectively

suffice for design of large wastewater treatment facility directly from bench scale units. No intermediate scale units are necessary. However, a similar situation does not exist for anaerobic fixed film fixed bed reactors. Scanty information is available for substrate, biomass and biogas kinetics for various types of detailed experimentation to formulate the scale up anaerobic fixed film bioreactors.

Studies were undertaken to formulate scale up equations for an anaerobic fixed film fixed bed reactor system. Important aspects of anaerobic biological treatment, *viz.* substrate utilization, biomass production and biogas generation, were used to determine various parameters relating to physical configuration of the reactors such as media depth, reactor diameter, and values or some important parameters such as particle size, etc. [Table 1.5].

These generalized equations have been developed, each accounting for one biological aspect, ie, substrate utilization, biomass and bio gas generation. Depending upon the shape of the reactor, rectangular square or circular, the last terms relating to

Table 1.5: Development of Modified Scale-up Equation for Anaerobic Fixed Film Fixed Bed Reactor.

$$\frac{Si-Se}{Si} K.\theta = K_1 \left[\frac{D}{\nu} . \frac{\mu_{max}}{Kd} . \frac{X}{Ks} . \frac{ç}{Si} . Y.K.\theta\right]^a \left[\frac{Di}{ht}\right]^b \left[\frac{ht}{dt}\right]^c \left[\frac{h}{ht}\right]^d$$

$$\frac{Si - Se}{Si} K.\theta = K_1 \left[\underbrace{\frac{D}{\nu} . ç . \theta}_{I} \quad \underbrace{\frac{\mu_{max} . X.Y}{Kd}}_{II} \quad \underbrace{\frac{K}{Si.Ks}}_{III}\right]^a \underbrace{\left[\frac{Di}{ht}\right]^b \left[\frac{ht}{dp}\right]^c \left[\frac{h}{ht}\right]^d}_{IV}$$

I - Hydraulics; **II** - Microorganisms; **III** - Substrate; **IV** - Physical configuration

$$\frac{Si - Se}{Si} = \frac{K_1}{K} \left[\frac{D. .\mu_{max}.X.Y.K}{\nu. Kd. Ks}\right]^a \frac{(\theta)^{a-1}}{Si^a} \frac{Di^b. h^d}{dp^c(ht)^{b+d-c}}$$

$$= \frac{K_1. K_2}{K.(ht)^{b+d-c}} . \frac{(\theta)^{a-1}}{si^a} . \frac{Di^b}{dp^c} . h^d$$

$$\text{where, } K_2 = \left[\frac{D . ç .\mu_{max} . X.Y.K}{\nu. \quad K_d . K_s}\right]^a$$

θ can be expressed as $\frac{Vv}{Q} = \frac{p \times \pi/4 \times Di^2 \times h}{Q}$ where, Vv = void volume
p = voids ratio

$$= \frac{K_1.K_2}{K.(ht)^{b+d-c}} . \left[\frac{p . \pi .Di^2. h}{4.Q}\right]^{a-1} . \frac{Di^b. h^d}{dp^c. Si^a}$$

$$= \frac{K_1.K_2}{K.(ht)^{b+d-c}} . \left[\frac{p\pi}{4}\right]^{a-1} . \frac{Di^{2(a-1)}. Di^b. h^{a-1}. h^d}{Q^a. dp^c. Si^a} . Q$$

$$\eta = K_3 . \frac{Di^{2a+b-2} . h^{a+d-1}}{(QSi)^a (dp)^c} . Q$$

$$\text{where } K_3 = \frac{K_1 . K_2}{K. (ht)^{b+d-c}} . \left[\frac{p\pi}{4}\right]^{a-1}$$

or

$$Si - Se = K_3 \frac{Di^{2a+b-2}. h^{a+d-1}}{(Qsi)^{a-1} . dp^c}$$

physical configuration can be changed to involve, length, breadth and media depth calculated for each equation pertaining to one aspect and whichever gives the maximum value that is to be adopted for design purposes.

The scale-up equations developed is mainly to use laboratory scale data to design corresponding pilot plant or full scale similar reactors.

2.1 Modified Scale-up Equation

The generalized scale-up equations based on substrate conversions are again modified and reconstructed for further simplification and to each the specific case of anaerobic fixed film fixed bed reactor, particularly cylindrical type. Total variables to be used in the equation are: substrate concentration (S_i), hydraulic retention time (HRT, θ), bio reactor diameter (D_i), media depth (D_p), and media depth (h). As the system is fixed film type without solid recycle, it was difficult to control the solids detention time (θ_c) and, therefore, θ_c is particularly eliminated for mathematical simplification purposes.

$$\text{Velocity (V)} = \text{Flow (Q)}/\text{A} \qquad (17)$$

Where, A is the net opening area at any cross section of reactor.

$$Q = \frac{\textit{Voids volume } (V_v)}{\theta} \qquad (18)$$

Voids volume and not the gross reactor volume of the media is used for calculation of HRT(θ). Assuming that the porosity is same throughout the media height or at any cross-section in the reactor, the cross-sectional area of the openings in the media is the same, implying

$$V_V = \text{(area) ht} \qquad (19)$$

Where, ht is the total media height.

Therefore,

$$\text{Velocity (V)} = V_v/(\text{Area} \ \ \theta) \qquad (20)$$

In the present investigation, scale-up equation is developed keeping substrate concentration (S_i) as constant. Therefore, the scale-up equation developed is valid for S_i considered in the equation. Three equations are developed to cover the wide range of S_i that is generally found to be valid in tannery wastewaters.

$$\eta = \frac{\text{Constant } D_i^{(2a+b\ 2)}\, h^{(a+d-1)}}{(QS_i)^a\,(dp)^c} . Q \qquad (21)$$

Where a, b, c and d are exponents

Q- Influent flow to the reactor (m^3/day)

η = efficiency of substrate COD removal

In any design problem, Q, S_i, and desired efficiency η are known. Media particle size (d_p) is predicted based on the media selection studies in the laboratory. D_i and h are the only unknowns in the equation. To calculate bioreactor diameter (D_i), media length (h) can be assessed based on the experience and topography of the area, reactor diameter D_i can be predetermined to facilitate the calculation of bioreactor height (h).

The above scale-up equations have been developed for standard conditions of pH–7.0, temperature - 35° C. The rate constants have been developed and the effects of pH and temperature have been determined for tannery wastewater and are presented in Table 1.6.

Table 1.6a: Characteristics of Combined Tannery Wastewater Used for Laboratories Studies.

Sl.No.	*Parameter*	*Magnitude*		
		Concentration of Tannery Waste		
1.	COD	1800	3600	6500
2.	pH	7.5	7.2	7.0
3.	Alkalinity	660	1200	1760
4.	Volatile Acids	88	136	317
5.	Ammonical nitrogen	103	106	109
6.	Total solids	11071	12731	15480
7.	Total volatile solids	2772	4107	6318
8.	Total suspended solids	35	35	40
9.	TOC	748	1517	2476
10.	Sulphate	657	845	2116
11.	Chloride	1758	2061	2614
12.	Sulphide	12	15	23
13.	Tannin	24	45	91
14.	Total phosphate	14	17	22
15.	Total nitrogen	140	144	150
16.	Chromium	14	16.5	22

All values are expressed in mg/L except pH.

Table 1.6b: Details of Laboratory Scale Anaerobic Fixed Film Fixed Bed Reactor.

Reactor No.	*Reactor Diameter, (D_1, mm*	*Media Size, (d_p), mm*	*Media Height, (ht), cm*
I	150	32	85
II	150	25	85
III	150	12	85
IV	150	6	85
V	100	6	85
VI	50	6	85

2.2 Experimental Set-up

The characteristics of combined tannery wastewater for experimental studies are given in Table 1.6a. Two fixed bed reactors were operated in series. The configuration and media of both the bioreactors are same. The physical configuration, media size and media height are presented in Table 1.6b. The selection of media (refractory brick) is based on extensive laboratory studies and is referred elsewhere. Hydraulic retention times (HRT) were varied from 0.5 to 05.0 days. Temperature and pH were varied from 20° to 45°C and 6.0 to 8.0 respectively in order to determine temperature and pH dependency upon substrate removal rate constant, specific growth rate and decay coefficient for micro-organisms. Table 1.7.

Table 1.7: Experimental Values of Various Important Parameters at Temperature 35°C and pH-7.

Sl.No.	*Parameters*	*Magnitude*	
1	**Properties of Wastewater**		
	–Viscosity, cm^2/s	0.007251	
	–Diffusivity, cm^2/s	1,273 x 10^{-5}	
	–Density, kg/m^3	1010	
2.	**Kinetic Constants and other related parameters (Temp. 35°C; pH-7)**		
		Reactor I	*Reactor II*
	–Substrate reaction rate (K), day^{-1}	2.1812	1.2846
	–Maximum specific growth rate ($_{max}$), day^{-1}	0.274	0.178
	–Half velocity saturation constant, mg/L	100	75
	–Yield coefficient (Y)	0.0506	0.045
	–Decay coefficient (kd), day^{-1}	0.032	0.021
	–Biomass concentration in bio-film (X), mg/L	9000	7500
	–COD removal, per cent	66.58	42.91
3.	**Biogas**		
	–Total methane generation, $m^3CH_4/m^3.d$	1.5	0.6
	–Methane yield, $m^3CH_4/kgCOD_r$	0.30	0.32
	–Methane content, percent	62	64

2.3 Application of Modified Scale-up Equation

The scale up equations for various influent substrate concentrations are developed for first anaerobic reactor. These are expressed as:

Influent substrate Concentration: 6500 mg/L

$$(S_i - S_c) = \frac{2.218 \, . \, D_i^{0.7214} \, h^{0.8466}}{(QS_i)^{0.3318}} \tag{22}$$

Influent substrate Concentration: 3600 mg/L

$$(S_i - S_c) = \frac{1.0352 \,.\, D_i^{0.362}\, h^{0.6827}}{(QS_i)^{0.1513}} \tag{23}$$

Influent substrate concentration; 1800mg/L

$$(S_i - S_c) = \frac{0.5395 \,.\, D_i^{0.3075}\, h^{0.6577}}{(QS_i)^{0.1235}} \tag{24}$$

These scale-up equations are developed on the basis of bench scale studies. Performances data for bioreactor-II has not been used in these equations. Unlike bioreactor-I for which substrate concentration (S_i) can be kept constant, S_i is a dependent variable for bio reactor-II and cannot be controlled or kept constant if the reactors are to be operated in series. Because of these inherent problems (for series type reactors), an indirect method was evolved to determine reactor dimensions for reactor-II which is summarized as follows:

- Calculate diameter and height for any given conditions of flow (Q), influent substrate concentration (80, media size dp), and removal efficiency ("I") using any of the above referred equations.
- Calculate volumetric loading rate for reactor-I in terms of kgCODr/m^3.d
- Determine loading rate reactor-II in terms of kg COD/d based on laboratory scale data (Figure 1.1).
- The effluent substrate calculation is influent substrate concentration for reactor-II and is, therefore known the wastewater flow rate (Q) for reactor-II in series with reactor-I will be same and is, therefore known. The desired removal efficiency for reactor-II is known before hands calculate the loading rate of kg CODr/d for reactor-II.

Conclusion(s)

It has been shown in this investigation that with both radial and pitch blade surface aerator, the oxygenation capacity is enhanced by increases in rotational speed, tank size, the number of blades, the aerator diameter and submergence level and optimum speeds and submergence have been indicated. In case of rotational speed, the optimum condition was given by

DN^c = constant

and the optimum submergence was obtained when the entire blade height immersed in the water.

In general, it may be said that a general approach to the problem of oxygenation capacity, the discharge rate of the aerator and mixing time based on the use of dimensionless groups provides a powerful technique for dealing with a very complex situation in aerational manner, though here it has not been possible to obtain a general data correlation which is independent of the specific physical design features

X-Axis = $kgCOD_r/m^3.d$ for reactor I Y Axis = $kgCOD_r/m^3.d$ for reactor II.

Si for Column–I: 6500 mg/L

Temp: 35 degree C

Media size: 32 mm

Column dia: 150 mm

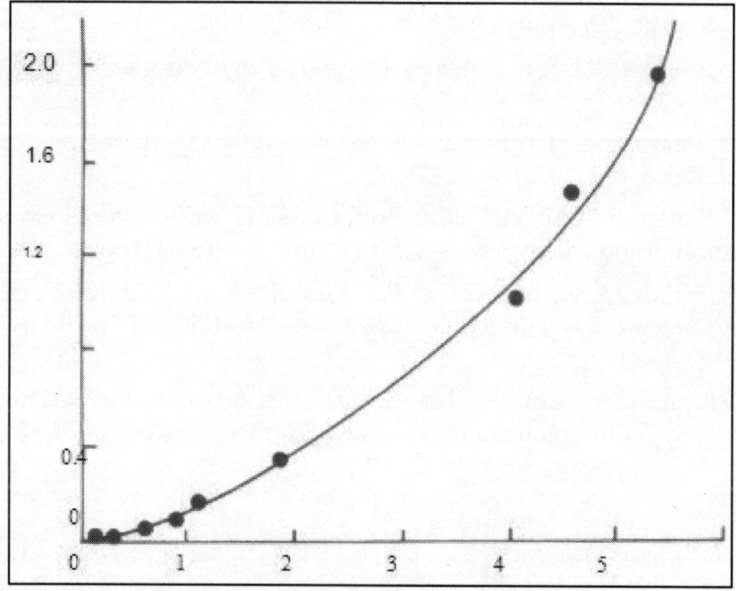

Figure 1.1: Loading Rate of Reactor I vs. Loading Rate of Reactor II.

of the mixing equipment because the flow regime depends on almost every property and dimension of the fluid and agitating system. Empirical correlations of the data, specific to the systems investigated here have been developed however, and it will be of interest to ascertain whether or not these can be applied, in a possibly modified form, to other designs of aerator, data on which are lacking. One further aspect which would appear to be of immediate interest is the effect of two or more surface aerators operating simultaneously and it is considered that data on such a system could give an useful insight into interaction phenomenon in particular and scale-up of aeration systems in general.

The final design equation based on dimension analysis for anaerobic fixed film fixed bed reactor is given by:

$$\eta = \frac{\text{Constant } D_i^{(2a+b\ 2)}\ h^{(a+d-1)}}{(QS_i)^a\ (dp)^c} . Q$$

References

1. W.W. Eckenfelder, Engineering aspects of surface aerator design. 22nd Purdue University Conference, Purdve (1967).
2. V.D. Emde, Entwurf Bau and Betiet von Abwassereingungasantga Wierer Mitterlvn gen. Wass. Abswas. Gewass. Halft 4, Wien (1969).
3. S.N.Kaul and V. Raman, Ind.J. Envir. Hlth 18, 169 (1976).
4. J.L. Cleasby and E.R. Beaumann, "Oxygen efficiency of bladed rotor;, JWPCF 40, 412 (1972).
5. K.H. Kalbskopf, " Flow velocities in aeration tanks with mechanical aerators;, Wat. Res. 6, 413 (1972).
6. S.N Kaul and T. Nandi, Treatment of tannery wastewaters with recourse to chromium and energy recovery, NEERI Internal Report (1994).
7. L.Szpyrkovicz, N. Jeremi and F.Z. Grandi, Application of electrochemical oxidation processes for tanning wastewater treatment, Toxicol Environ. Chem 44, 183-2-2 (1994).
8. S.N.Kaul and S. Shanta, Studies related to support matrix for fixed film fixed bed reactors, V. Environmental Science and Engineering (A-21), 133-170 (1994).

Chapter 2

Application of Fluidized Bed Reactor for Recovery of Chemicals from Pickling Wastewater

S.N. Kaul[1] and T. Nandy[2]

[1]*Principal, MIT College of Engineering, Pune*

[2]*Scientist, NEERI, Nagpur*

ABSTRACT

This chapter represents the application of full-scale fluidized bed technology with recourse to recovery of value-aided chemicals from the spent liquor generated from steel pickling processes in the steel manufacturing industry, achieving Zero Effluent Discharge [ZED]. Optimization of the process variables of a full-scale fluidized bed reactor for the recovery of hydrochloric acid [HCl] and iron oxide [Fe_2O_3] are discussed.

Keywords: *Pickling: Waste pickle liquor, Hazardous, Fluidized bed, Resource recovery, Hydrochloric acid, Iron oxide.*

Introduction

Pickling is the process of chemically removing oxides and scales from the surface of hot-rolled steel by the action of water solutions of inorganic acids. Depending on the type of product being processed, various acids such as sulfuric, hydrochloric, nitric, hydrofluoric, and phosphoric may be utilized either individually or in combination, in both batch and continuous processes. Until 1950, sulfuric acid accounted for more than 90 per cent of the pickling acid: however, in recent years an increasing number of steel mills have shifted to hydrochloric acid offers several advantages over sulfuric acid as follows: [Rennhack 1974; Azad 1976; Negro *et al.*, 1994]; [a] produces a more even and cleaner surface, [b] requires lower liquid temperatures, and [c] can be regenerated.

The pickling process includes a series of hydrochloric acid solutions at varying concentration and temperatures, and hot rolled steel is passed through it. This process removes fine scale, and bright the clean surface is produced. Iron oxide dissolved in acid are present in the form of insoluble chlorides, and spent acid is discarded which forms wastewater stream commonly referred to as "Waste Pickle Liquor". The pickled steel is then rinsed with water by sequential immersion in rinse tanks to remove all acid and dissolved iron. The rinse water is also discarded as wastewater alters certain washings. The following equations illustrate the chemical reactions occurring during the pickling process.

$$Fe_2O_3 + 6\,HCl \longrightarrow 2\,FeCl_3 + 3H_2O \quad (1)$$

$$FeO + 2\,HCl \longrightarrow FeCl_2 + H_2O \quad (2)$$

$$2\,FeCl_3 + H_2 \longrightarrow 2\,FeCl_3 + 2HCl \quad (3)$$

Waste pickle liquor from the pickling bath contains free hydrochloric acid and iron chlorides, which are considered as a source of environmental pollution. Around 0.13Kg. of the pickle liquor is generated per kg. Of steel processed. Because of their pollutant content and corrosiveness, these liquids appear in the list of hazardous wastes in all legislations [Negro et. Al, 1994]. The discharge of untreated waste pickle liquors into surface water causes water pollution due to its high chloride and iron contents, besides lowering the pH of the receiving stream. Dissolved oxygen of water is also lowered due to hydrolysis of iron salts to iron hydroxides. Discharge into land results in building up of iron and chlorides in soil, making it unfertile, and further polluting the groundwater.

To combat environmental pollution due to discharge of pickling liquors, two main options are available, which include end-of-the-pipe treatment, and recovery of value-added chemicals leading to zero effluent discharge.

End-of-the-pipe treatment of waste pickle liquor consists of neutralization, and/or deep-well injection. Neutralization of waste pickle liquor using lime and aeration is the most commonly adopted treatment process in various steel industries. Ferrous iron in the spent acid gets precipitated as ferrous hydroxide during neutralization, and subsequently converted to ferric hydroxide during aeration. Ferric hydroxide sludge separated is then dewatered through a vacuum filter or sludge-drying beds, where it is oxidized during drying. The neutralized wastewater is evaporated in open solar evaporation ponds. Usually, a large land area is required for solar drying of the neutralized wastewater. Also, an increase in chloride concentration attenuates the evaporation process. Removal of accumuln pressure lated salts causes problems in its disposal due to its hygroscopic nature. The overflow of wastewater from ponds may also contaminate the ground and surface water during monsoons (Gurnham, 1965; Kumar *et al.*, 1998: Kumar *et al.*, 1999; Nandita *et al.*, 2000).

Deep-well injection of waste pickle liquor is another method of treatment and disposal of pickling liquors. This method has now become obsolete due to severe environmental problems associated with it. Continuous disposal of pickle liquor causes sealing of permeable strata of the wells by deposition of reaction products, thereby reducing the capacity of the wells. The possibility of serious pollution of groundwater also cannot be ruled out [Rudolfs, 1997].

Resource recovery techniques that include roasting, pressure oxidation, ion exchange, and oxi-precipitation processes are also available for recovery of value-added chemicals from waste pickle liquors [Negro *et al.*, 1994].

The iron oxide obtained in roasting processes can be considered as a high quality raw material due to its high purity with a chloride content of 0.05-0.1 per cent and can be used in different industries, for example, as components for ferromagnetic materials [ferrites and high power magnets], as raw material for the production of iron dust [electrodes, use in chemical sector, etc.] and as an additive in the manufacture of different products [magnetic strips, colorings, abrasives, bricks, glass, cosmetics, etc.] In pressure oxidation processes, the capital and operational costs ae very high, and the purity of regenerated acid is only 18-20 per cent. Ion exchange process does not extract the ferrous iron in recoverable form, and environmental pollution problems are not fully solved. The oxi-precipitation process can only be used for liquors when the pickling agent has been recovered. This process obtains similar iron elimination rate than other processes, and yields an iron by product for its use, without further transformation, in other industries like electronic, painting, etc. [Negro *et al.*, 1994].

This paper highlights the application of a full-scale fluidized bed reactor achieving Zero Effluent Discharge [ZED] with recourse to recovery of value-added chemicals [hydrochloric acid and iron oxide] from the waste pickle liquor and optimizing the operational parameters.

Materials and Methods

Application of a full-scale fluidized bed reactor system was assessed and a steel coils, to achieve zero effluent kg/d of hot-rolled steel coils, to achieve zero effluent discharge of waste pickle liquor with recourse-to-resource recovery of value-added chemicals. The reactor system was optimized through evaluation of the reactor performance by varying the operating parameters, which included the reactor temperature and the waste pickle liquor feed rate, and also the optimization of air velocity for fluidization of the reactor bed.

Details of Fluidized Bed Reactor System

The full-scale fluidized bed reactor system understudy consisted of the following unit processes.

1. Fluidized bed reactor of size 2 m diameter and 2.5 m height.
2. Cyclone, including dust re-circulation system
3. Venturi, in which hot waste gases are cooled down from approximately 850° to 100° C. During this process, the normal pickling acid is strongly concentrated due to evaporation of water.

4. Separator, in which the waste gases coining from the venturi are separated from the venturi liquid.
5. Absorber, including scrubber, in which the hydrochloric acid is recovered by absorption of gaseous HCl in water.
6. Fume stack, which sucks off the gases developed in the reactor and keeps the entire system below atmospheric pressure. The residual inert is led through the stack in to the atmosphere.

Figure 2.1 shows the schematic of the full-scale fluidized bed reactor system under study, and the details are presented elsewhere [NEERI Technical Report, 2000].

Process Reaction

The fluidized bed media consisted of iron oxide particles, which are kept in a turbulent state by the combustion gas in the reactor; the spent liquor is pumped via a venturi concentration circuit. The high temperature maintained in the fluidized bed reactor results in the conversion of $FeCl_2$ into Fe_2O_3 and HCl. Thus, the reverse of the pickling reaction takes place in the reactor as follows:

$$2\,FeCl_2 + 2\,H_2O + ½O_2 \longrightarrow Fe_2O_3 + 4HCl \qquad [4]$$

About 95 per cent of the reaction takes place on the surface of oxide particles and only 5 per cent in the gap. While the in oxide reacting on the surface of the oxide particles contributes to the growth of oxide particles, the iron oxide formed in the gas phase generates fine iron oxide dust, which is carried out of the reactor with the combustion gases, in the cyclone, this dust is again removed from the gas and sent back to the reactor. The iron oxide formed is constantly drawn off by means of a vibrating cooling trough. This facilitates maintaining the constant height of the fluidized bed in the reactor.

The hot reaction gases from the reactor are cooled in the venturi to approximately 100°C. The finest dust particles carried with the gases are captured at the same time making use of dissipated heart to evaporate the spent liquor. A split stream is taken out of this venturi circulation system and conveyed to the reactor.

The gas current passes to the absorber and the hydrochloric gas is absorbed adiabatic ally by water in a reversible flow, achieving a solution 16-17 per cent hydrochloric acid.

Experimental Evaluation

Waste pickle liquor generated from steel processing was analyzed for pH, acidity, total iron, chlorides, free hydrochloric acid, and other parameters as per Standard.

Similarly, rinse water from the process was also analyzed for various parameters. Table 2.1 presents the physico-chemical characteristics of waste pickle liquor and rinse water used for the experimental study.

Propane gas was used as the fuel, and air was used as the oxygen supplement and also to maintain the bed in a fluidized condition. The propane gas requirement was in the range of 49.8–52.8m^3/h.

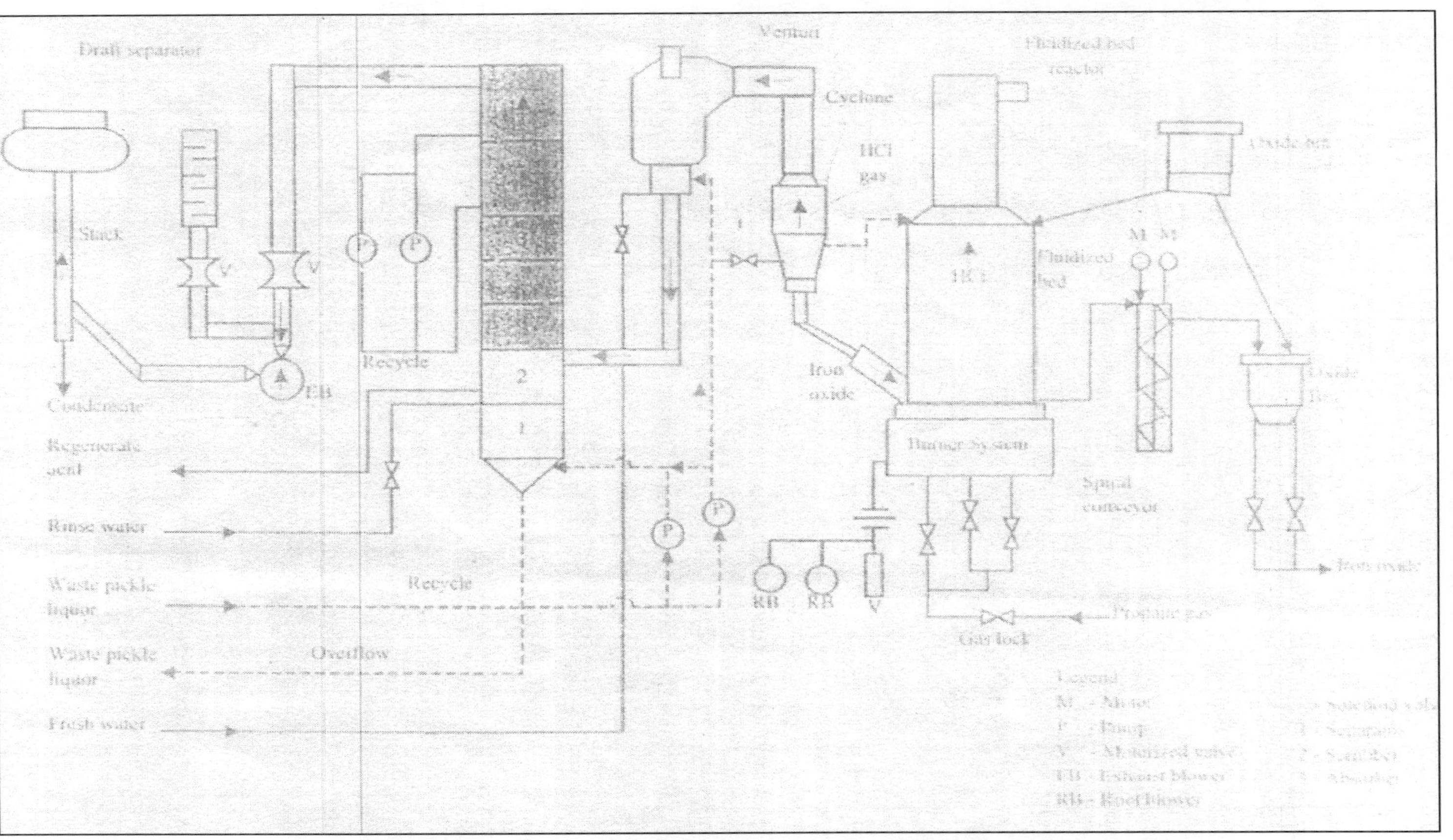

Figure 2.1: Schematic of the Full Scale Fluidized Bed Reactor System.

Table 2.1: Physico-chemical Characteristics of Waste Pickle Liquor and Rinse Water.

Parameters	*Concentration**	
	Waste pickle liquor	*Rinse water*
pH	<1	1 - 2
Acidity	21,000–25,000	15,000–20,000
Chlorides	88,000–100,000	42,000–50,000
Free HCl	3.3–4.2	0.6–0.9
Total iron	1,50,000–1,80,000	37,500–40,000
Total dissolved solids	90,000–110,000	45,000–60,000
Suspended solids	400–500	125–250
COD	26,500–28,400	5,500–6,000
Specific gravity	1.28–1.35	1.03–1.05

* All values are expressed in mg/l, except pH and specific gravity.

Methods [American Public Health Association, 1995].

To optimize the rector system with respect to temperature for recovery of acid, the studies were initially carried out by varying the temperature at a constant pickle liquor feed rate. The temperature was varied between 400 and 1,000°C, while the pickle liquor feed rate was kept at design value of 1,600 kg/h. [200 kg/h, m^3 of reactor volume]. The rinse water was used to absorb the HCl gases generated in the reactor, and the feed rate of rinse water was maintained at 2,400 kg/h [300 kg/m^3, h]

Optimizing the reactor temperature studies were carried out to assess the optimal pickle liquor feed rate at the optimum operating temperature achieved. The feed rate was varied between 900 and 1700 kg/h [112.5–212.5 kg/m^3 h] and the temperature was maintained in the range of 850 to 900° C.

To assess the effect of airflow rate on expansion of the bed and pressure gradient across the bed, studies were carried out by varying the airflow rate from 200 to 1.900 m^3/h. The pressure at various heights of the reactor was measured using pressure gauges installed. Fluidized systems are extensively used because of the high rate of heat transfer and uniform temperature within the bed, and the heat transfer coefficient was calculated based on the expression as depicted hereunder [Dow and Jakob]

$$hD/k = 0.55\,[D/L]^{0.65}\,[D/d]^{0.17}\{(1\text{-}e)\,Ps\,Cs/epCp\}^{0.25}\,(DG/\mu A)^{0.80} \quad (5)$$

Results and Discussion

Effect of Temperature on Recovery of HCl and Fe_2O_3

Conversion reaction of $FeCl_2$ to Fe_2O_3 and HCl occurs at temperatures above 450°C; hence, the reactor performance was assessed at temperatures varying from 400 to 1000°C, and the results are presented in Table 2.2. The effect of temperature on recovery of acid and iron oxide is depicted in Figure 2.2. Perusal of the figure indicates that hydrochloric acid and iron oxide recovery of 88.8 per cent (112.5 kg/m³ h) and 88.6 per cent (14.2 kg/m³ h) respectively, are achieved at a temperature of 500°C with a waste pickle liquor feed rate of 1,600 kg/h (200 kg/m³ h). The increase in acid recovery is marginal (only 9.6 per cent), with the increase in temperature from 500 to 1,000°C at a constant feed rate. Also, in case of iron oxide, the recovery is only around 8.2 per cent, with an increase in temperature from 500to 1,000°C. Operating the reactor system at temperatures below 500°C, a decrease in recovery of both acid and iron oxide is observed.

Table 2.2: Effect of Reactor Temperature on Recovery of Hydrochloric Acid and Iron Oxide at Constant Pickle Liquor Feed.

Rector Temperature (°C)	*Value-Added Chemicals Recovered*			
	Hydrochloric Acid		*Iron*	*Oxide*
	kg/h	*kg/m³.h*	*kg/h*	*kg/m³.h*
400	600	75.0	90	11.25
500	900	112.5	114	14.25
600	930	116.2	118	14.75
700	950	118.7	120	15.00
800	970	121.2	122	15.25
900	990	123.7	124	15.50
1000	1000	125.0	125	15.62

Waste pickle liquor feed: 200 kg/m³. h(1600 kg/h).

Value-Added Chemicals from Waste Pickle Liquor

Studies indicate an optimum operating reactor temperature of 500°C in terms of recovery of acid and iron oxide. However, temperatures in the range of 800-900°C are required to yield a dense grain of iron oxide resisting attrition, the reason being that under turbulent conditions, the granules of iron oxide developed are so small in size (in the range of 0.5-0.7mm) that they are transported by the gas stream out of the reactor. Hence it is necessary to agglomerate these granules to such a size that they are retained in the reactor in a fluidized condition. The agglomeration of iron oxide presupposes a molten state in the course of the reaction, and therefore, temperature in the reactor needs to be maintained above the melting point of ferrous chloride, *i.e.*, above 675°C. However, studies indicated temperatures in the range of 800-900°C to obtain a dense grain of iron oxide in the size of 4-5mm to retain them in the reactor in a fluidized condition. Thus operating the reactor system at temperatures of 800-900°C, recovery of HCl and Fe_2O_3 in the range of 96-97.7 per cent (121.2-123.7 kg/m³ h) and 94.2–95.6 per cent (15.2–15.5 kg/m³ h), respectively, are achieved with a pickle liquor feed rate of 200 kg/m³ h.

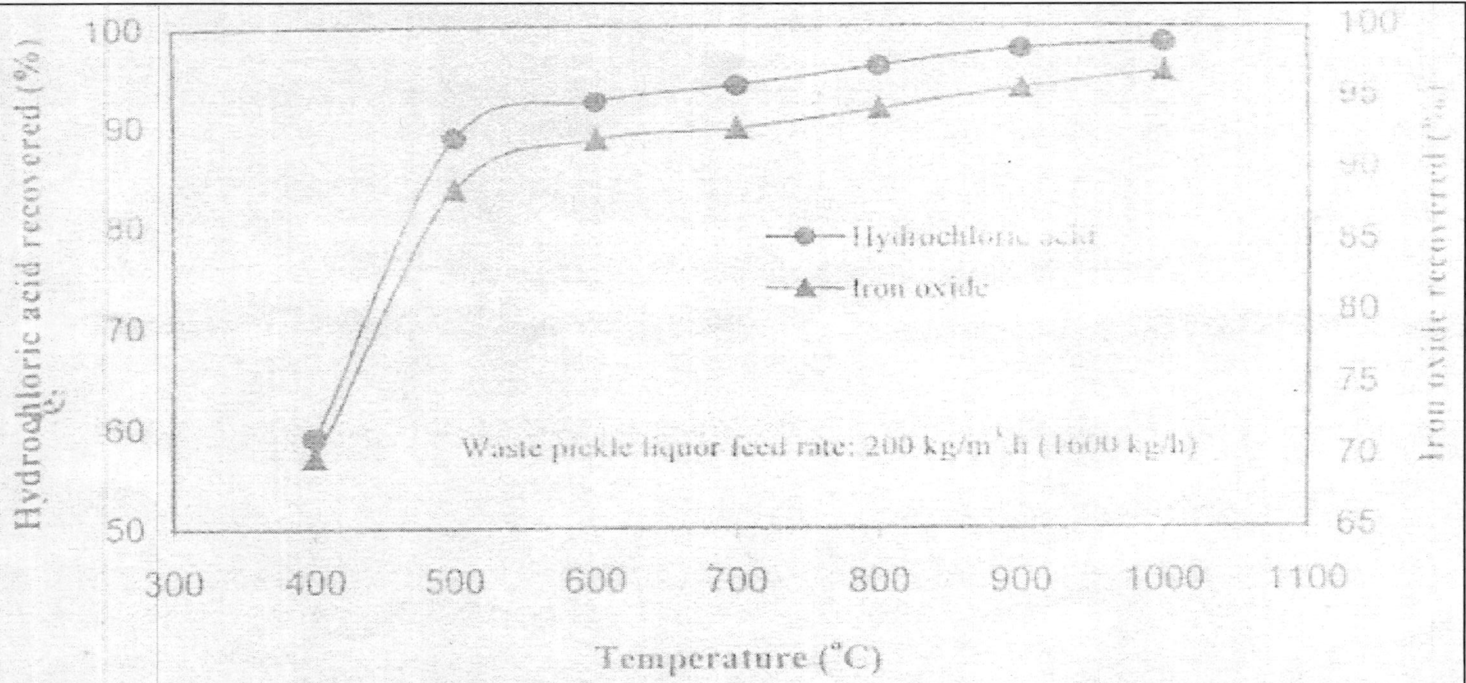

Figure 2.2: Effect of Fluidized Bed Reactor Temperature on Recovery of Hydrochloric Acid and Iron Oxide.

Effect of Waste Pickle Feed Rate on Recovery of HCl and Fe_2O_3

The fluidized bed reactor was also assessed by varying the waste pickle liquor feed rate operating the system at a constant temperature range of 850-900 °C. Table 2.3 presents the performance of the system at varying feed rates. The effect of waste pickle liquor feed rate on the recovery of acid and iron oxide, as presented in Figure 2.3 depicts a steady recovery of around 99 per cent hydrochloric acid and 98 per cent iron oxide with an increase in the feed rate from 112.5 to 187.5 kg/m³ h. However, a decrease in the recovery of both acid and iron oxide is observed upon further increasing the feed rate.

The data on the performance of the system at optimum operating conditions are presented in Table 2.4. A feed ratio of 187.5 kg/m³ h (1500 kg/h) ami reactor temperature in the range of 850 - 897°C was found optimum with recovery of HCl and Fe_2O_3 in the range of 125-128.7 kg/m³ h (1,000–1030 kg/h) and 15.9 -16.2 kg/m³ h (127-130 kg/h), respectively. Power consumption (average) in the process was found to be 0.0015 kWh per kg of HCl recovered.

Table 2.3: Effect of Waste Pickle Liquor Feed on Recovery of Hydrochloric Acid and Iron Oxide at Consult Reactor Temperature.

Waste Pickle Liquor Feed		*Value-Added Chemicals Recovered*			
		Hydrochloric Acid		*Iron Oxide*	
(kg/h)	*kg/m³·h*	*(kg/h)*	*kg/m³·h*	*(kg/h)*	*kg/m³·h*
900	112.5	640	80.00	80	10.00
1,050	131.2	750	93.75	93	11.62
1,150	143.7	810	101.25	102	12.75
1,300	162.5	900	112.50	113	14.12
1,500	187.5	1030	128.75	129	16.12
1,600	200.0	990	123.75	124	15.50
1,700	212.5	950	118.75	117	14.62

Reactor temperature: 850-900°C.

Value Added Chemicals from Waste Pickle Liquor

Effect of Air Velocity on Fluidization of Bed

The plot of superficial velocity of air induced and pressure gradient across the bed as monitored is shown in Figure 2.4. The initial part of the curve indicates the fixed bed, and as the velocity increases, the pressure gradient passes through a maximum value and then reduces within increase in velocity, because the porosity of the bed increases. This declining part of the curve indicates the fluidized bed. The decrease in pressure gradient continues until the velocity is high enough for transport of the material to take place. At this point, the pressure gradient starts to increase again, which indicates the transport of the bed.

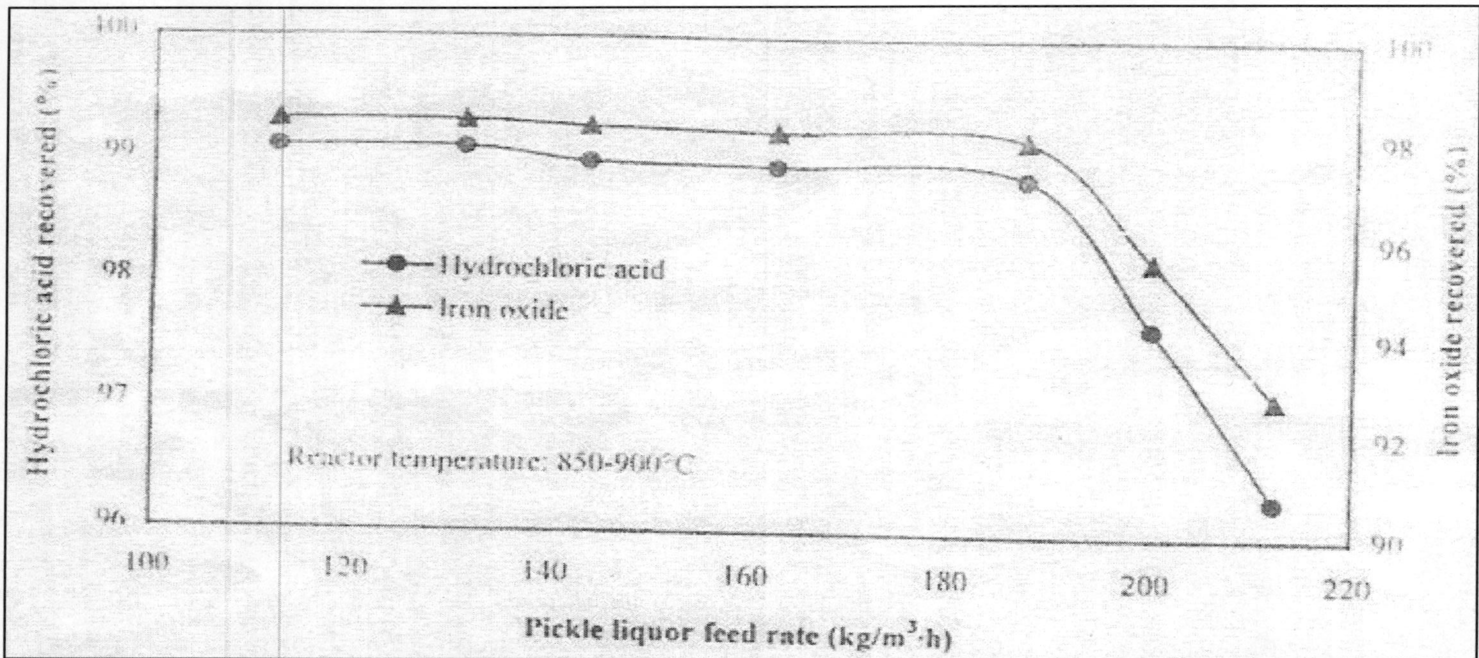

Figure 2.3: Effect of Waste Pickle Liquor Feed on Recovery of Hydrochloric Acid and Iron Oxide at Constant Reactor Temperature.

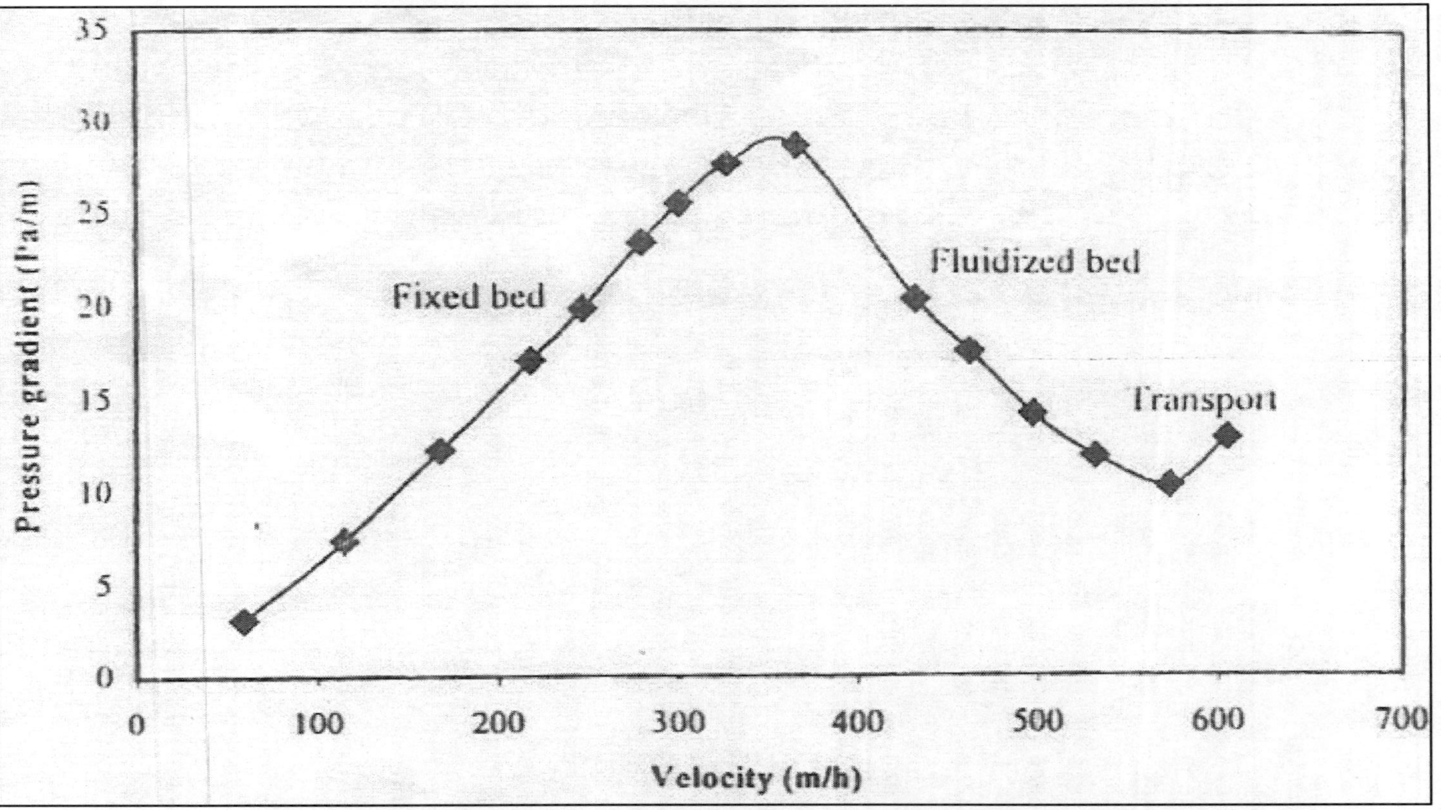

Figure 2.4: Plot of Superficial Velocity of Air vs. Pressure Gradient in Reactor Bed.

Table 2.4: Performance of Fluidized Bed Reactor at Optimum Operating Conditions.

Day	Waste Pickle Liquor Feed (kg/m³ h)	Reactor Temperature (° C)	Value-Added Chemicals Recovered*				
			Hydrochloric Acid		Iron oxide		Energy Consumption
			(kg/h)	(kg/m³ h)	(kg/h)	(kg/m³ h)	(kWh/kg HCl Recovered)
1.	187.5	853-893	1030	128.7	130	16.25	0.00155
2.	187.5	855-864	1020	127.5	130	16.25	0.00152
3.	187.5	854-884	1010	126.2	129	16.12	0.00142
4.	187.5	854-897	1000	125.0	127	15.87	0.00156
5.	187.5	850-867	1020	127.5	130	16.25	0.00153
6.	187.5	856-869	1000	125.0	129	16.12	0.00168
7.	187.5	853-892	1000	125.0	128	16.00	0.00144
8.	187.5	867-889	1030	128.7	129	16.12	0.00151
9.	187.5	858-883	1020	127.5	128	16.00	0.00141
10.	187.5	851-892	1010	126.2	129	16.12	0.0158

* Data presents average of 8 h monitoring each day.

In the case of gas-solid fluidized systems, two separate phases–a dense phase and a lean phase–are formed, and a condition of "aggregative fluidization" is produced. Froude Number ($u^2/g\ d$), where u is the velocity of air flow at which fluidization takes place, d is the diameter of the particles, and g is the acceleration due to gravity) gives a criterion from which the type of fluidization taking place can be predicted. At values greater than unity, aggregative fluidization takes place (Coulson and Richardson, 1991).

In the present study, the Froude Number determined ranges between 4 and 6 indicating aggregative fluidization. Due to intimate mixing and high degree of turbulence, heat transfer within the bed was found to be rapid, and heat transfer coefficient determined based on equation (5) is 305 W/m² K, and therefore uniform temperature was quickly attained throughout the system.

Conclusions

The application of a full-scale fluidized bed reactor for recovery of value-added chemicals from waste pickle liquor achieved an efficiency of 99 per cent in terms of acid (HCl) recovery, achieving Zero Effluent Discharge (ZED). In addition, iron oxide (Fe_2O_3) was also recovered with very high purity (chloride content of 0.05–0.1 per cent). The recycling of recovered acid has resulted in the reduction of HCl requirement in the pickling section, achieving savings of 55.3–60.2 per cent in a steel plant of a capacity of 1000 X 10^3 kg/d of hot-rolled steel coils. Based on the studies, the following conclusions are drawn:

1. The reactor temperature of 500°C was found optimum for the recovery of hydrochloric acid and iron oxide. However studies indicated that operating

temperatures in the range of 800 - 900°C is necessary to obtain dense grain of iron oxide in the size of 4-5 mm so as to retain them in the reactor in fluidized condition, and hence, the reactor temperature must be maintained between 800 and 900 °C to obtain a dense grain of iron oxide resisting attrition.

2. Steady recovery of both hydrochloric acid and iron oxide of around 99 and 98 per cent respectively, was achieved with an increase in the feed rate of waste pickle liquor from 112.5 to 187.5 kg/m³ h (900 1500 kg/h). However, recovery of both acid and iron oxide dropped upon further increasing the feed rate. A feed rate of 187.5 kg/m³ (1500 kg/h) and temperatures in the range of 850-900°C were found optimum with recovery of HCl and Fe_2O_3 of 125-128.7 kg/m³ h (1000–1030 kg/h) and 15.9–16.2 kg/m³ (127-130 kg/h), respectively. The strength of the recovered acid was found to be in the range of 15.5–17 per cent. Average power consumption was found to be 0.0015 kWh per kg of HCl recovered.
3. The Froude Number determined ranges between 4 and 6 indicating "aggregative fluidization" achieved in the reactor and heat transfer within the bed was found to be rapid, as indicated by a very high heat transfer co-efficient (305 W/m² K), thereby attaining quick uniform temperature throughout the system.

Nomenclature

h = Heat transfer coefficient (W/m² K)
k = Thermal conductivity of the air (W/m K)
d = Diameter of the particle (m)
D = Diameter of the reactor (m)
L = Depth of the bed (m)
e = Voidage of the bed
P_s = Density of the solid (kg/m³)
P = Density of the air (kg/m³)
C_s = Specific heat of the solid (J/kg K)
C_p = Specific heat of the air constant pressure (J/kg K)
μ = Viscosity of the air (Pa s)
G = Mass rate of flow of air (kg/s), and
A = Cross-sectional area of the bed (m²).

References

1. AMERICAN PUBLIC HEALTH ASSOCIATION, AMERICAN WATER WORKS ASSOCIATION. Water environment federation. (1995). *Standard Methods for the Examination of Water and Wastewater. 19tg ed, Washington, DC.*

2. AZAD, H.S. (1976), *Industrial Wastewater Management Handbook* New York; McGraw-Hill Book Co.

3. COULSON, J.M., and RICHARDSON, J.F. (1991). *Chemical Engineering* Vol. 4, London: Pergamon Press.

4. DOW, W.M. and JAKOB, M. (1951). Heat transfer between a vertical tube and a fluidized air solid mixture. *Chem. Engg. Prog.* 47, 120-126.

5. GURNHAM, C.F. (1965). *Industrial Wastewater Control. New York*: Academic Press.

6. KUMAR, S.M., CHARE, N.Y., VAIDYA, A.N., AND BAL A.S. (1998). Recovery of acid from pickling liquors, *Environ. Eng. Sci.* 15(4) 259-263.

7. KUMAR, S.M., VAIDYA, A.N., and BAL, A.S. (1999). Treatment options for the management of waste pickling liquor. *J/ntl Assoc. Environ. Manage 26*. 35-39.

8. NANDITA, D., KUMAR, S.M., CHARE, N.Y., VAIDYA., A.N. and BAL., A.S. (2000). Recovery of Chemicals; from waste liquor–A case study J.Av.wr. Environ. Manage 27, 189-194

9. NEERI technical report (2000). *Waste Management in a Steel Industry*, Nagpur, India.

10. NEGRO, C, LATTORREE, R. DUFOUR, J., FORMOSO. A., and LOPEZ M.F. (1994). Treatment and recovery of pickling liquors. *J. Environ. Sel Helth* A29(9), 1899-1926.

11. RENNHACK R. (1974). Novel developments in the regeneration of acid pickling liquors and other chloride bearing solutions. In P.E. Wulfinghoff, E., *Wastes–Solids Liquids and Gases*, New York; Chemical Publishing Co., p. 311-319

12. RUDOLFS W. (1997). *Industrial Wastes–Their Disposal and Treatment*. Bikaner: Allied Scientific Publishers, India

Chapter 3

Compatibility of Continuous Flow in Comparison with Windrow, Batch and Semi-Continuous Vermireactor System

S.N. Kaul[1], *A.B. More*[2], *P. More*[3] *and H. Gunjal*[4]

[1]*Principal, MIT Pune*
[2]*AP, Department of Civil Engineering, ICOER, Pune*
[3]*Sr. Lect., SCOE, Pune*
[4]*M.Tech. Student, BATU, Lonere*

ABSTRACT

Taking into account the current scenario of municipal solid waste management, efforts on efficient decentralized waste processing facility are on war-footing so as to avoid environmental degradation because of mismanagement of municipal solid waste in the city. Continuous flow vermicomposting system– is one of the technological options to Land filling for organic solid waste management is proposed in this chapter. To reduce the burden on centralized method it will be strongly recommended to use a treatment technology that will be effective to carry out the scientific treatment at or nearby source. In the cities there is rapid generation of the organic waste from hotels, institutes, public places, canteens, residential areas etc. It is necessary to develop a continuous flow vermireactor and has been emerged on prime basis. Landfill, open dumping and incineration are showing adverse environmental impact, it is therefore required to minimize the use of these systems and to act towards sustainable solution using end of pipe system which compels to conclude with minimization of solid waste at source without impact on environment and then by on human being. To treat the waste at the source a system needs to be developed which will prove to be a significant and sustainable solution to current environmental deteriorating conditions. Along with safe disposal of waste if there is some energy based product like vermiculture and vermicasting generation, then it will be more beneficial.

The vermicomposting technology which is the base of the processes would be given an enhancement with vermireactor by generating vermicastings which can be served as manure for fields and gardening and other related purposes. Complete decomposition of the organic matter resulted from enclosed method overcomes the problems faced by the open methods such as windrow, batch and semi continuous system. The open composting have some problems of flies, odour, and the main issue of the space requirement which would be render in this method to have the mobility and minimum space requirement. Thus enclosed continuous flow vermireactor will give new dimensions to the conventional vermicomposting and there will be increased awareness among the citizens about solid waste management practices on a decentralized basis and the city will be benefited by a solution to wet waste disposal management. So it will prove to be a best possible option of utilization of organic solid waste for culture generation.

Keywords: *Vermicasting, Vermiculture, Continuous flow vermireactor, Windrow, Batch system, Solid waste management, End of pipe system.*

Introduction

Indian Waste Sector

Estimated quantity of waste generated in India is about 27.4MT per year and per capita per day is at least 200-250g considering differences in economic zone and estimated to increase by 1.33 per cent annually. This enormous increase in MSW will have significant impacts in terms of the land requirement, treatment and disposal facilities for disposing this waste. The density of waste is between 330-560 kg/m^3. Waste collection in Indian cities ranges from 50 to 90 per cent. Out of the total collected waste, on an average 94 per cent is dumped on land and only 5 per cent is composted.

The municipal solid waste is governed under the rule of India's Municipal Solid Waste (Management and Handling) Rules 2000, which deals with the proper and scientific, collection, segregation, and disposal of biodegradable and non biodegradable solid waste within the limits of jurdication, in compliance with the rule.

Treatment Methods for Organic Municipal Solid Waste

The necessary steps to minimize the cost with sustainable technology are elaborated in the Chart 3.1.

Risk Factors Associated with Solid Waste Treatment

Compost

For capital cost considerations, compost plants adopt windrow method as a norm, disregarding its inherent odour nuisance potential. This emerges as a major risk factor since the host community finds it extremely damaging for its socio-economic and psychological well being.

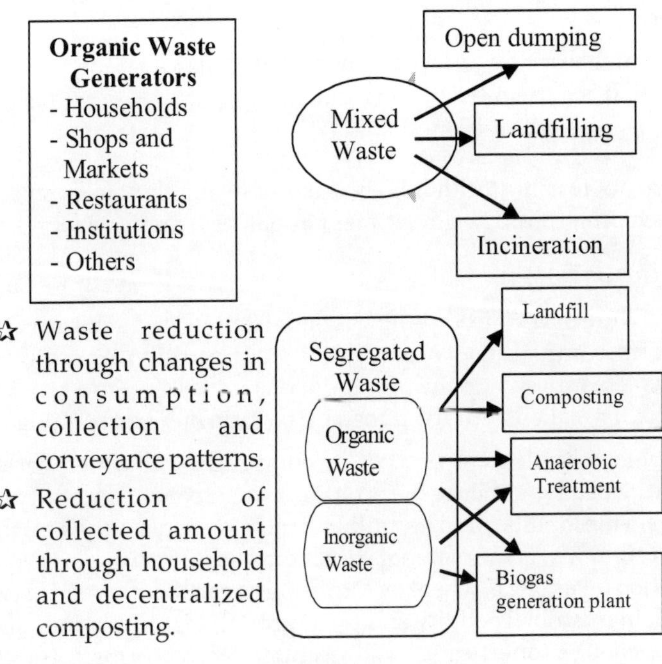

Chart 3.1

Bio-methanation

Bacteria involved in bio-methanation reactors are highly sensitive to temperature variations. This requires insulation and efficient instrumentation for real time monitoring of operating parameters. Preference for 'low cost' systems does not allow incorporation of these features while ambient temperature in several parts of the country varies from 5°C in winter to 45°C during summer. As a result biological process gets disrupted.

Open Dumping and Land Filling

These are the centralized methods and require collection and transportation of the waste to the respective sites. Further there is the problem of land deterioration, greenhouse gas emission, leachate formation, groundwater pollution, wind borne paper, plastics, odor, etc, traffic of large vehicles around the site, erosion of waste and/or cover soil, vermin/vector control.

Incineration

The recoverable energy is low and the cost is high. In many cities, incineration has met strong opposition because of emission of highly toxic dioxins and other pollutants. It is not feasible for many cities in developing countries, and cannot be regarded as the main option for waste treatment in the region.

Animal Feed

In general this option can only be expected to play a minor role in organic waste management. It is more restricted to rural areas.

Recycling

It is also not feasible technologies as it is unavailable at local level and therefore it is not much preferred. The government initiative is required for the purpose.

Vermicomposting

Vermiculture or vermicomposting is derived from the Latin term vermis, meaning worms. Vermicompsting is a process of consumption of organic material by earthworms. This speeds up the process of decomposition and provides a nutrient-rich end product, called vermicompost, in the form of 'worm castings'.

It is a biotechnological process of composting in which certain species of earthworm are used to enhance the process of waste conversion and produce a better product. It is a mesophilic process, utilizing micro-organisms and earthworms which are active at 10-32°C. As the material passes through the earthworms gut a significant transformation takes place, where by the resulting earthworm casting (worm manure) that are rich in microbial activity and plant growth regulators and are fortified with the pest repellent properties as well. Actually earthworms through a biological alchemy are able to transform garbage into gold. It is a faster process than composting.

Vermicompsting can be conducted all year-round, providing environmental conditions remain within acceptable limits. For increased efficiency, care should be taken to ensure that organic feedstock and conditions allow worms to reproduce successfully and to withstand moisture and climatic fluctuations. Given suitable conditions, vermicomposting appears to provide a relatively straightforward solution to the management of compostable organic wastes. For centuries, earthworms have been used as a means of decomposing wastes and improving soil structure. The breeding and propagation of earthworms and the use of its castings has become an important method of waste recycling throughout the world.

Processes Involved in Vermicomposting

It is the bio-oxidation of organic matter involving the joint action of earthworms and micro-organisms. The processes involved are:

Physical

- Aeration
- Fragmentation of substrate
- Dispersion of ingested material

Chemical

- Digestion of organic matter
- Increase or decrease of nutrients

Biological

- ✫ Stimulation/Inhibition of micro-organisms
- ✫ Modification of microbial communities.

Types of Vermicomposting Systems

Windrows

This system takes into account the availability of large land and other appropriate technology for operating the whole system. These systems are extensively being used for used both in the open and under cover.

Wedge System

This is a modified type of windrow system where one can easily harvest the vermicompost without disturbing the earthworms. In this system organic materials are applied in layers against a finished windrow at a 45 degree angle.

Bed and Bin System

In this system bins are used to breed and harvest the vermicompost and also in some cases beds are made on the ground for the same purpose. This method is labor intensive but is much easier to handle and is widely.

Reactor System

Reactor systems have raised beds with mesh bottoms. Feed stocks are added daily in layers on top of the mesh or grate. Finished vermicompost is harvested by scraping a thin layer from just above the grate, and then it falls into a chamber below. These systems can be relatively simple and manually operated or fully automated with temperature and moisture controls. For maximum efficiency, they should be under cover.

Factors that may be considered for selecting the appropriate vermicomposting technology for a project include: amount of feedstock to be processed; funding available; site and space restrictions; climate and weather; state and local regulatory restrictions; facilities and equipment on hand; availability of low-cost labors etc.

Project Description

Concept of Continuous Flow Vermireactor

Vermireactor uses the low strength organic (municipal solid) waste as a raw material to be feed in and it gets shredded, mixed and optimal operating conditions (oxygen, temperature, water etc.) are prevailed within the bioreactor where in the earthworms carrying out the vermicomposting utilizing the feed and then it grows and multiplies therein. The system progresses in a continuous flow pattern as all the environmental conditions will be prevailed in order to yield efficient quality of vermicasting and vermiculture in less amount of time and it requires less space than the conventional method and it is having mobility.

Continuous Flow Systems

The continuous flow system developed can be adopted for many mid-scale operations. The efficiency savings offered by the continuous flow design increases with the amount of material processed. This system design is almost ubiquitous in commercial mid to large-scale vermicomposting systems. Each of these systems uses a relatively deep top-fed container, in which the composting mass sits upon a raised floor made from a widely spaced wire mesh. Worms are added to the system and food waste is added gradually, layered with bedding material. The system is continually fed until the bin is nearly full. The worms generally move upward through the feedstock/bedding layers and vermicompost is harvested from below by scraping or cutting a thin layer of finished material from just above the grill using a rake or a manually or hydraulically-operated blade.

Continuous flow systems offer several advantages for medium to large scale composting operations. They are relatively straightforward to construct and operate. They are labour-efficient in terms of operation and harvesting finished material. They avoid the need for expensive equipment associated with technical 'in-vessel' systems and the turning and screening of windrowed material. Continuous flow vermicomposting designs are arguably the most efficient systems available, in terms of time and labor savings. However, regardless of efficiency or ease of operation, there is no design that eliminates the need for careful monitoring and good system management, which may require considerable initial experimentation and familiarization. One of the vertically continuous flow vermireactor designed by ***Dr. Clive Edwards*** at Rothampstead experimental station is shown in Figure 3.1.

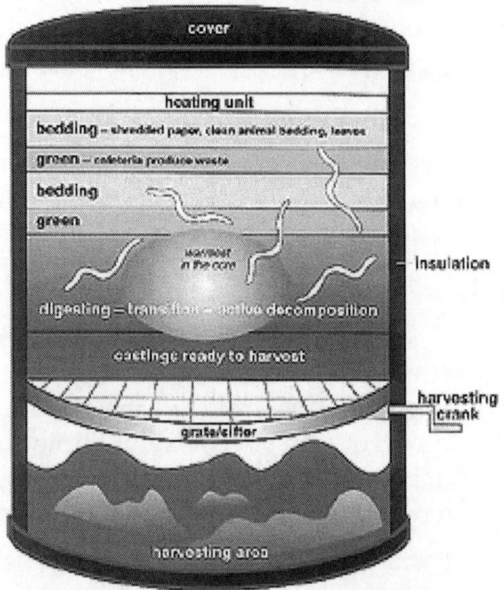

Figure 3.1: Continuous Flow Vermireactor, (designed by Dr. Clive Edwards)

Maintaining Continuous Flow

This system maintains an ongoing flow of vermicompost that is easily removed from the system without disrupting the worm activity or requiring complex or time-consuming harvesting methods. The worms most often used in vermicomposting are usually referred to as "surface feeders." Earthworms are oxygen breathing, moisture-loving animals that require organic material to be bacterially active before they eat it. In their natural environment, this is usually top few inches of soil or surface organic litter, such as leaves. In any system with a free flow of oxygen, monitored moisture level and abundant supply of decomposing organic material, earthworms may spread throughout the material unless the system is carefully managed. Earthworms may therefore be found anywhere within the continuous flow systems which meets their requirements.

Feeding Rates

The precise loading rate (at which raw feedstock can be added to a worm bed to encourage the worms to concentrate at or near the surface) will vary depending on the feedstock being used, temperature, moisture levels and the density of the worm population. Proper loading rates should be adopted so that new feedstock is not added until the majority of the previously added feedstock has been decomposed. Otherwise this can lead to a build-up of unprocessed material within lower layers. The worms will then spread into all the available food areas. Worm movement in the lower levels of a flow-through system often causes vermicompost to drop through the mesh floor before it has been sufficiently decomposed. Also, when the system is harvested, worms remaining in the material will fall through with the vermicompost and will either need to be separated using labour-intensive screening methods, or will be lost to the system.

Most operators of continuous flow systems find that frequent additions of thin layers of feedstock (1"-2" deep spread across the surface) produce the best results. Feedstock is sometimes mixed with bulking agents like compost, shredded leaves, cardboard, paper or straw or covered with an equally thin layer of these materials. Paper products are a preferred feedstock for earthworms, as they provide an easily accessible and digestible form of carbon.

Discussions

Methodology

Methodology is explained with the help of flowchart shown below;

1. Analysis of generated solid waste (Before, during and after treatment)

↓

2. Defining of composting zones

↓

3. Setting up fabricated continuous flow vermireactor

↓

4. Devising of solid waste management system

↓

5. Defining of end of pipe system

↓

6. Trial on digitally controlled vermireactor

↓

7. Application of cost benefit analysis to the model

Step 1

The municipal solid waste was collected from ward that is on decentralized basis and it was then segregated mechanically (inorganic such as plastic, clothes, glass, metal etc. and organic such as leftover food, vegetable waste, paper, fruits and leaves of the plant from garden). Then it was cut into small pieces to reduce the size and allowed to dry for some time. Then the waste is mixed properly to determine weight, volume and bulk density of the municipal solid waste for seizing the reactor. Characteristic study of the waste constituents present in the municipal solid waste should be done at site as well as in laboratory.

Step 2

In this the bioconversion rate and reaction time for vermicomposting will be found out by carrying out a pilot plant study of the reactor system (close vermicomposting) and based on it defining of the composting zones favorable for the continuous flow vermireactor will be demarked. Also study of the optimal design parameters (pH, Temperature, Moisture and Dissolved Oxygen) should be carried for efficient vermicomposting in the reactor system.

Step 3

Setting up of fabricated enclosed vermireactor to treat the solid waste on continuous basis. During the whole process qualitative and quantitative analysis of the municipal solid waste is done (before, during and after) the vermicomposting process in order to study the performance of the continuous flow vermireactor process. Determining of the kinetics of the process is major point of the reactor system.

Step 4

Solid waste management system deals with an integrated approach for collection, storage, transportation and disposal of low strength organic solid waste at source and with zero waste discharge facility through enclosed method. As it conserves time, space and energy required for treatment of solid waste by source reduction. This technology manages the solid waste generated at source by providing inbuilt facility for all the aspects of solid waste management.

Step 5

The continuous flow vermireactor is designed to provide the end of pipe treatment to organic municipal solid waste. Herein there is a complete decomposition of organic waste through scientific method. The system prevents direct emission of greenhouse gas emission into the atmosphere. Also it prevents entry of the leachate into the groundwater source by providing the system of collection. Thus organic load on landfill and open dumping is abolishing by conserving the source for manure production and vermicast generation.

Step 6

Application of digital control system to vermireactor for montoring and maintaining the favorable environmental conditions for efficient generation of vermiculture and vermicasting within the desired time span. That will help to run the reactor automatically without obstruction. Digitally controlled system will give new dimensions to vermicomposting system by enhancing it to achieve sustainability.

Step 7

Cost Benefit Analysis is that environmental economic tool that helps to assess the feasibility factor for the model by studying the benefits as against the conventional method of vermicomposting. The superiority of the advancement offered by this process is analyzed for the sustenance of the concept or method in the treatment.

Utilization of Vermicasting

Landscapes, Gardens and Nurseries

1. It reduces water consumption.
2. It improves the quality and appearance.
3. It reduces the bulk (as compared to FYM).
4. It eliminates the need for weeding operations.
5. It increases the shelf life of flowers.
6. It improves germination of seeds in nurseries.

Soil and Water Conservation

1. It helps in stabilization of earthen structures through better vegetative growth.
2. It promotes grass production in degraded open grass lands.
3. Enables reclamation of waste lands for plantation and cultivation.

Forest Development

1. Helps in early stabilization of transplants.
2. Enables early regeneration of partly degraded forest areas.
3. Increases overall vegetation in hilly regions.

Waste Management

1. Helps recycle segregated solid wastes from diverse sources.
2. No odor and destruction of pathogens. Also non toxic.
3. Environmentally Sustainable process.

Ecological Renewal

1. Helps in plantation of trees in extremely adverse conditions. Also improves growth.

Others

1. Cost effective pollution free valuable end product.
2. Established on site and scalable to suit any volume

Conclusion

The compatibility factor associated with the continuous flow vermireactor is the technological innovation that renders the problems coupled with the conventional methods of vermicomposting. It helps to achieve source reduction of organic waste generated from the different strata of the city. As it is a sustainable solution to the intensifying problem of solid waste in the city.

Vermiculture provides an amicable solution in the field of solid waste management. Bioconversion and the by-products obtained do not disturb the ecological balance. The manure obtained can replace the chemical fertilizers, which affects the soil as well as the water system. The pollution load on the environment can be reduced considerably by adopting this eco-friendly method of bio conversion.

Continuous flow vermireactor system is an enclosed biological reactor that provides favorable environmental conditions that catalyses the kinetic processes involved in the reactor. The favorable conditions are maintained by monitoring the performance and environmental parameter required for operating and maintenance of the vermicomposting process. Continuous flow vermireactor will cope up with time consumed by other conventional as it fasten the process. Also there will be a check on emission of landfill gases.

References

1. A report on 'India: Waste Sector', by Ministry of Environment and Forests, Government of India.
2. A report on 'Urban Organic Waste- From Hazard to Resource', IGES White Paper.
3. Asit Nema, report on 'Risk factors associated with solid waste treatment, Technology options in the Indian context', Foundation Green -Ensys, pp 1-8.
4. Dr. C. Visvnathan, Dr. Ing. Josef Trankler, Dr. Kurian Joseph, Dr. Nagendran, 'Vermicomposting as an Eco-tool in Sustainable Solid Waste Management".
5. Manuel Aria, Maria Gomez-Brandon, Cristina Lazacno, Jorge Domeinguez, Department de Ecoloxia Bioloxia Animal, Universidade de Vigo, Spain,' "earthworms and micro-organisms disentangling black box of vermicomposting', CODIS 2008.
6. Literature Review of worms in "Waste management" Vol.1, 2007. Recycled Organics unit.
7. A report on "Vermicompost production systems".
8. Worm casting application by George Hahn, California Vermiculturist.

Chapter 4

Study of UASB Reactor System as a Pretreatment Facility for Hydrogenated Oil Industry Wastewater

Sunita Shastry[1], *Tapas Nandy*[1], *S.R. Wate*[1] *and S.N. Kaul*[2]

[1]*National Environmental Engineering Research Institute, Nehru Marg, Nagpur, Mh., India*

[2]*Principal, MIT College of Engineering, Paud Road, Kothrud, Pune*

ABSTRACT

In the present study, a laboratory scale model of upflow anaerobic sludge blanket reactor system was investigated for pretreatment of wastewater from a hydrogenated vegetable oil industry in India. For 2-22 kgCOD/m^3.d variation in the substrate loading rate, 55-97 per cent of COD removal efficiency and 0.24-0.34 m^{3}CH$_4$/kgCOD$_r$ of specific methane yield was achieved.

Introduction

Hydrogenated vegetable oil, a substitute for butter oil, is extensively used in India. It forms a dietary supplement amongst all categories of population. The vegetable oil industry generally falls under small to medium scale though quite a few large scale industries also exist. Wastewater generated from hydrogenated vegetable oil industry is of great concern due to its potential toxicity in the form of fats both saturated and unsaturated, falling in line with other industries, *viz.* oil refining, dairy, food processing and slaughter house wastes contributing to fats.

Due to strict legislation for protection of environment, it is imperative to arrive at appropriate treatment technology for the wastewater emanating from this industry.

Experience with fatty wastewater through anaerobic route has been indicated positively (Ozturk, 1999).

Therefore in the present study, UASB reactor system which has found wide applications both for industrial and domestic wastewater (Hulshoff Pol *et al.*, 1986) has been experimentally investigated as a pretreatment facility to evaluate and assess its applicability for biogas recovery from hydrogenated oil industry wastewater.

Materials and Methods

The laboratory model reactor was fabricated using perpex glass column. The reactor dimensions comprised of 220 mm diameter and 1300 mm depth with total holding capacity of 50 litres (Figure 4.1). Gas solid liquid separator was provided as described by Lettinga (1980).

Deflection plates inclined at 45° were provided to prevent escape of biogas into settling zone. The three phase separator was fixed at height of about 800 mm from bottom. Outlet system was provided with V-notch to achieve uniform outflow (Shastry

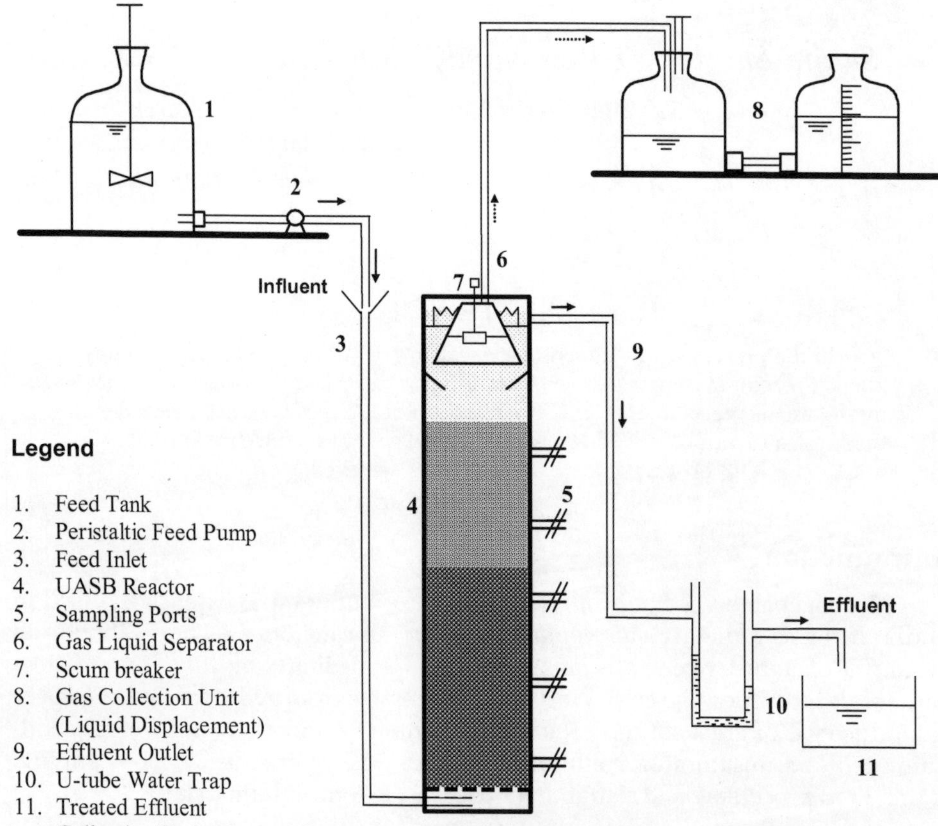

Figure 4.1: Experimental Set-up of UASB Reactor Used for the Study.

et al., 2004). Scum breaker was incorporated in the separator for breaking the scum. Biogas was collected in aspirated bottle through liquid displacement technique.

The wastewater used for study was collected from hydrogenated oil industry in Nagpur, India. Wastewater samples were collected after oil trap. Due to its alkaline nature (pH: 10.8-11.3), the wastewater was neutralized using 1N sulphuric acid (H_4SO_4). Phosphates being adequate, nitrogen was added to meet the nutrient requirement of C:N:P of 100:2.5:0.5.

All samples were collected hourly and composited for 12 hours and analysed for various parameters. All the analyses were performed as per Standard Methods (APHA *et al.*, 1995). The biogas collected was also analysed for methane, carbondioxide and hydrogen sulphide.

Results and Discussion

The wastewater characteristics used for the study is presented in Table 4.1. The BOD and COD concentrations were in the range of 1650-3400 mg/L and 3850–8290 mg/L, respectively. The BOD/COD ratio of 0.41-0.43 indicate biodegradable nature

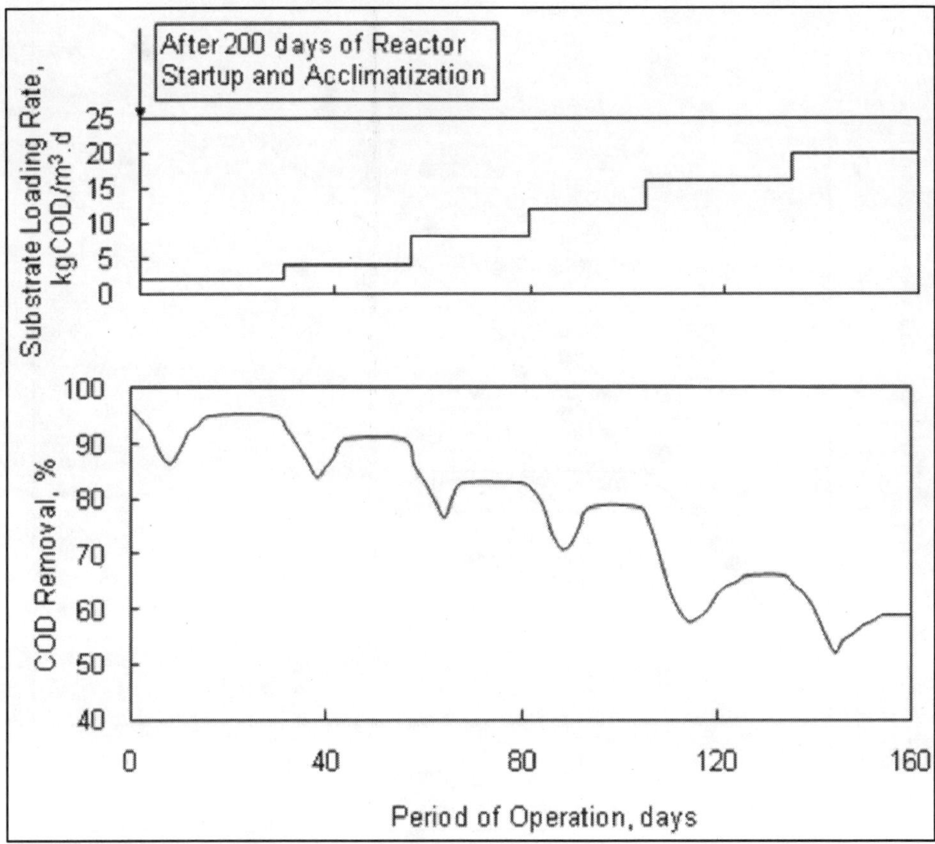

Figure 4.2: Reactor Performance for Step Increase in Loading Rates.

of wastewater with fats contributing about 1380-1650 mg/L making it amenable to biological treatment.

During the reactor startup, the sludge bed and blanket had VSS concentration in the range of 18000-19800 and 920-1815 mg/L, and subsequently increased to 38980-42540 mg/L and 4900-18430 mg/L, respectively.

The studies were carried out at neutral pH 7.0 and temperature 35°C. The performance of the reactor for step increase in loading rates 2 to 22 kgCOD/m^3.d is shown in Figure 4.2 which indicate fall in removal efficiency from 95 per cent to 60 per cent. For each kg of COD removal 290-338 l methane was produced. The conversion rates achieved fall in range with that achieved by Ubay *et al.* (1992) for olive oil mill wastewater using UASB reactor system.

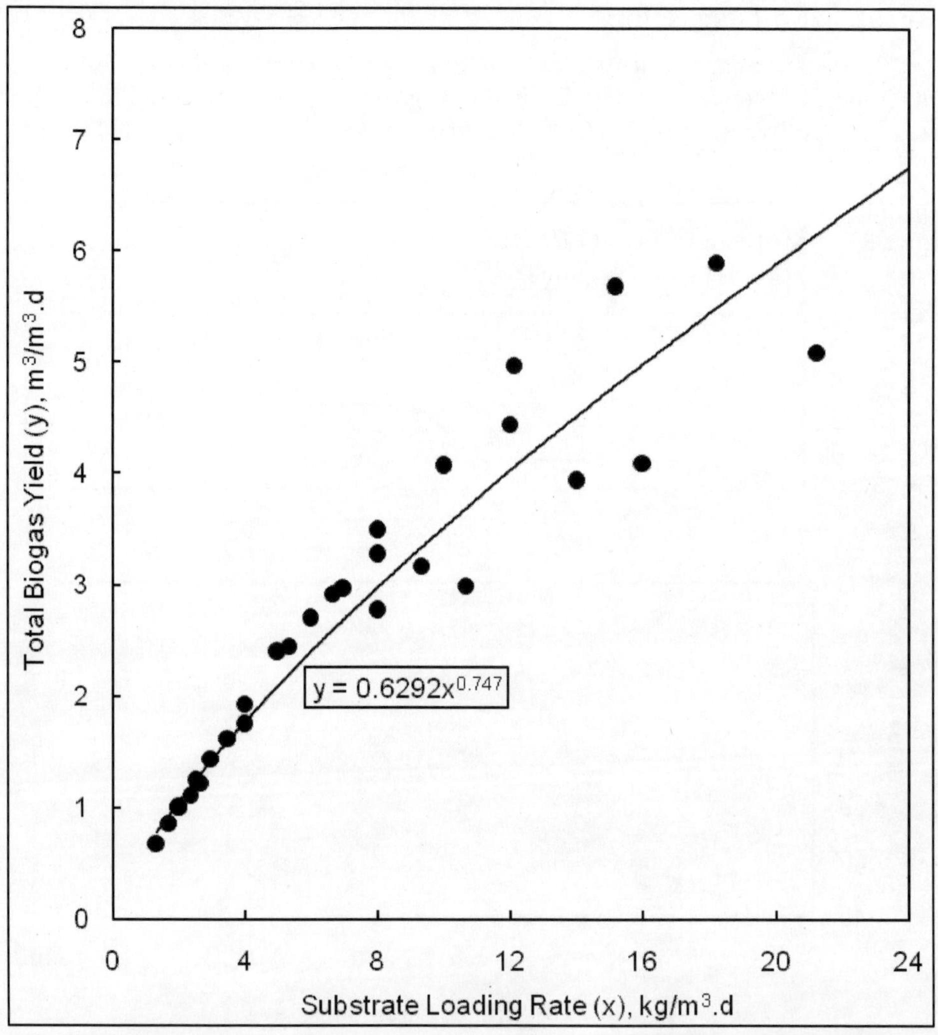

Figure 4.3: Plot of Total Biogas Yield as a Function of Substrate Loading Rate.

Plot of biogas yield and specific methane yield at various substrate loading rate are presented in Figures 4.3 and 4.4, respectively. The plot clearly indicates that methane yield increases with mass of substrate removed per unit volume per day of reactor system. However, the yield marginally decreases at higher loading rates.

The biogas analysis revealed that methane content decreased with increase in organic substrate loading rate, and the methane content varied from 53-66.7 per cent. Observations also revealed reduced methane content at low HRT, which is in agreement with Kaul *et al.*, 1994. The possible fall in methane content may be due to high organic acid yield during acid generation, resulting in activation of buffering action causing more carbon di oxide in off gases.

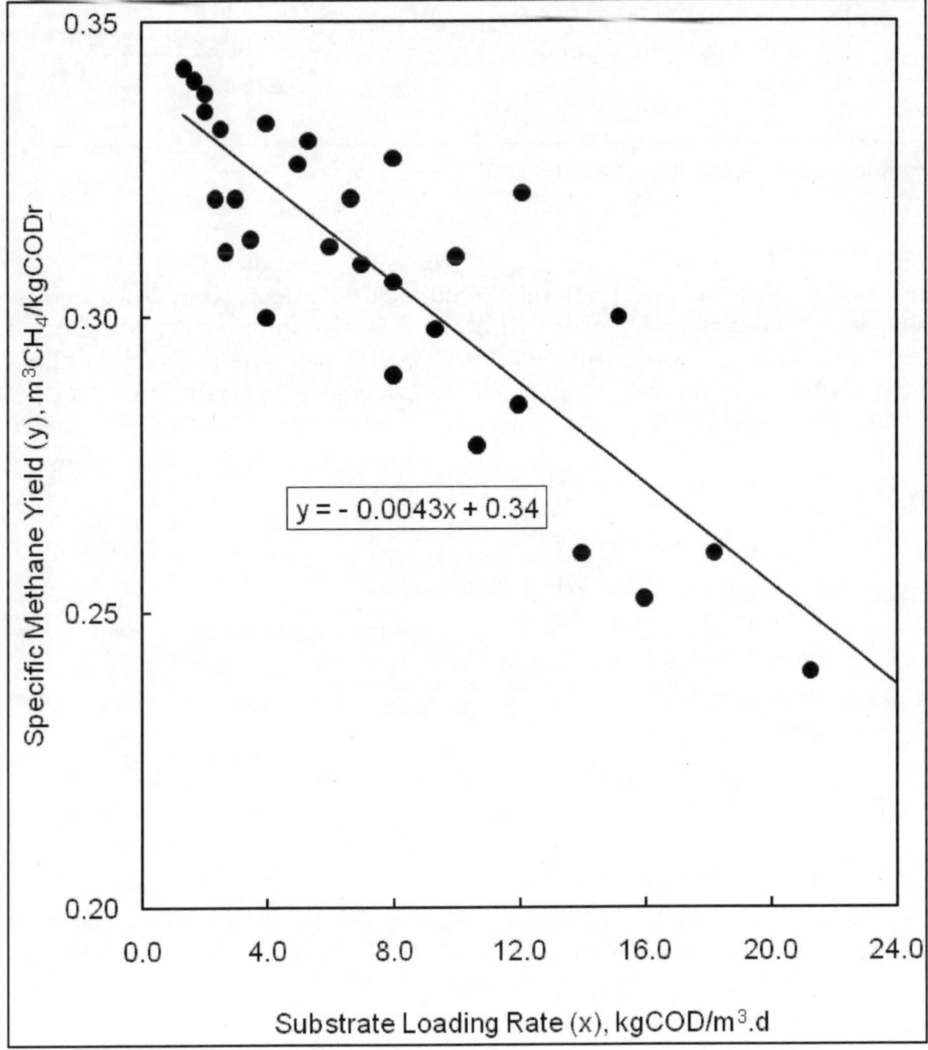

Figure 4.4: Plot of Specific Methane Yield as a Function of Substrate Loading Rate.

Table 4.1: Physico-chemical Characteristics of Wastewater Used for Study.

Sl.No.	Parameters	Concentration (Range)
1.	Temperature, °C	41-48
2.	pH	10.8-11.3
3.	Alkalinity	3750-4180
4.	Suspended solids	1500-1900
5.	Total dissolved solids	5870-6310
6.	BOD	1650-3400
7.	COD	3850-8290
8.	Total fatty matter	1380-1650
9.	Total kjeldhal nitrogen	17.2-18.8
10.	Phosphates	50.5-60.8
11.	Nickel	0.46-0.52
12.	Zinc	1.16-1.20

All values except pH and temperature are in mg/l.

Conclusion(s)

Evaluation of the system performance indicated the potential of upflow anaerobic sludge blanket reactor system for the treatment of medium strength hydrogenated vegetable industry wastewater with recourse to methane recovery of about 0.3-0.32 m^3 per kg substrate removal. Substantial organic load removal in terms of COD in the range of 80-85 per cent is achieved at loading rate of about 9-10 kgCOD/ m^3.d.

References

1. APHA, AWWA, WPCF (1995). "Standard Methods for Examination of Water and Wastewater", APHE, WPCF, New York.
2. Hulshoff, L. *et al.* (1986). "New Technologies for Anaerobic Wastewater Treatment", Wat. Sci. Tech. Vol.18, No.12, 4153.
3. Kaul, S.N. and Nandy, T. (1994). "Biogas Technologies". State of the art published by NEERI, Nagpur, India.
4. Lettinga, G. and Vinken, J.N. (1980). "Feasibility of the Upflow Anaerobic Sludge Blanket (UASB) Process for the Treatment of Low Strength Wastes", Proceedings of the 35th Industrial Waste Conference, Purdue University, 1980. Ann Arbor Science, Ann Arbor, Michigan.
5. Ozturk, I., San, H.A. and Eroglu, V. (1990). "Pilot and Full Scale Treatability Studies on Wastewaters from an Edible Oil Refining Industry", 44th Purdue Industrial Waste Conference Proceedings, 769-776.

6. Shastry S. (2004) "Treatment of Vanaspati Industry Wastewater using Upflow Anaerobic Sludge Blanket Reactor System" Doctoral Thesis submitted to Nagpur University, Nagpur, India.

7. Ubay, G., Özturk, I. and Ubay, E. (1992). "Atiksu Tipi ve Konsantrasyonunun Anaerobik Camur Özellikleri Üzerindeki Etkileri". SKKD, 3 (1), 141-146.

Chapter 5

Adsorption of Methylene Blue from Aqueous Solutions by an Economically Viable Indigenous Activated Carbon

Y.C. Sharma[1], Uma[1], S.N. Kaul[2] and S.N. Upadhyay[3]

[1]*Department of Applied Chemistry,*
[2]*MIT college Engineering and Technology, Kothrud Road, Pune*
[3]*Department of Chemical Engineering and Technology,*
Indian Institute of Technology (Banaras Hindu University), Varanasi

ABSTRACT

Adsorptive separation of a hazardous cationic dye, methylene blue by adsorption on activated carbon prepared from an agricultural waste material has been investigated. Activated carbon was prepared by pyrolyzing rice husk, an agro waste material. The activated carbon was found to possess both the micropores and mesopores and it displayed a surface area of 180.50 m^2/g. The surface morphology, surface functional groups and thermal behaviour of the activated carbon were investigated by Fourier Transformation Infrared (FTIR) Spectroscopy, X-ray diffractometry(XRD) and thermo-gravimetric analysis(TGA) respectively. The rice husk activated carbon(RHAC) was used for the removal of MB from aqueous solutions. The removal increased from 82.75 to 93.20 per cent by decreasing the initial concentration of dye from 100 to 60 mg/l. The process of removal followed first order kinetics and the value of rate constant of adsorption was found to be 1.41×10^{-2} min^{-1}. Adsorption data were fitted in different adsorption isotherm equations and their constants were calculated. Values of the thermodynamic parameters namely change in free energy, ΔG°, enthalpy, ΔH°, and entropy, ΔS° for the removal of the dye were also determined.

Keywords: *Activated carbon, Adsorptive separation characterization, Isotherm, Methylene blue.*

Introduction

Water pollution in general and the pollution of water by dyes/colours in particular has been a matter of concern for scientific workers since long time[1-4]. Solubility of a number of dyes in water is quite high and their removal from water is a challenge for the scientists. Many of the dyes produce toxic intermediate products during process of removal [5, 6] which are highly toxic. Textile and pulp and paper industries produce large amount of colored effluents. The color and other non-biodegradable constituents of the effluents pose serious environmental problems and because of their toxicity, they are also controlled by regulations. Methylene blue (MB) is a toxic dye and has been known to cause eye burns which sometimes result in permanent damage to vision in man and animals. It also causes difficulty in breathing, profuse sweating, mental problems, pain and other disorders [7-9]. Precipitation, coagulation, flocculation, adsorption, and filtration are the conventional methods used for treatment of colored effluents[10,11,12]. There is no single treatment method which is capable of adequate treatment of the color effluents. Adsorption is a method which has been widely reported for the removal of dyes from colored aqueous solutions and dye stuff effluents. Activated carbon has been frequently applied for decontamination of colored effluents [13] but its high cost limits its application for large scale treatment of wastewater in developing countries. To reduce the cost of treatment, significant number of non-conventional adsorbents have been used for the removal of colors from effluents but they offer a poor efficiency of removal. Present work has been devoted to the preparation of activated carbon from rice husk, an agricultural by product and its subsequent application for the removal of MB from aqueous solutions. Effect of important parameters namely contact time and initial concentration, effect of temperature and adsorbent dose on the removal has been studied. Extensive equilibrium studies have been carried out by fitting the data in various well known isotherm models.

Materials and Methods

Materials

Preparation of Carbon and Activated Carbon and their Characterization

Rice husk was collected from a rice mill, Varanasi, U.P. and was washed and dried overnight in a hot air oven. The dried husk was then subjected to carbonization at 450°C for 1 h using a tubular muffle furnace. A constant flow of 150 ml/min of high purity (99.99 per cent) nitrogen was maintained throughout the process of carbonization. The carbon formed was cooled to room temperature and washed with hot distilled water and 0.5 N hydrochloric acid until the pH of sample reached 7.0. The product was then dried in a hot air oven at 110°C overnight.

The primary carbon was impregnated with 20 per cent zinc chloride ($ZnCl_2$) solution for 24 h. Ratio of primary carbon and zinc chloride solution was1:1(w/w).The mixture was then dehydrated in an oven at 110°C for 12 h to remove moisture and activated under similar conditions that for carbonization but to a final temperature of 650°C for 2 h. The activated product was cooled under nitrogen gas flow to room temperature and was washed with hot distilled water and 0.10 M hydrochloric acid until the pH of washings reached 7.0. The activated carbon was then dried in a hot air oven at 110°C, ground and sieved to obtain desired particle size (150 μm) and stored in desiccators and was used for adsorption studies.

Micromeritics ASAP 2020 surface area analyzer was used to measure the surface area of the adsorbent. All absorbance measurements were recorded by UV-visible spectrophotometer (Spectronic 20, Bausch and Lomb, USA) at 663 nm. The surface area and pore size distribution of the activated carbon was determined by its N_2 adsorption isotherm by using Brunauer Emmett Teller (BET) method[14]. The FTIR spectral absorption studies were carried out by infrared spectrometer (Simadzu/ 8400S) to determine the presence of functional groups on the surface of adsorbent. The spectral wavelength was covered from 0-4500 cm^{-1}.

X-ray diffraction analysis of adsorbent was carried out using X-ray Diffractometer (Scifert and Co., Model ID–3000) to determine the phase characterization of RHAC. The powder sample for XRD was prepared by putting a thin layer of powder smeared over a glass plate. Goniometer speed was maintained at 1° min^{-1} and orientation was kept from 0° to 100°. Size of the RHAC was determined by using XRD result and the diffraction peaks from Scherrer formula:

$$X_s = 0.9\lambda / FWHM \cos\theta \tag{1}$$

where, X_s, crystallite size (nm); dikudd, wavelength of monochromatic X-ray beam(nm); and dikudd =0.154056 nm for CuKα radiation. FWHM: full width at half maximum for the diffraction peak under consideration (rad), and θ, the diffraction angle (degree).

Determination of pH_{zpc}

Determination of pH_{zpc} was done to investigate the surface charge of activated carbon. For the determination of pH_{ZPC}, 0.01 M NaCl was prepared and its pH was adjusted between 2.0 to 12.0 by using NaOH/HCl. 50 ml of 0.01 M NaCl was taken in the 250 ml Erylenmeyer flasks and 0.20 g of activated carbon was added to the solutions. These flasks were kept for 48 hrs and the pH of the solutions was measured by using pH meter[15]. Graphs were plotted between 'pH_{final} vs $pH_{initial}$' (with and without adding adsorbent)(Figure not shown). The point of intersection of the curves of 'pH_{final} vs $pH_{initial}$', which lies at 6.14 gives the value of pH_{ZPC} of activated carbon. The RHAC carried negative charged surface in solution which has significant role for adsorption of methylene blue[16].

Acidity and basicity of adsorbent can be understood by following two reactions:

$$RHAC–OH + H^+ \rightarrow RHAC–OH_2^+ \tag{2}$$

and

$$RHAC–OH \rightarrow RHAC–O^- + H^+ \tag{3}$$

The equations 2 and 3 show the nature of surface of RHAC to be basic and acidic respectively.

Results and Discussion

Surface Characterization of Adsorbent

Surface Area and Porosity

The specific surface area, the pore size distribution and the pore volume of rice husk activated carbon were determined by standard methods[17]. The pore analysis of RHAC showed the dominance of micropores having an area of 151.97 m^2/g, out of the total BET surface area 180.50 m^2/g.The BJH pore area was found to be 28.53 m^2/g for pore size range of 17-3000Å.The adsorption pore size was 22.52 Å whereas the BJH average pore size was 37.48 Å.

The surface area of RHAC was determined by volumetric adsorption of nitrogen at 77 K by the activated carbon. The system (instrument) provides data for determination of the monolayer adsorbed amount, apparent specific surface area and pore volume[18]. The sample (RHAC) was degassed for 12 h under vacuum at a temperature of 350 °C to remove any impurities prior to analysis. The adsorption of nitrogen under different partial pressures was determined. The mesopores and micropores can be defined by the hystereses in loops during adsorption at relatively high relative pressures (P/Po). The calculation of surface area, microporosity, mesoporosity and pore volume were carried out by standard methods[17].

FTIR Analysis

IR spectrum of rice husk gives an information of chemical structure and surface functional groups of the samples. Changes in the chemical nature of the cellulosic material rice husk were expected to occur in the process of carbonization and activation. Elimination of some elements in form of water and other species are responsible for these changes [19]. These changes were evidence of the formation of structures containing multiple carbon–carbon bonds, as well as the elimination of originally present oxygen and hydrogen atoms. The intense absorption band with maximum at about 2916.60 cm^{-1} can be attributed to the stretching vibrations of C-H bonds in alkanes. The band at 2291.69 cm^{-1}corresponds to vibration in alkanes and alkyl groups whereas the band near 1583.39 cm^{-1} is attributed to the vibration in aromatic group. The band at 1083.04 cm^{-1} is a less intense band and is related to the è(C-O) stretching vibration of the bonds in esters, ether or phenol group. The weak absorption bands at 777.28 cm^{-1} indicates γ (C-H) vibration in benzene derivative. The absorbance peak at 458.13 cm^{-1} was due to the bending vibration of siloxane bonds.

XRD Analysis

X-ray diffraction patterns for RHAC after thermal degradation in nitrogen or in air medium were recorded. The sample (RHAC) was amorphous, although a broad diffused peak was observed at about $2\theta=6°$ and about $2\theta=16°$ suggesting that silicon of rice husk is bonded with organic material and these peaks indicate the presence of amorphous form of carbon. Minor components identified in activated carbon are quartz, and alumina and major components are calcium silicate and silicon oxide. Diffraction peak of crystalline carbon was not observed. It was seen by the X-ray diffractrograms that the activated carbon has a heterogeneous surface [19].

TGA-DTA Analysis

Reaction bonding behavior of the RHAC has been obtained through TGA-DTA curves. Thermo gravimetric analysis of the substance is directly dependent on the decomposition temperature of its various oxides and functional groups. Interactions of surface groups with oxidizing gases are generally constant at lower temperature, whereas at higher temperature the surface groups decompose producing CO(150-600), CO_2(350and above), water vapor and free hydrogen (500 and above) [20]. As can be seen from the TGA curve, there is maximum weight loss, ~ 55 per cent at 450 °C under oxidizing atmosphere. However, ~ 5 per cent weight loss was recorded at 50-100 °C range. In oxidizing atmosphere pyrolysis of the adsorbent takes place and shows maximum degradation in the form of CO, CO_2, water vapour and volatile matters in gaseous form. An obvious weight loss between 50 to 600 °C is consistent with the analysis of TGA curve. There is one exothermal peak near 410 °C in the DTA cirve of RHAC. In addition, the peak intensity of rice husk activated carbon is found to rise once as temperature increases from 350 to 400 °C, while the peak intensity of RHAC has not obviously changed as temperature increases from 100 to 350 °C.This temperature zone is quite constant.

Effect of Contact Time and Initial Concentration on Removal of the Dye

The variation of removal of dye with contact time at different initial concentrations ranging from 60-100 mg/l. The removal of MB acquires a maximum at the time of equilibrium, *viz.*, 100 min. and becomes gradual there after. This indicates that after acquiring maxima, there is no significant change in the extent of adsorption. The removal increased from 82.75 to 93.20 per cent by decreasing the initial concentration of methylene blue from 100 to 60 mg/l. The removal (per cent) of the dye is higher in lower concentration ranges and this finding has industrial significance.

The effect of different doses of RHAC for the removal of MB was also carried out. The removal increased from 86.75 to 99.83 per cent by increasing adsorbent dose from 0.40 to 0.60 g for a solution of 50 ml containing MB. As expected, higher removal of dye was obtained at increased adsorbent dose. The higher removal at increased dose may be attributed to larger availability of active sites.

Effect of Temperature

Temperature is an important parameter for all adsorptive removal processes. It influences the adsorption of methylene blue also in present system. The effect of temperature was investigated in the temperature range 30 to 50 °C. The experimental results show that the removal of MB increased from 93.2 to almost 100 per cent (99.16 per cent) by increasing the temperature from 30 to 50 °C. Most of the adsorption processes are governed by exothermic processes, but the present studies lie in few examples of endothermic adsorption.

Activation energy, E_a, for removal of MB onto RHAC can be calculated by using Arrhenius equation [4]:

$$\ln k_1 = \ln A - E_a/RT \tag{4}$$

Where, A is the frequency factor (min^{-1}), k_1 = rate constant for the adsorption, E_a, activation energy (kJ mol^{-1}), T = Temperature (K), and R, the gas constant, 8.314 kJ mol^{-1} K^{-1}.The value of E_a can be calculated by the slope of graph 'ln k vs 1000/T' (Figure not shown). The activation energy for the present system was found to be 1.83 kJ/mol. Low activation energy indicates the process of removal to be governed by physical adsorption in present studies.

Kinetic Studies

The kinetics of removal of MB by RHAC was studied with respect to its different initial concentrations. For evaluating the adsorption kinetics of MB, the pseudo-first-order kinetic model has been used and found to fit the experimental data. The kinetic modeling of the removal of methylene blue on rice husk activated carbon was investigated by Lagergren's model [21]:

$$\log (q_e - q) = \log q_e - (K_{ad}/2.303)t \quad (5)$$

Where, q_e and q (both in mg/g) are amounts of methylene blue adsorbed at any time and at equilibrium, respectively, and K_{ad} (min^{-1}) is the rate constant of adsorption. The straight line plots of 'log (q_e- q) vs t' confirm that the process of removal is governed by first-order kinetics. The value of rate constant of adsorption was determined by slope and was found to be 1.41 $x10^{-2}$ at 30°C and its values were in increasing trend with temperature. The pseudo-first-order shows straight lines for most initial concentrations indicating suitability of this model.

Equilibrium Studies

Equilibrium studies for the present studies were carried out by fitting the equilibrium data in Langmuir, Freundlich, Temkin and Dubinin-Radushkevich isotherm equations and the constant were calculated. However, selection of an isotherm equation depends on the nature and type of the system.

Langmuir Isotherm

The Langmuir model assumes that uptake of methylene blue ions occurs on a homogeneous surface of the adsorbent by monolayer adsorption. The Langmuir equation is expressed by the following expression [22]:

$$C_e/q_e = 1/Q^o b + C_e/Q^o \quad (6)$$

Where, C_e (mg/l) is the equilibrium concentration of the solute (mg/l), q_e is amount adsorbed at equilibrium (mg/g), and Q^o (mg/g) and b (l/mg) are constants related to the adsorption capacity and energy of adsorption, respectively. A plot of C_e/q_e vs C_e gives a straight line.

For Langmuir isotherm the essential characteristics can be expressed in terms of a dimensionless equilibrium parameter (R_L) [21,23]:

$$R_{L =} 1/1+KC_o \quad (7)$$

Where, K is the Langmuir constant (l/g) and C_o is the initial dye concentration (mg/l).The value of R_L indicates the type of the isotherm to be either unfavorable, linear or favorable. If $R_L > 1$, it follows an unfavorable adsorption, a value 1.0 of R_L

indicates a linear, R_L lesser than 1.0 a favorable and a value of R_L equal to zero suggests an irreversible adsorption. Values of R_L at different temperature show favorable nature of adsorption for methylene blue on rice husk activated carbon.

Freundlich Isotherm

Adsorption data for the removal of MB by RHAC was fitted to the linear form of Freundlich isotherm:

$$\log x/m = \log K_f + 1/n \log C_s \quad (8)$$

Where, x/m is the amount adsorbed per unit mass of the adsorbate, C_s the equilibrium concentration, and 1/n and K_f are constants. The constant K_f is related to the degree of adsorption, and '*n*' provides the tentative estimation of the intensity of the adsorption. The values of the parameters were determined from the straight line plots of 'log x/m vs C_s'. K_f and n were calculated from the slopes and intercepts of the straight line plots. It may be noted that the values of K_f and n increase with an increase in temperature for all the concentrations indicating that adsorption is favorable at higher temperature [24,25].

Tempkin Isotherm

Tempkin isotherm exhibits heat of adsorption and the adsorbent–adsorbate interaction. It is characterized by uniform distribution of binding energy. According to Tempkin isotherm the adsorption in the layer decreases by the coverage due to interactions between adsorbate and adsorbent [26]. The Tempkin isotherm equation is given as:

$$q_e = RT/b.\ln A\, C_e$$

$$q_e = B_1 \ln A + B_1 \ln C_e \quad (9)$$

Where, B_1 =RT/b, T, absolute temperature (K), R, Gas constant (8.314J/mol K), A= Equilibrium binding, q_e=amount adsorbed face(mg/g), and C_e= Concentration of dye in aqueous solution at equilibrium. The value of B_1 and A have been calculated by slopes and intercepts of the plots respectively from a graph 'q_e vs ln C_e'. b is related with the heat of adsorption. Increasing trend of value of B_1 indicates the endothermic nature of adsorption reaction.

Dubinin- Radushkevich Isotherm

For the analysis of high degree of rectangularity, a well known equation, Dubinin-Radushkevich isotherm equation was applied to investigate the nature of adsorption of MB by RHAC[27]. The linear form of Dubinin - RadushKevich isotherm equation can be expressed as:

$$q_e = q_m \exp(-K\ ^2)$$

$$\ln q_e = \ln q_m - K\ ^2$$

$$= RT\ln(1+1/C_e) \quad (10)$$

The value of q_m and K can be obtained by the intercept and slope of the graph of 'ln q_e vs ^{2}'.

E, the mean free energy of sorption per molecule of the sorbate (kJ/mol) involved in the process can be calculated by D-R isotherm equation as follows:

$$E = (1/\sqrt{2K}) \quad (11)$$

The mean adsorption energy (E) gives information about chemical and physical nature of adsorption. The straight line plots indicate that adsorption follows D-R isotherm in this study. The magnitude of 'E' determines the type of adsorption process [27].

Thermodynamic Study

Thermodynamic study helps to decipher any reaction in a better way. The thermodynamic parameters namely, change in standard enthalpy (ΔH^o), standard entropy (ΔS^o), and standard free energy (ΔG^o), were calculated for the process of removal of MB onto RHAC at 30, 40 and 50°C, respectively. The following well known equations were used for the determination of these parameters[28]:

$$K_c = C_{ae}/C_e \quad (12)$$

$$\Delta G^o = - RT \ln K \quad (13)$$

$$\Delta H^o = R\,(T_2T_1/T_2\text{-}T_1).\ln(K_2/K_1) \quad (14)$$

$$\Delta S^o = (\Delta H^o - \Delta G^o)/T \quad (15)$$

Where, R (1.987Kcal/mol) is the universal gas constant, and T (K) is the absolute temperature. C_e is the equilibrium concentration of methylene blue in the solution (mg/l) and C_{ae}(mg/l) is the amount adsorbed on adsorbent at equilibrium. The values of K_c increased by increasing the temperature, which indicates the endothermic nature of the process of removal. K, K_1 and K_2 are equilibrium constants at temperatures T, $_{T1}$ and T_2 respectively. The values of K increased by increasing the temperature, which indicates the endothermic nature of the process of removal. Increasing trend of adsorption capacity with increasing temperature shows the nature of adsorption of methylene blue on rice husk activated carbon is endothermic. It also indicates that weak adsorptive forces between active sites of RHAC and dye molecule.

Conclusions

A mesporous/microporous activated carbon has been prepared from rice husk. FTIR, and XRD were used to characterize the activated carbon prepared. The surface of RHAC is found to be amorphous and heterogenous. Value of rate constant of adsorption shows that the rate of removal of MB by RHAC is significant. The negative values of ΔG^o indicate spontaneous nature of the adsorption of MB onto RHAC. The study confirms that rice husk activated carbon can be used as an alternative for more costly adsorbents used for dye removal in wastewater treatment processes.

References

1. P. Antonio, K. Iha, M.E.V. Sua´ rez-Iha. Kinetic modeling of adsorption of di-2-pyridylketone salicyloylhydrazone on silica gel, J. Colloid Interface Sci. 307 (1) (2007) 24-28.

2. R. Jain, S. Sikarwar. Adsorptive removal of Erythrosine dye onto activated low cost de-oiled mustard, J. Hazardous Mat. 164(2009) 627-633.

3. N. Feng, X. Guo, S. Liang. Adsorption study of copper (II) by chemically modified orange peel, J. Hazardous Mat. 164(2009) 1286-1292.

4. I.A.W. Tan, A.L. Ahmad, B.H. Hameed. Adsorption isotherms, kinetics, thermodynamics and desorption studies of 2,4,6-trichlorophenol on oil palm empty fruit bunch-based activated carbon, J. Hazardous Mat. 164(2009) 473-482.

5. V. Fierro, V. Torne´-Ferna´ndez, D. Montane´, A. Celzard. Adsorption of phenol onto activated carbons having different textural and surface properties, Microporous Mesoporous Materials 111 (1–3) (2008) 276-284.

6. A. Bahdoda, S. El Asria, A. Saoiabia, T. Coradinb, A. Laghzizila. Adsorption of phenol from an aqueous solution by selected apatite adsorbents: Kinetic process and impact of the surface properties, Water Res. 43(2009) 313-318.

7. G. Liu, J. Ma, X. Li, Q. Qin. Adsorption of bisphenol from aqueous solution onto activated carbons with different modification treatments, J. Hazardous Mat. 164(2009)1275-1280.

8. B. Royer, N. F. Cardoso, E. C. Lima, J. C.P. Vaghetti, N. M. Simon, T. Calvete, R. C. Veses. Applications of Brazilian pine-fruit shell in natural and carbonized forms as adsorbents to removal of methylene blue from aqueous solutions—Kinetic and equilibrium study, J. Hazardous Mat.164(2009) 1213-1222.

9. C. Namasivayam, D. Kavitha. Adsorptive removal of 2,4-dichlorophenol from wastewater by low-cost carbon from agricultural solid waste: coconut coir pith, Sep. Sci. Technol. 39(2004)1407-1425.

10. J.H. Weisburger. Comments on the history and importance of aromatic and heterocyclic amines in public health, Mutat. Res. 506 (2002), 9-20.

11. P.K. Malik. Use of activated carbons prepared from sawdust and rice husk for adsorption of acid dyes: a case study of Acid Yellow 36, Dyes Pigments 56(2003) 239-249.

12. S.T. Ong, C.K. Lee, Z. Zainal. Removal of basic and reactive dyes using ethylenediamine modified rice hull, Bioresource Technolology 98(2007) 2792-2799.

13. A. Aziz, M S. Ouali, El H. Elandaloussi, L.C. De. Menorval, M. Lindheimer. Chemically modified olive stone: A low-cost sorbent for heavy metals and basic dyes removal from aqueous solutions, J. Hazardous Mat. 163(2009) 441-447.

14. E.P. Barret, L.G. Joyner, P.P. Hanlenda. The determination of of pore volume and area distributions in porous substances, J.Am.Chem.Soc. 73(1951)373-380.

15. M. Namdeo, S.K. Bajpai. Investigation of hexavalent chromium uptake by synthetic magnetite nanoparticles, EJEAFChe. 7 (7) (2008.) 3082-3094.

16. N. El. Emad, J. A. Stephen, M. W. Gavin. Adsorption of basic dyes from aqueous solution onto activated carbons, Chemical Engineering J. 135 (2008)174-184.

17. S. Brauner, P.H. Emmet, E. Teller. Adsorption of gases in multimolecular layers, J. Am. Chem. Soc. 60(1938) 309-319.

18. K. Sujirote, P. Leangsuwan. Silicon carbide formation from pretreated rice husks, J. Mater.Sci.38(2003) 4739-4745.

19. S.D. Genieva, S.Ch. Turmanova, Turmanova, A.S. Turmanova, L.T. Vlaev. Characterization of rice husks and products of its thermal degradation in air or nitrogen atmosphere, J. athermal Calorimetry 93(2) (2008)387-396.

20. I.D. Mall, V.C. Srivastva, N.K. Agarwal. Removal of orange-G and methyl violet dyes by adsorption onto bagasse fly ash–kinetic study and equilibrium isotherm analyses, Dyes Pigments 69 (2006) 210-223.

21. G. McKay, M. El Guendi, M. Nassar. Equilibrium studies during the removal of dyestuffs from aqueous solutions using bagasse pith, Water Res. 21 (12) (1987) 1513-1520.

22. D.C.K. Ko, D.H.K. Tsang, F.J. Porter, G. McKay. Applications of multipore model for mechanism identification during the adsorption of dye on activated carbon and bagasse pith, Langmuir 19(2003) 722-730.

23. T.W. Weber, R. K. Chakkravorti. Pore and solid diffusion models for fixed–bed adsorbers, AIChE J.20(1974) 228-235.

24. G. Walker, L. Weatherley. Adsorption of dyes from aqueous solution—the effect of adsorbent pore size distribution and dye aggregation, Chemical Engineering J. 83 (2001) 201-206.

25. H. Freundlich. Uberdie adsorption in lunsungen, J. Phys. Chem. 7(1985)387-470.

26. M. Hosseini, S.F.L. Mertens, M. Ghorbani, M.R. Arshadi. Asymmetrical schiff bases as inhibitors of mild steel corrosion in sulphuric acid media, Mater. Chem. Phys. 78(2003) 800-809.

27. M.M. Dubinin, L.V. Radushkevich. Equation of the characteristic curve of activated charcoal, Chem. Zentr.1(1947) 875-878.

28. Y.C. Sharma, Uma, S.N. Upadyay. Removal of a cationic dye from wastewaters by adsorption on activated carbon developed from coconut coir, Energy and Fuels 23(2009)2983-2988.

Chapter 6
Biofiltration

Kiran Singh[1], R.S. Singh[1], B.N. Rai[1] and S.N. Kaul[2]

[1]Department of Chemical Engineering and Technology, Indian Institute of Technology (Banaras Hindu University), Varanasi
[2]CSIR-NEERI, Nagpur

Introduction

Several control technologies are used for the destruction and recovery of VOCs (Volatile Organic Compounds) emitted from industries. In general treatment of VOCs bearing waste gases can be done in two different ways - source control and emission control. Source control involves the reduction of emissions through substitution of raw materials, recycling, and its reduction. However, these measures may reduce the quality of the product or increase costs. Secondary control involves treatment of the waste gas after it has been produced. Conventional air pollution control technologies such as absorption, adsorption, condensation, thermal oxidation and catalytic oxidation are generally not cost effective for the treatment of VOCs bearing dilute waste air streams. Besides this, the above treatment technologies produce hazardous by-products also and require the addition of chemicals or fuels that may require further treatment or disposal, thereby creating additional environmental problems. A brief account of their salient features are summarized in Table 6.1.

Biofiltration, though the oldest of biological methods of waste control, is now emerging as a sustainable alternative for the treatment of air contaminated with VOCs. In biofiltration the polluted air is passed through a bed of packing media covered with a layer of aerobic micro-organisms. The micro-organisms are immobilized on the surface of the packing media. The VOCs get transferred from air to the water adhering to the bacterial growth on the media and then get biologically metabolized (Upadhyay and Kumar, 2004). Biofiltration is energy efficient and cost

Table 6.1: Characteristics, Advantages and Drawbacks of the Main Air Pollution Control Technology (APCT) Applied to VOC Elimination (Delhomenie and Heitz, 2005).

Technology	*Principle and Characteristics*	*Performances and Limitations*	*Costs/($m^3 \cdot h^{-1}$) Air (In US $)*
Adsorption	**Transfer of VOCs to a porous solid phase, fixed or fluidized** ☆ Materials: activated carbons, zeolites and polymers ☆ Example: Activated carbon adsorbs 10–30 per cent VOCs on weight basis ☆ Doubled installations:adsorption–desorption cycles ☆ Operating temperature<50–60°C: ignition risks may be present	☆ Conversion: 90–99 per cent ☆ Possible recovery of VOCs (desorption) ☆ Can accept variations of flow rates and shutdown periods ☆ Treatments for adsorbentregeneration are required ☆ Moisture level of effluents: <50 per cent ☆ High pressure drop ☆ Bed poisoning problems withcertain VOCs	Investment: $15–120 Operation: $10–35
Incineration	**Thermal oxidation of VOCs** ☆ 760 < Temperature < 1200 °C ☆ 0.3 < Residence time < 2 s ☆ VOCs concentration < 25 per cent of explosion limit ☆ Required O_2 level ~10 per cent	☆ Conversion: 98–99.5 per cent ☆ Possible energy recovery ☆ Elimination of halogenated or sulfurated VOCs with adequate, additional equipment ☆ High investment and operating costs ☆ Toxic by-products: CO, NO_x, dioxins, furans ☆ Efficiency decreases for low VOCs concentrations	Investment: $10–450 Operation: $20–150 (depends on the quantity of recovered energy)
Catalytic oxidation	**Thermal, catalytic oxidation of VOCs** ☆ 300 < Temperature 650 °C ☆ 0.07 < Residence time < 1s ☆ Catalysts: noble metals (Pt, Pd, Rh) on supports (alumina, other ceramics), or metal oxides (Cu, Ti, Ni, Mn, etc.) ☆ Catalyst life time: 2–5 years ☆ VOCs concentration:far lower than the explosion limit ☆ Required O_2 level ~2 per cent	☆ Conversion: 90–99 per cent ☆ Less energy required than incineration and less toxic by-products ☆ Catalyst deactivation problems (clogging, poisoning, overheating) ☆ Disposal of spent catalyst ☆ Combustion by-products	☆ Fixed catalyst Investment: $20–250 Operation: $10–75 ☆ Fluidized catalyst Investment: $35–220 Operation: $15–90
Absorption	**Transfer of the VOC to a liquid phase** ☆ Plate tower, bubble column, packed tower, atomizer ☆ Solvents: water (withadjusted pH), high boiling-point hydrocarbons, amines, etc. ☆ Counter-current operationincreases the VOC transferrate	☆ Conversion: 90–98 per cent ☆ Possible recovery and vaporization of the dissolved VOC with downstream treatments ☆ Inadequate for VOCs of lowsolubility ☆ Production of wastewater	Investment: $15–70 Operation: $25–120

Contd...

Table 6.1–*Contd...*

Technology	*Principle and Characteristics*	*Performances and Limitations*	*Costs/($m^3 \cdot h^{-1}$) Air (In US $)*
Condensation	**Liquefaction of highboiling-point VOC (>38 °C) via cooling and/orcompression** ☆ Cooling/cryogenic systems:water (5°C), brine (→ –35°C), liquid nitrogen (→ –185°C)	☆ Conversion: 50–99 per cent ☆ A recovery and vaporization way well adapted to saturated VOCs only ☆ Disposal of condensates ☆ Problems of frost deposits	Investment: $10–80 Operation: $20–120
Membranes	**Separation of gas mixtures through semi-permeable membranes** ☆ Materials: polymers (hollowfibers, silicones), porous ceramics ☆ Gas flow compressed beforemembrane separation	☆ Conversion: 50–98 per cent ☆ VOC are concentrated 5–100 times, and vaporization (recycle) possible ☆ Selective membranes, resistant to halogenated VOC ☆ High pressure drop ☆ High operating pressures ☆ Membrane cleaning required	Not available
UV/ photochemical oxidation	**Complete oxidation byoxygenated oxidizers (O_3, $_{H2}O_2$), and initiated by UVradiation** ☆ Possible utilization o fphotocatalysts (TiO_2, Fe_xO_y, etc.) ☆ Operating temperature: ambient	☆ Conversion: 90–98 per cent ☆ Moderate energetic costs ☆ Inadequate for halogenated VOCs ☆ Deposit of oxidation by-products on the catalysts surface (cleaning) ☆ Complex systems	Not available
Biotechnologies (detailed later)	**Biocatalytic oxidation of VOCs** ☆ Three configurations:biofilters (most frequent), biotrickling filters and bioscrubbers ☆ Biocatalysts: micro-organisms (bacteria, fungi) ☆ 30 s < Residence time < several min ☆ Operating temperature: 20–40°C ☆ Filter-bed life time: 3–5 years	☆ Conversion: 80–95 per cent ☆ Moderate installation and operating costs ☆ Low maintenance ☆ Strict control of biological parameters (pH, temperature, moisture level, nutrients, etc.) ☆ Large space required for biofilters ☆ Pressure drop problems	Investment: $10–70 Operation: $3–10

effective and also does not produce toxic end-products. The degradation of VOCs by biofiltration is very effective at low concentrations, *i.e.*, less than 1000 ppmv. This technology has been successfully used for removing a wide range of VOCs including mal-odorous pollutants such as ammonia, mercaptans, and sulphurous compounds. There have been many commercial biofilters using compost-type material installed in Europe and the United States over the past 20 years. Their applications to date also include several industries like, chemical and petrochemical industry, oil and gas industry, synthetic resins, paint and ink, pharmaceutical industry, waste and wastewater treatment, soil and groundwater remediation, and odor abatement applications. Biofilters and biotrickling filters are nearly similar and the basic difference between them is that in biotrickling filter nutrients are continuously trickled through the porous bed but in biofilters nutrient circulation is occasional. Therefore, biofiltration technology is cheaper than biotrickling filters.

Bioscrubber is essentially a two stage unit in which absorption occurs in first stage and biodegradation occurs in the other in which microbes are suspended. Bioscrubbers are usually applied for the treatment of readily soluble VOC (alcohols, ketones etc.) having concentration less than 5 g m^{-3}. The major problem associated with bioscrubbers is the generation of excess sludge and liquid waste. The guide lines for selecting treatment technology based on concentration and flow rates are given in Figure 6.1.

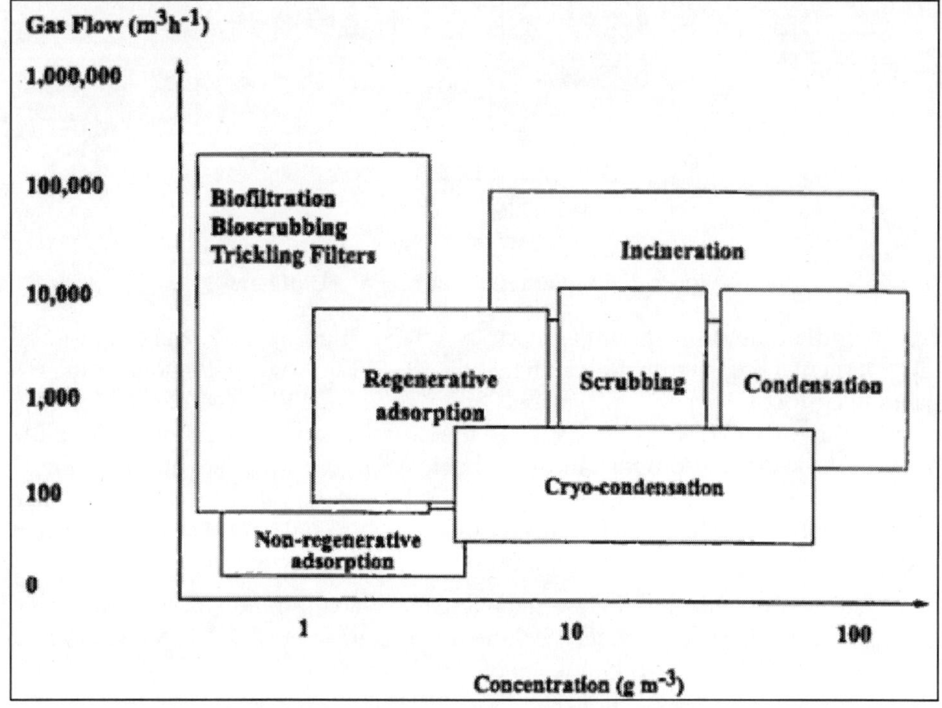

Figure 6.1: Guidelines to Choose VOCs Control Technologies, Based on Flow Rate and Concentration.

Biofiltration Technology

For the treatment of malodours compounds biofiltration process was proposed for the first time in the year 1923. Since 1950's the biofiltration process is being used to treat odorous compounds emitted from wastewater treatment plants. This process is also used to remove odour generated from solid waste processing, animal rendering plants, composting operations, and breweries. For the treatment of hazardous air pollutant biofiltration process has been widely and successfully used in Germany and the Netherlands since 1970's. In US, biofiltration process was first used to treat odours resulting from H2S gases in sewage works. There are more than 600 biofilters operating both in Germany and Netherlands and the technology is also widely spreading in United States. A schematic diagram of a typical biofilter is shown in Figure 6.2.

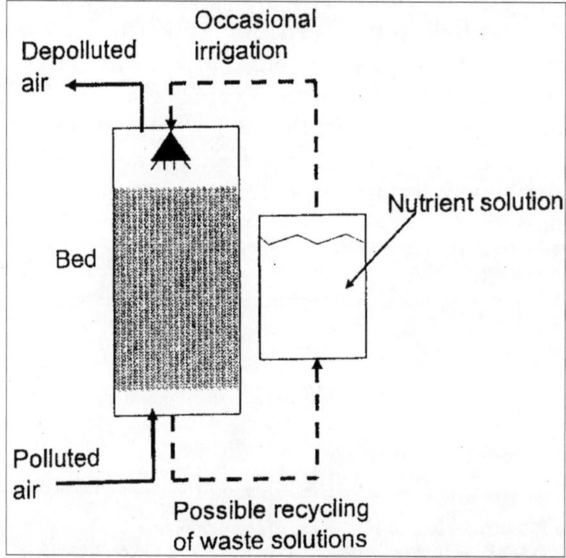

Figure 6.2: Schematic Diagram of a Biofilter.

In India a few research centers like BITS, Pilani, IIT Roorkee, National Environmental Engineering Research Institute (NEERI), Nagpur, Regional Research Laboratory (RRL), Trivendurm, Banaras Hindu University, Varanasi, have been doing work in this area. Giri *et al.* (2010) studied the degradation of dimethyl sulfide in the biofilter packed with compost and wood chips. Singh *et al.* (2006) and Kiran *et al.* (2010), examined various filter media like wood charcoal, yellow gram stalk, maize cobs and polyurethane foam. Raghuvanshi and Babu, (2009) used a mixture of coal and matured compost for the degradation of methyl ethyl ketone.

Many researchers have reviewed the research on biofiltration (Leson and Winer, 1991; Bohn, 1992; Singhal *et al.*, 1996; Swanson and Loeher, 1997; McNevin and Barford, 2000; Cohen, 2001; Pandey *et al.*, 2002; Groenestijin and Kraakman, 2005; Delhomenie and Heitz, 2005). Leson and Winer (1991) reviewed the history and current status of biofiltration discussed the principles and applicability of biofilters for a wide range of specific emission and sources. Bohn (1992) summarized the work done by other researchers and also provided the brief description of commercial biofilters running successfully at various locations in Europe and United States.

Bio-filtration Process

Biofiltration is nothing but bio-oxidation of contaminated air by beds of organic or inorganic media containing micro-organisms that have the ability to minimize pollutants aerobically. Bio-filtration utilizes a supported media such as lavarock, peat, compost, agricultural by-products, soil, etc. for microbial growth to remove odours and organic contaminants for air streams. The filter bed is kept wet by humidifying the influent air to maintain sufficient moisture for maintaining a biologically active layer (biofilm) surrounding the media. The contaminated air stream is passed through the packing media and the oxygen present in the contaminated air helps in maintaining aerobic conditions within the biofilter unit. As the contaminated gas passes through the filter bed, the gaseous contaminants are transferred from gas phase to water phase and get absorbed in the biofilm layer. The contaminants are now available to the micro-organisms as food and energy source to support microbial life and growth. This degradation process is an aerobic heterogeneous biochemical process controlled either by mass transfer or biochemical reaction or by both. The removal efficiency of biofilters varies with media characteristics (*e.g.* particle size, surface area, and porosity), carbon and energy source, nutrients, pH and temperature. The removal of pollutants during biofiltration process largely depends upon the presence of micro-organisms in the filter bed capable of biodegrading the contaminant of interest. The capacity of micro-organisms to degrade pollutants decreases with the increase in inlet load.

Biotransformation process can be expressed as:

$$\text{VOCs} + O_2 \xrightarrow{\text{Micro-organism}} \text{Biomass} + CO_2 + H_2O + \text{Heat} \quad (1)$$

The biodegradation of VOCs present in air stream begins with the transfer of pollutants from gas phase to aqueous phase and then to biofilm. These pollutants then get consumed by the microbial cells as the source of carbon and energy and get converted to CO_2, water and new biomass. Figure 6.3 shows the mechanism of biofiltration.

The transfer of pollutants from gas phase to aqueous phase is not a rate limiting step. The partition between gas and aqueous phase is generally described by Henry's law, which is

$$P = HC_w \quad (2)$$

Where, p is the partial pressure of contaminant in the gas phase (Pa), C_w is the equilibrium concentration of contaminant in the water phase (mol/l_{water} or g/l_{water}), and H is Henry's law constant (Pa l/mol or Pa l_{water}/g).

Pollutants with large Henry's law constants are difficult to degrade and require longer residence times than those with small Henry's law constant. Generally the elimination capacity of a biofilter declines with increasing Henry's law constant.

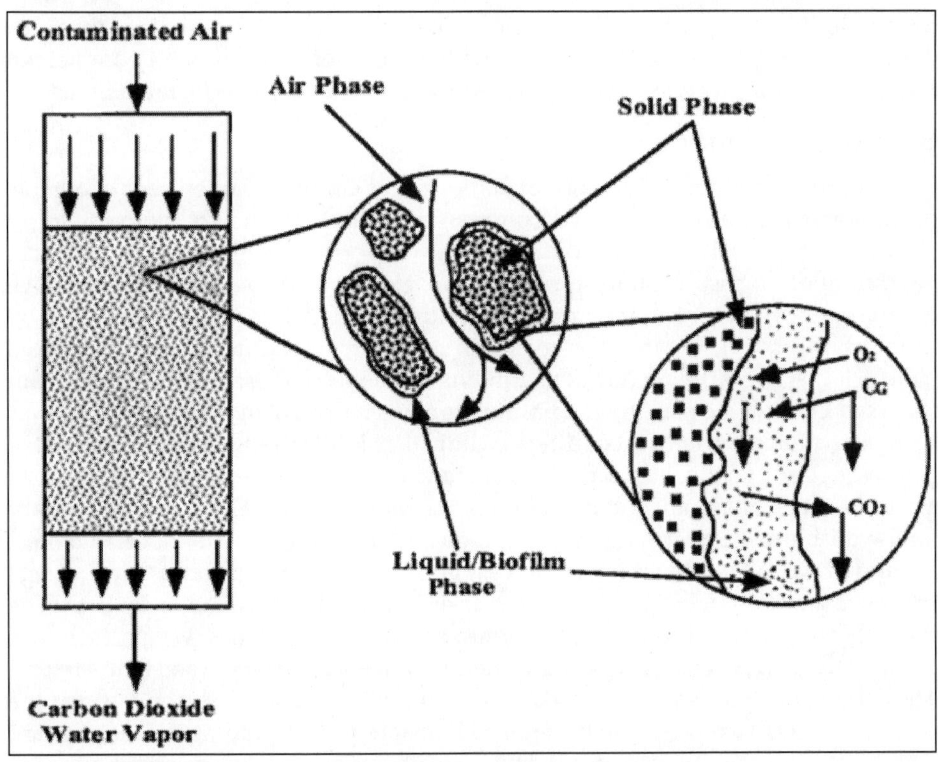

Figure 6.3: Internal Mechanisms of a Biofilter.

Contaminated air (CG) passes through the filter bed medium (compost, peat, soil, etc.) with (O_2) and sorbs into a microbial biofilm/liquid phase attached to the filter medium. Microbes convert the contaminant to carbon dioxide (CO_2) and water. (Devinney *et al.*, 1999).

The reversible adsorption of pollutant on the surface of filter media can be described by an equilibrium isotherm (Linear Freundlich, Langmuir, etc.) The transport of pollutants in the biofilm occurs by molecular diffusion following Ficks law.

The destruction of contaminants occurs in the biofilm, containing immobilized micro-organisms, surrounding the filter media. The micro-organisms degrade the gaseous contaminants and produce water, carbondioxide, some inorganic acids, intermediate products from the incomplete biooxidation of VOC, such as acetic acid, and microbial biomass. The development of biofilm is influenced by several factors such as gas flow rate and type of filter media. Filter media supports the biomass and provides moisture, nutrients, etc. for the microbial reaction. The biofilm thickness usually varies from tens of micrometers to more than 1 cm, with an average value of 1 mm or less. The microbial activity increases with the thickness of the biofilm, up to a level termed as the 'active thickness'. Above this value, the diffusion of nutrients becomes a limiting factor, thus differentiating the 'active' biofilm from the inactive one. Once the rate of adsorption of pollutants becomes faster than the rate of oxidation, the biofilter becomes overloaded, which decreases the removal efficiency.

Microbiology

Micro-organisms are the most critical component of the biofilter, as because of the microbial metabolism, the adsorbed pollutants are degraded into carbon dioxide and water and cell biomass. In the biofilters around 1 billion micro-organisms are present per gram of organic material. A complex microbiological community exists in the biofilter. A broad variety of bacteria, yeast, fungi, protozoa, nematodes, algae and other higher organisms are generally observed in biofilters. For the proper growth of micro-organisms and for the optimal contaminant degradation microbes require a suitable environment. The most important environmental factor for microbial function is the moisture in the contaminated air stream entering the biofilter and the filter media. Initially for evaluation of the microbial population dynamics cell counting method were used but now a days, molecular biological techniques like denaturating gradient gel electrophoresis (DGGE) of polymerase chain reaction (PCR) amplified 16S rDNA fragments, fluorescence in situ hybridization (FISH) and single-strand conformation polymorphism (SSCP) are applied. With the help of these techniques entire microbial community existing in the filter bed can be assessed and the most dominating microbe can be found out. In terms of biomass density, a biofilter contains between 10^6 and 10^{10} cfu^e of bacteria and actinomycetes (Delhomeni and Heitz, 2005).

The inoculation of the filter bed was done on the basis of the type of filter media used and the biodegradability of volatile organic compound. Generally natural filter media do not require any inoculation, since they have indigenous microbes. Most researchers preferably use packing media like compost and soil as they do not need any inoculation. During the course of biofiltration the microbes get acclimatized and then the most resistant microbial population dominates the bed. Highly soluble VOCs degrade very easily and a general source such as activated sludge having a wide spectrum of micro-organisms is often sufficient for inoculation. For treatment of the VOCs which are less soluble such as halogenated hydrocarbons like TCE etc, soil or water samples from site of contamination with the pollutant of interest or consortia that are enriched in the laboratory on the pollutants of interest, or pure cultures degrading the pollutant of interest are obtained either from culture collection or isolated from the mixed consortia, or leachate of the biofilter treating the pollutant of interest. By using such sources for inoculation of the filter bed the acclimation time can be minimized. Fungi generally grow at a slower rate in comparison to the bacteria but they are capable of degrading a broad variety of pollutant and can withstand with more adverse conditions. For toluene degradation bacterial consortia was used by many researchers (Singh *et al.*, 2010; Kiran *et al.*, 2010). But now a day's fungi have also used as an inoculum for toluene degradation by many researchers (Gracia-Pena *et al.*, 2001; Maestre *et al.*, 2007; Aizpuru *et al.*, 2005). Micro-organisms can survive for fairly long periods when the biofilter is not loaded up to 2 months, if sufficient nutrients are available from the filter material. Steele *et al.* (2005) observed difference between initial communities and community after acclimation due to restructuring of microbial communities after acclimation. The organisms which are capable of removing different type of odorants and VOCs are listed in Table 6.2.

Table 6.2: Micro-organisms Identified during the Degradation of VOC by Biofiltration.

Pollutant	*Micro-organisms*
Benzene	*Pseudomonas* sp.
Benzene	*Alcaligenes xylosoxidans*
Benzene	*Cladosporium sphaeraspermum, Exophila lecaniicorni, Phanerochaete chrysosporium*
BTX	*Phanerochaete chrysosporium*
Butylacetate *chrysosporium*	*Cladosporium resinae, Exophiala lecanii-corni, Phanerochaete*
Diethyl ether	*Graphium* sp. *ATCC 58400*
Ethylbenzene	*Phanerochaete chrysosporium*
Ethylacetate	*Rhodococcus fascians*
Methylisobutylketone	*Phanerochaete chrysosporium*
Phenols and Chlorophenols	*Pseudomonas putida, Pseudomonas fluorescence, Azotobacter, Flavobacterium, Bacillus* sp., *Achromobacter* sp., *Rhodococcus* sp., *Xanthomonas, Aspergillus, Penicillium, Neurospora, Deberomyces, Pichia, Trichosporon, Oospora*
Styrene	*Exophiala jeanselmei, Rhodococcus pyridinovorans, Phanerochaete chrysosporium* ATCC 34541
Toluene	*Pseudomonas pseudoalcaligenes, Paecilomyces variotii* CBS115145, *Cladosporium sphaerospermum, P. putida*
Trichloroethane	*Pseudomonas putida, Methylosinus trichosporium* OB3b, *Burkholderia cepacia* G4, *Pseudomonas mendonica* KR-1, *Nitrosomonas europaea, Pseudomonas putida* F1
Xylene	*Pseudomonas pseudoalcaligenes, Phanerochaete. chrysosporium*
Hexane	*Aspergillus niger, Fusarium* sp.
α-pinene	*Pseudomonas* sp., *Nocardia* P18.3*; Bacillus pallidus* BR425
Sulphurous odourants	*Hyphomicrobium* sp., *Thiobacillus thioparus, Thiobacillus ferroxidans, Pseudomonas* sp., *Xanthomonas* sp., *Aeromonas* sp., *Basidomycetes* sp.,
Formaldehyde	*Gliocladium deliquesecens, Paecilomyces variotii*

Source: Delhomenie *et al.*, 2002; Pandey *et al.*, 2002; Kennes and Veiga, 2004.

Pollutants are not always degraded independently by micro-organism. Several sparingly soluble VOCs such as high chlorinated aliphatics (*e.g.* trichloroethylene (TCE)) are degraded by biofiltration using co-metabolism *i.e.* certain other easily degradable hydrocarbon (*e.g.* toluene) helps in degrading TCE. On the other hand certain pollutant in mixed stream works as inhabitant for another one. Continuous exposure of microbes to high pollutant concentrations during biofiltration processes makes them less effective as the bacteria lose their ability to degrade pollutant. This also enhances the growth of non-pollutant degrading bacteria. In the biofiltration process change in the filter bed temperature during the degradation of VOCs indicates the intensity of microbial activity.

Factors Affecting Biofilter Performance

Nature of Pollutant

Pollutant's solubility and biodegradability have a great influence on the overall efficiency of biofiltration. The adsorption of pollutants into the biofilm largely depends on their solubility. High water solubility, low vapor pressure and hence low Henry's law constant are the desirable properties. Alcohols, aldehydes, ketones and some simple hydrocarbons are highly to moderately water soluble. On the other hand chlorinated hydrocarbons are less water soluble. The degradation of slightly soluble compounds at low concentrations in the biofilter coloumn is limited by the mass transfer rate into the biofilm. However, soluble compounds introduced at higher concentrations, are easily absorbed into the biofilm, but in this case biodegradation rate is the limiting factor. To increase the solubility of hydrophobic pollutant now a days some additives like silicone oil is being added.

Biofilter Media

The ability of biofilter to remove pollutant largely depends on the selection of the packing material as biofilter medium. The packing medium can briefly be described as solid phase on which adhesion of the micro-organisms takes place, resulting in the development of a biofilm around it, which grows as a result of pollutant degradation. The choice of biofilter medium is crucial in order to maintain long-term efficient biofilter operation. The desirable properties which are considered during the selection of an appropriate packing material are its specific surface area, density, porosity, pH, water holding capacity, buffering capacity, and elemental composition. Presence of these properties ensures intrinsically active biofilter medium with large effective surface area, minimum pressure drop, and uniform distribution of air throughout the bed without channeling. Biofilter medium also serves as living space and reserve of substrate for micro-organisms, as humidity reservoirs and as mechanical support for the maintenance of the internal structure of the filter bed. Filter materials can be broadly classified into two categories *i.e.* natural and synthetic filter materials. Soil, compost, peat, wood chips, biomass, coal are the most commonly used natural filter materials and Pall rings, granular activated carbon (GAC), polyurethane foam, polystyrene, etc. are commonly used synthetic media. Natural materials show compaction problem and may deteriorate over long term operations so some inert materials like poly vinyl alcohol, ceramic, perlite, vermiculite, glass beads, and polypropylene are blended with them to increase their strength. Nutrient supply is also a critical parameter for the microbial activity, so nutrients must be added during the operation. A synthetic filter material PVA/peat/KNO_3 composite bead are also developed which contains nitrogen and phosphate bearing compounds as nutrients.

Natural Filter Media

These filter media have most of the desirable properties required for achieving good removal efficiency such as the presence of intrinsic nutrients, indigenous microflora, large specific surface area (from 300 to 1000 $m^2 m^{-3}$), large water retaining capacity, low bulk density, ease of availability and low cost. However, there are some disadvantages also associated with the use of natural medias. Most of these packing materials have a common disadvantage of being strongly subjected to aging phenomena, resulting in bed shrinkage. This problem can be prevented by adding bulking inert agents such as polystyrene sphere, lava particles, glass beads, porous clay and ceramic.

Compost

Compost is the most widely used biofilter medium because it has most of the desirable properties required for a good medium. A wide variety of raw materials and processes are used to produce compost based media. These include municipal solid waste, industrial sludge, garden waste, animal manures and food and agricultural residues. The only disadvantage of compost is compaction of the bed, aging effect and problem of stability during long operation. Compost inherently contains broad spectrum of microbial population so inoculation is not necessary. But for rapid system start up sometimes addition of microbial inocula may be useful. There is also little requirement of external supply of nutrient in compost as compared to other materials used as biofilter media. Active compost has more available nitrogen than finished compost, but temperature in active compost may exceed 60°C. However, the airflow through the biofilter will normally reduce the temperature into mesophilic range. In order to overcome the limitations of compost bed biofilter now a days compost are used in combination of some other natural and synthetic media.

Agro Wastes

Use of agricultural waste as a filter media in biofilter column is very attractive from cost and environmental point of view. In recent years, there has been increasing trend towards more efficient utilization of agro-industrial residues. There are six characterized agricultural by-products as potential biofilter media- peanut shells, rice husk, coconut shells, cane bagasse, maize stubble and yellow gram stalk.

Peat, Coal and Soil

Peat is also an excellent medium for biofilters. Its low pH–3-4 is an attractive feature. It provides more surface area, better flow characteristics and less pressure drop during operations. Major problem associated with peat bed is moisture control. It also requires supplemental nutrient and buffering. Soil is a cost effective packing medium having indigenous microbial population. It is structurally stable and permits use of deeper beds. Its hydrophilic nature permits ease of moisture control. The problems associated with soil media are high pressure drop and its solubility in water. It is suitable for situations having large land area.

Synthetic Filter Media

These man made packing materials such as activated carbon, polyurethane foam, pall rings, and composite beads are some examples of the materials which have been successfully used as biofilter media. The synthetic media have several advantages over natural media such as high strength, low pressure drop, durability, better flow characteristics, large specific surface area, and easy removal of biomass. The disadvantages of these media are long acclimation period and need for inoculation. External supply of nutrient is also required but research is going on to develop new media with nutrients added on to their surface using appropriate physico-chemical techniques.

Granular Activated Carbon

During the last decade, GAC has been used successfully as a good packing material because of the advantage of rapid pollutant adsorption, large surface area and porosity and the resultant improved the degradation capacity of biofilters.

Polyurethane Foam

Polyurethane foam has been found to be very useful media in biofiltration. Due to open structure and high porosity, low density, easy inoculation procedure and ability to absorb water, the pressure drop over the packing remains low at relatively high gas velocity. The large specific area, less aging effect, and physical strengths make this media suitable for long time operation.

Pall Rings

Pall rings are used as biofilter packing medium which prevents media-bed plugging. However, startup is relatively long presumably because of poor biofilm establishment on the surface. The relatively low specific surface area is also a disadvantage for achieving a high elimination capacity. Pall rings are made of plastic and so its use as packing material may be beneficial because of its stability, low cost and high porosity.

Temperature

Temperature regulation is very important in biofiltration as the biological activity is temperature-dependent. General operating temperature range is 14-40°C, which is suitable for the growth of mesophilic micro-organisms. Biofilter performance is in general limited by the biological reaction rate or by the mass transfer rate. The temperature has a different effect on either of these limitations; consequently, effect of temperature on the performance varies depending on the particular application. These effects may counterbalance each other. The influence of temperature on microbial activity has been well studied and in the suboptimum range (15-35°C) it can in general be described by the Arrhenius equation. The temperature beyond the optimum range will cause a decline in the performance and even cause death of the micro-organisms. However this is favorable for thermophilic organisms. The metabolic rate of the thermophilic bacteria increases with temperature, doubling with each 10°C increase up to 65-70°C. Due to increased metabolic activity of microbial population higher removal efficiency and elimination capacities can be achieved. During the thermophillic operation of biofiltration less biomass is produced and thus clogging of bed does not occur and thus the pressure drop through the bed is minimized. While a warmer biofilter generally supports more active organisms, the physico-chemical effects of higher temperatures are usually unfavourable. For most gases with the rise in temperature the Henry's law coefficient increases; so that less of the compound will be dissolved in the water. This will reduce the sorption of the compound. Therefore, the transfer of the contaminant from the air to the micro-organisms will be less effective.

Moisture Content

Among the other parameters affecting biofilter performance (filter bed material, additional nutrients, temperature, pH, etc.), water availability within a biofilter must be carefully controlled to ensure high microbial activity. The micro-organisms need minimum water content for optimal growth and activity. Too little moisture content causes drying of the bed and leads to channeling, resulting in non-uniform gas distribution. Also the micro-organisms get deprived of water which causes significant reduction in biodegradation rate. Excess water has been shown to cause a reduction of biofilter porosity and formation of anaerobic zones. Along with this it will also increase the backpressure due to the reduced void volume, and channeling of the gas within the bed. Loss of water from the bed occurs due to (i) evaporation to an unsaturated influent gas stream, (ii) bio-oxidation caused by bed heating, (iii) metabolic water requirements, and (iv) heat exchange with the environment. Excessive moisture content reduces the transfer of oxygen and thus decreases the biological activity. To maintain optimum moisture content of the filter bed, the polluted air stream should be humidified before introducing in the biofilter column. For maintaining the moisture content it is often advantageous to have downflow biofilter. Since most of the drying occurs at the entrance to the filter bed, drying at the top is easily handled through direct water addition and flow from the top to the bottom.

pH Value

Most micro-organisms prefer a specific pH-range. Thus changes in the pH of the filter material will strongly affect their activity which in turn will affect the performance of the biofilter. The micro-organisms involved in the degradation of VOCs generally grow in the pH range of 6.5-7.5. The pH in compost filters is typically between 7 and 8.5, a range preferred by bacteria and actinomycetes. Highly acidic condition generally inhibits microbial activity and most bacteria favor neutral conditions. In low pH biofilters fungi are to be the dominating species. Sulfur oxidizing bacteria, however, are well adapted to acidic conditions. During microbial degradation, acidic by-products may be produced. Therefore, to maintain the pH of the filter bed, buffering agents may have to be added. Slaked lime ($Ca(OH)_2$, lime (CaO), limestone ($CaCO_3$), crushed oyster shell ($CaCO_3$), dolomite [$CaMg(CO_3)_2$], marl (soil + $CaCO_3$), and slag ($CaSiO_3$) are used to neutralize the acidic condition. In addition to controlling pH by buffering materials, periodic washing or back washing of biofilters are also done.

Pressure Drop

For economical operation of biofiltration process the pressure drop across the biofilter column should be minimum since an increase in pressure drop requires more enrgy and can result in air channeling through the media. Pressure drop across the bed depends on the volumetric flow rate of gaseous stream, media porosity and bed depth. Pressure drop is directly related to the moisture content in the media and the media pore size. Increased moisture and decreased pore size result in increased pressure drop. Consequently, filter media selection and watering is critical to biofilter performance and energy efficiency. Due to accumulation of biomass, pressure increases which can be minimized by periodic washing of the bed with water. This will help removing the excess biomass from the filter column and thus the proper aeration of the filter bed can be maintained.

Acclimation

Biofilters require an acclimation period following startup to reach optimum pollutant removal capacity. This period generally ranges from days to weeks depending upon system and pollutants being treated. Low loading rates may be necessary during this period. During acclimation period appropriate pollutant-degradation enzymes get induced within the microbial population and micro-organisms grow to established the biofilm on the media surface. The acclimation period can be shortened by inoculating the filter with organisms already acclimated to and actively degrading the pollutants present in the gas stream. Cultures isolated from already running biofilters or activated sludge plants can be used for inoculation purposes.

Operation under Transient Conditions

Most operating biofilters are subjected to varying feed conditions due to process shut downs, startup and surges in influent pollutant concentrations. During process shut down or low inlet pollutant concentrations, microbial population starves for the food and undergoes endogenous decay, resulting in decreased pollutant degrading activity. The longer the period of non-operation, longer is the duration of acclimation after restart-up to restore full activity. If the process shut down and transients are too frequent, micro-organisms adapt to the situation and restart-up acclimation periods is shorter. Continuation of air flow through the filter during process shut downs helps in maintaining aerobic conditions to reduce destruction of aerobic micro-organism film. Unstable or transient feed conditions during startup are the most detrimental because they can resist the formation of microbial growth and result in excessively long acclimation periods. Sudden increase in influent pollutant concentration is also detrimental to biofilter performance, leading to pollutant breakthrough, toxic inhibition or even death of the micro-organisms in the biofilter. In biofilters treating high loads the biomass will grow rapidly if nutrients are available. This may sometimes lead to clogging and pressure drops. To overcome such a situation backwashing is done.

Gas Flow Rate and Pollutant Loading Rate

Gas flow rate is one of the most important hydrodynamic parameters in the biofiltration process because it quantifies the amount of pollutant entering the biofilter column per unit time. The increase in gas flow rate will increase the inlet load and consequently the diffusion flux through the biofilm. It will also decrease the time of contact between the pollutant and the microbial growth on the filter material. The decrease in the pollutant residence time will cause a reduction in the removal efficiency. The high gas flow rate is also responsible for stripping water from bed which tends to desiccate the biofilter. Direction of flow may also have a role in biofiltration. Better distribution of nutrients may be under downflow condition but energy requirement to operate biofilter will be higher. The typical range of residence time is 30-60 s. The gas flow rate through the bed is an important parameter in deciding the loading rate of the biofilter. The typical range for the gas flow rate through the bed is 0.3-1.6 m^3/min-m^3.

Oxygen Supply

Oxygen supply is necessary for the operation of biofilters because the predominant micro-organisms used in biofiltration are aerobic, and require oxygen for metabolism. Aerobic heterotrophic bacteria present in filter beds require at least 5-15 per cent oxygen at the inlet gas stream to survive. The oxygen concentration is in most polluted gases is in general several order of a magnitude higher than the pollutant concentration. However, because of the low oxygen water solubility, biofilter performance may be limited by mass-transfer of oxygen into the biofilm and/or by diffusion in the biofilm. The elimination capacity is higher when partial pressure of O_2 is higher in gas thus demonstrating that at normal oxygen content, VOC removal is limited by oxygen availability.

Nutrient supply

Micro-organisms utilizes VOCs as their food source by bio-oxidizing it to CO_2 and water. In order to maximize the pollutant elimination capacity, nutrients like nitrogen, potassium, phosphorus, calcium, zinc, selenium, iron, manganese, magnesium, sulfur and sodium should be present at a concentration high enough to maintain an active, growing culture. Nutrient solution is also added to the biofilter column for attaining the optimum moisture content. In general, the nutrient requirement is qualitatively the same as the elemental composition of biomass. Although organic packing materials contain the necessary nutrients for the growth of micro-organisms, but the performance of toluene-degrading biofilters improves when an extra source of ammonium and phosphate is provided to the reactor. Therefore a good packing material for VOC treatment should have low inherent nitrification activity level and high slow-releasing nitrogen content rather than high soluble nitrogen content.

Bed Depth

The depth of biofilter medium ranges from 0.5 to 2.5 m. A depth of approximately 1m appears to be most common, to allow sufficient residence time while minimizing filter land area requirements. Some manufacturers recommend the use of multiple layers of biofilter media, since these will need less land area for high loading rates. Very high filter depths are restricted due to the fact of high pressure head and potential for compaction of bed at the bottom which further increases pressure drop and causes channeling.

Waste Gas Pre Treatment

The microbial communities in biofilters can be poisoned by the presence of toxic contaminants, excessive concentration of the contaminant, or excusion in environmental conditions like pH, temperature, and moisture content. In order to meet the basic requirements for optimal operation of the biofilter, waste gas conditioning is often required. A sufficient supply of oxygen and humidity, and an acceptable range of temperature and pH levels in the filter bed, are indispensable for the survival of the microbial community present in the bed. High particulate loads in the waste gas can adversely affect the operation of a biofilter by clogging the air distribution system and the filter material itself. Pretreatment options can include humidification for temperature and humidity control and/or the use of devices for particulate control, such as wet scrubber or a wet electrostatic precipitator.

Clogging and Channeling

In the biofilter column clogging and channeling occurs due to the overgrowth of biomass generally at the higher loading rates, as a result the performance of biofilter is altered due to less availability of specific surface area required for better transport of oxygen, water and VOCs. Therefore, the excess biomass should be periodically removed from biofilter columns for long term operation. Several techniques such as backwashing is used but it is found that this technique is only suitable in case of small-scale biofilters having cross-sectional areas of 0.004-0.015 m^2. Modified backwashing combined with air sparging or chemical treatment can also be done to avoid clogging and channeling. Agitation of filter media and change in mode of operation of biofilter from the upflow to downflow can help in reduction in excess growth of micro-organism.

Mathematical Modeling

Mathematical modeling is an important tool to improve the process design and performance of the biological treatment systems. The pioneer and most commonly used model was proposed by Ottengraf and van den Oever (1983). They proposed a steady-state model and used the concept of a bed of solid filter particles coated with a biologically active aqueous layer called biofilm. They assumed a plug flow model for the flow of contaminated gas through the filter bed. Oxygen was assumed to be in excess, thus oxygen limitation on kinetics was ignored. The biofilm thickness was assumed to be small compared to the diameter of the support particles and a constant value for the film thickness was used throughout the reactor despite the changes in the VOC concentration along the biofilter column. Pollutants are removed from the gas phase by absorption followed by simultaneous diffusion and a zero order reaction in the biofilm. Analytical solution of this model was presented in terms of dimensionless concentration. Biokinetics of biofilter can be explained with the help of most commonly used modified Monod model. Monod model is assumed for growth as a function of existing concentrations of biomass and the concentrations of contaminant:

$$\frac{dX_{act}}{dt} = \mu X_{act}, \quad \mu = \frac{\mu_{\max} C_{bf}}{K_s + C_{bf}}; \quad \frac{dX_{act}}{dt} = Y \frac{dC_{bf}}{dt}$$

Where, X_{act} is the biomass density, μ is the growth rate, μ_{max} is the maximum value of the growth rate, K_S is the Monod or half-saturation constant, and Y is the biomass yield. For high values of C_{bf}, the growth rate is constant, and growth rate is approximated by zero-order kinetics and similarly for low values of C_{bf}, growth is linear with contaminant concentration, and hence first-order kinetics may be presumed. However, when the model includes sufficient detail to show biodegradation rates as a function of depth within the biofilm, concentrations will range from the Henry's equilibrium value at the surface of the biofilm to zero at the maximum depth of penetration, so it is likely that both regimes will be encountered and the full form of the Monod equation will be needed. Often the appropriate values for K_S and μ_{max} are uncertain. Both values are strongly dependent on the conditions under which they are determined and most data in the literature are from experiments performed on micro-organisms in stirred, well-aerated suspensions, rather than in biofilms and are highly variable. Thus, these parameters are often fitted to the biofilter data obtained with the experiment being modeled.

Design and Operating Factors

Operation and performance of biofilter is generally reported in terms of removal efficiency, or pollutant elimination capacity as a function of pollutant loadings of gases or the empty bed retention time (EBRT). These terms are defined as:

$$\text{Removal Efficiency (RE)} = \frac{C_{in} - C_{out}}{C_{in}} \times 100 \text{ (per cent)}$$

$$\text{Elimination capacity (EC)} = \frac{C_{in} - C_{out}}{V} \times Q \quad (\text{g m}^{-3}\text{h}^{-1})$$

$$\text{Empty Bed Retention Time (EBRT)} = \frac{V}{Q} \ (s \ or \ \text{min})$$

$$\text{Pollutant loading (L)} = \frac{C_{in}}{V} \times Q \ (\text{g m}^{-3}\text{h}^{-1})$$

Where, C_{in} and C_{out} are the inlet and outlet pollutant concentrations, respectively, V is the packed bed volume (m^3) and Q is the air flow rate ($m^3 h^{-1}$).

Most of the researchers used EBRT as the parameter to evaluate the performance of biofilter at different gas flow rates. For a constant volume of biofilter bed, EBRT and gas flow rate have linear relationship. In most of the operational biofilters, the EBRT ranges from 15 seconds to several minutes. Generally EBRT considered in the range of 0.5 for readily degradable compounds to 10 minutes for recalcitrant compounds.

Selection of EBRT value for biofilter also depends on other operating conditions such the the VOCs concentration and biodegradability level. VOCs concentrations introduced beyond the toxicity threshold value tend to inhibit the microbial growth and activity. The actual appropriate bed height depends on media type and expected pressure drop.

A typical elimination capacity versus pollutant loading curve is shown in Figure 6.4. It is usual to report the performance as a function of the load, *i.e.* product of inlet concentration and air flow rate, rather than the concentration. This enables comparison of systems of different sizes operating under different conditions.

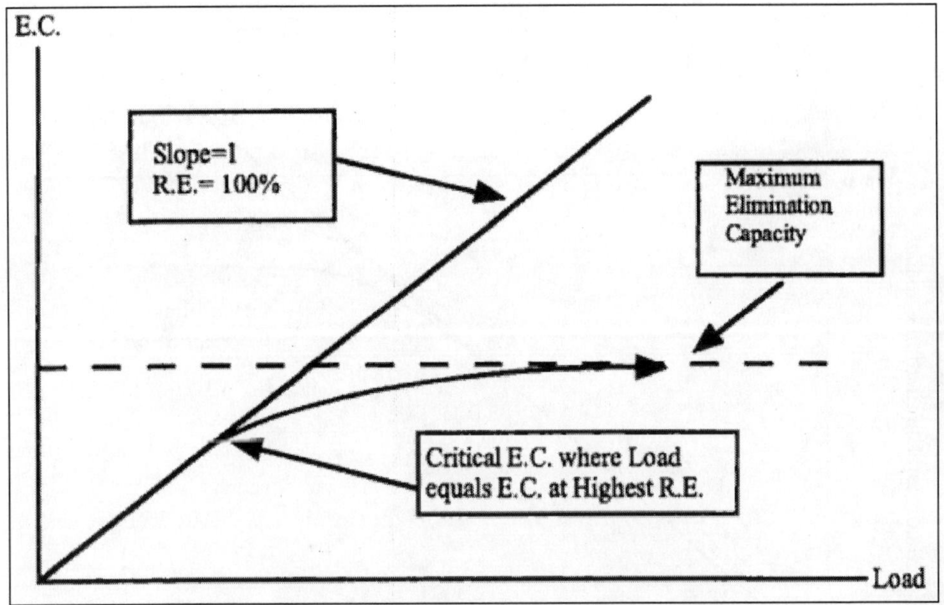

Figure 6.4: Typical Elimination Capacity vs. Load Curve.

Elimination capacity is always Equal to or less than the load. The ratio between elimination capacity and load is the removal efficiency of the system.

Figure 6.5 reveals that there are essentially three operating regimes:

1. Low loading, (also called first order regime): The elimination capacity and loading are identical and the pollutant is completely removed. The biofilter is operated well below its maximum elimination capacity. The performance increases proportionally with the loading.
2. Intermediate range: Breakthrough of the pollutant occurs with higher inlet concentration or higher air flow rates, the elimination capacity increases, but to a lesser extent than the loading.
3. High loading, (also called zero order regime): The biofilter is operated at its maximum elimination capacity. Increase in the loading doesn't result in further increase in elimination capacity, the removal efficiency decreases. As indicated on the figure, if the concentration is very low, the maximum elimination capacity is reduced accordingly. Above behaviour can also be explained with help of the controlling mechanism in the biofilter. At low loadings of pollutant, overall process is controlled by mass transfer and at high loading reaction in the biofilm becomes the controlling mechanism. For evaluating biofilter performance, one should consider both the maximum elimination capacity and the removal efficiency.

The relationship between off-gas flow rate, required residence time, and the corresponding reactor volume is the most crucial aspect in the biofilter design, since

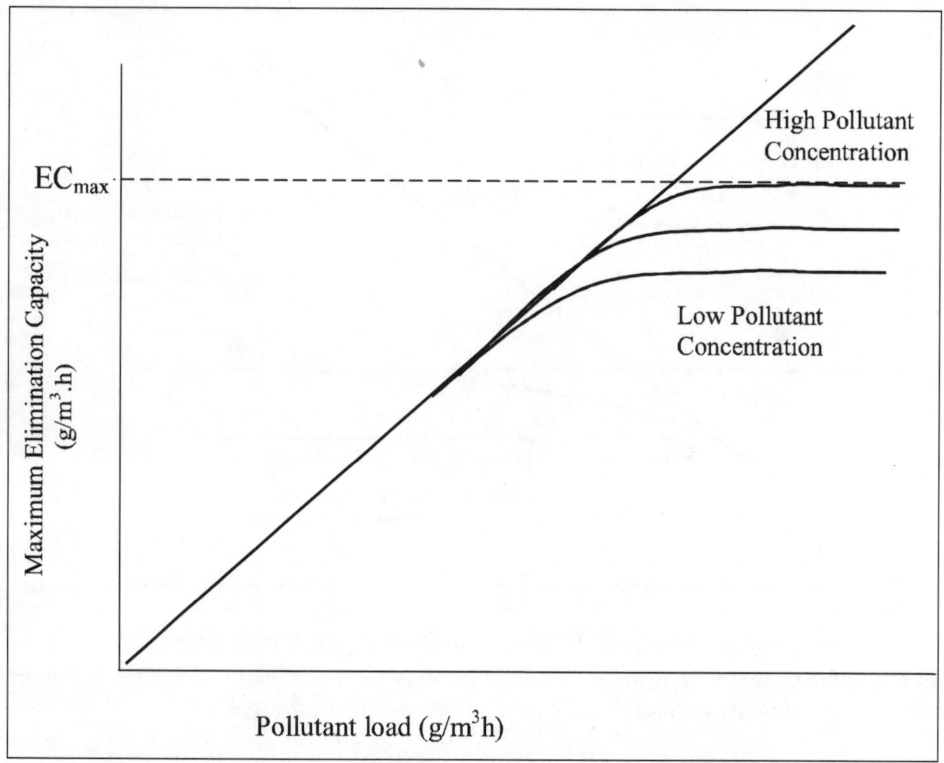

Figure 6.5: Elimination Capacity and Load Curve for a Typical Biofilter.

it strongly affects space requirements and capital cost of a biofilter. The elimination of a single pollutant in a well-functioning biofilter generally follows the concentration profile shown in Figure 6.6. Removal rates depend on the compound and the media type in the biofilter. It is reported in units of g/kg of dry media/day. The rate of removal is nearly linear with distance into the media or with Empty Bed Residence Time (EBRT) at higher concentrations. Generally, the lower-molecular–weight, less-complex compounds are more easily degraded and more quickly removed in a biofilter.

At lower concentrations, the rate of removal decreases and follows a power function. Lowering the polluted air flow rate (Case-B) by increasing the filter bed area increases the effective residence time and improves performance per unit of bed height, thus causing a steeper concentration profile. However, it also requires more filter volume per unit of air flow.

3. Advantages and Disadvantages of Biofiltration

Biofiltration has several advantages over conventional treatment technologies:

1. The main advantage of using biofiltration over other conventional control technologies are lower capital and operation and maintenance costs, low chemical usage, and no need of combustion source. Further, it does not

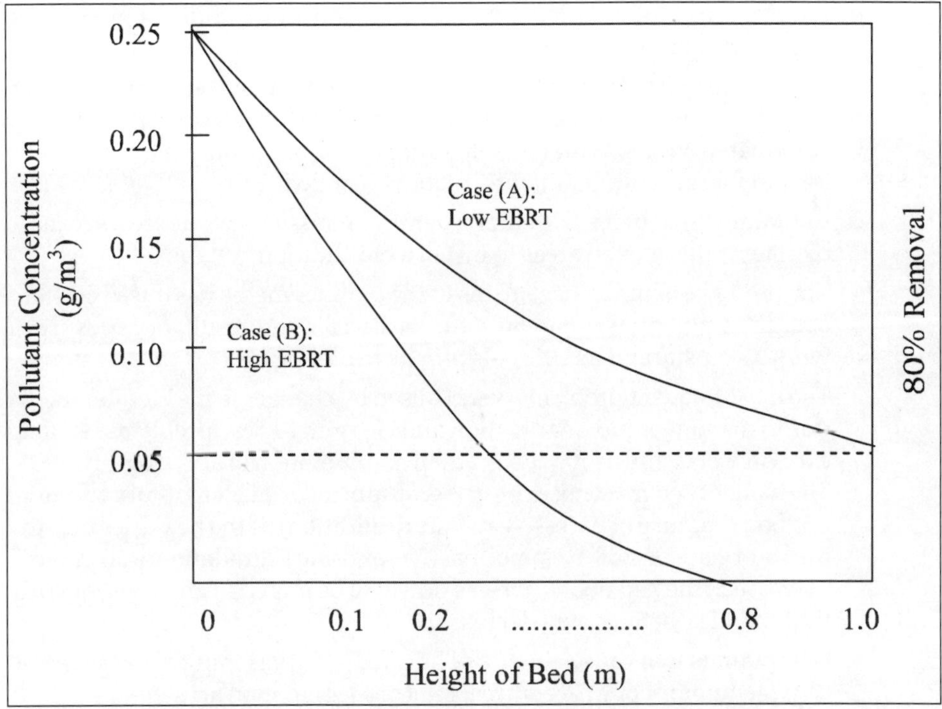

Figure 6.6: Removal of Pollutant in a Biofilter as a Function of Empty Bed Residence Time (EBRT).

transfer pollutants from one phase to another phase but it completely destroys the pollutants and produces non-harmful products.

2. Biofiltration units can be designed to physically fit into any industrial setting. A biofiltration unit can be designed in any shape and size. These can be in open field with the piping and delivery system underground or in-door. In addition, biofilters can be designed with stacked beds to minimize space requirements and multiple units can be run in parallel.
3. Biofiltration is versatile enough to treat odorants like hydrogen sulfide, toxic compounds, and VOCs simultaneously, from the emission of various industries. The treatment efficiencies of these constituents are above 90 per cent for low concentrations of contaminants (<1000 ppm) and large volumetric flow rates.
4. Different media, microbes and operating conditions can be used to enhance the performance of biofiltration system for many emission points.

Inspite of all their advantages, biofilters also have several disadvantages:

1. Fluctuating inlet stream characteristics can be detrimental to the biofilter's microbial population and overall performance and also risks violating emission discharge restrictions. The inability to maintain consistent

removal efficiency can be a major limitation unless full support of the regulatory authority and community is achieved.

2. Biofiltration cannot successfully treat some organic compounds, which have low adsorption or degradation rates. This is especially true for chlorinated VOCs. With the help of this technology only the biologically assimilable gaseous pollutants can be controlled.
3. Contaminant sources with high chemical emissions would require large biofilter units or open areas to install a biofiltration system.
4. The presence of micro-organisms in the biofilter media has raised concern over their potential release into the treated air and resultant exposure of workers on-site and individuals off-site to pathogens.
5. The most frequent problems were caused by changes in the biofilter media due to dry-out, rapid degradation and clogging. These problems result in excessive pressure drops and gradual accumulation of acidic by-products. The biofilter bed must not dry out (because of insufficient moisture addition) or flood (because of excessive moisture addition). If the bed dries out, the micro-organism will become inactive and biodegradation will rapidly decrease. If the bed floods, the gas flow will be impeded and consequently the biological activity will decrease.
6. Compounds that cause either a pH decrease or poisoning of bacteria will cause temporary or irreversible loss of biodegradation activity.
7. With the continuous supply of nutrient, biofilter accumulates large amount of biofilm, which leads to clogging. At present, clogging of the reactor over long run is one of the major obstacles for proper functioning. Clogging increases pressure drop across the reactor thereby decreases pollutant removal rate (Okkerse *et al.*, 1999; Cox and Deshusses *et al.*, 1999a). Clogging of air distribution systems, rapid corrosion of ductwork and concrete parts, emissions of odorous byproducts and flooding of media have also been observed.
8. Acclimation periods for the microbial population may take weeks or even months.

Observing the overall performance of different technologies available, it can be said that biofiltration has the capacity to serve as a viable VOCs control technology.

Conclusion

Biofiltration is an attractive technique for removal VOCs as well as odour from various emissions. The biological mechanism involved in the treatment of wastewater is time proved, the gaseous pollutant treatment are almost similar and forces a new engineering conception and has led new application in the field of environmental biotechnology. This new concept offers scope and challenges in the development of novel biofilter, support media, nutrient requirement, specific organisms and scale up to large systems. There are also potential possibilities in application of genetic engineering for development and exploration of more effective micro-organisms to degrade mixed streams of different types of pollutants.

References

1. Aizpuru, A., Dunat, B., Christen, P., Auria, R., Garcia-Pena, I., Revah, S., 2005. Fungal biofiltration of toluene on ceramic rings. J. Environ. Eng.-ASCE. **131**, 396–402.

2. Cohen, Y.,. 2001. Biofiltration - the treatment of fluids by micro-organisms immobilized into the filter bedding material: a review. Bioresource Tech. 77, 257-274.

3. Devinney, J.S., Deshusses, M.A., and Webster, T.S., 1999. Biofiltration of Air Pollution Control. CRC Press. Boca Raton. FL.

4. Delhomenie, M.C., Bibeau, L., Heitz, M., 2002. A study of the impact of particle size and adsorption phenomena in a compost-based biological filter. Chemical Engineering Science 57, 4999–5010.

5. Delhomenie, M.-C., Heitz, M., 2005, Biofiltration of air: a review, Critical Reviews in Biotechnology, 25, 53-72.

6. García-Pe a, E.I., Hernández, S., Favela-Torres, E., Auria, R., Revah, S., 2001.Toluene biofiltration by the fungus *Scedosporium apiospermum* TB1. Biotechnol Bioeng. 76, 61–69.

7. Groenestijn, J.M., van, Karaakman, N.J.R., 2005. Recent developments in biological waste gas purification in Europe. Chemical Engg. Journal 113 (2-3), 85-91.

8. Giri, B.S., Mudliar, S.N., Deshmukh, S.C., Banerjee, S., Pandey, R.A., 2010. Treatment of waste gas containing low concentration of dimethyl sulphide (DMS) in a bench-scale biofilter. Bioresource Technology 101 (7), 2185-2190.

9. Kennes, C., Veiga, M.C., 2004. Fungal biocatalysts in the biofiltration of VOC-polluted air. J Biotechnol 113, 305–319.

10. Kiran Singh, R S Singh, B N Rai and S N Upadhyay, 2010. Bio-filtration of Toluene Using Wood Charcoal as the Biofilter Media, Bioresource Technology, 101, 3947-395

11. Leson, G., Winer, A. M., 1991. Biofiltration: An innovative air pollution control technology for VOC emissions. Journal of The Air and Waste Management Association 41(8), 1045-1054.

12. McNevin, D., Barford, J., 2000. Biofiltration as an order abatement strategy. Biochemical Engg. Jl. 5, 234-242.

13. Maestre, J.P., Gamisans, X., Gabriel, D., Lafuente, J., 2007. Fungal biofilters for toluene biofiltration: Evaluation of the performance with four packing materials under different operating conditions. Chemosphere 67 (4), 684-692.

14. Ottengraf, S.P.P., van den Oever, A.H.C., 1983. Kinetics of organic compound removal from waste gas with a biological filter. Biotechnol. Bioeng. 25, 3089-3102.

15. Pandey, R.A., Padolay, K.V., Subbarao, T.V., Mudlier, S.N., 2002. Biological methods for deodorization of industrial emissions- an overview. Journal of I.A.E.M. 29, 158-175.

16. Raghuvanshi, S., Babu, B.V., 2009. Experimental studies and kinetic modelling for removal of methyl ethyl ketone using biofiltration. Bioresource Technol. 100, 3855-3861.

17. Singhal, V., Singla, R., Walia, A.S., Jain, S.S., 1996. Biofiltration–An innovative air pollution control technology for H_2S emission. Chemical Engg. World 31(9), 117-124.

18. Swanson, W.J., Loeher, R.C., 1997. Biofiltration: fundamentals, Design and Operations principles, and applications. Jl. of Environmental Engineering 123 (6), 538-546.

19. Steele, J.A., Ozis, F., Fuhrman, J.A., Devinny, J.S., 2005. Structure of microbial communities in ethanol biofilters, Chemical Engg. Jl. 113 (2-3), 135-143

20. Singh, R.S., Agnihotri, S.S., Upadhyay, S.N., 2006, Removal of toluene vapour using agro-waste as biofilter media, Bioresource Technol. 97, 2296-2301.

21. Singh Kiran, Singh R.S., Rai B.N., Upadhyay S.N., 2010. Bio-filtration of toluene using wood charcoal as the biofilter media, Bioresource Technology 101, 3947–3951.

22. Upadhyay, S.N. Kumar, V., 2004. Biological control of air pollutant, first edition. Consice Encyclopedia of Bioresource Technology. The Haworth Reference Press, New York.

Chapter 7

Biofiltration of Xylene Using Wood Charcoal as a Biofilter Media

***Kiran Singh*[1], *R.S. Singh*[1], *B.N. Rai*[1], *S.N. Upadhyay*[1] and *S.N. Kaul*[2]**

[1]Department of Chemical Engineering and Technology, Indian Institute of Technology (Banaras Hindu University), Varanasi

[2]CSIR-NEERI, Nagpur

ABSTRACT

Biofiltration of xylene vapors was investigated in a lab scale biofilter inoculated with *Pseudomonas putida* (MTCC 102), for 150 days. The biofilter was packed with wood charcoal and some glass beads were used for better air distribution. Acclimation of microbes was achieved within twenty days at an inlet concentration range of 0.04-0.08 g m^{-3}. Xylene vapors with a load ranging from 12 to 538 g m^{-3} h^{-1} were fed to the biofilter at various flow rates. Throughout the experimental period the removal efficiency was obtained in the range of 65-98.7 per cent. The maximum elimination capacity of 405.7 g m^{-3} h^{-1} was observed at an inlet load of 538.8 g m^{-3} h^{-1}.The pH of the packing media varied in the range of 8.2-6.8 during the operation period. Wood charcoal was found to be a very effective packing media for xylene biodegradation.

Keywords: *Biofiltration, Wood charcoal, Xylene, Inlet loading rate, Elimination capacity.*

Introduction

A large quantity of volatile organic compounds (VOCs) is released from various industrial sectors and processes. VOCs catalyze smog formation reactions, break the natural cycle of NOx, and deplete ozone by forming free radicals [1]. Emission control of VOCs has become one major concern in air pollution prevention since the 1990

Clean Air Act Amendments went into effect [2]. Conventional physical and chemical VOCs removal process is expensive, requires complex equipment and/or generates hazardous residues, so bioremediation and particularly biofiltration, is turning into a sustainable alternative for VOCs removal. Biofiltration is a low cost and non-generating hazardous residue, and shows an especial interest for its application in small industrial plants with low concentrations in the waste gas [3]. This technology has been successfully used to remove a wide range of pollutants such as VOCs, ammonia, mercaptans, and sulphurous compounds [4-7].

The biofilter consists of a filter bed that is usually comprised of both natural and inorganic media such as peat, compost, leaves, wood bark and/or soil. Main characteristics to consider upon the selection of an appropriate packing material are its specific surface area, density, porosity, pH, water holding capacity, buffering capacity and elemental composition [8]. The bed moisture is kept at a constant level by humidification of the influent air in order to maintain a biologically active layer surrounding the media known as the "biofilm" [9]. VOC- containing waste gases is transported to air/biofilm, interface, where they are absorbed into the biofilm and employed as carbon and/or energy sources by the micro-organisms. Being a heterogeneous biochemical process controlled either by mass transfer or biochemical reaction or both. The removal efficiency of biofilters varies with media characteristics (*e.g.* particle size, surface area, and porosity), carbon and energy source, nutrients, pH and temperature [10].

Among the VOCs that can be eliminated by biofiltration, xylene is considered as a relatively toxic pollutant. The current OSHA permissible exposure limit for xylene is 100 ppm as an 8-hr time weighted average concentration [36]. Xylene is an irritant of the eyes and mucous membranes at concentrations below 200 ppm and it is narcotic at high concentrations [37]. Acute exposure to high vapour concentration of xylene may cause chemical pneumonitis, hemorrhage into the air spaces, and pulmonary edema. Xylene is used as a solvent in printing, rubber, synthetic fiber, plastic insecticide, pesticides and leather industries and as a cleaner and paint thinner [11]. To date several biofilters and biotrickling filter using microbial consortia were used to remove xylene [11-18]. Several Xylene degrading bacterial and fungal strains are reported in the literature. Few of them are- *Pseudomonas putida* 39/D [20], *Xanthomonas maltiphilia* [21], *Sphingomonas strain* ASUI [22], *Rhodococcus* sp. Strain DK 17 [23], Transformed *E. coli* [24] and *Pseudomonas pseudoalcaligenes* [25]. Xylene degrading fungal strains are: *Cladophialophora* sp. Strain T 1(ATCCMYA-2335) [26] and *Phanerochaete chrysoporium* [27].

Elmrini *et al.*, 2004 [11] investigated the biofilter response to step and abrupt variations (shock loading effects) in inlet xylene concentrations. Elimination capacity and removal efficiencies as high as 61 g $m^{-3}h^{-1}$ and 96 per cent, respectively, were noticed within 24 h after any step change in loading rate, which revealed potential stability of biofilm to withstand shock loadings. Quinlan *et al.*, 1999 [28], investigated the biodegradation of synthetic and free phase mixtures of VOCs, containing benzene, toluene, ethylbenzene and isomers of xylene in a compost biofilter. Benzene and p-xylene was not essentially degraded as effectively as toluene and ethylbenzene.

Hwang *et al.*, 2007 [18], experienced nitrogen limitation during the treatment of p-xylene by using compost-based biofilters. They achieved an elimination capacity of 10 g p-xylene m^{-3} h^{-1} for 150 days by using food waste compost and concluded that a good packing material for VOC treatment should have a low inherent nitrification activity level and a high slow-releasing nitrogen content rather than a high soluble nitrogen content. Jorio *et al.*, 1998 [13] have shown a decrease in toluene removal due to the presence of xylene, while the presence of toluene, had negligible effect on the xylene.

Wu *et al.*, 2006 [15], simultaneously eliminated the mixtures of n-butyl acetate (n-BA), p- xylene and ammonia gas in a downflow biofilter. They achieved relatively low removal efficiency (40-100 per cent) for p-xylene because of its inhibition by the presence of high concentrations of n-BA. Heitz *et al.* (2000) [29] investigated the biofiltration of xylene vapours in a laboratory scale biofilter packed with peat mixed with structuring and conditioning agents and initially inoculated with a microbial consortium. The biofilter proved to be highly efficient in the removal of xylene at a gas flow rate of 0.4 m^3 h^{-1} corresponding to a gas residence time of 157 s.

The present paper describes the effects of various xylene inlet loads, gas flow rates, temperature and pH on the performance of the biofilter, using wood charcoal as a packing media.

Materials and Methods

Biofilter Setup

A schematic diagram of the experimental set-up used in this study is shown in Figure 7.1. The biofilter was made from cylindrical glass column with an inner diameter of 5 cm and an overall height of 80 cm. A 15 cm head space at the bottom was used for gas inlet and leachate collection, while a 15 cm space at the top was used for gas outlet and for nutrient supply. 50 cm of the column was packed with previously humidified equally sized, small cubical pieces of filter media and few glass beads were provided at the bottom. The filter media was supported on a stainless steel sieve plate that ensured homogeneous distribution of gas flow over the entire cross section of the filter bed. The biofilter column was provided with four sampling ports placed at a distance of 12.5 cm each to measure xylene concentrations, temperature and moisture content of bed. A valve is provided at the bottom of the column to collect the leachate. The biofilter was operated in upflow mode. Compressed and filtered air stream was dived into two parts. One part of the air stream was passed through the humidifier to assure a relative humidity value of at least 90 per cent. For producing xylene loaded air stream of desired concentration, the other part of the air stream was allowed to bubble through a 250 ml bubbler containing xylene. The flow of air stream into bubbler was controlled with a needle valve (Swagelock) and rotameter (Eureka, India). Rotameter and pressure control valve was used to maintain constant airflow to the biofilter column. Both the air stream was then mixed in a vertical rectangular mixing chamber made of perspex and then fed to the biofilter column. The nutrient solutions used during the experiment has the following compositions per liter of distilled water:

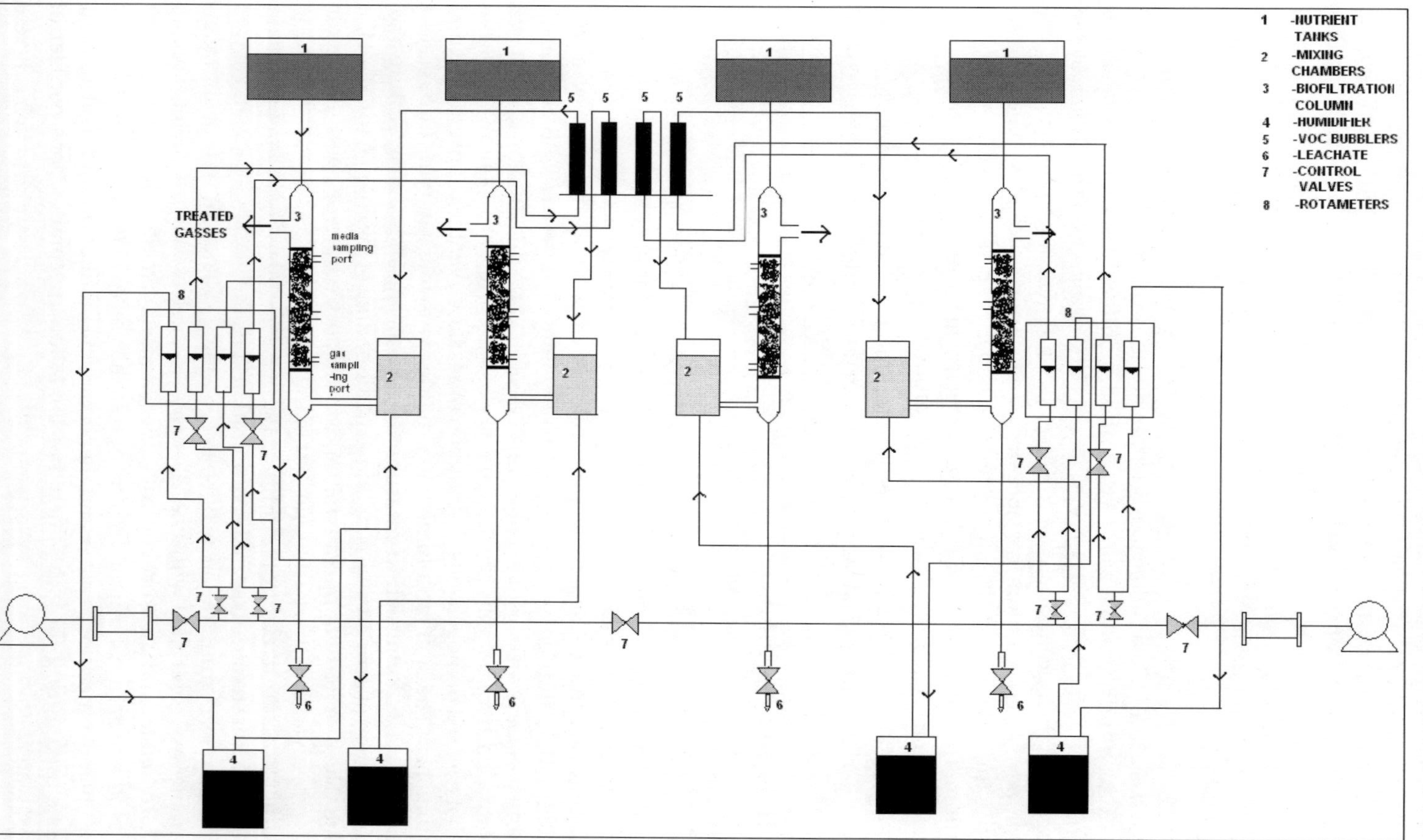

Figure 7.1: Experimental Setup.

KH_2PO_4- 4 g, $NaNO_3$- 4.5 g, $(NH_4)_2SO_4$- 1.5 g, $FeSO_4$- 0.28 g, $MgSO_4.7H_2O$- 0.35 g, $NaHCO_3$- 0.9 g, $MnSO_4$- 0.00152 g, $(NH_4)_2MoO_4$- 0.001 g and $CaCl_2$- 0.003g.

Filter Media

Wood charcoal was obtained from local market of Varanasi and cut into cubes of approximate 1cm size using hacksaw blades. Wood Charcoal cubes were dried in oven at 60^0C for two days and then cooled in a desiccator. Dry weight, Bed porosity, pH and moisture content of the filter media were calculated and found to be 0.6532 g, 68 per cent, 8.2, 65 per cent respectively. Chemical resistance study were carried out to check the resistance of wood charcoal sample against chemicals by placing wood charcoal samples in glass bottles containing pure xylene, tap water, distilled water, acidic solution and basic solution for two days. The results of visual observations and per cent recoveries were tabulated in Table 7.1. Specific surface area was measured by BET Surface Area and Porosity Analyzer and found to be 212 m^2/g. CHN content of the wood charcoal was measured with the help of CHN analyzer.The values of C, H, N content were 68.63 per cent, 3.65 per cent and 2.297 per cent respectively.

Table 7.1: Chemical Resistance Studies.

System	*Visual Effect*	*Per cent Recovery*
Pure xylene	Slightly colour change	88.3
Tap water	None	93.6
Distilled water	None	94.5
pH 4.6	None	87.4
pH 9.8	None	88.5

Analytical Techniques

During the experimental period temperature, pH of the leachate, pressure drop and moisture content were monitored daily. Temperature and moisture content was measured with Humidity and Thermometer meter (Model No. DT-615). Pressure drop across the bed was measured with the help of U-tube water manometer. pH was measured with digital pH meter. VOC concentrations in gas samples collected from different sampling ports of the biofilter column, in 25 ml glass samplers equipped with Teflon cap on one side, were measured using a gas chromatograph (Model-7610, Chemito gas chromatograph) equipped with a Chromatopack column (1/8" dia; 2m length; 80/100 mesh range) and flame ionization detector (FID). Nitrogen was used as the carrier gas. Temperatures of oven, injection port and detection port were 160, 175 and 200 °C respectively. The area of the chromatograms was monitored using the Iris 32 software. Gas samples were then injected into the FID gas chromatograph using 1ml SGE sampling syringe. Gas samples with known xylene concentration were used for the calibration.

Operation of Biofilter

Experiments were performed for a period of 150 days. The biodegradation of xylene vapour were investigated at various inlet concentrations and at four different flow rates *i.e.* 0.06, 0.12, 0.18 and 0.24 $m^3 h^{-1}$. During the whole experimental period constant composition of nutrient solution were supplied at the fixed rate (1 litre day^{-1}) periodically and the percolated liquid was recycled.

Micro-organism

The biofilter was inoculated with *Pseudomonas putida* procured from IMTECH, Chandigarh. The strain was then revived in 250-ml shake flasks containing 100 ml of nutrient broth. The flask was then incubated overnight in an incubator shaker at 26°C and 140 rpm. Overnight culture of *Pseudomonas putida* was used as inoculum for biofilter column.

Results and Discussion

The performance of the biofilter was evaluated in terms of (per cent) removal efficiency and the elimination capacity as a function of pollutant loading, which were estimated by the following equations:

$$RE = \frac{Cin - Cout}{Cin} \times 100$$

$$EC = \frac{Q(Cin - Cout)}{V}$$

$$IL = \frac{QCin}{V}$$

Where, Cin is the VOC concentration at inlet (g/m^3), Cout is the VOC concentration at outlet (g/m^3), V is the bed volume (m^3) and Q is the volumetric gas flow rate ($m^3 h^{-1}$).

Continuous Biodegradation Experiments

For examining the biofilter performance in the course of xylene degradation, operating periods were designed for 5 distinct phases as shown in Table 7.2. During the starting phase the biofilter was initially operated with an airflow rate of 0.06 $m^3 h^{-1}$ corresponding to an EBRT of 58.8 sec with an average inlet concentration range of 0.04-0.08 $g\ m^{-3} h^{-1}$ for 20 days. The results are shown in Figure 7.2. During this period the sufficient microbial growth was achieved in the biofilter and removal efficiency of about 99 per cent was obtained.

During phase I, flow rate was maintained at low level 0.12 $m^3 h^{-1}$ with corresponding EBRT of 29.4 sec. The xylene inlet concentration was varied between 0.1-0.8 $g\ m^{-3} h^{-1}$. Gradual increase in removal efficiency was observed. Maximum removal efficiency of 98 per cent was achieved on day 38th and on the same day the inlet xylene concentration was increased from 0.6 to 0.7 $g\ m^{-3}$but very slight variation in removal efficiency was observed. Further the inlet concentration of xylene was increased to 0.8 $g\ m^{-3}$ in the end of phase I.

In phase II the gas flow rate was increased and maintained at 0.18 $m^3 h^{-1}$ with corresponding EBRT of 19.6 s and the xylene inlet concentration was varied between

Table 7.2: Experimental Scheme for Continuous Xylene Degradation Experiments.

Nature of Biofilter Operation	*Flow Rate (m^3h^{-1})*	*EBRT (Sec)*	*Inlet Concentration Range (g m^{-3})*	*RE (per cent)*	*Operating Time (Days)*
Acclimatization	0.06	58.8	0.04-0.08	98-99	20
Phase I	0.12	29.4	0.1-0.8	98	26
Phase II	0.18	19.6	1-1.9	91	24
Phase III	0.24	11.76	1.6-2.2	75	20
Phase IV	0.24	11.76	1.2-1.8	75-70	30
Shut down phase	–	–	–	–	11
Restart	0.18	19.6	0.5-1	90-93	19

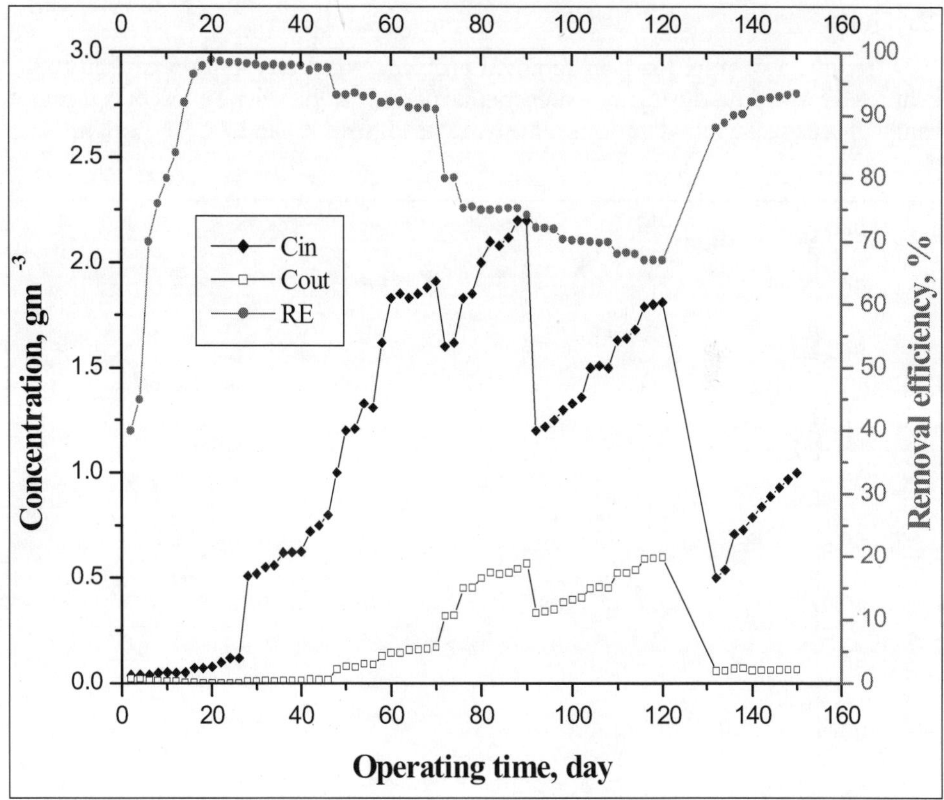

Figure 7.2: Biofilter Performance with Change in Inlet Xylene Concentration.

1-1.9 g m^{-3}. The removal efficiency was slightly decreased and then stabilized to 91 per cent in the end of phase II. In phase III xylene inlet concentration was further increased up to 2.2 gm^{-3} h^{-1} and flow rate was increased to 0.24 m^3 h^{-1} with corresponding EBRT of 11.76 s. During this phase the xylene removal efficiency suddenly decreased to 75 per cent. This can be attributed to the fact that for higher gas flow rates, the gas residence time in the biofiltration column is smaller. Consequently, the contact time between the polluted air stream and the degrading micro-organisms is smaller and the biodegradation ability of the filter bed diminishes leading to lower values of the removal efficiency of the contaminant [11].

In phase IV, the gas flow rate was remained unchanged at 0.24 $m^3 h^{-1}$ but the inlet concentration was kept low. Here also the removal efficiency was found to decrease further due to smaller residence time. The biofilter was then remained shutdown for ten days. In phase V, biofilter was operated again at the same flow rate as it was in phase II. The xylene inlet concentration was kept between 0.5-1 g m^{-3}. Biofilter responded very quickly to changes in flow rate and inlet concentrations and achieved nearly the same level of removal efficiency.

Effect of Inlet Loading Rate on Removal Efficiency and Elimination Capacity

The effect of inlet load on removal efficiency and elimination capacity was evaluated for 90 days during the experimental period for the xylene inlet concentration range of 0. 04–2.2 g m^{-3} at various air flow rates Figures 7.3 and 7.4. The xylene inlet

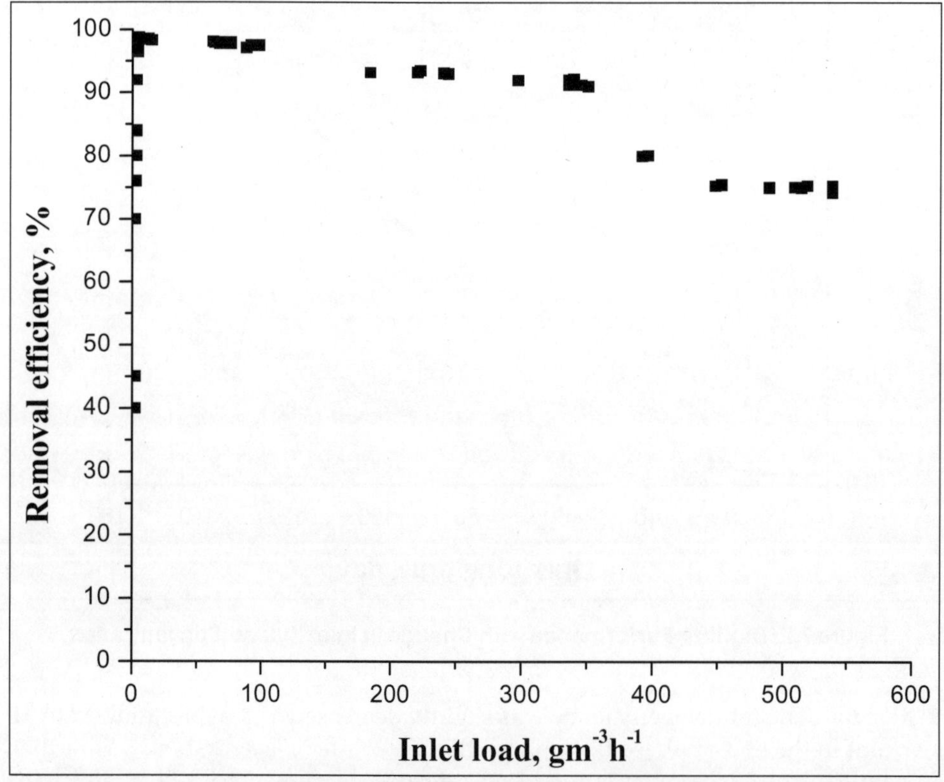

Figure 7.3: Influence of Inlet Xylene Load on the Removal efficiency of the Biofilter.

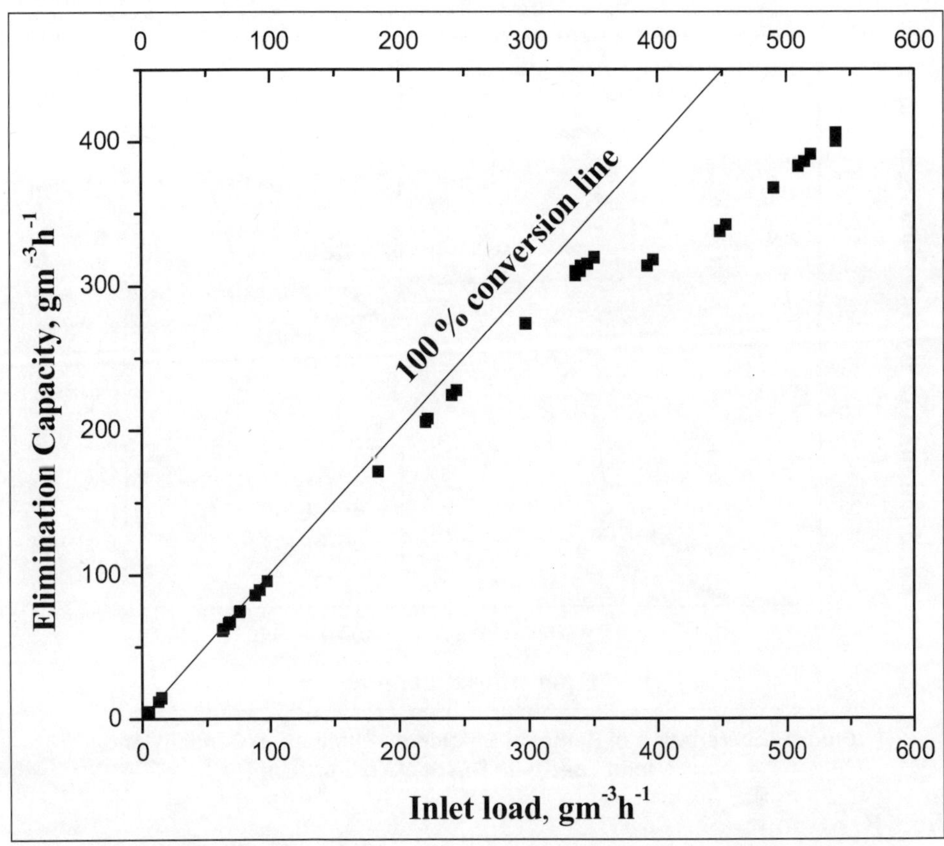

Figure 7.4: Influence of Inlet Xylene Load on the Elimination Capacity of the Biofilter.

load during the experiment, after the acclimation period was in the range of 12 g $m^{-3}h^{-1}$ to 538 g m^{-3} h^{-1}. Initially in phase I and II, the removal efficiency was found to increase with increase in inlet loading. About 98 per cent removal efficiency was obtained up to xylene load of 98 g m^{-3} h^{-1}. On further increase in inlet loading, removal efficiency gradually decreased. On the other hand a near linear relationship was observed between elimination capacity and inlet load up to the load of 250 g m^{-3} h^{-1}. Beyond this value, the elimination capacity increased slowly and reached to its maximum capacity of 405.7 g m^{-3} h^{-1} at an inlet load of 538.7 g m^{-3} h^{-1}. The values obtained for both removal efficiency and elimination capacity was better than the values reported by other researchers [11, 16-18].

Variation of removal efficiency, elimination capacity and inlet load with respect to time is shown in Figure 7.5. The removal efficiency, elimination capacity and inlet load versus time plot intersect corresponding to an inlet load of 375 g m^{-3} h^{-1} and this inlet load may be considered as the optimal operating point for the biofilter.

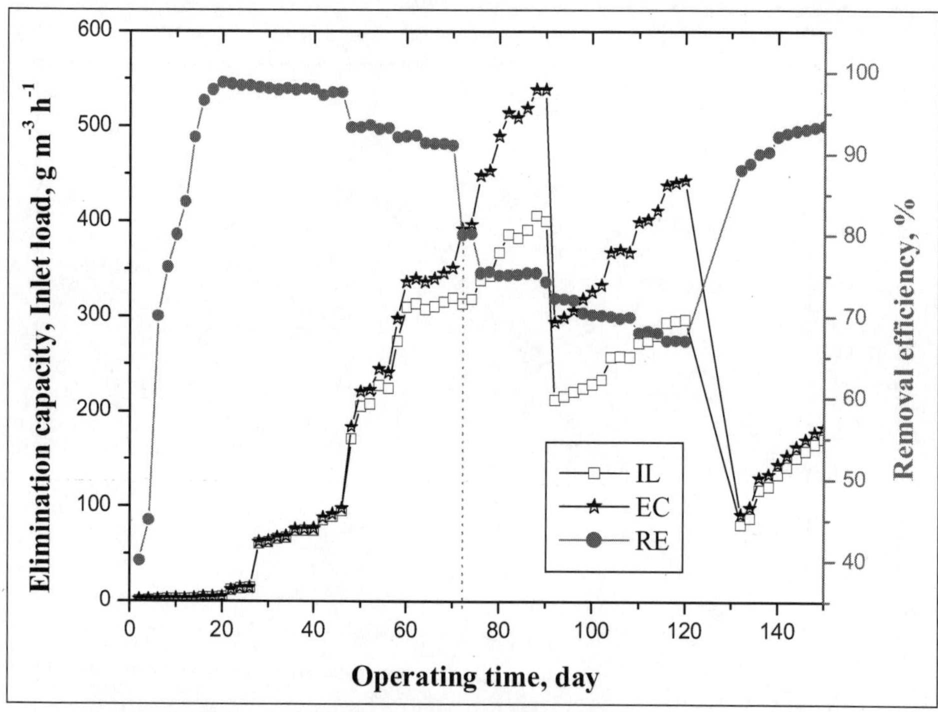

Figure 7.5: Variation of Removal Efficiency, Elimination Capacity and Inlet Load with Respect To time.

Bed Behaviour

Figure 7.6 shows the variation of pH and temperature of the bed with time. The filter bed temperature was monitored throughout the experimental period and was found to be in the range of 33.5-19.8°C. A difference of 2.5°C was observed between the bed temperature and ambient temperature. The intensity of microbial activity in the filter bed strongly depends on the biofilter temperature but in this study it was found that with the decrease in temperature the elimination capacity was increased. This may be explained by the fact that the biofilter was inoculated by *Pseudomonas putida* and the optimum temperature range for the growth of *Pseudomonas putida* is 25-30°C [30-32]. The bed behaviour confirms the presence of *Pseudomonas putida* in the xylene biofilter also. The pH of the bed was initially 8.2 and thereafter a slight decrease in pH with respect to time was observed. This may be due to the formation of carbonic acid in the biofilter. During the whole experimental period the pH was found to be in the range of 8.2–6.8, suited for the growth of *Pseudomonas putida* [33-34]. Filter bed moisture content was found in the range of 70-90 per cent throughout the experimental period.

The filter bed pressure drop is another key aspect of biofilter performance. It affects the energy consumption of the blower, which contributed most to the operation cost [35]. Negligible pressure drop was observed during the whole experimental period. No significant deterioration and compaction was observed in the present

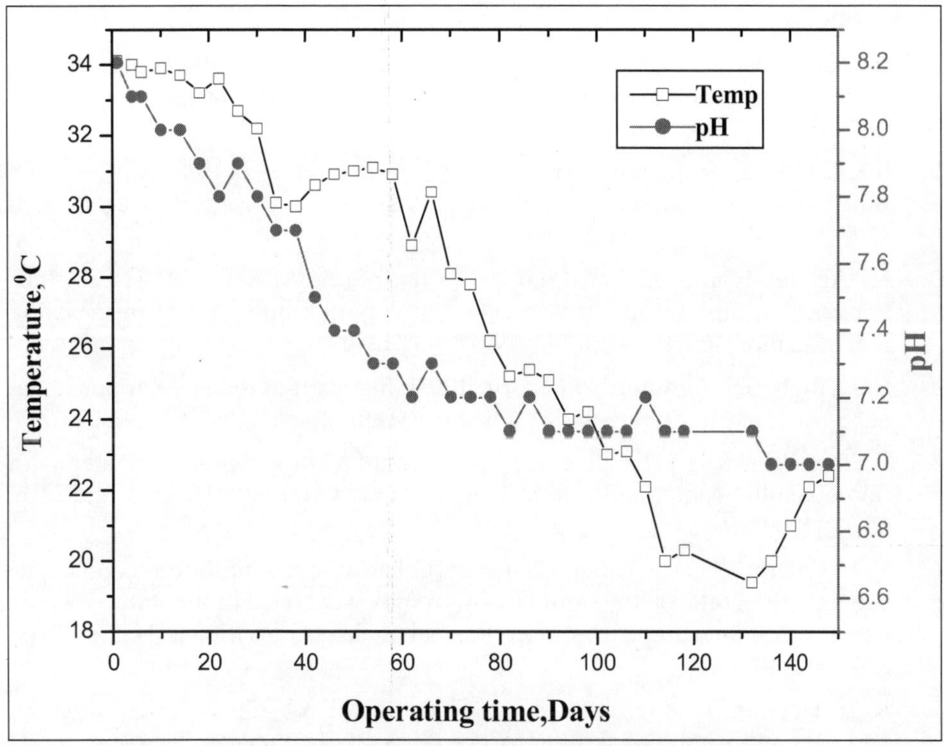

Figure 7.6: Variation of pH and Temperature of the Bed with Time.

study which proofs that wood charcoal is a very effective filter media for xylene biodegradation.

Conclusion(s)

Biofiltration of xylene vapors was investigated in a laboratory scale upflow biofilter packed with wood charcoal and inoculated with *Pseudomonas putida* (MTCC 102), for 150 days. The results obtained during the experimental periods proofs wood charcoal to be very effective filter media for xylene biodegradation. This media was found to have good mechanical stability and shown negligible pressure drop indicating the presence of uniform biofilm of constant thickness during the whole experimental period. Therefore its use in biofiltration process is very economical. Throughout the experimental period the removal efficiency was obtained in the range of 65-98.7 per cent. About 98 per cent degradation was achieved up to inlet loading of 98 g m^{-3} h^{-1}. Beyond this value the removal efficiency decreased with increase in inlet load. Elimination capacities increased linearly up to inlet loading of 250 g m^{-3} h^{-1}. The maximum elimination capacity of 405.7 g m^{-3} h^{-1} was observed at an inlet load of 538.8 g m^{-3} h^{-1}. The biofilter responded well during high loadings (300–538.8 g m^{-3} h^{-1}) and a removal efficiency of 80 per cent was achieved. After the shutdown of 11 days the biofilter easily recovered its performance and again shown a removal efficiency of 93 per cent.

References

1. S.B. Corchonoy, Atkinson, Kinetics of the gas phase reactions of OH and O_3 radicals with 2-carene 18. cineole, p-cyniene and terpinolene, Environ. Science Technol. 49 (1990) 1497-1502.

2. S.P.P. Ottengraf, Exhaust gas purification. In: Rhem, H.J., Reed, G. (Eds). Biotechnology. VCH Verlagsgesellschaft, Weinheim, Germany.8 (1986) (12) 425-452.

3. F.J. Alvarez-Hornos, C. Gabaldon, V. Martinez-Soria, M. Martin, P. Marzal, J.M. Penya-roza, Biofiltartion of ethylbenzene vapours: Influence of the packing material, Bioresource technology. 99 (2008) 269-276.

4. R.S. Singh, S.S. Agnihotri, S.N. Upadhyay, Removal of toluene vapour using agro-waste as biofilter media, Bioresource Technology. 97 (2006) 2296-2301.

5. E. Smet, P. Lens, H.V. Langenhove, Treatment of waste gases contaminated with odorous sulfur compounds. Critical Reviews in Environmental Science and Tech. 28(1), (1998) 89-98.

6. M.M. Galera, E. Cho, E. Tuuguu, S.-J. Park, C. Lee, W.-J. Chung, Effects of pollutant concentration ratio on the simultaneous removal of NH_3, H_2S and toluene gases using rock wool-compost biofilter, Journal of Hazardous Materials. 152 (2008) 624-631.

7. K.-L., Ho, Y.-C. Chung, Y.-H. Lin, C.-P. Tseng, Microbial populations analysis and field application of biofilter for the removal of volatile-sulfur compounds from swine wastewater treatment system, Journal of Hazardous Material. 152 (2008) 580-588.

8. H.L. Bohn, Biofilter media, Proceedings of the 89th Annual Meeting and Exhibition of the Air and Waste management Association, Nashville, Tennessee, June 23-28, Air and Waste management Association., Pittsburgh, Pennsylvania(1996).

9. R.J. Abumaizar, K. Walter, H.S. Edward, Biofiltration of BTEX contaminated air streams using compost-activated carbon filter media. Journal of Hazardous Materials. 60 (1998) 111-126.

10. S.N. Upadhyay, V. Kumar, Biological control of air pollutant, first edition. Consice Encyclopedia of Bioresource Technology. The Haworth Reference Press, New York. (2004).

11. H. Elmirini, N. Bredin, A. Shareefdeen, M. Heitz, Biofiltration of xylene emissions: bioreactor response to variations in the pollutant inlet concentration and gas flow rate, Chem. Eng. J. 100 (2004) 149-158.

12. Bibeau *et al.*, 1998

13. H. Jorio, L. Bibeau, G. Viel, Effects of gas flow rate and inlet concentration on xylene vapors biofiltration performance, Chem. Eng. J. 76 (2000) 209-221.

14. T.A. Gloria, R. Sergio, L.O. Ricardio, Hydrodynamic characterization of a trickle bed air biofilter. Chem. Engg. J. 113 (2005) 145-152.

15. D. Wu, X. Quan, Y. Zhang, Y. Zhao, Long-term operation of a compost-based biofilter for biological removal of n-butyl acetate, p-xylene and ammonia gas from an air stream, Biochemical Engg. Journal. 32 (2006) 84-92.

16. L. Li, J.X. Liu, Removal of xylene from off-gas using a bioreactor containing bacteria and fungi, International Biodeterioration and Biodegradation. 58 (2006) 60-64.

17. E. Jeong, M. Hirari, M. Shoda, Removal of p-xylene with Pseudomonas sp. NBM21 in biofilter, Journal of bioscience and Bioengineering. 102 (2006) 288-287.

18. J.W. Hwang, S.J. Jang, E.Y. Lee, C.Y. Choi, S. Park, Evaluation of composts as biofilter packing material for treatment of gaseous p-xylene, Biochem Engg. Journal. 35 (2007)142-149.

19. V. Saravanan, N. Rajamohan, Treatment of xylene polluted air using press mud-based biofilter, Journal of Haz. Mat.(2008). Article in press.

20. D.T. Gibson, V. Mahadevan, J.F. Davey, Bacterial metabolism of para- and meta-xylene: oxidation of the aromatic ring, J. Bacteriol., 119 (1974) 930-936.

21. J. Jung, D. Kafkewitz, Toluene and xylene degradation by a denitrifying strain of Xanthomonas maltophilia with limited or no oxygen, Chemosphere. 32 (1996) 1843-1850.

22. M. Bramucci, M. Singh, V. Nagarajan, Biotransformation of both p- and 2,6-dimethylnaphthalene by xylene monooxygenas cloned from a *Sphingomonas* isolate. Appl. Microbiol. Biotechnol. 59 (2002) 679-684.

23. D. Kim, Y. Kim, J. Jung, G.J. Zylstra, Y. Kim, S. Kim, E. Kim, Regioselective oxidation of xylene isomers by Rhodococcus sp. Strain DK 17. FEMS Microbiol. Lett. 223 (2003) 211-214.

24. T. Maruyama, H. Iida, H. Kakidani, Oxidation of both termini of p- and m-xylene by *Escherichia coli* transformed with xylene monooxygenase gene, J. Mol. Catal. B Enzym, 21 (2003) 211-219.

25. F.X. Prenafeta-Boldu, J. Vervoort, J.T.C. Grotenhuis and J.W. van Groenestijin, Substrate interactions during the biodegradation of BTEX hydrocarbons by the fungus *Cladophialophora* sp. Strain T1, Applied Environmental Microbiology. 68 (2002) 2660-2665.

26. Y.-S. Oh, S.,-C. Choi, Selection of suitable packing material for biofiltration of toluene, m- and p-xylene vapours, The Journal of Microbiological Society of Korea. 38 (2000) 31-35.

27. Y.-S. Oh, S.,-C. Choi, Y.-K. Kim, Degradation of gaseous BTX by biofiltration with *Phanerochaete chrysosporium*, The journal of Microbiol. 36 No. 1 (1998) 34-38.

28. C. Quinlan, K. Strevett, M. Ketcham, VOC elimination in a compost biofilter using a previously acclimated bacterial inoculum, J. Air Waste Manage. Assoc. 49 (1999) 544-553.

29. M. Heitz, H. Jorio, L. Bibeau, G. Viel, Effects of gas flow rate and inlet concentration on xylene vapors biofiltration performance. Chemical Engineering Journal. 76 (2000) 209-221.

30. H. Shim, B. Hwang, S.-S. Lee, S.-H. Kong, Kinetics of BTEX biodegradation by a coculture of *Pseudomonas putida* and *Pseudomonas fluorescens* under hypoxic conditions, Biodegradation. 16 (2005) 319-327.

31. N.C. Choi, J.-W. Choi, S.-B. Kim, D.-J. Kim, Modelling of growth kinetics for *Pseudomonas putida* during toluene degradation, Applied Microbiol Biotechnol. 81 (2008) 135-141.

32. S. Roy, J. Gendron, M.-C. Delhomenie, L. Bibeau, M. Heitz, R. Brzezinski, *Pseudomonas putida* as the dominant toluene-degrading bacterial species during air decontamination by biofiltration, Appl Microbiol Biotechnol. 61 (2003) 366-373.

33. Y.C., Chung, C., Huang, C.P. Tseng, Biodegradation of hydrogen sulphide by a laboratory-scale immobilized *Pseudomonas putida* CH11 biofilter, Biotechnol. Prog. 12 (1996) 773-778.

34. S. Moller, A. R. Pedersen, L.K. Poulsen, E. Arvin, S. Molin, Activity and three-dimensional distribution of toluene-degrading *Pseudomonas putida* in a multispecies biofilm assessed by quantitative in situ hybridization and scanning confocal laser microscopy, Applied and Environmental Microbiology. 62 No. 12 (1996) 4632-4640.

35. C. Ying-Chien, L. Yu-Yen, T. Ching-Ping, Removal of high concentration of NH_3 and coexistent by biological activated carbon (BAC) biotrickling filter, Bioresource Technology. 96 No. 16 (2005) 1812-1820.

36. www.osha.gov

37. www.aiha.org

Chapter 8

Activated Carbon: An Adsorbent for Pollution Control

Vinay K Srivastav[1], Jitendra K. Pandey[2] and D.R. Saini[3]

[1]HEMRL, Pune
[2]Max Plank, Germany
[3]National Chemical Laboratory, Pune

ABSTRACT

Activated carbons are one of the materials which are widely used in removal of various kind of contamination from air (gases) and water (liquids). Its structure, chemical and physical characteristics, production techniques have been presented. The influence of physical parameters and type of chemical and how these parameters affect its surface activities, pore size, and pore size distribution are discussed. The effect of structure on kinetics and mechanism of adsorption has also been outlined.

A wide range of natural, bio, bio waste and petroleum (rich) carbonaceous resources used for manufacturing of activated carbon are briefed. The popular characterization techniques like porosimetry, sorptometry, topography, surface area, pore size distribution and nature of the isotherms are reviewed in short. The application various kind of activated carbon like powder, granular, fiber and shaped product used for purification, decolorizing of water, solvent recovery, military uses, nuclear industries, air purification, domestic uses, precious metal, catalysis and surfactant recovery have been outlined and illustrated with few examples.

Introduction

The removal of contaminants out of liquids and gases is a process which has received considerable attention throughout the twentieth century. However, the interest in this topic is not confined to more recent years. In 1550 BC the Egyptians used charcoal to purify oils and early ocean going ships used to stored drinking water in wooden barrels, the inside of it was charred.

Activated carbon (AC) is a predominantly amorphous solid that has an extraordinarily large internal surface area and pore volume. These unique characteristics are responsible for its adsorptive properties, which are exploited in many different liquid- and gas-phase applications. AC is an exceptionally versatile adsorbent because the size and distribution of the pores within the carbon matrix. It can be controlled to meet the needs of current and emerging markets. Engineering requirements of specific applications are satisfied by producing AC in the form of powders, granules, fiber and shaped products. Through choice of precursor, method of activation, and control of processing conditions, the adsorptive properties of products are tailored for applications as diverse as the purification of potable water and the control of gasoline emissions from motor vehicles etc.

There are several type adsorption material used for fluids and gases purification and separation. Basically they can be classified in two broad categories based on physical structure like amorphous and crystalline, polar and non polar, hydrophilic and hydrophobic etc. Properties and application of few amorphous and polymeric adsorbent are given in Tables 8.2 and 8.3. Similarly, Table 8.6 shows polarity of commonly used amorphous adsorbent. As such AC is non polar however it often exhibit polarity depending on substitutions.

Functionalized acrylic and styrene based polymeric adsorbent are also used in large number of applications because they can be shaped in sheet, film, hallow fiber etc.

Manufacture and Processing

Commercial AC products are produced from organic materials that are rich in carbon, particularly coal, lignite, wood, nut shells, peat, pitches, agriculture by product, bio waste and coke etc. The choice of precursor is largely dependent on its availability, cost, and purity, often the manufacturing process and intended application of the product are also required considerations. Manufacturing processes fall into two categories, physical (thermal) activation and chemical activation. The effective porosity of activated carbon produced by thermal activation is the result of gasification of the carbon at relatively high temperatures. The porosity of products is further improved by chemical dehydration reactions occurring at significantly lower

Table 8.1: Composition as Percentage of Organic Mass of Barely Straw.

Constituents	% mass
Cellulose	31 -34
Pentosan	24-29
Lignin	14-15
Slicia	3-6
Ash	5-7

Table 8.2: Properties and Applications of Amorphous Adsorbents.

Adsorbent	Pore diameter, nm	Particle density, g/cm^3	Specific area, m^2/g	Applications
activated carbon (large pore)	1–10^3 (broad range)	0.6–0.8	200–600	water purification, sugar decolorizing
activated carbon (small pore)	1–10	0.5–0.9	400–1200	removal of light organics
carbon molecular sieve	0.4–0.5, 10–10^2 (bimodal)	0.9–1.0	100–300	air separation (N_2 production)
silica gel (high area)	2–10	1.09	800	general purpose
silica gel (low area)	10–50	0.62	300	general purpose
activated alumina	2–10	1.2–1.3	300–400	desiccants

Table 8.3: Properties and Applications of Polymeric Adsorbents.

Type of polymer	Representative commercial product	Properties	Applications
sulfonated styrene–divinylbenzene copolymers[a] with various degrees of cross-linking	Dowex-50 Amberlite IR120B	pore diameter, porosity, density, etc, vary with degree of hydration or dehydration	sugar separations[b], eg, various fructose, glucose
macroreticular sulfonated styrene–divinylbenzene	Diaion HPK-25	porosity ~0.33, microparticles ~80-μm diameter	removal of NH_3 or light amines[c]

[a] Ion exchanged to Ca^{2+} form.
[b] Liquid-phase operation.
[c] Gas or liquid phase.

temperatures Partial gasification is done by use of mild oxidizing agent such as carbon dioxide, steam or fuel gases at around 800^{0}C.Most of the activated carbon are either non polar or slightly polar depending on the treating agent used.

Thermal Activation Processes

Activation is controlled gasification done by an oxidising gas such as steam, CO_2 or air, at elevated temperatures, typically 800-1100°C.

Thermal activation occurs in two stages: thermal decomposition or carbonization of the precursor and controlled gasification or activation of the crude char. During carbonization, elements such as hydrogen and oxygen are eliminated from the precursor to produce a carbon skeleton possessing a latent pore structure. During gasification, the char is exposed to an oxidizing atmosphere that greatly increases the pore volume and surface area of the product through elimination of volatile pyrolysis products and from carbon burn-off, Carbonization and activation of the char are generally carried out in direct-fired rotary kilns or multiple hearth furnaces, fixed bed, but fluidized-bed reactors have also been used. Materials of construction used for chamber are steel and refractories, designed to withstand the high temperature conditions, *i.e.*, >1000 ^{0}C and are inert in activation processes Cheremisinoff and Ellerbusch (1978)

The production techniques required number of unit operation. Unit operations order, sequence, duration and condition require modification depending on precursor composition like per cent of cellulose; lignin and element like C, H etc. in order to tailored the desired product for specific application. The constituents of few precursors are shown in Tables 8.1 and 8.4. For example production of activated carbon from bituminous coal requires pulverization, briquetting, oxidation and devolatization while coconut shells did not require all these operations. To obtained the higher active product coconut shell require to charred slowly prior to activation of char and acid wash for low ash content. However few basic steps are common for most of the precursors.

Table 8.4: Chemical Composition of Few Raw Materials.

Per cent	*Almond Shell*	*Apricot Stone*	*Hazelnut Shell*	*Walnut Shell*
Carbon	50.3	50.5	51.4	49.0
Hydrogen	6.05	6.03	5.95	5.75
Ash	0.31	0.78	0.49	0.44
Moisture	8.71	8.1	7.7	11

The activating agent employed, in physical processes, has been investigated by many authors. Steam is preferable to carbon dioxide and much better than air for activating many chars. Utilising steam for activation requires a high temperature of activation to provide rapid oxidation, however, temperatures above 1000°C impair adsorptive power. Step wise production process is given below:

Step-1Analysis of precursor composition is first and essential step in order to select the higher carbonaceous starting raw material especially from agricultural by product, bio or vegetable waste before pulverization and cleaning.Number of precursor require demineralization prior to gasification/pyrolysis in order to reduce the ash content. Chemical compositions of various precursors are shown in Table 8.1 and 8.4.

Step-2 Precursor material is carbonized/pyrolysed usually in absence of air which liberate volatiles non carbon elements like H, O, N etc from the precursor resulting in formation of non-graphitisable char with a rigid carbon skeleton, made up of aromatic sheets and strips. Each carbonization condition (Time and Temperature) result in a rich carbon content sorbent with unique adsorption properties and higher Iodine number. Each precursor requires specific carbonization condition. Zulkamain *et al*. (1993).

Both these step are required to be performed prior to activation. For quality assurance product of each step requires through examination in term of performance and properties before proceeding further.

Several workers have shown that surface area and porosity increased with increasing temperature and time Gergova *et al.* (1994).It has also been reported that there is optimum condition for each precursor beyond which the yield decrease with time Gamer and Packer (1980).However, porosity development is mainly in the form of micro and mesopores. Rodriguez-Reinoso and Molina-Sabio (1 992) reported that activation with CO_2, opens and widens microporosity and that at high temperatures the exterior of the particles are rapidly burnt off. Increased porosity is accompanied by a parallel increase in surface area. Lu and Chung (1996), Sollars *et al.* (1990) showed that the rate of carbon followed linear relationship with activation times employed and the optimum activation conditions are 900^{0}C for 400 minutes, forming a meso-microporous carbon from waste tires for good phenol adsorbent.Comparing the effectiveness of steam and CO_2 as activating agents is difficult because factors other than the oxidising gas influence the final product. Smisek and Cerney (1970) state that carbonaceous gasification by steam and CO_2, occurs according to an endothermic reaction and therefore can be accurately controlled. Although both reactions precede endo thermally occur at different rates. Park *et al.* (1996) showed that steam activation was about three times more rapid than carbon dioxide activation. Pashchenko *et al.* (1996) also reported a decrease in reaction rate with CO_2, activation on carbon containing wastes, result in nearly 2 times less than that of steam.

The resultant pore structure is also dependent on the activator employed. Zietek *et al.* (1996) showed that pore volumes increased with both activators but that steam activation resulted in microporous structures and CO_2 resulted in larger micropores and mesoporous structures.Number of researcher have used oxygen as an activating agent. A clear picture is yet to be established and require lot more work in this regard due to involvement of large number of parameter including precursor composition.

Chemical Activation Processes

In contrast to the thermal activation of coal, chemical activation is generally carried out commercially in a single kiln. The precursor, usually wood, is impregnated with a chemical activation agent, typically phosphoric acid, inorganic additives, metallic chlorides and the blend is heated to a temperature of 450–700^0 C.These additives degrade and dehydrate the cellulose present in the precursor.For lignin sulphuric acid, phosphoric acid, zinc chloride are added.Vegetable matter are carbonized at around 900^{0}C for higher carbon per cent. Chemical activation agents reduce the formation of tar and other by-products, thereby increasing carbon yield. In case of sawdust.sawdust is first impregnated with concentrated phosphoric acid and fed to a rotary kiln, where it is dried, carbonized, and activated at a moderate temperature. To comply with environmental pollution regulations, the kiln off-gases are treated before discharge to the atmosphere. The char is washed with water to remove the acid from the carbon, and the carbon is separated from the slurry. The filtrate is passed to an acid recovery unit. In few allied plants it is not recycle but used manufacture of fertilizer. Where ever it is necessary, the pH of the activated carbon is adjusted, and the product is dried. The dry product is screened and classified in the required size range for specific granular carbon applications.

Carbon yields as high as 50 per cent by weight of the precursor by chemical treatment and 30 per cent by physical treatments are reported Like physical treatment chemical treatment are also required to be modified the desired product. Effect of various chemical treatment on activated carbon in term of surface area and pores volume are shown in Table 8.5.

Table 8.5: Effect of Various Chemical Treatment on Activated Carbon.

Chemical Additives	*Surface Area (BET)* m^2g^{-1}	*Pore Specific Volume* $cm_3g^{-1} \times 10^{-2}$
10 per cent $ZnCl_2$	198	1.1
20 per cent $ZnCl_2$	557	2.3
10 per cent $Fe_2(SO_4)_3$	70	5.4
20 per cent $Fe_2(SO_4)_3$	243	1.8
10 per cent $FeSO_4$	89	5.2
20 per cent $FeSO_4$	412	2.8

The chemical and physical nature of ACs are dependent on the precursor, the oxidising agent employed, the temperature of activation and the degree of activation.

In dry chemical activation often inorganic materials, such as chalk, lime, sulphuric acid, calcium chloride and zinc chloride are incorporated into precursors prior to pyrolysis to enhanced carbon percentage and pore structures to result in large surface areas. In wet-chemical processes, phosphoric acid, chloride salts of magnesium, iron and aluminum, sodium carbonate, and sodium and calcium hydroxide are most popular agent. The only common trait between all these chemicals is their strength as dehydrating agents. In chemical activation not only the type of chemical but treatment time, temperature and quantity of chemical utilized are equally important role. $ZnC1_2$help in increasing the yield of the product by preventing pyrolysis of the products, such as tars forming also tends to cause hydrogen and oxygen atoms to be stripped away as water, rather than as hydrocarbons or as oxygenated organic compounds, leaving the carbon skeleton largely untouched. KOH activation increase the surface area and pore volume of active carbons according to the relative amount of KOH incorporated into the precursor. Otowa *et al.* (1995) Allen and Balasundaram (1995) reported that the best product can be obtained with iron nitrate in case of lignite. Hayashi *et al.* (1996) have found that K_2CO_3 works effectively as an activator for bean curd refuse. Many other chemicals have been used as potential activation agents, the main ones are: ammonium salts, borates, boric acid, calcium oxide, cyanides, ferric and ferrous compounds, hydrochloric acid, manganese dioxide, nickel salts, nitric acid and sulphur. US patent number 4,149,994 (1979) describes a chemical process without carbonisation. Activation is performed by adding polar compounds to coal powders, containing non-polar groups bonded with polar groups, and drying at a temperature of 120 ^{0}C. Such carbons perform well in wastewater treatment.

Chemically modified carbon with increase amounts of N, Cl, or O have microporus surface that exhibit basic polar (N) or acidic (CL)groupos.Which is expected to effect the adsorption and de-adsorption of the material.

Table 8.6: Common Adsorbent Based on Pore Size and Polarity.

Surface Polarity	*Pore Size Distribution*	
	Narrow	*Broad*
Polar	Zeolites (Al rich)	Activated alumina silica gel
Nonpolar	Carbon molecular sieves silicalite	Activated carbon

AC Regeneration

It is often required for it reusability. In general, both the processes (physical and chemical) involved more than one step. However chemical activation can be conducted at 200-800^{0}C while in contrast to physical require (Typical- 800-1100^{0}C) this in itself results in different bum-off rates of the carbon skeleton but often it is reported that the carbon yields in chemical activation are greater than those for physical activation. As a result of this temperature difference, chemically activated carbons lack the semi-conductor-redox properties that are seen in physically activated carbons. This is a result of restriction of formation of graphite zones due to the low temperatures used in chemical processes. It is often reported that physical activation resulted in smaller surface areas, smaller yields and significant pore differences, while chemical activation result in more uniform micro and mesopore size distribution. The formation of wide micropores and narrow mesopores are also dictated by activation technique. However, the ramifications and possibilities are further heightened when gasification of these carbons, with steam or CO_2 etc., is considered.

Leboda *et al*. (1996) showed that calcium incorporated into a variety of precursors, by ionic exchange or in solution as calcium acetate. It acts as a catalyst in porosity development. It was found that in the presence of $CaCO_3$, steam activation caused changes in both the micro and mesoporous structure.

Precursor Materials

Large number of low cast precursors like normal every day waste material, agriculture by product, bio waste old tyres have been used for it production. Bagasse, Kelp and seaweed, Beet-sugar sludges, Lampblack, Blood Leather waste, Bones, Lignin, Carbohydrates, Lignite, Cereals Molasses, Coal Nut shells, Coconut shells, Oil shale, Coffee beans, Peat, Corncobs and corn stalks, Petroleum acid sludge, Cottonseed hulls, Petroleum coke, Distillery waste, Potassium ferrocyanide residue, Fish, Pulp-mill waste, Blue Dust, rice hulls, Fruit piths, Rubber waste, Graphite Sawdust and Wood are also used.

The extent to how successful a precursor will be for producing AC is often gauged by its ash content. Flynn *et al*. (1988) stated that, due to its high ash content, Northern Ireland lignite was unsuitable for producing activated carbon. This problem can be overcome by demineralising, with mineral acids, *to* remove the majority of inorganic matter which ashes on carbonisation. Demineralisation is a chemical modification process and is often performed in industrial processes prior to pyrolysis and gasification. US patents 4,149,994 and 4,149,995 by Murty (1979 a, b).

Other Manufacturing Processes

Different chemical activation processes have been used to produce carbons with enhanced adsorption characteristics. ACs of exceptionally high surface area of 3000m ^{2}/g have been produced by the chemical activation of carbonaceous materials with potassium hydroxide. ACs are also produced commercially in the form of cloths, fibers, and foams generally by chemical activation of the precursor with a Lewis acid such as aluminum chloride, ferric chloride, or zinc chloride. Activation are further improved by microwave or plasma treatment and electromagnetic irradiation.

Forms of Activated Carbon Products

To meet the engineering requirements of specific applications, ACs are produced and classified as granular, powdered, fiber or shaped products. Granular activated carbons are produced directly from granular precursors, such as sawdust and crushed and sized coconut char or coal. The granular product is screened and sized for specific applications. Powdered activated carbons are obtained by grinding granular products. Shaped AC products are generally produced as cylindrical pellets by extrusion of the precursor with a suitable binder before activation of the precursor AC fiber are produced from pitch or PAN precursor by high speed sinning, extensive mode carbonization, graphitization and activation.

Adsorption capacities of various chemicals by AC are shown in Table 8.7 and properties of few commercially available ACs are given in Table 8.9.

Physical and Chemical Properties

The structure of activated carbon is best described as a twisted network of defective carbon layer planes, cross-linked by aliphatic bridging groups. X-ray diffraction patterns of activated carbon reveal that it is non graphitic, remaining amorphous because the randomly cross-linked network inhibits reordering of the structure even when heated to 3000^{0}C. This property of activated carbon contributes to its most unique feature, namely, the highly developed and accessible internal pore structure. The surface area, dimensions, and distribution of the pores depend on the precursor and on the conditions of carbonization and activation. Mechanical integrity of ACs are also desirable in number of applications

Pore sizes are classified by IUPAC in to three broad classes:

- ✰ Micropores - pore width <2 nm)
- ✰ Mesopores - pore width 2–50 nm)
- ✰ Macropores - pore width >50 nm)

The pore structure is mainly classified into two categories (a) closed pores and (b-h)open pores. The open pores consist of several types: (b)through pores are open at

Table 8.7: Adsorption Capacities of Activated Carbon.

Compound	Adsorption capacity, mg/g	Compound	Adsorption capacity, mg/g
bis(2-ethylhexyl)phthalate	11,300	guanine	120
butyl benzyl phthalate	1,520	styrene	120
heptachlor	1,220	1,3-dichlorobenzene	118
heptachlor epoxide	1,038	acenaphthylene	115
endosulfan sulfate	686	4-chlorophenyl phenyl ether	111
endrin	666	diethyl phthalate	110
fluoranthene	664	2-nitrophenol	99
aldrin	651	dimethyl phthalate	97
PCB-1232	630	hexachloroethane	97
β-endosulfan	615	chlorobenzene	91
dieldrin	606	*p*-xylene	85
hexachlorobenzene	450	2,4-dimethylphenol	78
anthracene	376	4-nitrophenol	76
4-nitrobiphenyl	370	acetophenone	74
fluorene	330		74
		1,2,3,4-tetrahydronaph-thalene	
DDT	322		
2-acetylaminofluorene	318	adenine	71
α-BHC	303	dibenzo[*a*,*h*]anthracene	69
anethole	300	nitrobenzene	68
3,3-dichlorobenzidine	300	3,4-benzofluoranthene	57
2-chloronaphthalene	280	1,2-dibromo-3-chloropropane	53
phenylmercuric acetate	270	ethylbenzene	53
hexachlorobutadiene	258	2-chlorophenol	51
γ-BHC (lindane)	256	tetrachloroethylene	51
p-nonylphenol	250	*o*-anisidine	50
4-dimethylaminoazobenzene	249	5-bromouracil	44
chlordane	245	benzo[*a*]pyrene	34
PCB-1221	242	2,4-dinitrophenol	33
	232	isophorone	32
1,1-dichloro-2,2-bis(*p*-chloro- phenyl)ethane			
(DDE)		trichloroethylene	28
acridine yellow	230	thymine	27
benzidine dihydrochloride	220	toluene	26
β-benzenehexachloride	220	5-chlorouracil	25
n-butyl phthalate	220	*N*-nitrosodi-*n*-propylamine	24
N-nitrosodiphenylamine	220	bis(2-chloroisopropyl) ether	24
phenanthrene	215	phenol	21
dimethylphenylcarbinol	210	bromoform	20
4-aminobiphenyl	200	carbon tetrachloride	11
2-naphthol	200	bis(2-chloroethoxy)methane	11
α-endosulfan	194	uracil	11
acenaphthene	190	benzo[*ghi*]perylene	11
	190	1,1,2,2-tetrachloroethane	11
4,4′-methylenebis(2-chloro-aniline)			
		1,2-dichloropropene	8.2
benzo[*k*]fluoranthene	181	dichlorobromomethane	7.9
acridine orange	180	cyclohexanone	6.2
1-naphthol	180	1,2-dichloropropane	5.9

Contd...

Table 8.7–*Contd...*

Compound	Adsorption capacity, mg/g	Compound	Adsorption capacity, mg/g
4,6-dinitro-o-cresol	169	1,1,2-trichloroethane	5.8
1-naphthylamine	160	trichlorofluoromethane	5.6
2,4-dichlorophenol	157	5-fluorouracil	5.5
1,2,4-trichlorobenzene	157	1,1-dichloroethylene	4.9
2,4,6-trichlorophenol	155	dibromochloromethane	4.8
2-naphthylamine	150	2-chloroethyl vinyl ether	3.9
pentachlorophenol	150	1,2-dichloroethane	3.6
2,4-dinitrotoluene	146	1,2-*trans*-dichloroethylene	3.1
2,6-dinitrotoluene	145	chloroform	2.6
4-bromophenyl phenyl ether	144	1,1,1-trichloroethane	2.5
p-nitroaniline	140	1,1-dichloroethane	1.8
1,1-diphenylhydrazine	135	acrylonitrile	1.4
naphthalene	132	methylene chloride	1.3
1-chloro-2-nitrobenzene	130	acrolein	1.2
1,2-dichlorobenzene	129	cytosine	1.1
p-chloro-*m*-cresol	124	benzene	1.0
1,4-dichlorobenzene	121	EDTA	0.86
benzothiazole	120	benzoic acid	0.76
diphenylamine	120	chloroethane	0.59
		N,N-dimethylnitrosamine	6.8×10^{-5}

Table 8.8: Parameters of Physical Adsorption and Chemisorption.

Parameter	Physical adsorption	Chemisorption
heat of adsorption $(-\Delta H)$	low, <1.5 or 2 times latent heat of evaporation	high, > 2 or 3 times latent heat of evaporation
specificity	nonspecific	highly specific
nature of adsorbed phase	monolayer or multilayer, no dissociation of adsorbed species	monolayer only may involve dissociation
temperature range	only significant at relative low temperatures	possible over a wide range of temperature
forces of adsorption	no electron transfer, although polarization of sorbate may occur	electron transfer leading to bond formation between sorbate and surface
reversibility	rapid, nonactivated, reversible	activated, may be slow and irreversible

two ends independent pores,(c) blind pores are open only at one end,(c) do not connect with other pores and crossing pores (d) connect with other pores.

From the viewpoint of pore shape, pores are classified as follows: cylindrical pores (e), conical pores (f), ink-bottle pores (g), slit pores (h), and so on. These help in specifying the pore structure.

In amorphous adsorbent like AC the specific area can be enhanced up to 3000 m^2/g.But with this high area carbon tend to lose its physical strength and this limits their usefulness in many practical applications. The high area materials also contain a large proportion of very small pores, which renders them unsuitable for applications involving adsorption of large molecules. The distinction between gas carbons, used for adsorption of low molecular weight, and liquid carbons, which are used for adsorption of larger molecules such as color bodies from the liquid phase, is thus primarily a matter of pore size.

Table 8.9: Properties of Few Commercial Activated Carbons (US).

Property	Manufacturer / Precursor / Product grade / Product form / Typical range	Gas-phase carbons			Liquid-phase carbons		
		Calgon Coal BPL Granular	Norit Peat B4 Extruded	Westvaco Wood WV-A 1100 Granular	Calgon Coal SGL Granular	Norit Peat SA 3 Powdered	Westvaco Wood SA-20 Powdered
particle size, U.S. mesh[b,c]	<4	12 × 30	3.8 mm	10 × 25	8 × 30	64%	65 – 85%
apparent density, g/cm^3	0.2–0.6	>0.48	0.43	0.27	0.52	0.46	0.34–0.37
particle density, g/cm^3	0.4–0.9	0.80		0.50	0.80		
hardness number	50–100	>90	99				
abrasion number					>75		
ash, wt %	1–20		6			6	3–5
BET surface area, N_2, m^2/g	500–2500	1050–1150	1100–1200	1750	900–1000	750	1400–1800
total pore volume, cm^3/g	0.5–2.5	0.8	0.9	1.2	0.85		2.2–2.5
CCl_4 activity, wt %	35–125	>60					
butane working capacity, $g/100\ cm^3$	4–14			>11.0			
iodine number	500–1200	>1050			>900	800	>1000
decolorizing index							
Westvaco	15–25						>20
molasses number							
Calgon	50–250				>200		
Norit	300–1500					440	
heat capacity at 100°C, $J/(g \cdot K)$[d]	0.84–1.3	1.05			1.05		
thermal conductivity, $W/(m \cdot K)$	0.05–0.10						

[a] Specific values shown are those cited in manufacturers' product literature (19). Typical ranges shown are based on values reported in the open literature.
[b] Unless otherwise noted.
[c] Approximate mm corresponding to cited meshes are mesh: mm—4: 4.76; 8: 2.38; 10: 2; 12: 1.68; 25: 0.72; 30: 0.59; 325: 0.04.
[d] To convert J to cal, divide by 4.184.

Adsorption capacity and rate of adsorption depend on the internal surface area and distribution of pore size and shape but are also influenced by the surface chemistry of the activated carbon. The macro porosity of the carbon is important for the transfer of adsorbate molecules to adsorption sites within the particle. Functional groups are formed during activation by interaction of free radicals on the carbon surface with atoms such as oxygen and nitrogen, both from the precursor. The functional groups make the surface of AC interactive which influence adsorptive properties. ACs are generally considered to exhibit a low affinity for water, which is an important property with respect to the adsorption of gases in the presence of moisture. However, the functional groups on the carbon surface can interact with water, rendering the carbon surface more hydrophilic. Surface oxidation, which is an inherent feature of activated carbon, results in hydroxyl, carbonyl, and carboxylic groups that impart an amphoteric character to the carbon, so that it can act as acidic or basic. The electro kinetic properties of an activated carbon product are therefore, important with respect to its use as a catalyst support. As well as it influence the adsorption of many molecules. The presence of specific groups on ACs helps in solvent recovery applications. In addition to surface area, pore size distribution, and surface chemistry, other important properties of commercial activated carbon products include pore volume, particle size distribution, apparent or bulk density, particle density, abrasion resistance, hardness, and ash content

Adsorption Mechanism

Adsorption starts with diffusion followed by physical, capillary action and chemisorptions. Analysis of the process is complex as all the process occurs simultaneously. Parameters involved in specific adsorption mechanism are given in Table 8.8. In physical adsorption the forces are relatively weak, involving mainly van der Waals (induced dipole–dipole) interactions, supplemented in many cases by electrostatic contributions from field gradient–dipole or–quadrupole interactions. By contrast, in chemisorption there is significant electron transfer, equivalent to the formation of a chemical bond between the sorbate and the solid surface. Such interactions are stronger and more specific than the forces of physical adsorption and are obviously limited to monolayer coverage. The differences in the general features of physical and chemisorption systems can be understood on the basis of the surface forces. ACs material is expected to have major surface adsorption and in minor way capillary action.

Transport in a macropore can occur by several different mechanisms, the most important are bulk molecular diffusion, Knudsen diffusion, surface diffusion, and Poiseuille flow. In liquid systems bulk molecular diffusion is generally dominant, but in the vapor phase the contributions from Knudsen and surface diffusion could be large or even dominant. The contribution from Poiseuille flow, ie, forced flow through the pore under the influence of the pressure gradient, is generally relatively minor as the pressure gradients are usually small.

In micro pores diffusion in which the pore diameter is not much greater than the molecular diameter the diffusing molecule never escapes from the force field of the pore wall. Under these conditions steric effects and the effects of nonuniformity in the potential field become dominant and the Knudse mechanism does not work. Diffusion occurs by an active process involving jumps from site to site, just as in surface diffusion, the diffusivity strongly dependent on both temperature and concentration. The true driving force for any diffusive transport process is due to the gradient of chemical potential concentration. This distinction is not important in dilute systems where thermodynamically ideal behavior is approached. However, it becomes important at higher concentration levels and in micropore and surface diffusion.

In Capillary condestion. equilibrium vapor pressure in a pore or capillary is reduced by the effect of surface tension. As a result, liquid sorbate condenses in a small pore at a vapor pressure that is somewhat lower than the saturation vapor pressure. In a porous adsorbent the region of multilayer physical adsorption merges gradually with the capillary condensation regime, leading to upward curvature of the equilibrium isotherm at higher relative pressure. In the capillary condensation region the intrinsic selectivity of the adsorbent is lost, so in separation processes it is generally advisable to avoid these conditions.

Adsorption Kinetics

In intrinsic adsorption process chemisorptions may be regarded as a chemical reaction between the sorbate and the solid surface, it is an activated process for which the rate constant (k) follows the familiar Arrhenius rate law:

$$k = k_0 e^{-E/RT} \quad (1)$$

Depending on the temperature and the activation energy (E), the rate constant may vary over many orders of magnitude. In practice the kinetics are usually more complex than expected, since activation energy generally varies with surface coverage as a result of energetic heterogeneity and/or sorbate-sorbate interaction. As a result, the adsorption rate is commonly given by the Elovich equation (2):

$$q = \frac{1}{k'} \ln(1 + k''t) \quad (2)$$

Where, k' and k" are temperature-dependent constants. In contrast, physical adsorption is a very rapid process, so the rate is always controlled by mass transfer resistance rather than by the intrinsic adsorption kinetics. However, under certain conditions the combination of a diffusion controlled process with adsorption equilibrium constant that varies according to equation 1 can give the appearance of activated adsorption. A porous adsorbent in contact with a fluid phase offers at least two and often three distinct resistances to mass transfer: external film resistance and intra particle diffusional resistance. When the pore size distribution has a well-defined bimodal form, the latter may be divided into macropore and micropore diffusional resistances. Depending on the particular system and the conditions, any one of these resistances may dominant or the overall rate of mass transfer may be determined by the combined effects of more than one resistance. Most of the system follows either first order or second order kinetics.

Adsorption Equilibrium

As per the Henry's Law the distribution of a sorbate between fluid and adsorbed phases is governed by the principles of thermodynamics. Equilibrium data are commonly reported in the form of an isotherm, which is a diagram showing the variation of the equilibrium adsorbed phase concentration or loading with the fluid-phase concentration or partial pressure at a fixed temperature. In general, for physical adsorption on a homogeneous surface at sufficiently low concentrations, the isotherm should approach a linear form, and the limiting slope in the low concentration region is commonly known as the Henry's law constant. The Henry constant is simply a thermodynamic equilibrium constant and the temperature dependence therefore follows the usual Van't Hoff equation as:

$$\lim_{p \to 0} \left(\frac{\partial q}{\partial p} \right)_T = K' = K'_0 e^{-\Delta H_0/RT} \tag{3}$$

In which - ΔH_0 is the limiting heat of adsorption at zero coverage. Since adsorption, particularly in the vapor phase, is usually exothermic, - ΔH_0 is a positive quantity and K'_0 therefore decreases with increasing temperature. A corresponding dimensionless Henry constant (K) may also be defined based on the ratio of adsorbed and fluid-phase concentrations:

$$\lim_{c \to 0} \left(\frac{\partial q}{\partial p} \right)_T = K = K_0 e^{-\Delta H_0/RT} \tag{4}$$

Since, for an ideal vapor phase, p = cRT, these quantities are related

$$K = RTK'; \; -\Delta H_0 = -\Delta U_0 + RT \tag{5}$$

Henry's law corresponds physically to the situation in which the adsorbed phase is so dilute that there is neither competition for surface sites nor any significant interaction between adsorbed molecules. At higher concentrations both of these effects become important and the form of the isotherm becomes more complex. The isotherms have been classified into five or six different types as shown in Figure-3. Isotherms for a microporous adsorbent are generally of type I; the more complex forms are associated with multilayer adsorption and capillary condition.

Langmuir Isotherm

Type I isotherms are commonly represented by the ideal Langmuir model:

$$\frac{q}{qs} = \frac{bp}{1+bp} \tag{6}$$

Where, *qs* is the saturation limit and b is an equilibrium constant which is directly related to the Henry constant (K_0 = b*qs*). The Langmuir model was originally developed to represent monolayer adsorption on an ideal surface, for which qs corresponds to the monolayer coverage. However, in applying this model to physical adsorption on a microporous solid, the saturation limit becomes the quantity of sorbate required to fill the micropore volume. This expression is of the correct form to represent a type I isotherm, since at low pressure it approaches Henry's law while at high pressure it tends asymptotically to the saturation limit. The equilibrium constant b = (K"/qs) decreases with increasing temperature (eq. 3); therefore, for a given pressure range, the isotherm approaches rectangular or irreversible form at low temperatures (large b) and linear form at high temperatures (small b). Although very few systems conform accurately to the Langmuir model, this model provides a simple qualitative representation of the behavior of many systems and it is therefore widely used, particularly for adsorption from the vapor phase. According to the Langmuir model the heat of adsorption should be independent of loading, but this requirement is seldom fulfilled in practice. Both increasing and decreasing trends are commonly observed. For a polar sorbate on a polar adsorbent (ie, a system in which electrostatic forces are dominant) a decreasing trend is normally observed, since the relative importance of the electrostatic contribution declines at high loadings as a result of preferential occupation of the most favorable sites and consequent screening of cations. In contrast, where van der Waals forces are dominant (nonpolar sorbates), a rising trend of heat of adsorption with loading is generally observed. This is commonly attributed to the effect of intermolecular attractive forces, but other explanations are also possible. In homologous series such as the linear paraffins the heat of adsorption increases linearly with carbon number.

Freundlich Isotherm

The isotherms for some systems, notably hydrocarbons on activated carbon, conform more closely to the Freundlich equation:

$$q = bp^{1/n} \qquad (n > 1.0) \tag{7}$$

Although the Freundlich expression does not reduce to Henry's law at low concentrations, it often provides a good approximation over a wide range of conditions. This form of equation can be explained as resulting from energetic heterogeneity of the surface. Superposition of a set of Langmuir isotherms with different b values (corresponding to sites of different energy) yields an expression of this form. In most of the cases Langmuir three parameter model works well. However, Langmuir with four to five independent parameters for better fitting and prediction is used by several workers.

Characterization of Activated Carbons

Sorbent Characterization

Activated carbons are materials which cannot be easily characterized simply by chemical analysis or structural techniques. Every carbon is unique with its physical and chemical characteristics due its parent material, pre and post treatments treatment and conditions, whether it chemical, physical or a combination of both. However, standard methods of characterization have been established in terms of pore classification, surface area or surface topography, in order to compare carbons and to assess their adsorptive potential.

Topography

Topographical *i.e.* stumes of the external surface of activated carbon gives an indication of the porous nature of the carbon's exterior. The external pores are generally large pores, normally mesopores and macropores depending on the parent material, and are important in terms of adsorption as they are transport pores allowing the passage of the adsorbing species, from the adsorptive, to the adsorbents internal porous structure.

Scanning electron microscopy and High-Resolution Transmission Electron Microscopy (SEM/HRTEM) is often used for surface investigations. SEM micrograph of carbonized barely strawis shown Figure 8.1 and carbonized and activated Coconut are show in Figure 8.2. Both the figure exhibit open shell structure with the thin shell wall thickness. Formation of the cell and enhancement of surface area mainly occur on activation as illustrated in Figure 8.2.

Porosimetry

Mercury porosimetry is used extensively as a means of characterizing porous media and powders. By far the most important aspect of mercury porosimetry is its use for determining pore size dmibution.

Mercury porosimetry works on the principle that, as mercury is a non-wetting liquid towards most substances (having a contact angle with mercury greater than 90"), it will not penetrate into openings without pressure being applied. If the solid to be investigated is placed in a container with a tapered capillary end, evacuated and filled to a predetermined level with mercury, when pressure is applied the mercury level will fall as it is pushed into the pores. The decrease in the mercury level is directly related to the pressure exerted and from this the volume-pressure porosimetric curve can be constructed Structural hysteresis or ink-bottle hysteresis arising from the non-cylindrical nature of pores in real porous bodies (cylindrical capillaries do not display hysteresis). Due to structural hysteresis, surface areas obtained from porosimetry will never be accurate.

Although there are lot of limitation and problems linked with the accuracy of porosimetry data, the method provides a characteristic of pore space geometry, this information combined with a comparative method can be used as a means of sorbent characterization.

Sorptometry

Gas sorption analysis and the interpretation of its physical and chemical findings is by far the most widely used and investigated method of sorbent characterization. In general, physisorption is used to determine the surface area, pore distribution and pore volume of a sorbent whereas chemisorptions is used to assess the area of any active surface component. Initially the physical adsorption (and desorption) isotherm is determined for the sorbent to be characterized.This is generally determined by nitrogen adsorption at the liquid nitrogen temperature of *77K*, however, other small molecule adsorptivities have also been used, *Ar* (*77K*), Kr (*77K*), Xe (*77K*), alkanes

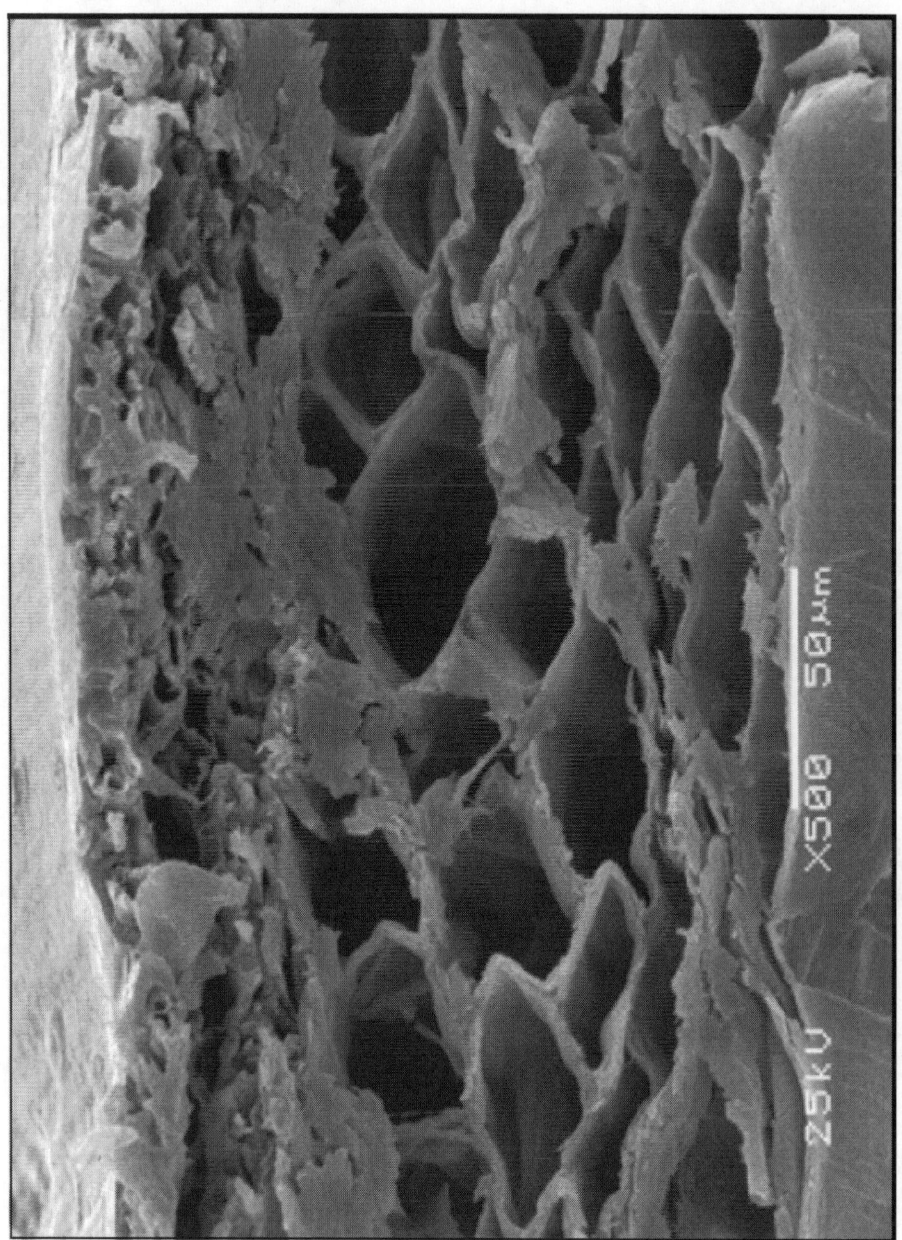

Figure 8.1: SEM for Inner Surface of Carbonzized Barley Straw at 200ºC with Magnification 500.

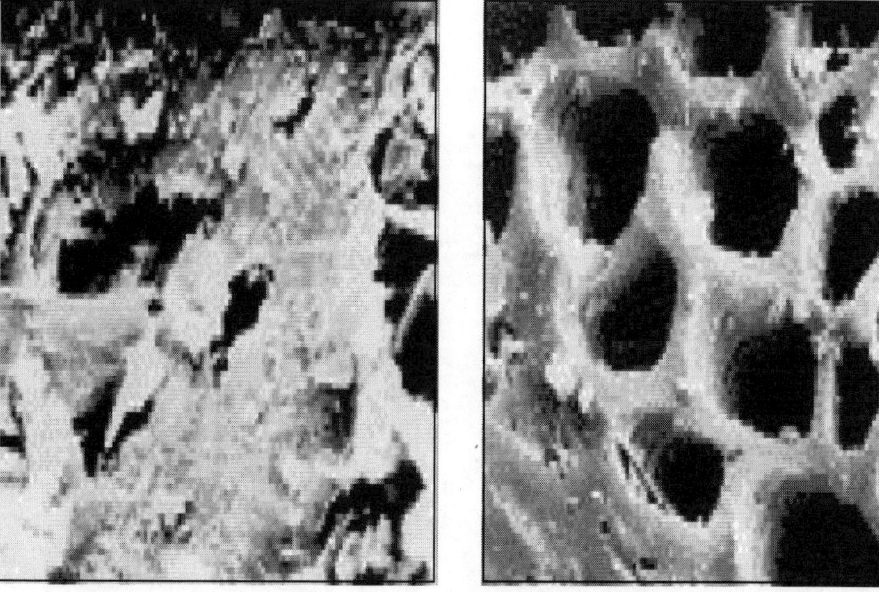

Carbonized Coconut Activated Coconut

Figure 8.2: SEM Micrograph of Carbonized and Activated Coconut.

(298K), *CO,* (195K) and 0, (90K). The shape of the isotherm provides an indication of the nature of the pore structures in the sorbent make-up.

Nitrogen adsorption data is a good source of information which can be used to yield information regarding the pore size distributions of the carbons. The application of these methods by Brunnauer *et al.* (1940) yield information on pore radii and pore size distribution for monolayer adsorption.

Isotherms

An isotherm is a pictorial representation of the equilibrium adsorption desorption process. It shows the equilibrium relationship, at constant temperature or pressure, and the adsorbed quantity. Brunauer *et al.* (1940) noted that although isotherms were different for all sorbents and sorbates investigated, common shapes of isotherms were observed.Six isotherm types are generally used as a basis for classification, and these are shown in Figure 8.3.

Type I- reversible isotherms are characteristic of micro porous solids as seen for example in activated carbons or zeolites. Uptake reaches a limiting value as *p/p,* tends to 1, indicating that uptake is dependent on micropore volume accessibility rather than surface area. Typical systems include nitrogen on charcoal at -183°C and many dyes on activated carbon 20°C.

Type II- isotherms are common to non-porous or macroporous solids. They indicate unrestricted monolayer-multilayer adsorption, such as the sorption of nitrogen on iron catalysts at 195°C.

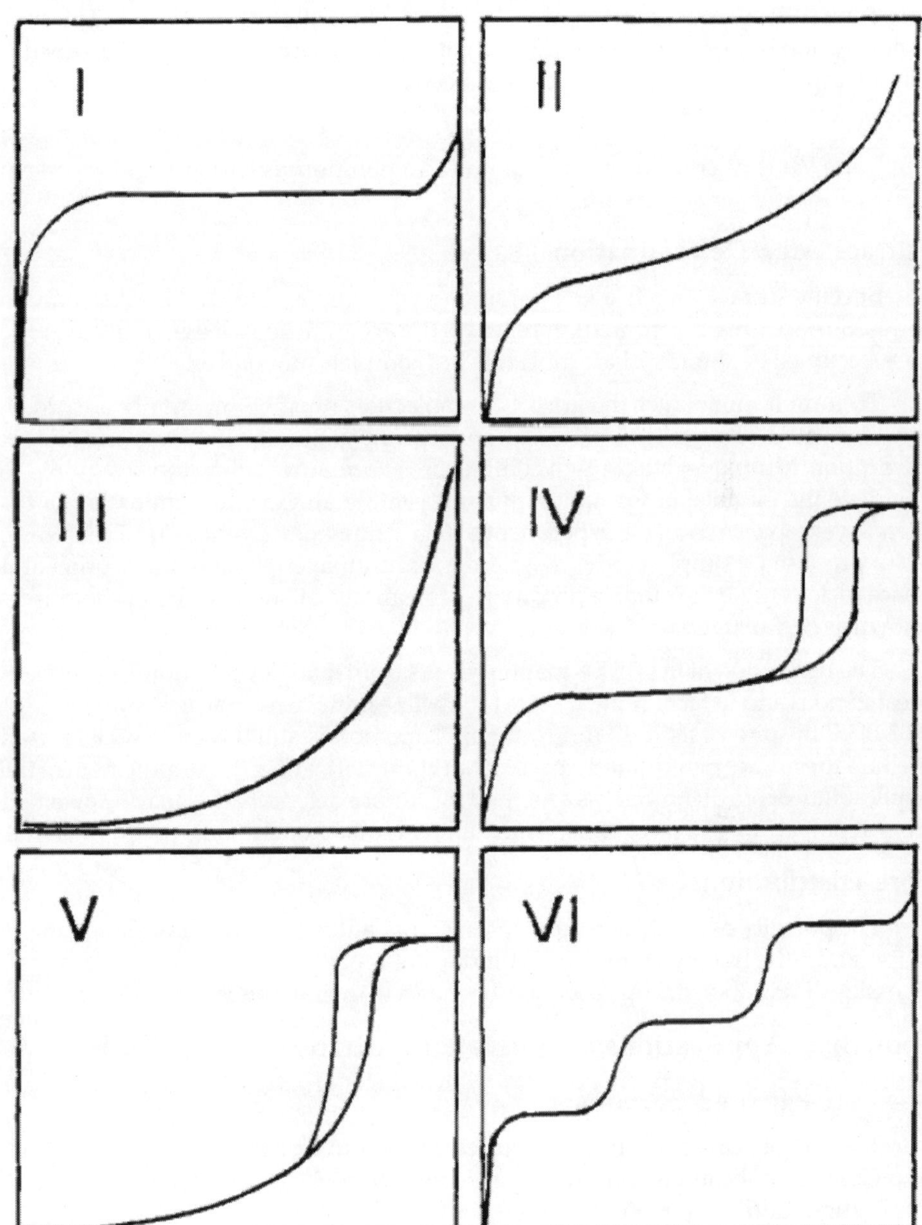

Figure 8.3: Isotherm Classification.

Type III- isotherms are uncommon and are usually associated with water vapors adsorption. Generally in this type of adsorption, the adsorbent-adsorbate interactions are weak as compared with the adsorbate-adsorbate interactions. An example is bromine on silica gel at-79 ^{0} C

Type IV-isotherms are recognised by the presence of a hysteresis loop. This type of adsorption is observed frequently with many mesoporous industrial adsorbents. The adsorption of n-decane on vycor has been used to study isotherm hysteresis.

Type V- is related to the Type III isotherm and is very uncommon.

Type VI- isotherms are seen with uniform non-porous surfaces and represent stepwise multilayer adsorption.

Surface Area Determination

Specific surface area is a key parameter which is used to characterize porous solids. Surface area determination relies on the accurate knowledge of the average area, occupied by the adsorbate molecule in a complete monolayer.

To form a monolayer the adsorbate molecules must be physically adsorbed, however, it is known that physical adsorption generally involves multilayer adsorption. Multilayer adsorption commences at pressures below those required to complete the monolayer formation thus preventing an exact determination of the monolayer capacity from the experimental data. Brunnauer, Emmett and Teller (BET, 1940) proposed a simple model to account for multilayer physical adsorption and hence to determine the monolayer capacity for surface calculation. This is popularly known as BET equation.

The derivation of the BET equation results from many over simplifications and idealizations and in fact its linear form is only linear in the relative pressure range of 0.05-0.35. In spite of its limitations, the BET method is still the most widely used method for surface determination in sorbent characterization. Some indication of the applicability of the BET model as a method for surface determination may be obtained from a comparison of data obtained with different sorbets on the same porous material.

Pore Distribution

The porosity of a solid is related to its texture and refers to the pore space in the material, *i.e.* the fraction of the bulk volume that is occupied by pore or void. space. Pore size distribution groups pores with respect to their diameter.

Industrial Applications of Activated Carbons

Uses of Activated Carbons

The uses of activated carbon for particular application requiring a specific grade. The various applications can broadly be split into two categories, (i) liquid-phase applications and (ii) gas-phase applications.

Liquid-Phase Applications

Activated carbons in liquid-phase applications differ from gas-phase carbons primarily in pore size distribution. Liquid-phase carbons have significantly more pore volume in the macropore range, which permits liquids to diffuse more rapidly into the mesopores and micropores. The larger pores also promote greater adsorption of large molecules, either impurities or products, in many liquid-phase applications.

Liquid-phase activated carbon can be applied either in a powder, granular or shaped form. The average size of powdered carbon particles is 15-25 micron, and they are most often used in batch applications. Granular or shaped-carbon particle size is usually 0.3-3.0 mm, and these carbons are more suitable for use in continuous flow systems.In liquid phase applications batch and continuous process can be utilized for efficiency and economic.

Batch Systems

In batch process stirred vessels are most often used in treating material with powdered activated carbon. The type of carbon, contact time, and amount of carbon vary with the desired degree of purification. The efficiency of activated carbon may be improved by applying continuous, counter-current, carbon-liquid flow with multiple stages.

Continuous Systems

Granular and shaped carbons are used generally in continuous systems where the liquid to be treated is passed through a fixed bed. Compounds are adsorbed by the carbon bed in the adsorption zone. As carbon in the bed becomes saturated with adsorbates, the adsorption zone moves in the direction of flow, and breakthrough occurs when the leading edge of the adsorption zone reaches the end of the column.

Gas-Phase Applications

Gas-phase applications of activated carbon include separation, gas storage, and catalysis. Although only about 20 per cent of activated carbon production is used for gas phase applications, these products are generally more expensive than liquid-phase carbons and account for about 40 per cent of the total value of shipments. Most of the activated carbon used in gas-phase applications is fiber, granular or shaped profiles.

These usually exploit the differences in the adsorptive behavior of gases and vapors on activated carbon, on the basis of molecular weight and size. For example, organic molecules with a molecular weight greater than about 40 are readily removed from air by activated carbon. It also used odour removal and gas purification.

Water Treatment

A major use of activated carbon is in the treatment of both potable and effluent water. Potable water primarily from rivers can be contaminated by sewage, effluent discharge, algae breakdown, etc. This can be rendered drinkable by chemical treatment, but it may be left with an undesirable taste or odour (Cheremisinoff, 1993). Activated carbons can be used as a final polishing treatment to remove such as tastes or odours. Activated carbon is never used in primary treatment, but is normally used in conjunction with a suitable chemical treatment. Industrial effluent containing toxic or harmful contaminants can be treated with activated carbon to avoid sea or river pollution. Also effluents from dyes, dairy, detergent manufacturers, refiners, electroplating and chemical manufacturers, and from manufacturers of insecticides, herbicides and fungicides, can effectively be treated with activated carbon. Activated carbon is also a very effective dechlorinator, and therefore is used by manufacturers of soft drinks and beers where water has been super chlorinated to render it sterile.

Decolourising

Activated carbon is first used in decolourising sugar. It is still extensively used for this application. It is also used as a decolouriser in the purification of edible oil, foodstuffs, pharmaceutical intermediates and chemicals such as esters, surfactants and alcohols. The wine and spirit industry uses activated carbon to ensure that white wine, vodka and bacardi, etc., reach the consumer with the correct appearance and flavour.

Solvent Recovery

Industries such as paints, adhesives, transparent film, printing, rubber, plastics, *dry* cleaning, chemicals and textiles, increasingly use hydrocarbon solvents during the manufacture of their products. These solvents must be recovered for both environmental and economic reasons. For this purpose activated carbons are particularly effective, and adsorb most solvents for any concentration above 1000 ppm. The method involves passing solvent-laden air through a bed of carbon until the bed is saturated. At this point the air stream is redirected to another bed and the first bed is stripped of the solvent by passing steam at 105 ^{0}C through the bed in the reverse direction to air flow. The solvent is then separated from the aqueous phase, and the carbon bed goes into a drying cycle to prepare it again for the solvent-laden air stream. Solvents recovered by this method include chloroform, carbon tetrachloride, acetone, pentane, methyl ethyl ketone, tetrahydrofuran, white spirit, benzene, toluene, xylene, petroleum and ether.

Military Uses

Activated carbon was first used in World War I to protect against chlorine gas. Most of the world's armed forces now use activated carbon to protect against attack by toxic gases, such as mustard gas, during conflicts. Other such gases include hydrogen cyanide, cyanogens chloride, phosphine and arsine. These carbons are put to use in personal gas masks, shelter filters and inlet filters for tanks and arm owed cars. A further use is in military suiting where combat uniforms contain a layer of carbon impregnated material under the outer cover. This protects the wearer from splashing of liquid gas after a toxic gas canister has exploded.

Nuclear Reactors

In the western world, all nuclear power reactors have activated carbon ventilators installed as a precaution against radioactive iodine leaks form the core or heat exchanger systems. Special carbons impregnated with potassium iodide or potassium tri-iodide are commonly used for the purpose. Also a potential problem with radioactive krypton and xenon was highlighted as a result of the Three Mile Island accident in 1979. To cope with this, special off-gas delay beds were designed so that passage of radioactive krypton and xenon through the bed could be delayed, until such time as the radiation hazard could have decayed to an acceptable level. These types of bed are now being fitted to many nuclear reactors.

Air Treatment

There are a large number of applications in which activated carbon is used to remove noxious contaminants from air streams. The common most include the ventilation of areas where large numbers of people congregate, such as airports, hospitals, submarines, office blocks and theatres. It is essential that toxic contaminants, which result from industrial activity, such as mercury vapour, sulphur dioxide and hydrogen sulphide, must be removed with maximum efficiency. In these instances a highly efficient impregnated carbon is usually employed.

Domestic Uses

In recent years there has been a large increase in the number of uses to which activated carbon can be put in the home. These include cooker hoods, fridge deodourisers, air purifiers, deep fat fiyer cartridges, Odour eater foot insoles and cigarette filters.

Precious Metal Recovery

Activated carbon can adsorb trace quantities of gold and silver from a cyanide solution. The process has been known since the1930, but it is only since the1970 that its full commercial implications have been recognised. Today its usage is widely practiced and is increasing at a rapid rate. The carbon-in-pulp (CIP) process for the recovery of gold in an agitated, counter-current, multistage extraction process has been developed and used successfully in recent years.

Catalysis

Many chemical reactions require a catalyst to improve efficiency, accordingly, in many cases activated carbon's large surface area is used as a support for such catalysts, thus further improving efficiency. Examples of its use include the production of chloro-fluoro carbons, terephthalic acid, vinyl chloride and sulphuryl chloride.

Miscellaneous Uses

Numerous relatively low-volume activated carbon uses make up a small proportion of carbon consumption. Small carbon filters are used in households for purification of tap water, oils and dyes. Numbers of organics are recovered by activated carbon in dry cleaning. Electroplating solutions are treated with carbon to remove organics that can produce imperfections when the thin metal layer is deposited on the substrate. Medical applications include removal of toxins from the blood of patients with artificial kidneys and oral ingestion into the stomach to recover poisons or toxic materials. Other uses include ELCD (evaporative loss control device) canisters, gas desulphurisation, de-ozonisation, condensate, de-oiling and carbon dioxide purification

Notations

b = Langmuir equilibrium constant

c = Sorbate concentration in fluid phase

E = Activation energy

$-\Delta H_0$ = Limiting heat of adsorption
K = Dimensionless equilibrium constant
K′ = Henry's law constant
k = Rate constant
k_0 = Pre exponential factor
k′, k" = Temperature dependent constant
p = Partial pressure of sorbate
q = Adsorbed phase concentration
q_s = Saturation limit in Langmuir expression
R = Gas constant
t = Time
T = Temperature
AC = Activated carbon
m^2/g = Square meter per gram

References

1. Allen, S.J. and Balasundaram, V. 1995. Contrasting adsorption exhibited by lignite-based activatedcarbons. J. Chem. Tech. Biotechnol., 64,261-268.
2. Ashmore, P.G.,(1963), Catalysis and inhibition of chemical reactions, Butterworths, London p. 163.
3. Brunnauer, S.H., Emmett, P.H., and Teller, E. 1940.Classification of isotherm shape J.Chem.Soc., 62.
4. Cheremisinoff, N.P. 1993. Carbon adsorption for pollution control. Prentice Hall, Englewood Cliffs, N.J.
5. Flynn, P.J., Allen, S.J., and McKay, G. 1988. Production of activated carbon from Northem Ireland lignite deposits. In the Proceedings of the Fifth Conference of the Irish Manufacturing Committee on Advances in Manufacturing Technology, Belfast, Northem Ireland.
6. Gamer, L.J., and Packer L.J. 1980. High grade granular activated carbons from Victorian brown coal. In the Proceedings of the Fifth International Conference on Coal Research, Dusseldorf, FRG, 2, September, 805-818.
7. Gergova, K., Petrov, N. and Eser, S. 1994. Adsorption properties and microstructure of activated carbonsproduced from agricultural by-products by steam pyrolysis. Carbon, 2, (4) 693-702.
8. Leboda, R., Skubiszewska-Zieba, J. and Bogillo, V.I. 1996. Change of porous structure of activated carbons during their gasification catalysed by calcium and steam activation. In the Proceedings of The European Carbon Conference 'Carbon 96', Newcastle, UK, 2, July, 612-613.

9. Lu, W. and Chung, D.D.L. 1996. Mesoporous activated carbon filaments. In the Proceedings of TheEuropean Carbon Conference ' Carbon 96', Newcastle, UK, 1, July, 26-27.
10. Otowa, T., Nojima, Y. and Itoh, M. 1996. Activation mechanism, surface properties and adsorption characteristics of KOH activated high surface area carbon. Fundamentals of Adsorption, Proceedings of the Fifth International Conference on Fundamentals of Adsorption, 709-716, Kluwer Academic Publishers.
11. Park B.B., Lee, Y.S., Shim, H.B., Ryu, S.K. and Rhee, B.S. 1996. Adsorption of NOx and SOX withactivated carbon fibrcs. In the Proceedings of The European Carbon Conference 'Carbon 96', Newcastle, UK, 1, July, 74-75.
12. Pashchenko, L.V., Khazipov, V.A., Shendrik, T.G. and Galushko, L.Y.A. 1996. The Reception of adsorbents from carbon containing waste. In the Proceedings of The European Carbon Conference 'Carbon 96', Newcastle, UK, 2, July, 433433.
13. Sircar. S., Golden, T.C., and Rao, M.B. 1996. Activated carbon for gas separation and storage. Carbon, 34.
14. Smisek, M. and Cemy, S. 1970. Active carbons-manufacture and properties and applications. Elsevier.
15. Zietek, S., Babel, K., Kielczewski, M. and Zahrzewski, R. 1996. The effect of carrier pyrolysis and activation conditions on the properties of Cr-Cu-Ag sorbents. In the Proceedings of The European Carbon Conference 'Carbon 96', Newcastle, UK, 1, July, 68-69. NO. 1 pp 1-12.
16. US patents 4,149,994 and 4,149,995 by Murty (1979 a, b).
17. US patent number 4,149,994 (1979).

Chapter 9

Assessment of the Environmental Exposure of Honey Bees to Neonicotinoid Insecticides coming from Corn Coated Seeds

Andrea Tapparo[1], Vincenzo Girolami[2], Luca Mazzon[2], Chiara Giorio[1], Matteo Marzaro[2] and Andrea Targa[1]

Università degli Studi di Padova

[1]Dipartimento di Scienze Chimiche, via Marzolo 1, 35131 Padova, Italy

[2]Dipartimento di Agronomia Ambientale e Produzioni Vegetali, Agripolis, 35020 Legnaro (Padova), Italy

ABSTRACT

Colony Collapse Disorder (CCD), the sudden loss of most honeybees of the hive represents a worldwide crisis. The consequent losses in crops productivity and plant pollination constitute a major emergency from both economical and ecological standpoints. Although on the causes of CCD several hypothesis have been advanced (parasitic mites, viruses, insecticides, etc.) up to the present no one has been clearly supported by experimental results. In several countries the phenomenon is mainly observed in the period of corn sowing (spring), that generally occurs using corn seeds coated with specific insecticides. As a consequence, seed-coating neonicotinoid insecticides that are extensively utilized in the corn crops have been blamed for CCD.

In this connection, two possible ways of environmental exposure and intoxication of honeybees to neonicotinoid insecticides were studied: 1) the atmospheric emission of particulate matter containing the insecticide by the sowing-machine and 2) the translocation of the systemic neonicotinoids from the coated seed to guttation drops of young corn plants. Quantitative measurements conducted both in laboratory and in the field demonstrated that both mechanisms of environmental diffusion of neonicotinoids can produce high exposure levels for bees, with lethal effects compatible with CCD phenomenon: the death manifests within few minutes if bees consume guttation drops or within few hours if they are exposed to particulate matter produced by the sowing machine.

Keywords: *Neonicotinoid insecticides, Guttation, Particulate matter, Honeybees, CCD.*

Introduction

The death of honeybees is a worldwide crisis of rapidly increasing proportions. USDA-ARS reports that in winter 2008, 36 per cent of America's (2.4 million) bee hives have been lost and European figures align on the same trend. The phenomenon is described as Colony Collapse Disorder (CCD) and it is characterized by a sudden disappearance of worker bees that do not return to the hive, with a loss of up to 60 per cent of the hives. CCD has been worldwide observed in the last years, with a rapidly increasing number of cases in Europe, USA and Australia. In these areas CCD mainly occurs at the beginning of spring [1] in the period of the extensive corn sowing. As most flowering plants' sexual reproduction depends on bee pollination, the problem constitutes a major emergency from both economical and ecological standpoints.

A lot of hypothesis were formulated to explain CCD. Besides parasitic mites, viruses [2] and sub-lethal doses of insecticides [3,4], the neonicotinoid insecticides were suspected to cause CCD. In effect they systemically translocate into plant tissues but the amounts detected in nectar or pollen are lower than 10 ppb, while higher doses (>40 ppb) are necessary for abnormal honeybee foraging behavior or for bee loss (>0.5 ppm) [5]. Up to the present, no one of the hypothesis on the causes of CCD was clearly confirmed or refused.

As for the toxic effects of insecticides on honeybees, particular attention was focused on the neonicotinoid systemic insecticides imidacloprid, clothianidin and thiamethoxam and on fipronil (a phenylpirazole, non-systemic insecticide) which are extensively used in corn crops utilizing seeds coated with the insecticide.

In this connection, two possible way of exposure (and intoxication) of honeybees to neonicotinoid insecticides were recently studied in our laboratory. The first one takes into account the emission of particulate matter containing the insecticide during the sowing of corn by the sowing machine. This mechanism was proposed for the first time by Greatti [1] but it has not been completely clarified and quantified yet.

The second one was recently postulated by Girolami [6] and it is based on the translocation of significant amount of the neonicotinoid insecticide from the coated seed to guttation drops of young corn plants. Guttation (from the latin *"gutta"* = drop) is the formation of small drops of xylem sap on the tips or along the edges of leaves. It is a physiological phenomenon occurring in many vascular plants, in particular grasses, water entering roots creates a slight pressure that forces it to rise and be exuded through the hydatodes at leaf margins. It's a common but often unnoticed phenomenon as confused with dew. Bees require intense drinking activity and water-fetching to the hive [7] and are known to collect water from guttations [8]. In corn crops, guttation solutions can persist drinkable into the crown cup of the young plants for the whole day.

In the present work the results of the quantitative study on both ways of exposure of honeybees to the seed coating insecticides used in corn crops are reported.

Materials and Method

Sampling and Analysis of Guttation Drops

Commercial corn seeds coated with different insecticides imidacloprid (Gaucho®, 0.5 mg/seed), clothianidin (Poncho®, 1.25 mg/seed), thiamethoxam (Cruiser®, 1 mg/seed) and fipronil (Regent®, 1 mg/seed), were sowed to collect guttation drops from emerging corn plants. In spring 2009 corn seedlings were grown in open field (spaced 20 cm within the row and 75 cm between rows). Tests were carried out also in laboratory (greenhouse) by sowing coated seeds in pots (diameter 15 cm, 2-5 plants per pot). A total of 6-8 pots for each insecticide were used and equal numbers were sown with uncoated seeds as control, or with seeds coated with fungicides (Fludioxonil, Metalaxyl).

Guttation drops from the leaves of corn plants were collected by a pipette from single plants or homogeneous groups of plants, every morning in the first 20 days after the seedling emergence. Samples were stored at 4 °C until the instrumental analysis.

The content of insecticides was measured by HPLC-DAD analysis by using conventional instrumentation [6]. More recently a rapid LC method based on high efficiency columns has been optimized. Rapid analysis were performed on a Shimazdu Prominence UFLC-XR instrument equipped with UV-Vis diode array detector (DAD) and a Shimadzu XR-ODS II (2.2 µm, 2x100 mm) analytical column under the following experimental condition: eluent flow rate of 0.4 mL/min, gradient elution (0-0.5 min: 70/30 per cent water/acetonitrile; 0.5-2 min, linear gradient to 100 per cent acetonitrile; 2-3 min, 100 per cent acetonitrile), 5 µL of injector volume, 45 °C of column temperature, multi-wavelength acquisition of the detector signal and analytes quantification at dikudd=215 nm for fipronil, dikudd=252 nm for thiamethoxam and dikudd=269 nm for clothianidin and imidacloprid. External calibration was performed by analysis of standard solutions in the 0.05–10 mg/L concentration range of analytes in water/methanol (50 per cent). The instrumental analysis were performed by direct injection of the guttation solutions, after filtration on a Millex HV 0.45 µm (Millipore) syringe filter.

Chemicals for the preparation of the standard solutions of fipronil, thiamethoxam, clothianidin and imidacloprid were from Fluka (Pestanal, purity >99.7 per cent for the three neonicotinoids and >97.5 per cent for fipronil). Methanol (VWR) and acetonitrile (Riedel de Haen) were of HPLC grade. Water was purified by a Millipore MilliQ equipment.

Characterisation of Particulate Matter Emitted by Sowing Machine

Two sampling campaigns were performed to characterize particulate matter emitted by the sowing machine. Commercial seeds coated with clothianidin (1.25 mg/seed) have been used in two different formulations: together with conventional seeds largely utilized in crops up to 2008, a new type of coated seeds with a thicker film was tested. The sowing machine was a Ribouleau mod. Monosem NG Plus and it worked normally for ca. one hour for each experiment. Total suspended particulate matter (TSP, 10-30 min of sampling time) was collected at the waste pipe of the fan of the pneumatic device that provides to the distribution of the seeds. The air flow rate at the waste pipe was estimated by using a Darcy anemometer (450-650 m^3/hour).

During the sowing both the particulate matter that falls down on the soil (dry deposition) and the atmospheric particulate matter present at the field margin (TSP and PM_{10}) were sampled. Two portable TSP samplers (SKC Universal), equipped with polycarbonate (SKC, 0.8 µm) or cellulose esters filters (SKC) were positioned at the fan drain of the sowing machine. Three TSP samplers (Zambelli, ZB1 timer) and a PM_{10} sampler (Zambelli Explorer plus), equipped with glass fibre filter (Whatmann) were utilized at the field margin or at a distance of 15 m along the wind direction. Cellulose esters filters (Carl Schleicher et Schull - mod. Selecta) with a diameter of 18.5 cm were used to collect dry depositions at the field margin.

All filters were stored in freezer before the laboratory analysis. They were then extracted with a measured amount of methanol (VWR) in an ultrasonic bath for 30 min. Sample solutions were analysed by the HPLC-DAD aforementioned procedures.

Results and Discussion

The results of instrumental analysis revealed that the corn plants from laboratory cultivation (greenhouse) produced guttation drops containing astonishingly elevated amounts of the insecticides coming from the original seeds. This effective translocation was constantly confirmed in the whole period in which guttation phenomenon occurs (during the first 20 days after the seedling emergence with a production of about 30-150 µL/day/plant). The concentration of the insecticides in the drops was in the range of 4-24 mg/L for thiamethoxam, 8-36 mg/L for clothianidin and 17-81mg/L for imidacloprid, with higher values observed in the guttation drops produced during the first days after the seedling emergence. In the next ten days a diminishing of the concentration of all the neonicotinoids in drops was observed (Figure 9.1). On the contrary the concentrations increase again in the last 10 days when the guttation phenomenon is considerably reduced and then it rapidly ends.

In an attempt to have a better simulation of the field conditions, the growing of only one seedling for each pot produced guttation drops containing similar concentrations of the insecticides as compared to the aforementioned samples. Similar levels of insecticides have been measured also in the guttation drops from field cultivation.

All guttation drops sampled from reference plants (both laboratory and field blank samples from non-coated seeds) contained levels of insecticides lower than the instrumental detection limits (LOD of the rapid LC procedure were 15 µg/L for

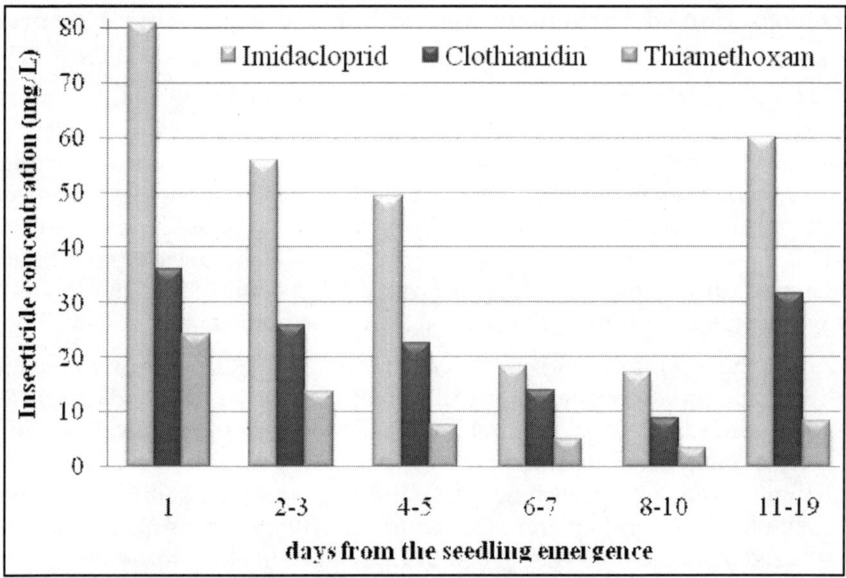

Figure 9.1: Concentration of Neonicotinoid Insecticides in Guttation Drops of Emerged Corn Plants (Greenhouse).

thiamethoxam, 10 µg/L for clothianidin and imidacloprid and 25 µg/L for fipronil). Fipronil, a non-systemic phenylpirazole insecticide that shows lower water solubility than neonicotinoids, was never revealed in guttation drops.

Preliminary results of the study seem to reveal that the content of insecticide in the seed influences the effective amount released, even if other factors can determinate the neonicotinoids content in guttation drops. High concentrations of imidacloprid (up to 200 mg/L) and thiamethoxam (up to 146 mg/L) were observed using seeds coated with 1.25 mg/seed of the specific insecticide.

By making a comparison between guttation drops collected from the top and the crown cup of the leaves, lower concentrations of insecticides were generally found in the samples collected from the crown cup. This fact is probably due both to the possible dilution of guttations by the dew and to degradative processes of the insecticide, for example photodegradation.

It is worth to notice that all these guttation solutions, if orally administered to honeybees, induce acute toxic effects within few minutes [6]. For instance the characteristic abdomen contraction and the irreversible wings block are well associated to the neurotoxicity of the neonicotinoids. Moreover the death of bees, after assumption of guttation drops containing such high neonicotinoids concentration, was always observed within fifteen minutes [6].

Literature [8] and direct beekeepers' observations report that guttation drops can be used by honeybees for their foraging especially in the early spring when they require intensive drinking activity and water-fetching for the hive [7]. As a consequence, corn guttations coming from neonicotinoids coated seeds may represent a possible way of exposure to lethal doses for bees.

As for the characterization and the quantification of the particulate matter emitted during the corn sowing by the sowing machine that releases fragments of seeds surface through the air flow generated in the pneumatic device of seeds distribution, experiments have been conducted in the field using commercial seeds coated with clothianidin (1.25 mg/seed) of two different formulations. The first one was used in crops up to 2008. The second one is a new type of seed-coating with a thicker film, which has been introduced in 2009 as it should be more resistant to the abrasion. Under normal sowing conditions, total particulate matter released in the atmosphere at the waste pipe of the pump of the sowing-machine has been sampled and analyzed. At the same time both the atmospheric particulate matter present at the field margin (TSP and PM_{10}) and the coarse particulate matter that falls down around the field were estimated.

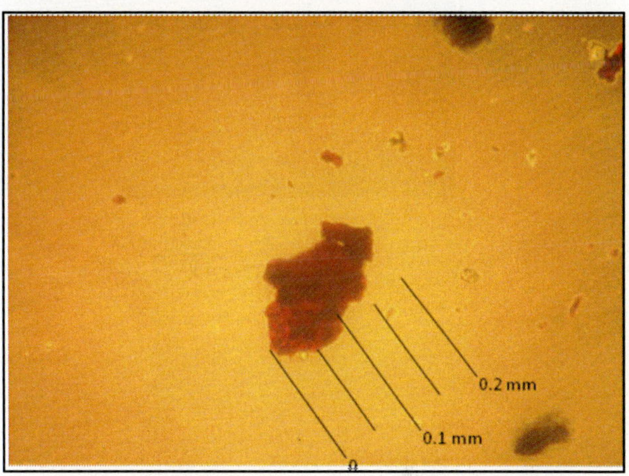

Figure 9.2: Optical Microscope Image of Coarse Seed Coating Particles.

At the end of 1 hour sowing the accumulation of red particles of the seed coating was well observed in the proximity of the waste pipe of the fan of the machine. These coarse particles (diameter lower than 1 mm, Figure 9.2) due to surface abrasion can be easily produced also by a simple manipulation of the seeds. Their fate is probably the rapid deposition on soil, with a very limited capability to be transported in the environment by the wind.

On the other hand, the presence of a significant tail of the dimensional distribution of these erosion particles that approaches the range of the fine particles (few micrometers) can be hypothesized. In effect, E-SEM-EDS examination of the sampled PM confirmed the presence of fine particles containing the insecticides (Figure 9.3). Of course, the environmental diffusion of these fine particles is expected to be higher than that associated to the coarse ones, and increased toxic effects toward bees can be expected.

From analysis of the sampled particulate matter, it was found that the concentration of clothianidin emitted from the air waste pipe of the machine, under normal sowing condition, is in the 3.6–2.9 mg/m^3 range. Emission factors for clothianidin were estimated in about 2.0 g/hour (or 1.2 g/ha) for the first kind of coated seed (employed until 2008), and in about 1.6 g/hour (0.9 g/ha) for the new seeds (available from 2009). In the latter case a loss of insecticide from the seed of ca. 1.1 per cent was estimated.

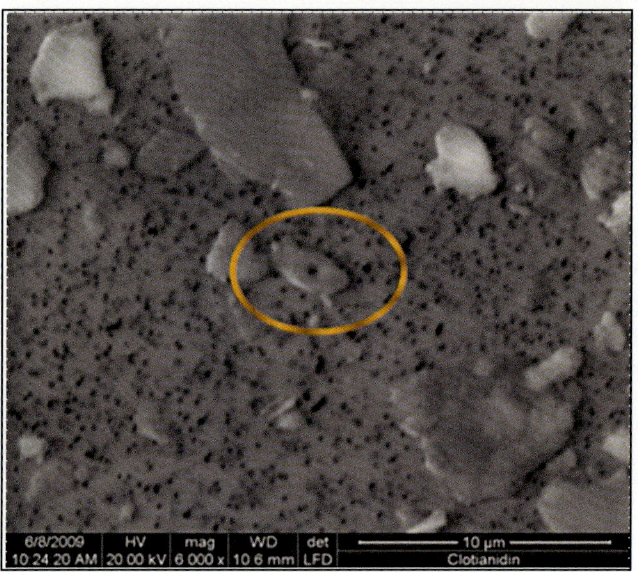

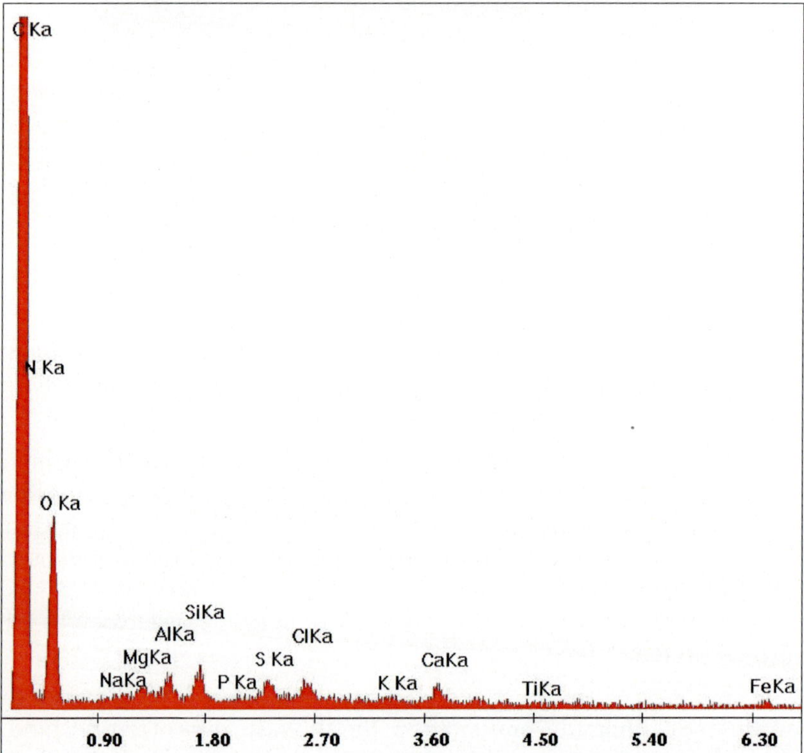

Figure 9.3: E-SEM Image of PM_{10} Particles Sampled during the Sowing of Corn Seeds Coated with Clothianidin. Among typical crustal particles (soil) a fine particle from the seed coating is detected by its X-ray emission spectrum (EDS, sulphur and chlorine signals of the insecticide).

This mass loss can be attributable mainly to coarse particles emission. In effect, the seeds coated with the new film of insecticide seemed to release a lower amount of coarse particles. Nevertheless, analytical measurements revealed that both kinds of coatings release large amounts of fine particles containing the insecticide. A complete dimensional characterization of the emitted particles is actually in progress.

The particles which fall down at the field margin (dry deposition) were estimated in about 300 µg/m^2 of insecticide for the first kind seeds, after ca. one hour of work activity of the sowing machine (about half when the new commercialized seeds were utilized). At the field margin the concentrations of clothianidin present in the total suspended particulate (TSP) were 0.10-0.24 µg/m^3. The insecticide concentrations found in the PM_{10} confirm a coarser dimensional distribution of the particles coming from the new seed coating, by levels of ca. 10 ng/m^3 of clothianidin. 60 ng/m^3 was the concentration of insecticide measured in PM_{10} during the sowing of the old seeds.

Preliminary tests of exposure of caged bees to the particulate matter present at field margin confirm the lethal effect of these powders, but with a longer time spans and a lower repeatability than those measured after assumption of guttation drops.

In conclusion, guttation drops and the particulate matter released during the sowing of corn seeds coated with neonicotinoid insecticides may represent two significant mechanisms of environmental diffusion of the neonicotinoids. Both ways are able to produce lethal exposure toward bees. The new proposed seed coating results more resistant in comparison to that used up to 2008 but it doesn't prevent the dispersion of large amounts of fine particulate matter containing the insecticide. The present quantification of the two exposure ways for bees to neonicotinoids seems to suggest the causal hypothesis that associates such insecticides to CCD, at least in the spring time during the period of the corn sowing.

References

1. Greatti M., Sabatini A.G., Barbattini R., Rossi S., Stravisi A. *Risk of environmental contamination by the active ingredient Imidacloprid used for corn seed dressing. Preliminary results.* Bull. of Insectology 56 (2003) 69-72.

2. Anderson D., East I. J., Cox-Foster D., Conlan S., Holmes E.C. *The latest buzz about colony collapse disorder.* Science 319 (2008) 724-725.

3. Desneux N., Decourtye A., Delpuech J.M. *The sublethal effects of pesticides on beneficial arthropods.* Annu. Rev. Entomol. 52 (2007) 81-106.

4. Johnson R., Evans J.D., Robinson G.E. Berenbaum M.R. *Changes in transcript abundance relating to colony collapse disorder in honey bees (Apis mellifera).* PNAS 106 (2009) 14790-14795.

5. Yang E.C., Chuang Y.C., Chen Y.L., Chang L.H. *Abnormal foraging behavior induced by sublethal dosage of Imidacloprid in the honey bee (Hymenoptera: Apidae).* J. Econ. Entomol. 101 (2008) 1743-1748.

6. Girolami V., Mazzon L., Squartini A., Mori N., Marzaro M., Di Bernardo A., Greatti M., Giorio C., Tapparo A. *Translocation of neonicotinoid insecticides from*

coated seeds to seedling guttation drops: A novel way of intoxication for bees. J. Econ. Entomol. 102 (2009) in press.

7. Kuehnholz S., Seeley T.D. *The control of water collection in honey bee colonies*. Behav. Ecol. Sociobiol. 41 (1997) 407-422.
8. Shawki M.A.A., Titera D., Kazda J., Kohoutkova J., Taborsky V. *Toxicity to honeybees of water guttation and dew collected from winter rape treated with Nurelle®*. Plant. Protect. Sci. 42 (2005) 9-14.

Chapter 10

Agricultural Waste to Energy: New or Old Agricultural Paradigm?

M. Bruna Zolin

Professor, University Ca' Foscari of Venice
Department of Economics, Italy

ABSTRACT

The large amount of agricultural waste generated in developed and developing countries is a problem for citizens and farmers who are obliged to get rid of it. However, agricultural waste as an energy source can represent a solution to a series of issues including the scarcity of fossil resources, their growing cost, their impact on the environment in terms of greenhouse emissions and climate change. From a policy perspective there are two main and important orientation: the minimisation of environmental impacts (reducing and preventing waste generation) and the reduction of resource use. Especially renewable energy, obtained by agricultural waste, could represent an aid to face the problems of isolation and resulting economic and social disadvantages in rural and remote areas of developed but, over all, developing countries, where their use can be a solution to provide electricity and services to resident populations in a sustainable way (both from an economic and environmental point of view) and so could contribute to improve the standards and quality of life in these areas.

Introduction

World energy consumption has been increasing more and more over the years; it is still now mainly satisfied by fossil fuels, raising a series of issues such as the scarcity of fossil resources, their growing cost, their impact on the environment in terms of greenhouse emissions and climate change.

A solution can be found in renewable energy, which is unlimited and non-polluting energy, it has a much lower environmental impact than conventional sources of energy and several other benefits such as stimulating the economy (through research activities, new technologies development, etc.) and creating job opportunities. Renewable energy, in particular, could represent an aid to face the problem of isolation and resulting economic and social disadvantages in rural and remote areas of developed but, over all, developing countries, where the use of non-conventional resources can be a solution to provide electricity and services to resident populations, in a sustainable way (both from an economic and environmental point of view) and so could contribute to improve the standards and quality of life in these areas.

More and more countries, in fact, are investing in renewable energy sources, even if the quota of worldwide energy consumption satisfied by them is still low (about 1.6 per cent in 2006), but it is worth underlining that renewable energy data could be underestimated (sometimes and in some areas, in fact, renewable energies are used by households to satisfy their domestic energy needs).

This paper analyses the role of renewable energy obtained from agricultural waste. Section two summarised the legislative framework comparing India and the EU. Section three analyses some stylised facts to discuss the possible implications for food supply of the production of biofuels that pose social (rising food prices) and environmental (deforestation) risks. A fourth section will analyse the use of agricultural wastes as an energy source. A final section will conclude.

Renewable Energies: Legislative Comparison

Renewable energy policies are very recent; they began to emerge at the end of eighties and especially in the past five years.

Policy targets for renewable energy exist in almost seventy countries worldwide, including all 27 European Union countries (REN21, 2007) and at least sixty countries have some type of policy to promote renewable power generation. These policies are founded on different basis (feed-in tariff; taxes; investments, energy certificates, etc.) as shown in the scheme 1, which considers the different typologies of renewable energy promotion policies in some states of particular interest.

Worldwide the most common renewable energy promotion policy is represented by feed-in law[1].

This policy was enacted for the first time by United States in 1978. Feed-in policies were next adopted in Denmark, Germany, Greece, India, Italy, Spain and Switzerland in the early 1990s. By 2007, at least 37 countries and 9 states/provinces had adopted such policies, more than half of which have been enacted since 2002. These policies have had the largest effect on wind power, but have also influenced solar power, biomass and small hydro development (Table 10.1).

1 This is an incentive structure to encourage the adoption of renewable energy through government legislation: the electricity utilities are obligated to buy renewable electricity at above market rates set by the government.

Table 10.1: Renewable Energy Promotion Policies in some States.

Country	*Renewable Energy Promotion Policies*					
	Feed-in Tariff	*Capital Subsidies, Grants or Rebates*	*Investments or Other Tax Credits*	*Tradable Renewable Energy Certificates*	*Public Investments, Loans or Financing*	*Public Competitive Bidding*
France	X	X	X	X	X	X
Germany	X	X	X		X	
Italy	X	X	X	X		
Spain	X	X	X		X	
United States	*	*	X	*	*	*
China	X	X	X		X	X
India	*	*	X		X	X

* Some states/provinces within these countries have state/province-level policies but there is no national-level policy.

Source: REN21, Renewable 2007 Global status report, our elaboration.

New energy solutions, in fact, strongly depend on appropriate tariff schemes.

Other common forms of policy support for renewable power generation include direct capital investment subsidies or rebates. Tax incentives and credits are also common ways of providing financial support to use renewable energy.

The Indian Framework

India has an ancient and strong tradition of energy planning and intervention programmes. Programmes for promoting renewable energy and for improving energy efficiency date back to the 1940s. In particular, during the 1980s Ministry of Non-conventional Energy Sources, MNES[3], focused on development, dissemination, and demonstration of various renewable energy technologies. Since 1993 emphasis has shifted towards policies, planning and institutional linkages to promote renewables within specific application sectors, and, in 2000, a complete policy for development of the sector was formulated by MNES, with the following broad objectives.

1. Meeting the minimum energy needs through renewable energies;
2. Providing decentralised energy supply in agriculture, industry, commercial and household sectors in rural and urban areas;
3. Producing 10 per cent of additional grid power generation capacity from non-conventional energy (NCE[4]) by 2012.

---

2 Re-christened as Ministry of New and Renewable Energy, since October 2006.

3 Conventional fuels include: fossil fuels (petroleum, coal, propane, & natural gas) and also, in some instances, nuclear materials such as uranium. Conversely, alternative fuels, also known as non-conventional fuels, are any materials or substances that can be used as a fuel, other than conventional fuels. Some well known alternative fuels include biodiesel, ethanol, butanol, chemically stored electricity (batteries & fuel cells), hydrogen, methane, natural gas, vegetable oil, biomass, & peanut oil. This means that a non-conventional energy is not necessarily a renewable energy.

The government provides many different types of incentives to encourage private investments in this sector, these include: providing budgetary resources by the government for demonstration projects; extending institutional finance from the Indian Renewable Energy Development Agency (IREDA) and other financial institutions for commercially viable projects, with private sector participation; promoting private investment through fiscal incentives, tax holidays, depreciation allowance, facilities for banking of power for the grid and remunerative returns for the power provided to the grid.

Several state governments also provide several incentives for the promotion of NCE, which could be summarised in three principal kind: authorisations[4]; fiscal incentives[5] and financial supports[6].

At national level, we can find some information about renewable energies and energy saving in the following documents:

1. *Electricity Act 2003:* "The State Commission shall discharge the following function (..): promote cogeneration and generation of electricity from renewable sources of energy by suitable measures for connectivity with the grid and sale of electricity to any person, and also specify, for purchase of electricity from such sources, a percentage of the total consumption of electricity in the area of a distribution licensee" (Section 86).
2. *National Electricity Policy 2005 and Tariff Policy 2006* stipulates that the share of electricity from non-conventional sources would need to be increased; considering the fact that it will take some time before non-conventional technologies compete, in terms of cost, with conventional sources, the commission may determine an appropriate deferential in prices to promote these technologies.
3. *National Rural Electrification Policies, 2006* establishes that for villages/ habitations where grid connectivity would not be feasible or not cost effective, off-grid solutions based on stand-alone systems may be taken up for supply of electricity. Where these also are not feasible and if only alternative is to use isolated lighting technologies like solar photovoltaic, these may be adopted.

The European Union Framework

With regard to energy, the role of the European Union is to drive sustainable processes, leaving the single Member States to deal with specific areas of intervention.

---

4 For example, industrial clearances are not required for setting-up of an NCE industry and no clearance is required from Central Electricity Authority (CEA) for power generation projects up to Rs 1,000 million.

5 For example, a five-year tax holiday is allowed for NCE power generation projects.

6 Promotion of medium, small, mini and micro enterprises for manufacturing and servicing of various types of NCE systems and devices; financial support is available to NCE industries for R&D projects in association with technical institutions; soft loans are available through IREDA for NCE equipment manufacturing; etc.

The European guidelines and support are more complex and articulated, yet, at the same time, more lenient, than national ones.

Currently, the main EU guidelines and regulations concerning energy and renewable energy could be summarized in the following documents:

1. European Parliament's directive, *Promotion of the use of energy from renewable sources*, 2009;
2. Communication from the Commission *Second Strategic Energy Review. An EU energy security and solidarity action plan*, 2008;
3. Communication from the Commission to the European Council and the European Parliament: *An energy policy for Europe*, 2007;
4. Communication from the Commission *Renewable Energy Road Map. Renewable energies in the 21st century: building a more sustainable future*, 2006;
5. Communication from the Commission *Action Plan for Energy Efficiency: Realising the Potential. Saving 20 per cent by 2020*, 2006;
6. *Green Paper on Energy Efficiency or Doing More with Less*, June 2005;
7. Directive 2001/77/CE on Electricity Production from Renewable Energy Sources.

The Communication from the Commission to the European Council and the European Parliament *An energy policy for Europe* (January, 2007) represents a strategic review of the European energy situation and introduced a complete set of European Energy Policy measures (the so called 'energy' package). The new European energy policy includes main topics and objectives fixed till then by the *Renewable Energy Road Map* (2006)[7], *Action Plan for Energy Efficiency* (October 2006)[8] and the *Green Paper on Energy Efficiency* (June 2005)[9]. The main established objectives, therefore, are:

1. Establishing the internal energy market, which means:
   - ☆ separation between the management of gas and electricity networks and production or sales activities;
   - ☆ harmonisation of competences and independence of energy at Community level, their reinforced collaboration and obligation to take into account the Community objectives;
   - ☆ reduction of energy poverty by developing an Energy Customers' Charter which will principally encourage the implementation of aid schemes for the most vulnerable citizens in the face of increasing energy prices.

---

7 Communication from the Commission to the Council and the European Parliament "Renewable Energy Road Map. Renewable energies in the 21st century: building a more sustainable future", COM(2006) 848 final.

8 Communication from the Commission Action Plan for Energy Efficiency: Realising the Potential. Saving 20 per cent by 2020, October, 2006.

9 Green Paper on Energy Efficiency or Doing More with Less, June 2005.

2. Ensuring a secure energy supply, minimising the EU's vulnerability concerning the external dependence and possible energy, through measures supporting strategic oil stocks and increasing security of electricity supply.
3. Reducing greenhouse gas emissions, through the so called initiative "20-20-20".
4. Developing energy technologies, by strengthening the European Strategic Energy Technology Plan which will cover the entire innovation process, from the initial research to entry onto the market[10].
5. Considering the future of nuclear energy.
6. Implementing a common international energy policy, in particular by strengthening the European Energy Charter[11], taking the initiative in an agreement on energy efficiency.

In 2008, with the proposal for a Directive on the promotion of the use of energy from renewable sources[12], the EU established a common framework for the promotion of energy from renewable sources and committed itself to completing the initiative "20-20-20".

In particular, the Directive proposal establishes that each Member State shall ensure that the share of energy from renewable sources, in gross final consumption of energy in 2020, is at least its national overall target for the share of energy from renewable sources in that year[13]. With regard to transports, each Member State shall ensure that the share of energy from renewable sources in all forms of transport in 2020 is at least 10 per cent of final consumption of energy in transport in that Member State. As far as biofuels are concerned, with effect from 2017, the greenhouse gas emission savings from the use of biofuels and other bioliquids shall be 50 per cent. After 2017 it shall be 60 per cent for biofuels and bioliquids produced in installations whose production has started from 2017 onwards. Finally, Member States shall introduce in their building regulations and codes appropriate measures in order to increase the share of all kinds of energy from renewable sources in the building sector.

---

10 This strategic plan will support the Seventh Framework Programme for Research , which foresees a 50 per cent increase in spending on research in the energy sector, along with the Intelligent Energy for Europe programme.

11 Council and Commission Decision 98/181/EC, ECSC, Euratom of 23 September 1997 *on the conclusion, by the European Communities, of the Energy Charter Treaty and the Energy Charter Protocol on energy efficiency and related environmental aspects.* The Energy Charter Treaty establishes a framework for international cooperation between European countries and other industrialised counties with the aim of developing the energy potential of central and Eastern European countries and of ensuring security of energy supply for the European Union. The Protocol on energy efficiency and related environmental aspects aims to promote energy efficiency policies that are compatible with sustainable development, to encourage more efficient and sound use of energy and to promote cooperation in the field of energy efficiency.

12 COM(2008)0019 – C6-0046/2008 – 2008/0016(COD).

13 These mandatory national targets are consistent with a target of at least a 20 per cent share of energy from renewable sources in the Community's gross final energy consumption in 2020.

Stylised Facts

Globalisation and the rising importance of key emerging economies are having diverse effects on world agricultural markets. The strong growth in demand of processed products as well as agricultural materials are originated mainly in developing countries and transitions economies.

The size of the population and the population growth rate are of a great importance for the analysis of food demand trend.

The growth rate, very high in Africa and Asia, affect the food consumption. Consumption patterns are also changing (Table 10.2).

Table 10.2: Population (in thousands).

	1997		*2006*		*1997-2006*
	Total	*Per cent*	*Total*	*Per cent*	*Variation (per cent)*
Africa	532,215	9.57	652,662	10.75	18.45
NorthAmerica	466,908	8.40	507,203	8.36	7.94
SouthAmerica	329,651	5.93	373,808	6.16	11.81
Asia	3,478,106	62.54	3,781,850	62.30	8.03
Europe	726,203	13.06	727,612	11.99	0.19
Oceania	28,568	0.51	27,278	0.45	–4.73
World	5,561,651	100.00	6,070,413	100.00	8.38

Sources: Our elaboration on United Nations Statistics Division data.

The largest growth variation occurs in the high value added product such as meat in developing and transition economies. As income and urbanisation growth, people move away from a diet consisting largely of staple food crops into a diet that includes more fish, meat, dairy products and eggs.

The growth of economy has improved living standards. According to World Bank, the emerging economies (India, China, and Brazil) have achieved an annual GDP growth over the last two decades between 5-10 per cent. Over the very the long term (1850-1970s), food/agricultural prices have been stable.

They have trended upwards in the 1970s and they stabilized again and fell at the end of that decade. Since 2002, the increase in the prices of all fuels and some metals exercised an upward pressure on many agricultural commodities. The linkages existing between crop and energy prices affect the supply of agricultural production through direct and indirect effect.

With a high percentage of household income spent on food, some governments have introduced measures within the food and agriculture sector to address food supply and tackle inflation.

Historically the objective of food security has been the key driving force behind policy and the provision of an adequate supply of affordable food is still a top priority today.

High international commodities prices have transmitted signals to farmers to allocate more resources and increase agricultural production. When measured at the global level, the cereal production responded with 11.47 per cent expansion, meat with 6.25 per cent.

However, not all the farms responded similarly as high world prices: expansion in Asia cereals production and utilization are negative (due to changing in diet), but in meat the rates growth more then 35 per cent (Tables 10.3–10.6).

Table 10.3: Meat Production (tonnes).

	1997		*2007*		*1997-2007*
	Ton/1000	*Per cent*	*Ton/1000*	*Per cent*	*Variation (per cent)*
Africa	10,222	4.73	12,495	4.37	18.19
NorthAmerica	38,196	17.66	46,226	16.18	17.37
Central America	4,766	2.20	6,806	2.38	29.98
SouthAmerica	22,755	10.52	32,491	11.37	29.97
Asia	82,931	38.34	128,135	44.85	35.28
Europe	51,540	23.83	52,437	18.35	1.71
Oceania	5,067	2.34	6,074	2.13	16.58
World	216,298	100.00	285,716	100.00	24.30

Source: Our elaboration on FAO Statistics data.

Table 10.4: Meat Consumption (tonnes).

	1993		*2003*		*1993-2003*
	Ton/1000	*Per cent*	*Ton/1000*	*Per cent*	*Variation (per cent)*
Africa	9,472	4.97	12,529	5.01	24.40
NorthAmerica	34,023	17.87	40,529	16.22	16.05
CentralAmerica	4,387	2.30	7,101	2.84	38.22
SouthAmerica	17,462	9.17	23,599	9.45	26.01
Asia	68,572	36.01	108,555	43.45	36.83
Europe	53,775	28.24	54,247	21.71	0.87
Oceania	2,730	1.43	3,277	1.31	16.67
World	190,422	100.00	249,836	100.00	23.78

Source: Our elaboration on FAO Statistics data.

Table 10.5: Cereal Statistics (million tonnes): Production.

	2002		*2008*		*2002-2008*
	Ton/1000	*Per cent*	*Ton/1000*	*Per cent*	*Variation (per cent)*
Africa	115,6	5.71	149,0	6.51	22.42
North America	335,1	16.55	457,0	19.98	26.67
Central America	35,7	1.76	41,8	1.83	14.59
South America	103,0	5.09	135,5	5.92	23.99
Asia	982,2	48.51	969,1	42.37	–1.35
Europe	433,5	21.41	500,5	21.88	13.39
Oceania	19,8	0.98	34,4	1.50	42.44
World	2,024,8	100.00	2,287,2	100.00	11.47

Source: Our elaboration on FAO Statistics data.

Table 10.6: Cereal Statistics (million tonnes): Total Utilization.

	2002		*2008*		*2002-2008*
	Ton/1000	*Per cent*	*Ton/1000*	*Per cent*	*Variation (per cent)*
Africa	156,1	7.56	200,1	9.09	21.99
North America	296,3	14.35	355,2	16.13	16.58
Central America	56,0	2.71	65,5	2.97	14.50
South America	97,8	4.74	120,0	5.45	18.50
Asia	1,070,6	51.87	1,030,2	46.79	–3.92
Europe	372,8	18.06	414,3	18.82	10.02
Oceania	14,9	0.72	16,4	0.74	9.15
World	2,064,2	100.00	2,201,7	100.00	6.25

Source: Our elaboration on FAO Statistics data.

In these countries, population continue to increase and domestic production growth is unable to meet food demand.

From the point of view of energy security, the growing demand for biofuel production could eventually further strengthen the domestic capacity and/or create a growth in imports of such commodities. Expanding the land resource, through for example land reclamation or an improvement in land quality, is theoretically possible only in developing countries, where this resource has not yet been completely exploited. However, in these countries, insecure property rights, land and labour distribution and allocation as well as inequality in the access to land and credit, all create large inefficiencies. As a consequence, the increase in the availability of arable land meets many constraints[14] (Table 10.7).

Table 10.7: Arable Land (1.000 Ha).

	1997		*2005*		*1997-2005*
	Total	*Per cent*	*Total*	*Per cent*	*Variation (per cent)*
Africa	196,060	14.11	213,119	15.00	8.00
North America	223,195	16.06	220,112	15.49	–1.40
Central America	31,004	2.23	30,936	2.18	–0.22
South America	104,939	7.55	108,324	7.62	3.12
Asia	492,675	35.45	511,458	35.99	3.67
Europe	294,356	21.18	280,073	19.71	–5.10
Oceania	41,936	3.02	51,458	3.62	18.50
World	1,389,812	100.00	1,421,169	100.00	2.21

Source: Our elaboration on FAO Statistics data.

From the developed countries' point of view, located mainly in temperate areas, and bearing in mind that ethanol and bio-diesel productions are not economically viable without public support, a different combination of technologies, oil crude prices and renewable resources policies could change the current situation. Because of the relatively high costs in the EU for both land and labour, biomass production is actually more expensive per unit of energy produced than that produced by available residues and wastes.

Energy Scenario

World primary consumption of energy (and thereby world primary production[15]) has been strongly increasing over the years and in 2006 it achieves 472,274 Quadrillion of Btu (+40 per cent in respect to 1980), whereas world production of energy amounts to 469,412 Quadrillion of Btu (+39 per cent in respect to 1980) (Figure 10.1).

---

14 According to FAO terminology, arable land is the land under temporary crops (double-cropped areas are counted only once) plus temporary meadows for mowing or pasture, plus land under market and kitchen gardens, plus land temporarily fallow (less than five years). The abandoned land resulting from shifting cultivation is not included in this category. Cultivated land is the sum of arable land and the land under permanent crops. Land under permanent crop refers to land cultivated with crops that occupy the land for long periods and need not be replaced after each harvest.

15 Primary production/Total production of primary energy: any kind of extraction of energy products from natural sources and conversion to a usable form is known as primary production. Primary production takes place when the natural sources are exploited, for example in coal mines, crude oil fields, hydro power plants or fabrication of biofuels. Transformation of energy from one form to another, such as electricity or heat generation in thermal power plants, or coke production in coke ovens, is not primary production.

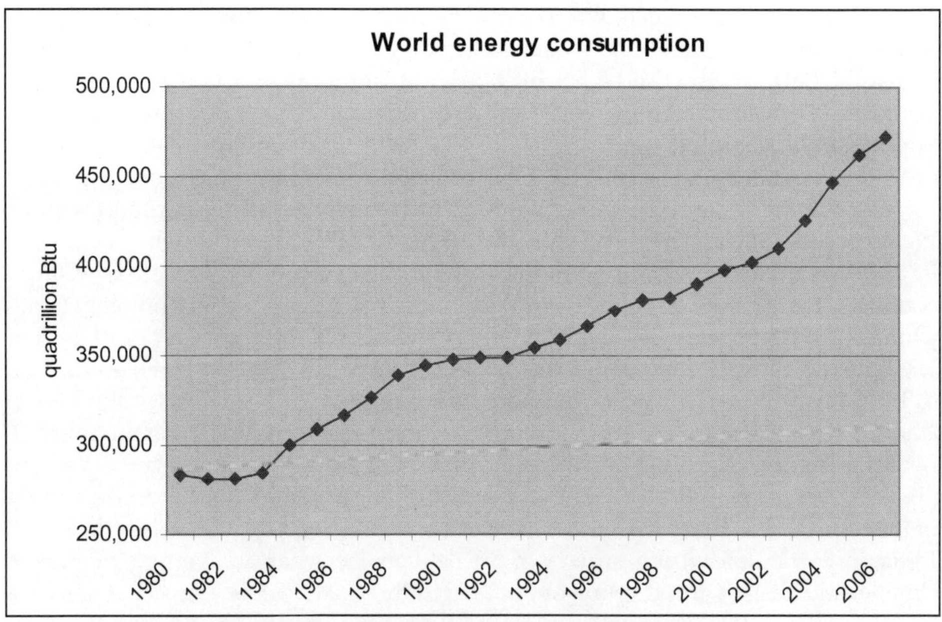

Figure 10.1: World Energy Consumption Trend.
Source: EIA (Energy Information Administration)–
International Energy Annual 2008, our elaboration.

Petroleum, dry natural gas and coal represent 86.2 per cent of supply and demand of energy; net hydroelectric power and net nuclear electric power represents 12 per cent, whereas renewable resources play a marginal role in the supply and demand of energy in terms of quantities (about 1.6 per cent), even if they show an high increase in the period 2000-2006 (+36 per cent if we consider geothermal, solar, wind and wood and waste electric power; +12 per cent if we consider biomass, geothermal and solar energy not used for electricity generation). However, renewable energy data, in our opinion, could be underestimated: biomass in some areas is traditionally used by rural and semi-rural households to satisfy their domestic energy needs and this does not appear in official statistics (Zolin, 2008).

Among the conventional energies, coal shows the highest increase in production and consumption, followed by dry natural gas. Related environmental benefits and the efficiency of dry natural gas make it an attractive energy producing fuel choice in many countries, however, higher petroleum and natural gas prices make coal the best choice in those nations where coal resources are available in sizeable quantities.

Net nuclear electric power increased by 7.6 per cent, in the period 2000-2006, and an increase in nuclear power production is expected in the near future in many countries (EIA, 2008)[16].

---

16 Cuts in nuclear power generation are expected only for OECD Europe, where several nations have programmed the elimination and non-replacement of some old reactors and others where nuclear power is forbidden.

Between 2000 and 2006, world primary energy production and consumption increased by 15.7 per cent. Energy consumption grew faster than population, which increased, between 2000 and 2008, by 9 per cent, but more slowly than GDP (gross domestic product), which increased in the same period by 33 per cent. As well as population, GDP growth (and GDP per capita), in fact, is a fundamental factor which can influence energy demand. When GDP grows, the population has more disposable income for more energy consuming products such as vehicles, electronics and heating. World population is growing at a rate of 1.16 per cent (78 million people) per annum in 2008, with strong variation among different States (population of non-OECD countries continues to grow most rapidly: India has a 1.6 per cent annual growth; China has a 0.6 per cent annual population growth rate but it has a population of 1.33 billion; Eastern Europe, by contrast, is showing a negative growth rate). According to ISEE (Institutional Sustainable Energy Exchange) data[17], totally, for the period 2006-2030, world population is estimated to increase at an annual average rate of 1 per cent. The rate of growth in world GDP, rather, is assumed to average 3.3 per cent per year over the period 2006-2030, reflecting the rising weight in the world economy of the non-OECD countries, where growth remains fastest. Nevertheless the crisis, India, in particular, with a growth of 6.4 per cent per year on average, overtakes China as the fastest growing major country in the 2020s. Europe, instead, shows a GDP growth rate lower, which amounts to 1.9 per cent per year. The volume of energy consumption will differ among countries and regions, depending on levels and changes in economic development, fuel and prices, governmental policies and other factors.

For the future, world energy demand (having in mind population and GDP trends) is expected to grow by 1.6 per cent per year (more than population pressure, less than the expected growth in GDP) on average from 2006-2030, that means a 45 per cent increase from 2006 levels. In this period, China and India account for 51 per cent of incremental world primary energy demand and, all together, non-OECD countries account for 87 per cent of the increase.

The EU's energy production and consumption are unsustainable due to their limited resources and their impact on the environment. Europe is a net importer of energy, while internal production levels are decreasing.

At European level[18], the production of primary energy showed a gradual decrease over the years. If we consider the period from 2000 to 2006, the decrease amounts to 7.5 per cent, shifting from 893 million toe[19] in 2000 to 1,831 million toe in 2006.

---

17 All data are available in the ISEE publication World Population and GDP growth, http://www.hart-isee.com/index.php?page=world-population-GDP-growth.

18 We consider EU25. Data used always consider EU25, also for the years before 1st May 2004.

19 For European energy production we use a different unit of measurement (toe - *tonne of oil equivalent*) instead of Btu (British thermal unit), since, in order to have a higher detail, we have used different databases (Eurostat for the EU and EIA for world and India).

Energy production for the EU25 has been in the main nuclear energy (about 30 per cent of total), which increased 4.5 per cent from 2000 to 2006, followed by coal and lignite (21.6 per cent of total) and natural gas (24.1 per cent of total) which, however, decreased 12.3 per cent and 16 per cent respectively in the period 2000-2006. Energy production from crude oil (13.4 per cent of European energy production in 2006) registered the highest decrease in this period, equal to 46 per cent.

In 2006, renewable energies represent 14.7 per cent of European energy production, with an increase of about 23 per cent compared with 2000. Biomasses and waste top the list of renewable energy production (about 69 per cent in 2006) followed by hydroelectric power (20.2 per cent), wind (5.8 per cent), geothermal (4.5 per cent) and solar (0.8 per cent). All renewable sources show a steady increase in the years from 2000 to 2006, with the exception of hydroelectric power (-17 per cent).

As far as European energy consumption is concerned, in the period 2000/2006 it increased by 5.8 per cent. Renewable energy registered the highest increase (+23 per cent), followed by gas (+10.2 per cent) and nuclear (+4.5 per cent). In 2006, oil, with 37 per cent of total European energy consumption has the first position, even if this increase from 2000 has been only by 2.2 per cent, it is followed by gas (24 per cent), coal and lignite (18 per cent) and nuclear (14 per cent). Energy consumption from renewable resources only amount to 7 per cent of total.

In 2006, India's production of primary energy amounted to 12,388 quadrillion of Btu (about 3 per cent of world energy production) and Indian energy consumption (17,677 quadrillion Btu) represents 4.4 per cent of world energy consumption. Despite the fact that India is the world's eleventh greatest energy producer and the world's sixth greatest energy consumer, it is a net energy importer (Sidhu, 2007).

Energy supplies for India are in the main coal energy (more than half of India's energy needs are met by coal), due to the vast availability of coal reserves (about 10 per cent of the world's total) and consequently to the high domestic production (8,203 quadrillion of Btu in 2006); followed by petroleum, which covers about 30 per cent of total Indian energy consumption, but more than 60 per cent of that oil is imported[20].

In the period 2000-2006, energy production increased by 20.6 per cent in India and consumption by 23.8 per cent. Natural gas has shown the fastest growth rate of any fossil fuel: consumption increased by 42.4 per cent in 7 years and production by 28 per cent. Production and consumption of hydroelectric power also increased by about 33 per cent, in this period, but renewable sources and in particular geothermal, solar, wind and wood and waste showed the highest increase, both, in terms of production and consumption (+69 per cent).

Nevertheless the EU and India represent different realities (India is an emerging economy, characterised by high population and GDP growth rates; Europe, instead, is a developed region, where population grows little or does not grow at all and GDP

---

20 In the period 2000-2006 petroleum demand increased by 19 per cent and at present India is the world's sixth largest oil consumer.

increases more slowly), pollution, scarcity of resources and uncertainty of energy supplies are shared problems. Data, in fact, show a common tendency by the side of the EU25 and India to invest in renewable energies, in particular in wind energy, which can represent a possible solution for a sustainable development.

Agricultural Waste Might be a Renewable Energy Source?

The potential agricultural wastes are primary such as crop residues (land managed for grain) and forest residues (land managed for timber/pulp and for fire protection). The secondary potential agricultural wastes are: manure and biosolids, mill wastes, black liquor (pulp by product) and urban wood wastes. They can produce a variety of transportation fuels: ethanol, hydrogen, diesel gasoline. Managing land for energy, food and fibber.

Few studies have been conducted on the utilisation of agricultural waste for composting and/or animal fodder. Most of the proposed solutions have not been implemented because they did not meet the basic sustainability conditions: social progression, economic improvements environmental protection.

The main problems mainly in rural areas are wastes, agricultural wastes. Some of agricultural waste are used as animal fodder and as fuel resources in very primitive ovens with health problems and damage to environment (air pollution problems). The five crops that have the highest account of waste are: rice, corn, wheat/barley (cereals), cotton and sugarcane. Many rural areas in the developing countries have not sewage treatment network.

In general a large amount of sewage is generated in the villages and due to the lack of a sewer system, sewage is drained into the nearest canal causing soil, water pollution as well as health problems.

The best solution is utilise all agricultural waste, sewage and municipal solid waste as source of energy, fertiliser, animal fodder and other products, according to the production, demand and need of the waste production.

Some compatible techniques could be located mainly in rural areas: briquetting, biogas, composting, animal fodder and other recycling techniques for solid wastes.

Briquetting process[21] (easy to use, transport and store) starts with the collection of wastes (rice and wheat straws, cotton and corn stalks, sugar cane waste and fruit branches) followed by size reduction, drying and compaction by extruder or press.

Biogas is the anaerobic fermentation of organic materials by micro-organisms under controlled conditions.

Composting is the aerobic decomposition of organic materials by micro-organisms under controlled conditions. It requires specific conditions such as temperature, characteristic of the raw material.

---

21 Briquettes were discovered durino the First and Second Wars for heat and electricity producion using semple tecnologies.

Animal fodder. Transforming wastes into animal foodstuff needs to transform the agricultural waste into a more edible forms, with a higher nutritional value and better digestibility. The mechanical method has some limits such as the high cost and unfeasibility for small farms.

Household municipal solid waste: consists of organic, paper, plastic, tin, aluminium, textile and glass waste and dust. The opportunities vary from a place to another and it is necessary to separate the waste that could be mixed with other derived from different sector.

Conclusion(s)

Renewable energy has been assuming an increasing importance in recent years as a possible solution to global problems such as the increasing energy demand (in particular in developing countries, where population and GDP are growing faster than elsewhere), the shortage of conventional energy resources, the uncertainty of energy supplies and the resulting increasing energy costs; as well as, of course, the problem of atmospheric pollution and climate change.

Agricultural wastes used as an energy source can represent a good solution.

In general, the volume of waste generated by economic activities is located in four main economic sectors: agriculture, industry construction and services and the main treatment types utilised are recovery and disposal. In the EU the industry and construction sectors generate the highest volume of waste, followed by services and agriculture. The main treatment types utilised are recovery and disposal. With the international community seeking to expand the development of different types of renewable energy, more farmers in developed countries are using engines to generate onsite power from biogas created from converted animal waste and other agricultural organic materials. For example, in the European Union (EU), where agriculture is responsible for 9 per cent of greenhouse gas emissions, countries are being encouraged to expand the use of cogeneration technologies to reduce various industrial source points of methane gas emissions—including agricultural operations. The waste of 2,500 cows, 15,000 pigs or 300,000 chickens can create energy to supply more than 900 EU homes. Although the practice is more widely accepted in Europe, a growing number of farmers in Central America are installing similar systems to generate onsite power from methane-rich biogas that is created from the breakdown of organic materials, including animal manure and energy crops. By producing biogas from animal manure as substitute for fossil fuels, any additional greenhouse gas emissions are avoided. Additionally, farmers could also benefit from the biogas generation's end-product: a high-quality, agricultural fertilizer with a positive impact on land on the local water bodies. The use of agricultural waste to produce biogas gives farms another way to reduce their operational costs and greenhouse emissions.

In fact, the pollution approach and the resource use approach can be considered as the main guidelines to define and carry on sustainable public waste management policy.

Agricultural wastewater treatment relates to the treatment of wastewaters produced in the course of agricultural activities. Agriculture is a highly intensified industry in many parts of the world, producing a range of wastewaters requiring a variety of treatment technologies and management practices.

Renewable energy, from agricultural waste, therefore, can represent an important development opportunity in rural areas, but it is necessary that territories and local populations are able to implement recovery/recycling operations.

In rural areas of developing or emerging countries, the traditional fuels, such as fuel wood and agricultural residues, are still the dominant energy sources with harmful impacts on the environment and on health. The policy for these areas has to provide access to modern energy production and energy from agricultural wastes could be an appropriate solution.

References

1. Antle, J.M. and Heiderbrink, G. 1995. Environment and development: theory and international evidence. *Economic Development and Cultural Change*, 43 (3), 603-625.
2. Aqeel, A. and Butt, M.S. 2001. The relationship between energy consumption and economic growth in Pakistan. *Asia-Pacific Journa*l, 8 (2), 101-110.
3. Berndes, G., Hoogwijk, M., van der Broek, R. 2003. The contribution of biomass in the future global energy supply: a review of 17 studies. *Biomass and Bioenergy*, 25, 1-28.
4. Carrosio G., Prestia A. (2007). Flows and fixities of the rural areas: mobilities of local communities and the energy question, presented at XXIInd Congress of the European Society for Rural Sociology, Wageningen.
5. Commission of the European Communities (2004). *The share of renewable energy in the EU*, Bruxelles.
6. Commission of the European Communities (2005a). *Biomass Action Plan*. Brussels.
7. Commission of the European Communities (2005b). *Green Paper on Energy Efficiency or Doing More with Less*, Bruxelles.
8. Commission of the European Communities (2006a). *An EU Strategy for biofuels*, Bruxelles.
9. Commission of the European Communities (2006b). Communication from the Commission Action Plan for Energy Efficiency: *Realising the Potential. Saving 20 per cent by 2020*; Bruxelles.
10. Commission of the European Communities (2006c). *Renewable Energy Road Map. Renewable energies in the 21st century: building a more sustainable future*. Brussels.
11. Commission of the European Communities (2007a). Communication from the Commission to the European Council and the European Parliament: *An energy policy for Europe*, Bruxelles.
12. Commission of the European Communities (2007b). *Intelligent Energy Programmes*, Bruxelles.

13. Commission of the European Communities 2008. *Second Strategic Energy Review. An EU energy security and solidarity action plan*. Brussels.

14. EIA (Energy Information Administration) 2007. *International Energy Outlook*, World Energy and Economic Outlook.

15. EIA (Energy Information Admistration) (2008). *International Energy Outlook*, World Energy and Economic Outlook.

16. European Commission (2005). Research project ExternE: Externalities of Energy, Methodology 2005 Update, Luxemburg, http://www.externe.info/.

17. European Parliament's and Council directive 2009. 2009/28/EC: *Promotion of the use of energy from renewable sources and amending and subsequently repealing Diresctives 2001/77/EC and 2003/30/EC*.

18. European Union and Council of European Union, Directive 2001/77/CE on *Electricity Production from Renewable Energy Source*, Brussels.

19. FAO 2007. *Crop prospects and food situation.*, Food and Agriculture Organization of the United Nations, Rome.

20. Foster, A.D. and Rosenzweig, M.R. 2003. Economic growth and the rise of forests. *Quarterly Journal of Economics*, 118 (2), 601-637.

21. Malthus, T. R. 1803. *An Essay on the Principle of Population.* Penguin Books Limited, 1988 edition.

22. Menanteau, O., Finon, D., Lamy, M. L. 2003. Prices versus quantities: choosing policies for promoting the development of renewable energy. *Energy policy*, 31 (8), 799-812.

23. MNES 2006. *Annual report 2005-2006*. Ministry of Non-conventional Energy Sources, New Delhi.

24. MNES (2008), *Annual report 2007-2008*. Ministry of Non-conventional Energy Sources, New Delhi.

25. OECD-FAO 2009. *Agricultural Outlook 2009-2018.* Organisation for Economic Co-operation and Development and Food and Agriculture Organization of United Nations.

26. Suri, V. and Chapman, D. 1998. Economic growth, trade and energy: implications for the environmental Kuznet curve. *Ecological Economics*, Elsevier.

27. World Bank 2005. *World Bank group progress in Renewable energy and energy efficiency: 1990-2004*. Washington D.C., USA.

28. Zolin M.B. (2008). Biomass energy: new opportunity for agriculture and marginal areas?, Proceeding INDO-ITALIAN International Conference on "Green and Clean Environment", Pune, India.

Web sources

29. F.O. Licht's 2008. *World Ethanol and Biofuels Report*. Available at: http://www.agra-net.com/portal/puboptions.jsp?Option=menu and pubId=ag072.

30. Food and Agriculture Organization of the United Nations 2008. *Production Statistics* (ProdStat). Available at: http://faostat.fao.org/site/526/default.aspx.

31. Food and Agriculture Organization of the United Nations 2008. *Population Statistics* (PopStat). Available at: http://faostat.fao.org/site/452/default.aspx

32. Food and Agriculture Organization of the United Nations 2008. *Consumption Statistics*. Available at: http://faostat.fao.org/site/345/default.aspx.

33. Food and Agriculture Organization of the United Nations 2006. *Statistical Yearbook*. Available at: http://www.fao.org/statistics/yearbook/default.asp.

34. Institute for Applied Systems Analysis 2007. *Biofuels and food security–Final Report*. Available at: http://www.opecfund.org/publications/pdf/BiofuelAndFood Security.pdf.

35. ISEE (2008). World Population and GDP growth, http://www.hart-isee.com/index.php?page=world-population-gdp-growth.

36. OECD-FAO. *AgLink-Cosimo database Statistics*. Available at: http://www.oecd.org/document/6/0,3343,en_36774715_36775671_40969158_1_1_1_1,00.html.

37. Ravindranath, N.H., Manuvie, R., Fargione, J., Canadell, J.G., Berndes, G., Woods, J., Watson, H., and Sathaye, J., 2008. *Greenhouse Gas Implications of Land Use and Land Conversion to Biofuel Crops*. Available at: http://www.globalcarbon project.org/global/pdf/pep/Ravindranath_2009_GHGs per cent 20implications per cent 20of per cent 20Biofuel per cent 20Crops.SCOPE.pdf.

38. REN21 (2007). Renewables 2007. Global status report, http://www.ren21.net/pdf/RE2007_Global_Status_Report.pdf.

39. Sidhu K. S. (2007). Non-conventional Energy resources, http://www.indiacore.com/bulletin/kssidhu-non-conventional-energy-resources.pdf.

40. World Bank. 2009. *Key Development Data and Statistics*. Available at: http://web.worldbank.org/website/external/datastatistics.

Chapter 11

An Innovative Approach in Remediation of Soils Contaminated by Hydrocarbons

Paolo Plescia and Maria Tocino Amor

C.N.R.-ISMN–Italian National Council of Research Institute for the Study of Nanostructurated Materials Via Salaria km 29,300–PO. BOX 10–00016 Monterotondo Station (Rome) Italy

ABSTRACT

This work shows the results of the experiments conducted on soils heavily contaminated by hydrocarbons (C10-C40 content equal to 52,000 ppm) and Polycyclic Aromatic Hydrocarbons (PAHs total content equal to 110 ppm) through a new flotation process. The process is achieved using absorption substances (polyurethane powders) that removes emulsions created inside the processing waters. The results of tests indicate a reduction of C10-C40 hydrocarbons concentration in soil at 802 ppm total (- 98 per cent) and a reduction in PAH concentration at 8.2 ppm total (- 92.5 per cent). Fine and very fine fractions (< 90 µm) are removed from clean soil and accumulated in the float fraction, which is separated by a secondary treatment. The fraction rich in hydrocarbons may be recovered and burned in a thermal power plant. The whole process is suitable for use in fixed or mobile plants.

Keywords: *Contaminated soils, Hydrocarbons, Polymeric adsorbents, Froth Flotation*

Introduction

The flotation is a separation method widely used in mining and, recently, is of interest in some remediation processes (Anderson *et al.*, 1993; Clifford 1993; Griffiths,

1995; Dai *et al.*, 1995). It is highly advantageous over other technologies for soil remediation, such as soil washing and thermal desorption, because it allows to separate simultaneously the organic fraction (contaminated organic) and the inorganic one (metals) (Griffiths, 1995; Dai *et al.*, 1995; Vaughan, 2000). Other Authors showed that the most important parameters in the flotation of soils contaminated with hydrocarbons, BTEX and PAHs are the concentrations of collectors, the pH and kinetic characteristics of the particles. We add that one of the fundamental problems in the separation of oils and paraffins from soils derive from emulsions dissolved in the water of process and from oily clay particles that precipitate on the surface of the clean sands and gravel as a result of electrical effects (Plescia *et al.*, 2002). Hence the need to remove the oil fraction that is emulsified with water during the agitation and subtract as much as possible colloidal soil fractions, which are the richest in oil. To remove the emulsions from the water of process is possible to use absorbent substances, which may retain the oils and at the same time be easily separated by flotation. In recent years, many polymeric substances were used in absorbing oils spilled in seawater. Our group has recently developed a new product based on polyurethane foam to absorb the oil from bilge waters (Faudree *et al.*, 1980; Ik-Shung *et al.*, 2003; Thompson, 2005). It is based on a low cost material, derived from the recycling of polyurethane foams from fridges and air conditioners. The polymer is lighter than water and has a high affinity with oils; it's hydrophobic and do not absorb water. The product, called "Polysolver", (www.arcobalenoa.it) has a very high absorbing capacity and can be subjected to mechanical compression for the recovery of mineral oils (Table 11.1).

Experimetnal Procedures

The soil used for the laboratory tests comes from an old refinery site in Liguria (La Spezia) and is composed of gravel, sand and clay in the ratio 20, 35 and 45 per cent. The mineralogy of this soil was derived from X-ray diffraction and TGA (thermogravimetry).

The sample contains 5.1 per cent of C_{10}–C_{40} and 28 per cent of water. The material oil content is such that makes it muddy, with a strong smell of hydrocarbons and sulfur compounds (Figure 11.1). There are also few fibers of amphibole asbestos.

The phases of treatment were:

1. Condition of the soil in the flotation cell for 5 min
2. Addition of 1 per cent of surfactant additive (SDS, Sodium Dodecyl sulphate)
3. Flotation for 5 min of oily rich particles
4. Addition of the additive adsorption
5. flotation and separation of floating foam
6. Collection and fractions' analysis.

The tests were performed using EPA method 418.1 for Total Recoverable Petroleum Hydrocarbons (TRPH) and EPA method 8270 for total PAHs.

Results Obtained

The first floating product is the richest fraction in hydrocarbons, followed by the polymer foam that has absorbed oil emulsions. Further skimming led to the release of fractions gradually lighter. The sink, which contains mainly sand and gravels is clean and free from oily phases. The results obtained from flotation tests, in the presence of absorption additives compared with the classical flotation, show improvements in quality as well as in the capacity of oil removal. The soil after flotation is visually clean, as is shown in Figure 11.2. The hydrocarbons are concentrated in the float phases. These can be separated into a lighter phase, with the polymer containing oil and other hydrocarbons and a heavy one, with oily clays. The chemical analysis shows thank the Hydrocarbon content (C_{10}-C_{40}) in the original sample (51000 ppm) has been reduced to 802 ppm, with a total reduction of 98.43 per cent. The clay has a C_{10}-C_{40} content equal to 48500 ppm, as the untreated soil. The polymer foam retains 57300 ppm of total hydrocarbons, practically all the oil remained in the emulsion (Table 11.2). The value of 802 ppm of Hydrocarbons in the soil after treatment is under the specified limit within the Italian and European legislation on "Contaminated Land". Therefore the cleaned soil can be reinstated at the same place of its collection. PAHs results are even more interesting: the PAH content is considerably reduced (to 92.5 per cent total PAH; 100 per cent Naphthalene; 78 per cent benzo(bk)fluoranthene). In terms of yield, the 81.8 per cent of the soil is restituted clean. The residues are constituted of clay contaminated with oil, about 13 per cent of the soil, and the polyurethane material with free oil.

Table 11.1: Physical Characteristics of Polyurethane Foam "Polysolver" Used in Flotation Tests.

Parameters	*Polysolver*
Physical form	Flocks and powder
Pore diameters	From 0.2 to 100 microns
Color	Light yellow
Density	0.16 g/cc
In water	Hydrophobic
Adsorption (in weight per cent)	420 per cent standard oil 320 per cent gasoline 300 per cent heavy oil
Specific surface	> 100 mq/g
Ignition temperature	310°C

Discussion of Results and Conclusions

The data analysis leads to the following conclusions:

- ☆ The flotation is extremely enhanced for contaminated soils with hydrocarbons and PAHs, if it is done in combination with absorption additives for the removal of the oil emulsions.

Table 11.2: Chemical Analysis of Contaminated Soil and Products after Flotation: (A) Untreated Soil, (B) Treated Soil, (C) Clays Fraction (Heavy floating fraction), (D) Polyurethane Foams (Light floating fraction).

	Soil mg/kg	tr. soil mg/kg	clay mg/kg	foams mg/kg
Hydrocarbons (C10-C40)	**51260,5**	**802,0**	**48500,0**	**57328,0**
PAH				
naphtalene	11,2	0,0	1,6	7,9
acenaphthylene	2,0	0,1	2,3	3,0
acenaphtene	3,5	0,1	1,9	3,6
fluorene	7,4	0,1	6,1	2,4
anthracene	18,1	0,2	2,2	27,4
phenantrene	1,7	0,1	0,8	3,0
fluoranthene	9,1	0,8	14,9	5,2
pyrene	12,9	0,5	19,0	6,7
benzo-a anthracene	4,5	0,4	5,2	4,5
crysene	6,9	0,8	10,1	11,3
benzo(b+k)fluoranthene	7,3	1,6	10,0	8,3
benzo(a)pyrene	5,0	0,6	8,6	4,5
indeno(1,2,3)cd-pyrene	0,0	0,0	0,0	0,0
dibenzo(a,h)anthracene	5,7	1,0	4,0	3,8
benzo(g,h,i)perylene	14,1	1,9	23,9	7,1
Total PAH	**109,4**	**8,2**	**110,6**	**98,7**

Figure 11.1: POLYSOLVER Image on Scanning Microscopy.

Figure 11.2: Soil Untreated (A), Clean Soil (B), Clays Rich Floated Fraction (C) and Polymer Rich Fraction (D).

- ☆ The clay fraction contaminated with hydrocarbons is easily detached from the sands and gravels, and thus aggregated to the float fraction.
- ☆ The hydrocarbons are quickly absorbed by the polyurethane surface and can be quickly separated. The complete separation of the fine fraction brings the presence of a concentrate containing clay, oil and heavy paraffins, and the adsorbent mixed with light oils. The oil concentrate can be further refined in order to extract and recover the oil or can be used within an incinerator.

References

1. R.A. Griffiths, "Soil-washing technology and practice", J. Hazardous Materials 40 (1995) 175–189.
2. W.C. Anderson (Ed.), "Innovative Site Remediation Technology: Soil Washing: Soil Flushing", American Academy of Environmental Engineers, Annapolis, MD, 1993.
3. J.L. Underwood, "Soil cleanup by in-situ surfactant flushing. III. Reclamation of multicomponent sodium dodecylsulfate solutions in surfactant flushing", Separation Sci. Technol. 30 (1995) 2277–2300.
4. Q. Dai, K.H. Chung, "Bitumen-sand interaction in oil sand processing", Fuel 74 (12) (1995) 1858–1864.

5. C. van Rijt, "Cleaning contaminated sediments by separation on the basis of particle size", Water Sci. Technol. 28 (1993) 283–295.
6. S.R. Clifford, "Removal of organic contamination from Buffalo River sediment by froth flotation", Miner. Metall. Processing 294 (1993) 195–199.
7. R. Varadaraj, "Decontamination of hydrocarbon containing substrates". US Patent c5,417,864 (Cl. 210–703), 1995.
8. M. Wilichowski, J. Werther, "Applicability of flotation in the washing of soil", Chem.-Ing.-Tech. 67 (1995) 760–763.
9. J. Zheng, C.C. Harris, P. Somasundaran, "A study of grinding and energy input in stirred media mills", Powder Technol. 86 (1996) 171–178.
10. J.E. Norman, O.C. Ralston, "Conditioning surface for froth flotation", AIME Technical Publication, 1074, 16, 1939.
11. Q. Dai, K.H. Chung, "Hot water extraction process mechanism using model oil sand", Fuel 75 (1996) 220–226.
12. D.A. Sabatini, R.C. Knox, J.H. Harwell (Eds.), "Surfactant - Enhanced Subsurface Remediation: Emerging Technologies", ACS Symp. Ser., 594, 1995 NATO/CCMS Pilot Study, n.219, 1998.
13. Pignatello JJ, Xing BS (1996), "Mechanisms of slow sorption of organic chemicals to natural particles" Environ.Sci.Technol. 30: 1-11.
14. Ribeiro, A., Villumsen, A., Réfega, A., Vieira e Silva, J. and Bech-Nielsen, G. (1998), 16th World Congress of Soil Science, Montpellier, France.
15. Vaughan D. J. and Wogelius R.A., "Environmental Mineralogy" EMU Notes in Mineralogy, EMU Press, 2000.
16. Thompson Timothy, "Foam Composite for absorption of Hydrocarbon and process for making same", PCT/US2005/002971, 2005.
17. Faudree, III, Thomas L., "Foamaceous hydrocarbon adsorption medium and method and system for making same" 1980, US Pat 4230566.
18. Ik-Shung Ahn, Heon-Sik Choi Ju-Young Kim, "Comparison of Nanosized amphiphilic polyurethane (APU) particles with SDS an anionic surfactant for the soil sorption and the extraction of phenanthrene from soil", Proceedings of Adsorption Science and technology, 2003.
19. Hayes D. Thomas, "Foam transport process for in-situ remediation of contaminated soils", Gas Research Institute UN Pat. No. 318462 filed on 1994-10-05.
20. Pizzagallo M. D.R., Napola A., Spagnuolo M., Ruggiero P. "Mechanochemical removal of organo-chlorinated compounds by inorganic components of soil" Chemosphere, 55 (2004), 1485-1492.
21. Plescia P., Gizzi D., Benedetti S., Caramuscio P., De Stefano L. and De Luca F. "Trattamento meccanochimico di suoli contaminati da olii pesanti", (2002): ACQUA and ARIA (Arti poligrafiche Europee SRL, A.N.E.S.), N° 3 Marzo 2002, 93–99.

Chapter 12

Dynamic Heavy Media Separation in Plastics Recycling Industry

Paolo Bozano[1], Paolo Bozzato[1] and Girolamo Belardi[2]

[1]*Ecomin S.r.l., Via Privata Galla n. 10, 16010 Serra Riccò, Genova, Italy*

[2]*Istituto di Geologia Ambientale e Geoingegneria (IGAG-CNR), Area della Ricerca di Roma RM 1–Via Salaria Km 29,300 C.P. 10, 00016 Monterotondo Stazione–Roma, Italy*

ABSTRACT

The results of the performance in the treatment of plastics of different composition and shape with a Tri-Flo™ equipped with a new feeding system for cylindrical separators (Dynafeed™), recently developed and patented, are presented. Separation tests were carried out with a PS (polystyrene)/PP (polypropylene) mixture, ABS (Acrylonitrile butadiene styrene) and PE (polyethylene)/PET (polyethylene terephthalate) mixture. It has been found that single particle sizes and shapes are separated by the Tri-Flo™ with high sharpness, with limited variations of the Ep in the range 0,004÷0,007 g/cm^3. The offset Δ_{50}, is dependent on particle size and shape. It is possible obtain an high grade and recovery of the concentrates. The results have enlarged the field of application of dynamic cylindrical DMS processes in the plastics recycling industry. The Tri-Flo, a multistage Heavy Medium (HM) cylindrical separator, can be ideal technology for the separation of plastics of different polymeric composition and of other light 'post-consumer' secondary materials because very sharp separation can be obtained with particles of different shape, it has high sink-float capacity and produces three products in one vessel simplifying plant layout and reducing costs at high efficiency.

Keywords: *Dynamic dense medium separation, Cylindrical cyclone, Plastics separation, Particles shape.*

Introduction

Plastics recycling has become an important part of commerce and environmental protection.

There are many kinds of plastics in wide use today including polyethylene (PE), polystyrene (PS), polypropylene (PP), and polyvinyl chloride (PVC).

Another common example is the polyethylene terephthalate (PET) plastic used in the production of soft-drink bottles. Many type of plastics can easily be reused as raw material for other plastics products. However, in some cases, it is necessary that the purity of the raw material be at last 99.9 per cent.

Gravity separation is an effective and economical method of separating materials by their density differences. In the DMS processes, both static and dynamic separators may be used. However, as a consequence of low density differences in DMS, low particle settling rates, static separators may be used only for large particles separations. Moreover, for plastic materials with low wetting capability, surface tension forces may hinder the separation or require a pre-wetting treatment,

using, if necessary, surface conditioning agents. Alternatively, a process of sink-float separation and flotation has been proposed (Dodbida G. *et al.*, 2002). Because of low settling rates in DMS processes, the use of dynamic separators may be more appropriate. As the mass forces are increased by centrifugal acceleration, the influence of surface forces due to viscosity or surface tension are reduced or eliminated. Dynamic separators such as the conical cyclone DSM cyclone and the multistage cylindrical separator Tri-Flo can be used. Because of the different particles shape resulting from the comminution of plastic materials, using the DSM cyclone did not yield acceptable results. Independent of their density, these particles are flat or filiform. Due to the large radial centripetal flow, in the DSM cyclone, flat and filiform particles are directed towards the float discharge.

The Tri-Flo™ is a proven technology, more than 16 Tri-Flo™ pilot plants and 30 Tri-Flo™ industrial plants have been supplied worldwide, mainly in mineral processing.

Experimental

Apparatus

The Tri-Flo™, reported in Figure 12.1, is a cylindrical multistage separator applied in difficult separations or where a 3-product separation is required.

The material is fed at atmospheric pressure separately from the medium. The first stage float material directly enters the second stage for a second separation. In a difficult 2-product separation only one Correct Medium (CM) is used and the second stage further washes the

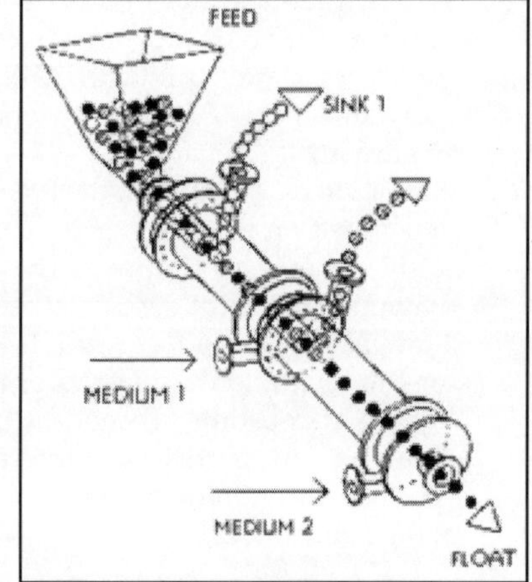

Figure 12.1: Tri-Flo™ Schematic View.

float material for improved separation sharpness. Two CMs are used in a 3-product separation and the second stage operates the low gravity cut.

A multistage rougher-scavenger or scavenger-cleaner approach looked most hopeful. Two advantages were expected from this. The first advantage stems from the fact that in a two stage separation the first stage removes the greater amount of sink particles, acting as a rougher, and thus allows the second stage to operate under more constant scavenger conditions. This becomes particularly important when there is a high proportion of sinks or when this proportion varies significantly.

The second is based on the fact that under otherwise similar conditions a two stage separation will be sharper than that of a single stage.

Partition curves in Figure 12.2 show that: the overall separation of a rougher cleaner circuit is improved if compared with a rougher only and that an abnormal load in the first stage causes some heavies to be misplaced in the first stage float product but the second stage, which is working in more stable conditions since the first stage smoothes out variations, can correct the situation.

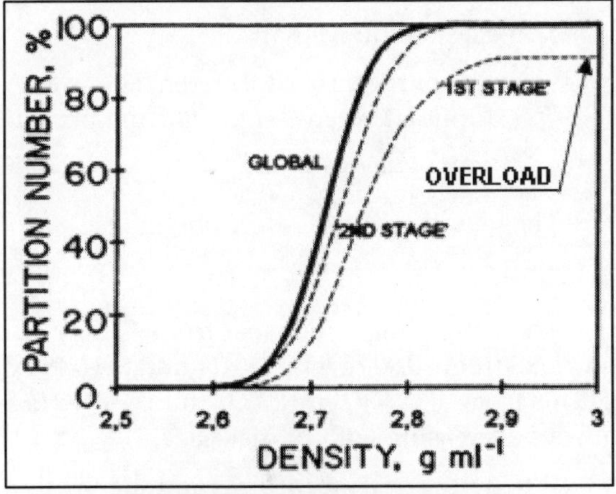

Figure 12.2: Partition Curves of a Double Stage Circuit.

Pilot scale tests have been performed with a two stage Tri-Flo 100 mm ID equipped with a new feeding system (Dynafeed™) that has been recently developed to spin the feed material before entering the separator, thus further improving the separation sharpness and the throughput capacity of cylindrical cyclones (Bozzato P. *et al.*, 2005).

First Tests with -8+4 mm Polystyrene/Polypropylene Mixture

The Tri-Flo™ was first tested by conducting separation tests on a mixture of polystyrene (1,04 g/ml) and polypropylene (0,9 g/ml) in the same proportion, using water as a medium. The particle size of the feed material was in the -8+4 mm range. The tests were performed to simulate a four-stage Tri-Flo™: after a first separation in the two-stage vessel, the float was treated again in the same vessel, obtaining four sinks and one float on the whole. In order to obtain a better separation, the first two stages were implemented with high back-pressure to produce clean polystyrene in the sinks. The successive stages were set with low back-pressure to obtain clean polypropylene in the float. The results were satisfactory, as all the polypropylene was collected in the final float while the polystyrene was collected in the four sinks with the following distribution: sink1 = 80 per cent; sink2 = 16 per cent; sink3 = 3 per cent; sink4 = 0.9 per cent.

Only 0,1 per cent of polystyrene was reported in the float after the four stages, so the polypropylene content of the float product was 99.9 per cent.

Additional Tests to Evaluate Influence of Particles Shape on Dynamic DMS Processes

Tests have been performed for studying the influence of particle shape and size on the separation in dynamic dense medium separation processes using compact, elongated and flat particles of ABS (acrylonitrile butadiene styrene) with different shape and size. Partition curves have been determined in dynamic dense medium separators (cyclones and Tri-Flo™) for each particle type using water and salt solution (water and sodium chloride) in the range density from 1 g/cm^3 and 1.12 g/cm^3.

These tests showed that:

- ☆ The particles with different shape and size have different "offset" values (difference between d_{50} and medium density) and different Ep values;
- ☆ This behavior greatly differs in cyclones and Tri-Flo™

The following results were obtained:

Tri-Flo™

The offset ranges: for cubes from -0,009 g/cm^3 (6mm) to +0,023 g/cm^3 (1,5mm); for plates from +0,009 g/cm^3 (36x6mm^2) to +0,051 g/cm^3 (1,5x12x12mm^3); for elongated prisms from -0,006 g/cm^3 (3x3x6mm^3) to +0,065 g/cm^3 (1,5x1,5x12mm^3). Offset is always decreasing with particle size.

When treating a mix of sizes and shapes, though the individual partition curves for size and shape have a low Ep, as a consequence of the offset (which is different for size and shape), the combined partition curves are worsened; the following data have been observed:

1. cubes — offset
   Ep=0,010 g/cm^3 — Δ_{50}=0,004 g/cm^3
2. elongated prisms — offset
   Ep=0,016g/cm^3 — Δ_{50}=0,023 g/cm^3
3. plates — offset
   Ep=0,011 g/cm^3 — Δ_{50}=0,024 g/cm^3
4. mix — offset
   Ep=0,014 g/cm^3 — Δ_{50}=0,018 g/cm^3

Conical Cyclone

Elongated and flat particles show very high offset values, and only for a few tracers it was possible to calculate a partition curve in the range of tested medium densities. Only positives offset values have been observed. It is very difficult to find a clear relationship between particle shape and size and resulting partition curve. The offset for cubes ranges from +0,021 g/cm^3 (intermediate particles: 3 mm) to +0,044 g/

cm^3 (bigger particles: 6 mm), for plates and elongated prisms it is difficult to identify a clear trend according to particle size, whilst there is a strong dependence of the offset from elongation. It was possible to observe very high positive offsets, in particular for very elongated (dikudd>3) and very flat (dikudd <0,5) particles; these particles entirely reported to float, with offset values greater than 0,9 g/cm^3.

As a general rule, the partition curves of individual sizes and shapes indicate that very sharp separations are obtained with both conical cyclones and Tri-Flo™. The Ep is in the order of 0,004÷0,007 g/cm^3. It has been found that there is a large variation in the offset Δ_{50} according to the particle elongation, and this effect is very much amplified with the conical cyclone.

The separation of compact particles as cubes may be sharper than the separation of elongated or flat particles. It appears that elongated particles give the worst separation (cutter mills producing a larger proportion of elongated particles should be avoided).

For the Tri-Flo™, the mix of all particles gives a partition curve having an offset Δ_{50}=0,018 g/cm^3 and Ep=0,014 g/cm^3. For the conical cyclone, the mix of all particles gives a partition curve having a worse Ep.

This is as a consequence of the different flow patterns and the low centripetal flow that characterise cylindrical separators such the Tri-Flo. It has been found that cylindrical cyclones with peripheral sink discharge have a tangential velocity similar to a centrifuge (the forced vortex prevails $v_t/r = c$). while in conical cyclones the free vortex with high shear rate prevails $v_t r^n = c$) (Ferrara, G. *et al.*, 1997; Chine', B. *et al.*, 1997).

It can be noted that in spite of the low Ep values obtainable with individual sizes and shapes, under the effect of the large offset values, the Ep of the mix may be worse than expected. Obviously, the results get better with a more restricted composition of particle size and shape and get worse with a larger composition.

Pilot tests with plastic tracers and plastic flakes

Additional tests have been performed in the two stage Tri-Flo 100 mm ID, using water as dense medium and treating the following materials (Figure 12.3):

1. Plastic tracers: PE/PET mixture; sub-spherical shape; size -4+3 mm;
2. Plastic tracers: PE/PET mixture; flat shape; size -15+10 mm;
3. Plastic flakes: PE/PET mixture from post-consumers bottles; various shape, recyclable material produced in a shredding plant; size -20+1.84 mm.

The purity grade obtained with plastic tracers has been of 100 per cent for both materials. the results with PE/PET mixture from post-consumers bottles are shown in the Table 12.1 and Figures 12.4 and 12.5.

Conclusions

It has been found that single particle sizes and shapes are separated by the Tri-Flo™ with high sharpness, with limited variations of the Ep in the range 0,004÷0,007 g/cm^3. The offset Δ_{50}, is dependent on particle size and shape.

Table 12.1: Results Obtained Treating Shredded Post-Consumers Bottles.

Medium Water	PET (g)	PE (g)	PET (cum.) (g)	PE (cum.) (g)	PET (%)	PE (%)	PET purity grade (%)	PET recovery (%)	PE purity grade (%)	PE recovery (%)
Size range 1										
FEED	53.65	439.80			10.9	89.1				
SINK										
1st treatment	35.08	0.25	35.08	0.25			99.3	65.4		
2nd treatment	11.65	2.04	46.72	2.29			95.3	87.1		
FLOAT										
1st treatment	18.57	439.55							95.9	99.9
2nd treatment	6.93	437.51							98.4	99.5
Size range 2										
FEED	82.72	435.96			15.9	84.1				
SINK										
1st treatment	67.58	1.82	67.58	1.82			97.4	81.7		
2nd treatment	11.67	1.71	79.24	3.53			95.7	95.8		
FLOAT										
1st treatment	15.14	434.14							96.6	99.6
2nd treatment	3.48	432.43							99.2	99.2
Size range 3										
FEED	164.48	343.38			32.4	67.6				
SINK										
1st treatment	156.78	0.13	156.78	0.13			99.9	95.3		
2nd treatment	6.55	0.42	163.33	0.55			99.7	99.3		
FLOAT										
1st treatment	7.71	343.25							97.8	100.0
2nd treatment	1.16	342.83							99.7	99.8
Size range 4										
FEED	278.21	272.29			50.5	49.5				
SINK										
1st treatment	252.81	0.18	252.81	0.18			99.9	90.9		
2nd treatment	21.06	0.24	273.87	0.41			99.9	98.4		
FLOAT										
1st treatment	25.40	272.12							91.5	99.9
2nd treatment	4.34	271.88							98.4	99.8

Size range 1 = -12.5 mm;

Size range 2 = -12.5+5.6 mm;

Size range 3 = -5.6+2.8 mm;

Size range 4 = -2.8+1.84 mm

Figure 12.3: a and b: Plastic Tracers: Sub-spherical and Flat Shape; c: Mixture from Post-consumers Bottles.

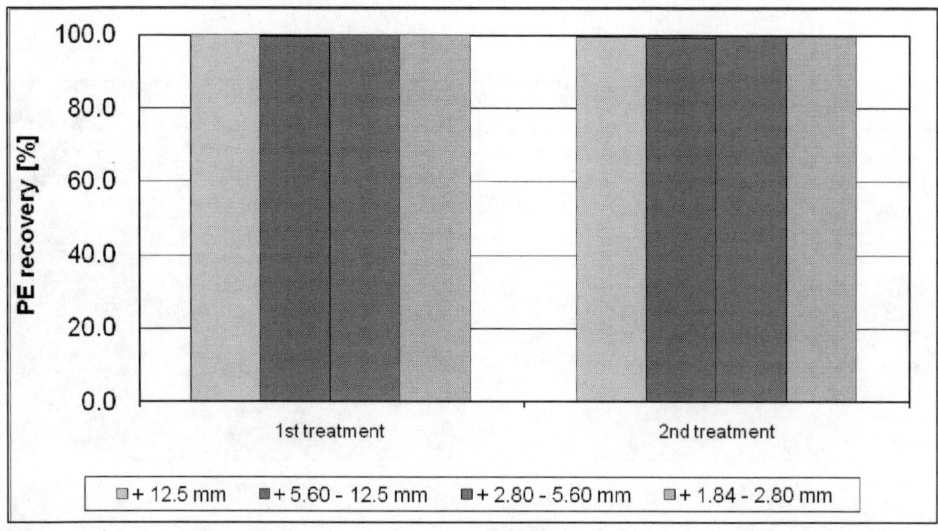

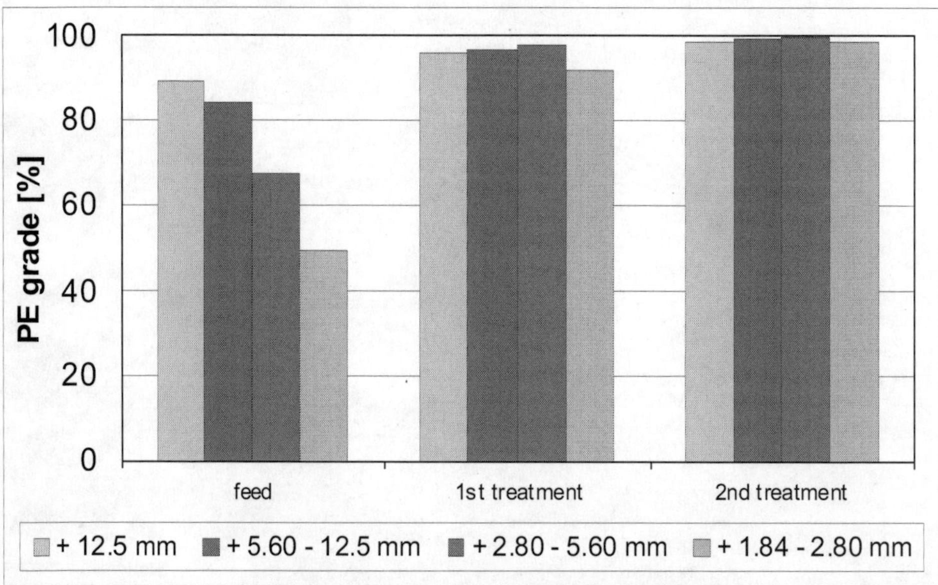

Figure 12.4: PE Recovery and Grade Obtained Treating Shredded Post-Consumers Bottles.

The results obtained so far demonstrate that in dense medium separation of plastics, the Tri-Flo™, is more suitable than conical cyclone. In fact due to the lower offset values, Tri-Flo™ gives better overall partition curves considering a mix of sizes and shapes as in plastic scraps occur.

It would be very important to get a picture of the complex behaviour of different shaped particles due to the flow patterns in dynamic separators; it may be of great

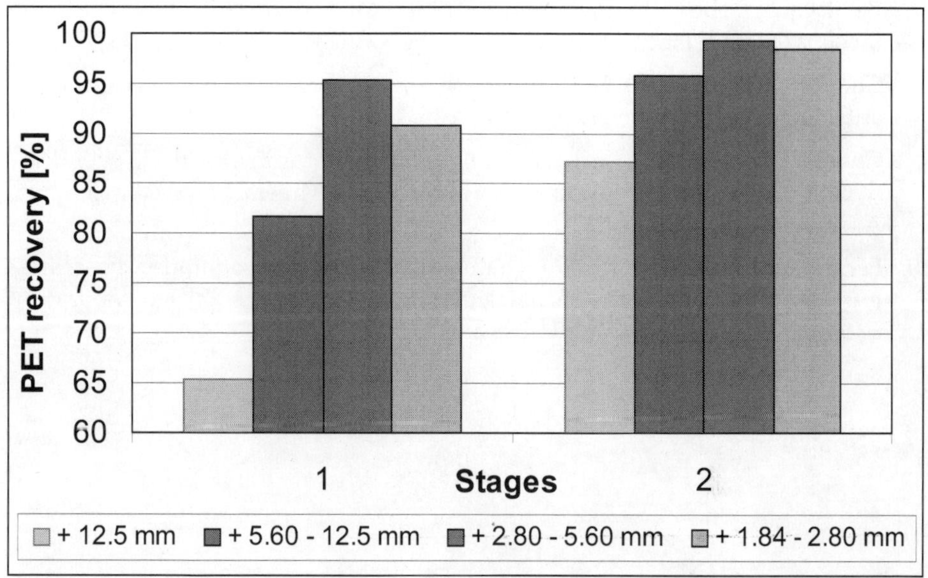

Figure 12.5: PET Recovery Obtained Treating Shredded Post-Consumers Bottles.

assistance in the analysis of dense medium processes, particularly when different particle shapes are involved (recycling industry).

References

1. Bozzato, P., Peng, F.F., 2006. "Multistage Dense Medium Circuit Analysis and Application", SME Annual Meeting, Preprint 06-075, Mar. 24-Mar. 30, St. Louis, MO.
2. Bozzato, P., Peng, F.F., Gu, Z., 2005. "Dense Medium Tri-Flo Separator–A review and New Development," SME Annual Meeting, Preprint 05-132, Feb. 27-March 3, Salt Lake City, Utah.
3. Barbee, C.C., Luttrell, G.H., Wood, C.J., Bethell, P.J., 2004. "Simulation of Heavy Medium Cyclone Performance," *Mineral and Metallurgical Processing*, Vol. **21** (4), pp. 38-42.
4. Luttrell, G.H., 2003. "Density Separations: Are We Really Making Use of Existing Process Engineering Knowledge?", *Advances in Gravity Concentration*, R.Q. Honaker and W.R. Forrest (Ed.), SME Inc., Littleton, CO, pp. 1-15.
5. Dodbiba G., Haruki N., Shibayama A., Miyazaki T., Fujita T., 2002, "Combination of sink–float separation and flotation technique for purification of shredded PET-bottle from PE or PP flakes", Int. J. Miner. Process. 65, 11–29.
6. Bevilacqua, P., Bozzato, P., Ferrara, G., 2000. "Influence of particle shape on dynamic DMS processes", Proceedings of the XXI IMPC, Rome, pp. A7 95-103.

7. Bevilacqua, P., Bozzato, P., Ferrara, G., 1997. "Analysis of dense medium circuits for separating plastics", Proceedings of the XX IMPC, Aachen, pp. 383-393.

8. Chinè, B., Ferrara, G., 1997. "Comparison between flow velocity profiles in conical and cylindrical hydrocyclones", Kona 15, 170–179.

9. Chinè, B., Ferrara, G., Concha, F., 1997. "Flow patterns in the Tri-Flo cylindrical dynamic separator", Proceedings, XX International Mineral Processing Congress, Aachen, Vol. 2, pp. 505–520.

10. Ferrara, G., Ruff, H.J., 1982."Dynamic dense medium separation processes–Further developments with particular reference to the Tri-Flo separator", *Erzmetall*, Vol. 35 (6-7), pp. 294-298 and 395-398.

Chapter 13

The Mechano-Chemistry for the Energy Recovery from Urban from Exhaust Batteries

Paolo Plescia and Maria Amor Tocino

C.N.R.-ISMN (National Council of Research–Institute for the Study of Nanostructurated Materials - Area Rome 1, Salaria Street km 29,300 Montelibretti (Rome) Italy

ABSTRACT

Annually the amount of urban solid wastes produced in Italy is approximately 26 million of tons, of which about 30 per cent is constituted by water and 10 per cent by inert and metal components. The remaining 60 per cent is a mixture of paper, plastic and putrescible materials with a calorific power at least equal to 2500 Kcal/kg. Nowadays the production of RDF (Refuse-Derived Fuel) does not represent an interesting alternative to the fossil combustibles because of its low calorific power and, more important, because of the presence in high amount of toxic elements, ashes and inert components. Based on past experience in mechanochemistry, a new technology has been carried out using a mechanochemical reactor able to induce in the treated materials an irreversible changing of the molecular structure, as well as a strong dimensional reduction in micro-spherule with diameter smaller than 300 μm, with an high increase of the calorific power of the waste (between 10 per cent and 40 per cent), followed by a decrease in the chlorine amount. Two series of mechanochemical reactors were developped, to convert the urban solid wastes in noble fuels (refuse derived fuels RDF).

Keywords: *Mechano-chemistry, Urban solid wastes, Chlorine, Metals.*

Introduction

Annually in Italy the amount of urban solid waste produced is equal to 28 million of tons, equivalent to over 500 kg pro capite (1). Without any efficient separated waste collection (< 10 per cent), the most part of waste is constituted for nearly 60 per cent by organics, paper, cardboard and plastic, with an high content of water (32 per cent) (Table 13.1). The common destination of wastes are the over 2000 dumps present in Italy. Over the past two years, a strong action of the Italian Government is forcing the creation of new incinerators, which solve only part of the problem and create many problems of pollution. Furthermore, the recent EU-regulations in terms of energy efficiency of incinerators (EU 98/2008, 2) require a quality of RDF much higher than at present. The main problems that lead to energy deficiency of RDF are:

- ✰ The water content
- ✰ The presence of heavy metals
- ✰ The presence of chlorinated substances both inorganic (salts) and organic (plastic with chlorine, solvents, disinfectants, detergents)

Table 13.1: Composition of Urban Solid Waste (Updating Corse of Energy from Refusals, Polimi, 2005)

Constituent	content in MR residue	composition			carbon content		LHV
		moisture	ash	volatile fraction	total	% renewable	
	% by weight						MJ/kg
paper & cardboard	24.5	14.0	5.0	81.0	37.6	100	13.22
Wood	6.0	22.0	1.5	76.5	37.6	100	13.87
Plastic	19.0	6.0	9.0	85.0	55.5	0	26.18
glass & inert material	3.5	2.5	95.0	2.5	1.0	0	-0.061
Metals	3.5	5.0	92.5	2.5	1.0	0	-0.122
organic fraction	31.5	70.0	9.0	21.0	9.6	100	1.719
fines	12.0	30.0	35.0	35.0	20.5	60	4.395
MR residue	**100**	**31.8**	**16.6**	**51.6**	**27.6**	**16.0**	**10.11**

volatile fraction % by weight of total	
C	27.6
Cl	0.64
H	3.49
O	19.7
N	0.15
S	0.06
Total	**51.6**

The presence of NaCl and plastics with chlorine (PVC) causes, during the combustion process between 250 and 800 °C, the development of hazardous chlorinated compounds, for instance PCDD and PCDF (3, 4). Regarding the preliminary separation, the reader is referred to the extensive specialized bibliography. For the preliminary separation of metals, the experience on our systems has been conducted with a magnetic separator over-belt and a separator of non-ferrous metals by means of eddy currents; this combination can recover metals of wastes with a very high efficiency. To dehydrate the waste have been proposed various solutions, but each necessarily lead to an abnormal increase in processing costs. In fact, if we separate the putrescible organic materials as manure (compost) could reduce the water content below 30 per cent (1, 2). The elimination of chlorine is another big problem. In fact it is not only the separation the PVC, which is accomplished relatively easily, but you must also remove the inorganic salts of chlorine, for the reason that they contribute to producing dioxins, hydrochloric acid and corrode all the metal structures of the incinerator (5, 6). To remove these components and get a good fuel it is necessary to conduct a "refining" of the waste, by using a pre-separation systems and an efficient method of processing and drying of the waste.

The Mechanochemical Reactors

To achieve a complete refining of urban waste has been devised a system configuration which, in addition to the separations of metals, introduce a micronization of wastes using high-energy mills, which we called "mechanochemical reactors". The mechanochemical reactor is based on an high energy mill, able to transmit a continuous stress to the materials to obtain a comminution, dewatering and some chemical reactions. Because of these large compression and shear stresses, the chemical structure of the materials are subjected to some important changes (7, 8). The mechanical energy transferred to the material turns partially into heat and the remaining energy causes macro and microscopic changes in the treated material:for example, the treatment for elimination of organic molecules from complex substances (for instance DDT, PCB etc.), the removal of chlorine from chlorinated polymers, the processing of asbestos to convert it in amorphous gel (9, 10, 11), the destruction of polycyclic hydrocarbons (12).The study conducted for a decade started on the physics of ring mills, machines that have the most important rate of friction (8, 9) and therefore suitable for micronize of cellulose and polymeric materials; in the same period it was developed several models of computation that have allowed to analyze energies that are generated in the mills and in this way to predict the timing of treatment (8, 9) (Figure 13.1). Unfortunately, for many technical problems it was impossible to scale up the ring mill to the industrial level and thus were studied some planetary mills, as the THOR mill, designed in 2006 and the latest hybrids mill called TRITOR (Figures 13.2 and 13.3). In the THOR mill, the electrical engine drives a central axis, which in turn it moves four "satellites", the jars. The summation of the movements makes it possible to move the balls inside the jars (about 100 kg of steel balls per jar) according to a rotation movement and causes a shear strain on the material subjected to grinding. After a development of two years, the THOR mill is now used industrially, with a rate of production of 500 kg/hour of highly refined fuel for cement plants and power stations. During the same period began the development of third generation of mechanochemical reactors, based on the combination between an attrition mill and a nutation mill; this project was called TRITOR. The mechanics of this new mill cause the reduction of sizes of the particles and contemporary loss of the water, for the contemporary action of intensive shear and pressure between the grinding balls and particles. This system will be used in the experimental production of fuel from residues of the scrapping of cars (car-fluff). To determine the energy used in mechanochemical reactors was used a direct technique based on the use of films for the determination of the pressure produced by Sensor Products (Pressurex®, Sensor Products Inc. 300 Madison Avenue, Madison, NJ 07940 USA).These films are imprinted with a color that has an intensity proportional to the level of the external pressure. It has thus analyzed the mechanisms of action of planetary mills and attrition mills on the materials and we have seen that the first reach very high levels of pressure (6 kbar), with a main shear stress, but they have a discontinuous level of pressure on the surface of the jar, while the latter have an higher level of shock action and a lower

Figure 13.1: The Laboratory Ring Mill.

Figure 13.2: The First and Second Prototype of the Industrial Mechano-chemical Reactor Based on Planetary Motion (THOR, Assing SpA and CNR ISMN, 1999-2007).

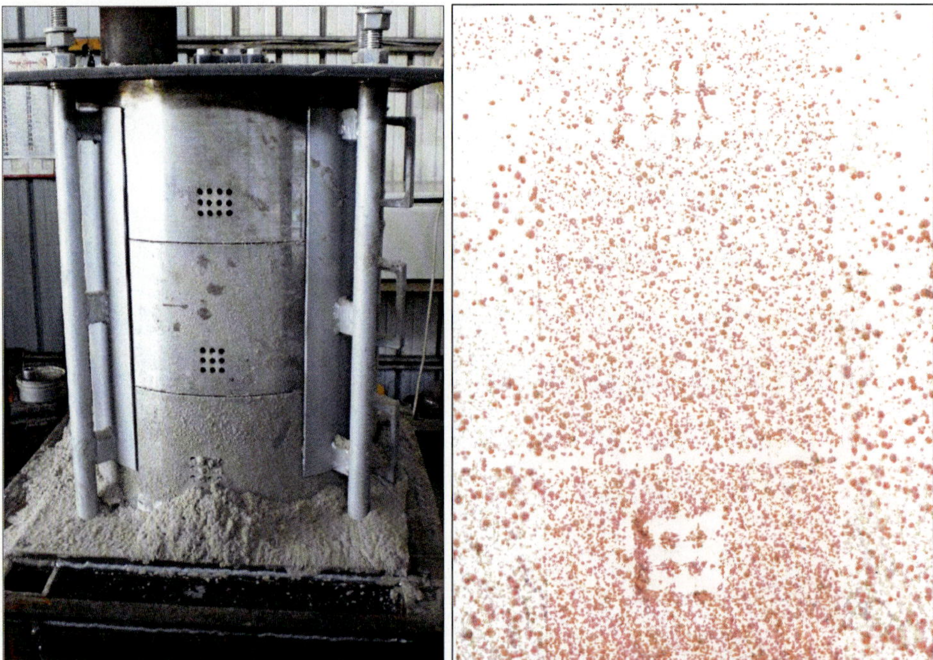

Figure 13.3: The TRITOR Mill (Arcobaleno.A and CNR ISMN, 2008-2009) and the Imprinting of Balls on the Pressure Sensor (max pressure 3.5 kbar).

level of shear stress but may act on the entire volume of the jar (Figures 13.3 and 13.4). From these considerations it is possible to choose a mechanochemical reactors for specific uses, and in particular:

- A planetary mill as THOR systems to treat cellulosic and putrescible wastes
- An attritor mills as TRITOR for treating plastic polymer.

Resume of Experimental Study and Results on RDF

The experimental phase has involved a lot of tests on samples of paper, wood, vegetables, fats and animal tissues. After the mechanochemical treatment the samples have been carefully characterized: chemical, microstructural, thermochemical and thermogravimetrical analyses in air and nitrogen have been made, to understand the behaviour during the combustion, before and after the milling process. Particularly significant are the results on the improvement of lower calorific values (LCV) and reduction of sizes of products (Table 13.2). For a simple mixture of paper, wood and PE polymers, with a starting LCV of 19 KJ/kg, the end LCV was 22 KJ/g. For a mixture of paper, wood, PET and 40 per cent of organic material (with a humidity equal to 70 per cent (L.C.V. = 13 KJ/g) the final LCV was 21 KJ/g. The material, after treatment, appears reduced to an impalpable dust (Figure 13.5). The treated RDF tends to a minor ignition temperature, as in Figure 13.6, where is represented the ignition temperature of some RDF vs. coal dust for cement kiln. Finally, the mechanical

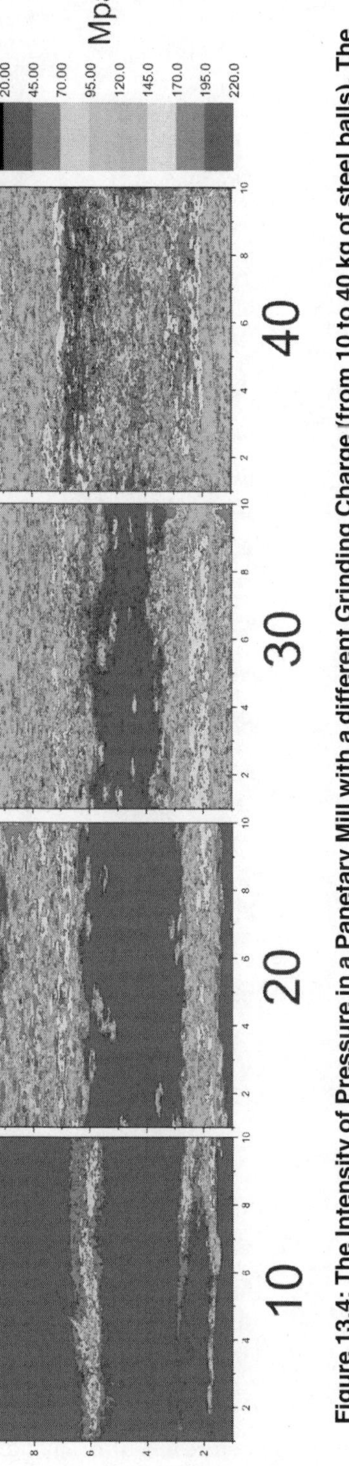

Figure 13.4: The Intensity of Pressure in a Panetary Mill with a different Grinding Charge (from 10 to 40 kg of steel balls). The intensity of color is related to pressure (max about 3.5 kbar).

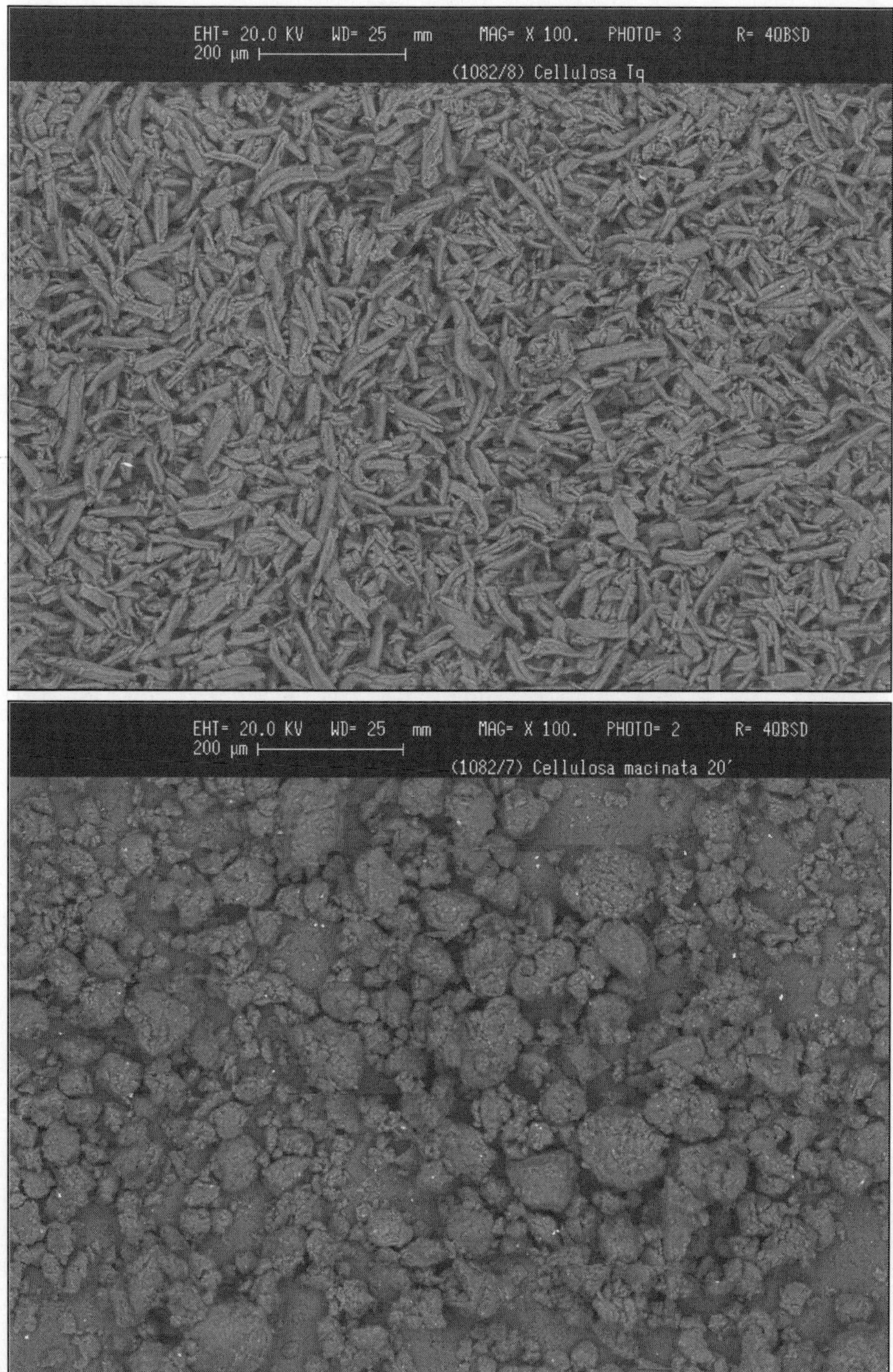

Figure 13.5: Cellulose Morphology before and after Milling Treatment (SEM images).

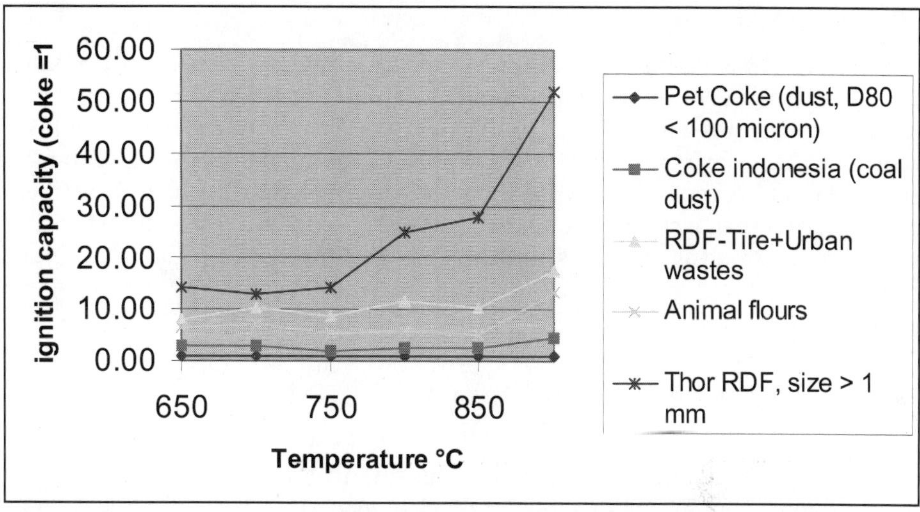

Figure 13.6: Ratio between the Ignition Capacity of Coal Dust with the Value Obtained from the RDF from different Sources, such as RDF THOR, Petroleum Coke, Animal Flours and RDF obtained by Mixing between Municipal Solid Waste and Tires.

action of compression and shear can achieve substantial dehydration (3 per cent of residual water starting from 35 per cent), a significant reduction of bacteria and a reduction of chlorinated organic substances. Chlorine, in particular, is extracted from the waste as hydrochloric acid and can be easily picked up if not immediately in contact with carbonates.

Table 13.2: Comparison between a Normal RDF and a Treated RDF (THOR).

Parameters	*Units*	*Not Treated RDF*	*Treated RDF*
Size	mm	> 50 mm	90 per cent < 20 μm
Heating Value	Kj/kg	12.000	22400
Moisture		35 per cent	2.9
Cl	On dry matter per cent	0.9	0.65
S	On dry matter per cent	0.6	0.14
Ashes	On dry matter per cent	20	17
Cr	mg/kg	150	92.3
Cu	mg/kg	300	98.2
Mn	mg/kg	400	41.3
Ni	mg/kg	40	6.8
As	mg/kg	9	1.1
Cd	mg/kg	7	0.1
Hg	mg/kg	5	0.6
Pb	mg/kg	200	38

Conclusion(s)

The experience on the study and construction of new mechanochemical reactors led to a wide range of products that can be usefully employed in the treatment of urban solid wastes and hazardous wastes. The power of mechano-chemical technology stems from its substantial simplicity of use and the possibility to make small plants and low environmental impact. Finally, a word to the cost of treatment: to bring a bad urban waste to a good fuel, they are needed on average 40 /ton. This fuel could be burned as a biomass, with practically the same LCV, but less ashes and emissions.

References

1. AA.VV "Energy from wastes " Polytechnic University, Milan, 2005.
2. European Union Regulation 98/2008.
3. AAVV "The relationship between Chlorine in Waste Streams and Dioxin emissions from Waste Combustor Stacks", American Society of Mechanical Engineers, New York, 1995.
4. P. Shepherd "Polyvinyl Chloride Plastics in Municipal Solid Waste combustion impact upon dioxin emissions",–Solid waste Association of North America, Maryland 1993.
5. Sheng-yong LU, Jian-hua YAN, Xiao-dong LI, Ming-jiang Ni, Ke-fa CEN and Hui-fen DAI "Effects of inorganic chlorine source on dioxin formation using fly ash from a fluidized bed incinerator" Journal of Environmental Science, 19, 6, 2007, pp 756–761.
6. Masaki Takaoka, Takashi Yamamoto, Atsuhiro Shiono, Nobuo Takeda, Kazuyuki Oshita, Tadao Matsumoto and Tsunehiro Tanaka "The effect of copper speciation on the formation of chlorinated aromatics on real municipal solid waste incinerator fly ash", Chemosphere 59, 10, 2005, pp 1497-1505.
7. V. V. Boldyrev, Mechanochemistry and Mechanical Activation of Solids, Imperial College Press; 1st edition (February 2006) 250 pp.
8. P. Plescia, G. Martinelli, S. Benedetti, D. Gizzi, "Mechanochemical dissociation of Calcium Carbonate: laboratory data and relation to natural emission of CO_2", Physics of the earth and Planetary interiors, 2004, vol. 142, i.3-4, pp 205-214.
9. G.Paoloni, P.Plescia, "Mechanochemistry as treatment method of inertization: applicability to the asbestos' wastes", Faculty of Chemical Engineering– University of studies "La Sapienza" Rome, Thesis 1999.
10. S. A. Rowlands, A. K. Hall, P. G. McKormick,"Destruction of toxic materials" Nature, vo. 367- 368, 1984.
11. S. Louiselle, M. Branca, G. Mulas, G. Cocco, Selective mechanochemical dehalogenation of clorobenzenes over calcium hydride", Env. Science and Technology vol. 31-1 pages 261-265, 1997.

12. Field L.D.; Sternhell S.; Wilton H.V. "Mechano-chemistry of some Hydrocarbons" Tetrahedron, 53, 11, 1997, pp. 4051 - 4062.

13. Q. Zhang, F. Saito, K. Shimme, S. Masuda, "Dechlorination of PVC by a mechanochemical treatment under atmospheric condition" J. Soc. Powder Technol. Jpn 36, pages 468-473 (1999).

Chapter 14

Application of Hydrometallurgical Process for Metals Recovery from Exhaust Batteries

***Stefano Ubaldini*[1], *Pietro Fornari*[1], *Ida De Michelis*[2] and *Francesco Ferella*[2]**

[1]Istituto di Geologia Ambientale e Geoingegneria (IGAG - CNR) Area della Ricerca di Roma RM 1 - Via Salaria Km 29,300 C.P. 10 00016 Monterotondo Stazione–ROMA

[2]Dipartimento di Chimica, Ingegneria Chimica e Materiali, Università degli Studi di L'Aquila, 67040 Monteluco di Roio, L'Aquila, Italy

ABSTRACT

The application of hydrometallurgical processes have been experimented starting from spent alkaline batteries. Main aim of the research was to develop a preliminary experimental work to recover selectively MnO_2 and Zn from leaching solution containing Mn and Zn, both at concentration of 10 g dm^{-3}, by electrolysis, with the scope to apply electrochemical technologies for Zn and Mn recovery from exhaust batteries. The results have demonstrated the technical feasibility of the application of the mentioned technologies; in fact, high recoveries of manganese and zinc dissolved in solution were achieved (97 per cent MnO_2 and 98.50 per cent Zn).

Keywords: *Exhaust batteries, Hydrometallurgy, Electrowinning, Zn, Mn.*

Introduction

This experimental work involves the utilisation of technologies well know in the framework of the mineral processing and the metal dissolution by hydrometallurgical processes for Mn–Zn recovery from wastes. The wastes were constituted from spent alkaline batteries.

Disposal of spent batteries constituted a serious environmental problem in terms of heavy metals content when these devices are disposed off in inadequate way (Vegliò *et al.*, 2003; Beolchini *et al.*, 2009).

Alkaline and zinc-carbon batteries represent about 80 per cent of the market. In particular, alkaline batteries, have an average content of manganese dioxide (MnO_2) of 37 per cent, while zinc content is of 16 per cent. The high Mn and Zn content justify the interest to recover the metals (Vatistas *et al.*, 2001). Main processes applied are the pyrometallurgical processes, that involve high energetic consumption and pollution problems, and hydrometallurgical processes, chipper and at low energetic cost. Hydrometallurgical processes are generally characterised by different steps of pretreatment followed by leaching and metal separation (Ubaldini *et al.*, 2000; Vegliò *et al.*, 2003).

In this work, hydrometallurgical processes have been applied starting from spent alkaline batteries. The experimental work was focalised on Zn and Mn electrochemical recovery from liquid synthetic solutions (Ubaldini *et al.*, 2008). The composition of the alkaline batteries utilised was of 35 per cent MnO_2 and 16 per cent Zn.

Main aim of the research was the development of a preliminary experimental work to recover selectively MnO_2 and Zn from synthetic leaching solution containing Mn and Zn, both at concentration of 10 g dm^{-3}, by electrolysis, with the scope to apply electrochemical technologies for Zn and Mn recovery from exhaust batteries.

The results have demonstrated the technical feasibility of the application of the mentioned technologies.

In all the operative conditions, a MnO_2 and Zn deposit at high grade of purity has been achieved; this fact was demonstrated from the results of the analyses conducted by diffractometric techniques.

Materials and Methods

Pre-treatment and Characterization

Spent alkaline batteries were manually dismantled. Dismantling products such as plastic films, ferrous scraps and paper pieces were discharged and mixture of the cathodic (manganese oxides) and the anodic (zinc oxides and electrolytic solution) materials in form of powder was extracted (Vatistas *et al.*, 2001). The powder was dried ground, using a ball mill, sieved and analysed by X-ray diffraction (XRD), utilising an automatic diffractometer Bruker mod. D8 Advance. The results show that the following main species are presents, in order of decreasing abundancy: Zincite (ZnO), Quartz (SiO_2), Manganese Oxide (Mn_2O_3), Calcium Manganese Oxide

($Ca_4Mn_2O_7$), Sylvite (KCl) and Potassium Oxide (KO_2). Quantitative chemical analysis was conducted by Inductively Coupled Plasma Mass Spectrometry (ICP-MS) utilising an Perkin-Elmer apparatus mod. Elan 500, with data station.

Results have showed that powder was composed mainly by 35 per cent MnO_2, 14 per cent Zn, 4.9 per cent K, 0.65 per cent Ni and 0.6 per cent Fe.

Iron represents the major impurity when Zn and MnO_2 are deposited in the electrowinning step, so it has been precipitated by an alkaline reagent.

Preparation of the Synthetic Solution

With the aim to develop a preliminary electrowinning study, synthetic solutions have been prepared utilising reagents of high analytical degree of purity (RPE Carlo Erba), taking account of the powder composition and of the concentration of reagents such as sulphuric acid and lactose, utilised for chemical leaching. Lactose was added to favour the manganese reduction (Ubaldini *et al.*, 2005). Distilled water was used for preparation and dilution of the stock solution.

The concentration of synthetic solution utilised was of 10 g dm^{-3} Mn and of 10 g dm^{-3} Zn, in H_2SO_4 0.5 M (Ubaldini *et al.*, 2005). Initial pH of the solution was of 0.15.

Electrowinning

A cylindrical glass laboratory cell of 200 cm^3 in volume has been used. The cell was equipped by a thermostatic water jacket connected with thermostatic bath, magnetic system for stirring and three electrodes: as cathode, a platinum mesh; as anode, a spiral of platinum wire; as reference, a saturated calomel electrode (SCE). During MnO_2 deposition, the working electrode (W) was the anode and the cathode was the counter electrode (C). During Zn deposition, the W was the cathode, while the anode was the C. In both processes, the reference electrode was constituted by SCE, to permit an electrolytic process at controlled potential.

The cell was connected to a potentiostat-galvanostat (AMEL mod. 555B), equipped with an instrument system to automatically control the process parameters at constant potential conditions. By analysing the residual solution with ICP-MS, the purity of the MnO_2 and Zn deposition was monitored, while the deposit was analysed by XRD.

Electrowinning experiments were carried out in potentiostatic operations at constant anodic potential and stirring conditions (see Table 14.1). The kinetics of MnO_2 and Zn deposition were studied by selected voltages (Ubaldini *et al.*, 2006).

In this way, the voltage between working and reference electrodes was automatically monitored.

The corresponding voltages (V) between working and counter electrodes (cell voltages) was measured with differential electrometer. Average experimental cell voltages were utilised to calculate the energy consumption during the tests.

The current flowing through the electrolytic cell was automatically integrated. To calculate the current efficiency and the average energy consumption, experimental data were utilised (Ubaldini *et al.*, 2006).

Table 14.1: Experimental Conditions for MnO_2 and Zn Electrodeposition.

Factors	*MnO_2*	*Zn*
Anode Vs. SCE (V)	1.2-1.8	1.45
Cell Voltage (V)	2.2-2.8	3.8
Current Int. (mA)	1000-1800	1200
Curr.Den. (mA/cm^2)	10-18	12
pH	0.47-0.59	5
Temperature (°C)	95	95
Time (h)	8	8
Agitation Cond.(rmp)	200	200

Results and Discussion

Synthetic solution prepared for electrowinning study was constituted from a manganese and zinc sulphate solution, together with other dissolved metals; for this reason solution has been purified before the electrowinning section. Purification has been carried out by precipitation-cementation (data not reported here).

Main observations consequent to the experimental texts for electrolytic selective recovery of MnO_2 from solution containing Zn (electrolytic kinetic of MnO_2 at potential controlled), have been the following:

- ✰ On the basis of data achieved during the preliminary electrodeposition texts, the experiments should be conducted at temperature of 95°C, at constant potential (Figure 14.1), at fixed value of 1.8 V (Anode - SCE) and corresponding cell potential (anode - cathode) of 2.6-2.8 V.
- ✰ After 120 minutes of electrolysis 97.35 per cent of MnO_2 has been recovered, with average faradic yield of about 48 per cent.
- ✰ Energetic consumption after 120 minutes was of 3.73 kWh/kg of MnO_2 recovered.
- ✰ Zinc remains quantitatively in solution.

By Zn electrolysis of a synthetic solution containing only Zn sulphate, after MnO_2 cathodic deposition, about 98.50 per cent of metallic Zn was achieved, with a faradic current efficiency of 72.43 per cent, after 2 hours of electrolysis.

A summary of the main results achieved during electrodeposition tests for selective recovery of MnO_2 and Zn from synthetic solution containing Mn e Zn, both of concentration of 10 g dm^{-3}, by electrolysis, is reported in Table 14.2.

Table 14.2: MnO_2 and Zn Electrowinning: Main Results.

Species	*Recovery (per cent)*	*η (per cent)*	*E (kWh/kg)*
MnO_2	97.35	48.00	3.73
Zn	98.50	72.43	4.28

η: faradic current yield; E: Energetic consumption.

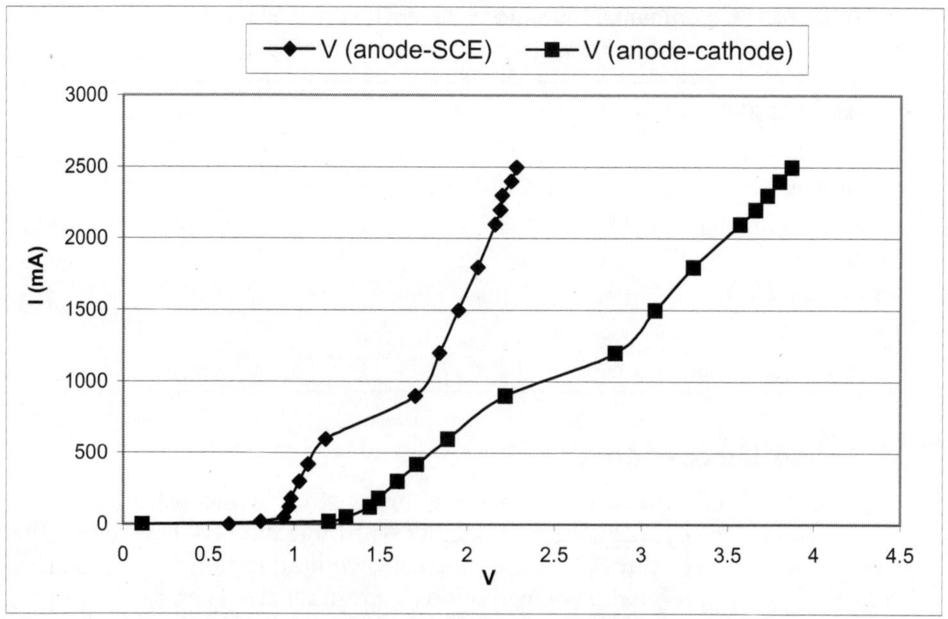

Figure 14.1: Measurement of the Decomposition Potential (Vd) of the Synthetic Solution Containing 10 g dm^{-3} Mn and 10 g dm^{-3} Zn.

Acknowledgements

This work was supported by Italian National Council Researches under CNR Project n. 6 - 1 - 132.48.2 and under project CNR RSTL n. 0042.0005. Moreover, the authors are grateful to Prof. F. Veglio for the suggestions given in the course of experimental activity and to Mr. R. Massidda and Ms. Emanuela Tempesta for helpful collaboration during the experimental work.

Conclusions

Results have demonstrated the technical feasibility of the application of electrochemical technologies for Zn and Mn recovery from exhaust batteries.

A quantitative recovery of manganese (over 97 per cent MnO_2) and zinc dissolved in solution (98.50 per cent Zn) has been achieved.

References

Beolchini, F., Fonti, V., Ferella, F., Ubaldini, S., Vegliò, F., 2009. "*Nickel, vanadium and molybdenum extraction from exhaust LC-finer catalysts by biohydrometallurgical technologies*", Proceedings of the 1st International Conference Biotechnologies and Metals, Edited by Jana Kadukova, Alena Luptakova, Maria Prascakova, Hedviga Horvathova, pp. 11-14, Kosice, Slovak Republic.

Ubaldini, S., Veglio', F., Toro, L., Abbruzzese, C., 2000, "*Combined bio-hydrometallurgical process for gold recovery from refractory stibnite*", Minerals Engineering, Vol.13, n.14-15, pp.1641-1646.

Ubaldini, S., Fornari, P., Abbruzzese, C., Vegliò, F., 2005, " *Electrochemical recovery of MnO_2 from solutions containing waste biomasses*", Proceedings of the 7th European Symposium on Electrochemical Engineering (EFCE), pp. 499-504, Toulouse, France.

Ubaldini, S., Abbruzzese, C., Fornari P., Luptakova, A., Prascakova, P., 2006,"*Electrowinning and Bioremediation processes for toxic metals removal from Acid Mine Drainage*", Acta Metallurgica Slovaca, 12, pp. 405-410.

Ubaldini, S., Abbruzzese, C., Fornari, P., Luptakova, A., Massidda, R., Vegliò, F., 2008. *Electrohydrometallurgical recovery of Zinc and Manganese from spent batteries*, Acta Metallurgica Slovaca, Vols.14, pp. 262-267.

Vatistas, N., Bartolozzi, M., Arras, S., 2001, "*The dismantling of spent alkaline line manganese dioxide batteries and the recovery of the zinc from the anodic material*", Journal of Power Sources, 101, 182 187.

Veglio', F., Quaresima, R., Fornari, P., Ubaldini, S., 2003, " *Recovery of Valuable Metals from Electronic and Galvanic Industrial Wastes by Leaching and Electrowinning*", Waste Management, Vol. 23, Issue 3, pp. 245-252.

Chapter 15

Pesticide Hazards, Monitoring and Control

H.S. Rathore

Applied Chemistry Department, Z.H. College of Engineering and Technology Aligarh Muslim University, Aligarh

ABSTRACT

Pesticides are any substance or a mixture of substances used for preventing, estroying, repelling, or mitigation of any pest. They may be biopesticides, antimicrobials, and pest control devices. Pests may be insects, bacteria, fungi, and viruses. The process of chemical crop-protection is profit-induced poisoning of the environment.The poisoning action of pesticides can be divided into four modes namely; (a) Physical Poisons, (b) Nerve Poisons, (c) Protoplasmic Poisons, and (d) Respiratory Poisons. Thus the pesticides affect every one *i.e.* they affect your health, your family's health and your community's health. A detailed description of pesticide hazards, monitoring and control is described in this chapter.

Keywords: *Pesticides, Hazards, Monitoring, Prevention, Control.*

Causes of Pesticides Hazards

Pesticide Stockpile

It is a common phenomenon around the world.Stockpiles are obsolete pesticides/ chemicals that can no longer be used for the desired purpose and so it requires disposal. The risk of contamination due to container leakage pose serious dangers to the community living in the area as well as to the passersby.

Pesticides Exposure

Farmers and workers are regularly exposed to pesticides. Children are also exposed. Non target objects are exposed in three ways;

(a) *Exposure Through the Skin:* It is through spills on clothing or directly on the skin by touching the treated flora and fauna, by mixing pesticides etc.

(b) *Exposure Through Breathing:* It is very common for workers who spray pesticides or for people as well as other animals stay nearby when the spray is carried out. Hence it is important to note that some pesticides have no distinguished smell.

(c) *Exposure Through Swallowing:* It occurs by eating or drinking contaminated food or water, by eating food with out washing hands or without changing clothes and by inhaling contaminated air.

In most cases, workers are not aware of hazards of pesticides and they handle them without proper precautions. Pesticides are poisons which enter our body through different pathways and once it is in, there is a very little chance of getting the pesticide out of the body system.

Hazard Based Classification of Pesticides

The World Health Organization (WHO) has classified the pesticide as follows:

Ia. Extremely Hazadous- colour code is RED.

Ib. Highly Hazardous-colour code is RED.

II. Moderately Hazrdous-colour code is YELLOW.

III. Slightly Hazadous-colour code is BLUE.

IV. Not Hazardous- colour code is GREEN.

This classification does not say anything about long-term chronic effects? For example, endosulfan, is classified as "MODERATELY HAZARDOUS", can cause severe long-term health effects such as cancer, mental disorders and endocrine disruption.

The Food and Agriculture Organization of the United Nations (FAO) recommends that WHO Ia and Ib pesticides should not be used in developing countries, and if possible II should also be avoided.

Lethal Dose of Pesticides

Class I: 2-3 Drops or one pinch

Class II: One tea spoon

Class III: Two tea spoons

Class IV: One cup or One glass

Acute Health Effects of Pesticides

Acute Toxicity

It is also known as acute poisoning. It occurs when the poisonous effects of pesticide are realized with in no time.

Chronic Toxicity

These are the poisonous effects which are realized after a long period of time say after months or some times after years of exposure to the pesticide.

Some acute health effects are headache, dizziness, nausea, chest pain, vomiting, skin rash mussel pain, excessive sweating, cramp, diarrhoea, difficulty in breathing, blurred vision, and ultimate death.

(a) *Local Acute Effects:* These are the effects which affect the part of the body that is in direct touch of pesticide. For example, dryness, burning, redness, and itching of eyes, throat, and skin, watering of eyes, and cough. The skin problems such as redness, itching, burning, rashes, blisters and discolouration. If fingernails and toenails turn black or blue the pesticide poisoning is confirmed. In the case of acute poisoning the nails will even fall off.

(b) *Systemic Acute Effects:* These effects occur by getting the pesticides in the body in- turn affects the whole body. Since the pesticide contaminates blood, all the sophisticated organs such as eyes, heart, lungs, stomach, intestines, liver, kidneys, muscles, brain, and nerves are affected.

(c) The symptoms of systemic poisoning and the quickness of poisoning depend on the period of exposure to the pesticide and the degree of acuteness of toxicity of the pesticide.

(d) Some well known pesticides which are causing acute skin diseases are given below:

- ☆ *Fungicides:* Benomyl, Captafol, Captan, Chlorothalonil, Copper Sulphate, Dicloran (DCNA), Dinobuton, Dinocap, Ditalimfos, Dithianon, Dyrene, Folpet, HCB, Maneb, Mancozeb, Organotins, PCNB, PCB, Sulfur, Thiophanate, Thiram, Triphenyltin, and Zineb.
- ☆ *Herbicides:* 2, 4-D, Alachlor, Allidochlor, Amitrole, Atrazine, Barban, Diquae, Glyphosate, MCPA, Nitralin, Paraquate, Phenmedipham, Propachlor, and Pyrazon.
- ☆ *Insecticides:* Arsenic, Cryolite, Dienochlor, Dinitrophenol, Naled, Parathion and Pyrethrins.
- ☆ *Fungicides:* Dazomet, Dichloropropene, Ethylene oxide, Metam Sodium and Methyl bromide.

Note: If rash appears on exposure to pesticide and it disappears on removing, it confirms that it is due to pesticide.

(e) Chronic Effects of Pesticides

1. *Neurotoxic Pesticides:* The pesticides which harm nervous system are called neurotoxins. Some pesticides are very harmful to the brain and nerves. Commonly known symptoms of organic brain diseases are severe memory problems, difficulty in concentration, changes in personality, paralysis, seizures, unconsciousness and coma.
2. *Liver Defects:* Liver breaks down the toxic chemicals into less harmful substances. The pesticide harms the liver that can lead to toxic hapatitis.
3. *Stomach Disorderness:* The pesticide poisoning is causing vomiting, stomach aches and diarrhea. Long exposure to pesticide is causing serious stomach problems such as loss of apatite. If one swallows the pesticide there will be terrible damage to the stomach. It is due to the damage of stomach walls.
4. *Immune System Defects:* Some pesticides have been found to disturb the immune system of our body. So the body enables to protect us from illness including cancer, and infectious diseases such as HIV, AIDS tuberculosis, cough and cold etc., allergic reactions etc.
5. *Hormones Disbalance:* Hormones are chemicals produced by organs such as the brain, thyroid, parathyroid, kidneys, adrenals, testes, and ovary. These chemicals control the important body functions. Some pesticides affect reproductive hormones causing decreased sperm production in the male and abnormal egg development in the female. Some pesticides can cause thyroid enlargement which could lead to thyroid cancer.
6. *Pesticides Exposure Causes miscarriage:* Mother's exposure to some pesticides during the first month of pregnancy can cause a miscarriage.
7. *Pesticides Exposure Causes Stillbirth:* Mother's exposure to pesticides during pregnancy is related to the occurrence of stillbirth Parents' exposure to pesticides before a baby is conceived can cause a stillbirth.
8. *Pesticides Exposure Causes Sterility in Men:* Some pesticides are known to cause sterility in man. Pesticides that harm sperm cells are called spermatotoxines or reproductive toxins.
9. A Husband Working With Pesticides Affects His Pregnant Wife Who Is Not Dealing With Pesticides.
10. Babies Are Exposed To Pesticides By The Breastfeeding.
11. *Pesticides Cause Birth Defects:* The U.S. Environment Protection Agency has the following list of some commonly used pesticides which are known or suspected to cause birth defects:

(a) *Herbicides:* Acrolein, Bentazine, Cyanazine, Bromoxylin, 2,4-D, Dinocap, Dinoseb, Diquat, Nitrofen, Picloram and 2,4,5-T.

(b) *Fungicides*: Benomyl, Captafol, Folpet, HCB, Mancozeb, Maneb, Tributyltin oxide, Tributyltin fluoride and Triphenyltin acetate.

(c) *Insectcides:* Avermectin, Chlordimeform, Endosulfan, Ethion, Phosmet, Methyl parathion, Mirex and Trichlorfon.

12. *Pesticides Cause Cancer:* There are some carcinogenic pesticides which are called Carcinogens. The chances of getting cancer are more if one is exposed to cancer causing pesticides than those who are not exposed to pesticides. U.S. Environmental Protection Agency (EPA) as published the following list of the pesticides which are know or suspected to cause cancer.

(a) *Insectcides:* Arsenic pentaoxide, Cadmium, Chlordane, Chlordimeform, DDT, Dichlorvos, Dieldrin, DDVP, Heptachlor, Dipropyl Isocinchomeronate, Ethoprop, Fonoxycarb, Lindane, Oxythioquinox, Para-dichlorobenzene, Propargite, Propoxur, Pyrethrins, Thiodicarb, and Trichlofon.

(b) *Fungicides:* Captafol, Captan, Chlorothalonil, Folpet, HCB, Maneb, Mancozeb, Iprodione, Metiram, Silica aeroge, Thiophanate-methyl and Vinclozolin.

(c) *Herbicides:*Acetochlor, Acifluorfen, Alachlor, Amitrole, Oxadiazon, Arsenic acid, Cacodylic acid, Diuron and Propyzamide.

(d) *Fumigants*: DBCP, EDB, Dichloropropane, Dichloropropene, Ethylene sodium, Metam sodium, and Propylene Oxide.

(e) *Others*: Arsenic trioxide, Chromic Acid, Creosote, Daminozide, Formaldehyde, Ortho-phenylphenol, Pentachlorophenol, Potassium dichromate, S, S, S-tributyl phosphorotrithioate and Sodium dichromate.

13. Malnutrition enhances the toxicity of the Pesticides: People who have insufficient food intake are prone to have higher risk of pesticides exposure. In rats, that were short on proteins the pesticide poisoning become much more:

    DDT- 4 times as toxic

    Carbaryl- 8 times as toxic

    Lindan- 12 times as toxic

    Endosulfan-20 times as toxic

    Captan-2100 times as toxic

14. Some mixtures of pesticides are more dangerous than their parent pesticides.

15. The following pesticides which are similar to Nerve Gas are more Acutely Toxic because they attack cholinesterase, a substance responsible for the proper functioning of our nervous system:

(a) *Organophosphates*: Azinophosmethyl, Demeton methyl, Dichlorvos (DDVP), Disulfoton, Ethion, Ethyl parathion (Parathion), Fenamiphos, Fensulfothion, Methamidophos, Methidathion Methyl parathion, Mevinphos, Phorate, Sulphotepp and Terbufos.

(b) *Methylcarbamates*: Aldicarb, Carbofuran, Fomentanate, Methomyl, Oxamyl and Propoxur.

Monitoring of Pesticide Poisoning

1. *Symptoms of Mild Poisoning by Nerve Gas Type Pesticides:* (i) Abdominal pain, (ii) Blurry vision, (iii) Chest pain, (iv) Diarrhoea, (v) Dizziness, (vi) Excessive sweating, (vii) Headache, (viii) Muscle pain and cramp, (viii) Nausea and vomiting and (ix)Increased secretion from the eyes, nose and mouth.
2. *Symptoms of Moderate Poisoning by Nerve Gas Type Pesticides:* (i) Confusion, (ii) Problem in walking,(iii) Problem in concentration, (iv) General weakness, (v) Muscle twitching and (vi) Small pupils.
3. If the gradual increase in poisoning, symptoms will be as follows: (i) Sleeping trouble, (ii) Nightmares and (iii) Restlessness
4. *Symptoms of Severe poisoning by nerve gas type pesticides:* (i) Unconsciousness, (ii) Involuntary urination and defecation, (iii) Coma, (iv) Small pupils (marked miosis), (iv) Blue color of lips and nailsbeds (cyanosis), (v) Difficulty in breathing, (vi) Convulsions and (vii) Death if severe poisoning in not recognized and treated in time.
5. Symptoms of Poisoning by Methyl Bromide:

   (i) Headache, (ii) Dizziness, (iii) Nausea, (iv) vomiting, (v) Shakiness, (vi) Walking problem and (vii) Convulsions.
6. Symptoms of Poisoning by Organochlorines:

   (a) Symptoms of Acute Poisoning:

   (i) Nausea, (ii)Vomiting, (iii) Headache, (iv)Dizziness, (v) Agitation/ Confusion, (vi) Muscle spasms, (vii) Muscle weakness and (viii) Convulsions, (b) Symptoms of Chronic Poisoning: (i) All above, (ii) Pallour, (iii) Skin lesions, (iv) Sterility, (v)Heart problems, (vi)Liver problems and (vii) Loss of consciousness.
7. *Symptoms of poisoning by parquet:* (i) Irritation and injury to eyes, skin, nose and throat, (ii) Skin and throat ulcers, (iii) Nose Bleeding, (iv) Failure of liver and kidney and (v) Scarring of lungs causing death from suffocation.
8. *Symptoms of poisoning by endosulfan:* (i) Damage of red blood cells, thyroid, kidneys, liver, muscles and the developing foetus, (ii) Hepatotoxic, (iii) Genotoxic, (iv) Mutagenic, (v) Clastogenic, (vi) Tumour promoter and (vii) inhibits immune function.
9. *Cholinsterase test:*The level of cholinesterase goes down by poisoning of organophosphate or carbamate pesticides *i.e.* nerve gas type pesticides. There are two tests one for cholinesterase in red blood cells and another for cholinesterase in plasma. The results are as follows:

   (i) 100 per cent of base line-Normal for you

   (ii) 75 per cent of base line -Mild poisoning–Usually no symptoms

   (iii) 50 per cent of base line-Moderate poisoning-May have symptoms

   (iv) 25 per cent of base line–Severe poisoning–Usually have headache, dizziness, nausea, vomiting, sweating etc.

   (v) 0 per cent of base line–Unconscious-Can die if treatment is not given.

Control of Pesticides Hazards

1. Take Caution
   (a) Don't handle pesticides with your bare hands
   (b) Don't smoke when handle pesticides
   (c) Don't face the wind when spraying pesticides
   (d) Don't consume pesticides–treated fruits/food
   (e) Don't eat any thing before washing your hands
   (f) Don't store pesticides in home specially in places where foodstuffs and stored
   (g) Don't use pesticides for your own treatment such as treatment of head lice
2. If you feel that you are poisoned by pesticide
   (a) Stop working with the pesticide immediately.
   (b) Change your clothes and remove the pesticide by washing with water and soap
   (c) Inform your family members, co-workers etc. to take you for medical care
   (d) Consult the trained technical manpower in this area
   (e) Follow the occupational health and safety regulations for pesticide exposure in agricultural workers etc.
3. If pesticide splashes in eyes
   (a) Open eyelids with your fingers and rinse eye/eyes immediately with running water at least for 30minutes.
   (b) Never use eye drops because it is not antidotes.
   (c) Consult heath worker as soon as possible. Take container of the pesticide with you to show it to the health worker.
4. If pesticide spills on the skin
   (a) Wash off the pesticide with soap immediately.
   (b) Remove contaminated dresses.
   (c) Consult the health worker, keep the pesticide container with you to show the health worker.
5. If some one swallows pesticide
   (a) Lay down the person with the head lower than body. Turned to one side to prevent aspiration in case of vomiting.
   (b) Clean the mouth and eyes or bathe the patient if possible.
   (c) Help the person to vomit by tickling the back of the patient's throat. Vomiting is dangerous for unconscious or convulsing or the patient suffering by heart diseases.

(d) If patient wants water give him a lot of water, avoid milk alcohol or any other foodstuffs.

(e) take the patient to the health worker do't forget to take pesticide container.

6. The best control is to make available the well trained health worker, round the clock.
7. Drugs that can be used is atropine. It is an antidote for the treatment of poisoning by organophosphate and methylcarbamate. It is a drug that counteracts the effects of a poison. Another drug is pralidoxime (2-PAM) that is recommended as an antidote for some organophosphates' poisoning. It is effective only if it is given soon after the poisoning occurs. Drugs should be given by the trained Health worker in the prescribed procedure.
8. Improvc thc cconomic condition of the farmers

(a) Offer the farmers "some thing extra" through government–supported prices or subsidies for their commodities.

(b) Provide the increase in net farm income from free trade in which they have the highest prices and lowest cost.

(c) In India, just after independence, cooperative agricultural farms were developed to give training and demonstrations to the farmers. The farmers were also paid to compensate their loss of time in training. This process may be revived again to provide training for the proper use of pesticides.

References

Books

1. AK De. Environmental Chemistry,4th ed., New Age International (P) Limited, New Delhi, 2003.
2. ASY Chau, BK Afghan, JW Robinson (Eds.).Analysis of Pesticide in Water, Vols. I, II, III.CRC Press, Boca Raton, FL., 1981.
3. CA Edwards (Ed).Environmental Pollution by Pesticides, Plenum Press, New York, 1973.
4. JL Tadio(Ed).Analysis of Pesticides in Food and Environmental Samples, CRC Press, 2008.
5. J Stenersen. Chemical Pesticides:Mode of Action and Toxicology, CRC Press, 2004.
6. Leo ML Nollet, Hamir S. Rathore (Eds). Handbook of Pesticides:Methods of Pesticide Residues Analysis, CRC Press: Taylor and Francis Group, 2009.
7. Michael A. Kamrin (Ed.).Pesticide Profiles:Toxicity, Environmental Impact, and Fate, CRC Press:Taylor and Francis Group, 1997.
8. S Yedia, AK Dixit. Abatement of Pestcide Pollution: Removal of Organochlorine Pesticides from Water Environment, Narosa, New Delhi, 2004.

9. Y Ming–HO. Environmental Toxicology: Biological and Health Effects of Pollution. 2nd edn., CRC Press, Boca Raton, FL, 2005.

Reviews

1. Q-B Chen, YL Ji. Safety evaluation and risk management for modern chemical pesticides. Yunnan Nongye Daxue Xueboo 20(1):99-106, 2005.
2. R. Pandey. Pesticides and sterility. Everyman's Science XXXVIII (2):84-86, 2006.

Papers

1. C Dharmani, K Joya. Epidemiology of acute organophosphate poisoning in hospital emergency room patients.Rev Environ Health 20(3):215-232, 2005.
2. K. Haerkoenen. Pesticides and induction of aneuploidy in human sperm. Cytogenet Genome Res 111(3-4):278-283, 2005.
3. N Xue, X Xue, Z. Jin. Screening of 31 endocrie-disrupting pesticides in water and surface sediment samples from Beijing Guanting reservoir. Chemosphere 61(11):1594-1606, 2005.
4. SS Nielsen and BA Mueller, AJ De Roos, HMA Viernes, FM Fariri, H Checkoway. Risk of brain tumors in children and susceptibility to organophosphorous insecticides: The potential role of paraoxonase (PONI). Environ Health Perspect 113 (7):909-913, 2005.
5. YC Yuan, HC Chen, YK Yaun. Sublethal effects of paraquate and malathion on the freshwater shrimp, *Macrobrachium nipponense*. Acta Zool Taiwan 14(2):87-95, 2004.
6. YJ Wu, L Yang Wli. Multitarget toxicity of organophosphorus pesticides. Huanzing Yu Zhiye Yixue 22(4): 367-370, 2005.

Chapter 16

Environmental Assessment of Fly Ash from some Thermal Power Stations for Reclamation of Mined Out Areas

Gurdeep Singh

Professor and Head
Centre of Mining Environment,
Department of Environmental Science and Engineering,
Indian School of Mines, Dhanbad

ABSTRACT

Coal occupies an important position in energy sector in India since India has vast reserves of thermal grade coal. Combustion of coal to generate electricity produces large volumes (>120 million tonnes) of coal combustion residues (CCRs) particularly fly ash as Indian coals contain high ash. These fly ash dominated CCRs accumulates in on-site piles and ash ponds leading to serious environmental problems, particularly contamination of ground and surface waters due to leaching of trace elements. There is a widespread concern to enhance the utilisation of fly ash considerably. One of the major areas of utilisation of fly ash can be as back fill material for mine reclamation. This study was carried out to evaluate the long-term leaching of trace elements for assessing the high volume utilisation of fly ash in mined out areas reclamation.

The leaching characteristics were evaluated through various shake (ASTM 30 Days, TCLP, SLP) and open column percolation experiments. It has been established that open percolation column experiments are the only leaching methodology to evaluate the actual leaching behaviour of trace elements from coal/fly ashes in real world situations. Overall the leaching characteristics envisaged that fly ash generally is environmentally benign as there does not occur significant leaching of trace elements as determined though long-term open-column leaching experiments.

Environmental assessment through open percolation column experiments carried over three (03) years established the fly ash from various thermal power stations as environmentally benign material. Field studies from actual ash ponds also validate this observation. High volume fly ash utilisation in mine reclamation is gaining momentum on the basis of initial success achieved in some areas. Overall this study indicates that fly ash can be disposed and utilised in mined out areas in an environmentally acceptable manner.

Keywords: *Fly ash, Trace elements, Leaching, Shake test, Open column percolation experiments, Mine reclamation.*

Introduction

Electricity is an essential need of any industrial society and no nation can progress without adequate supply of power. Growth in its demand during the past decades has been phenomenal and has outstripped all projections. In India also, there has been impressive increase in the power generation from a low capacity of 1330 MW in 1947 at Independence to about 81,000 MW at end of March 1995 (Trehan *et al.*, 1996) and at about 1,20,000 MW at the end of August 2005 (India Energy Forum, 2005). However, despite this substantial growth there still remains a wide gap between demand and supply of power which is expected to worsen in the years and decades to come. We already are experiencing shortage of nearly 8 per cent of the average demand and 16.5 per cent of peak demand. Moreover, with the quantum jump expected in demand for power in the future due to rapid industralisation and changing life styles of populace as a result of economic liberation, shortages shall further increase unless immediate steps are taken to increase power production. It has been estimated that 1,40,000 MW of additional power would be required to meet such demand by the end of 10th five year plan *i.e.*, year 2007 A.D.(Trehan *et al.*, 1996).

India has vast reserves of coal and it is expected that this shall remain as prime source of energy in the early part of 21st century. Presently, thermal power stations account for about 70 per cent of installed generation capacity of 81000 MW (Sampath *et al.*, 1996). In a pulverised coal fired thermal power plant about 40-55 per cent ash is produced. About 20 per cent falls down due to gravity and is removed as bottom ash and the remaining fly ash. This fly ash is collected by mechanical and/or electrostatic precipitator. As per the available estimates the production of coal ash in India including both fly ash and bottom ash is about 120 million tonnes per annum which is likely to touch 170 million tonnes per annum by 2012 A.D. Thus, the utilisation of fly ashes has drawn considerable concern and attention of scientists, technologists, environmental groups, government, regulators etc.

Studies of trace elements and the elements present in fly ash are distributed into the fractions of the fly ashes based on volatilisation temperature (Bahor *et al.*, 1981). It is found that elements appear to partition into three main classes.

1. Elements that are not volatilised and reported equally in both fly ash and bottom ash. These elements include Al, Ba, Ca, Ce, Co, Cu, Fe, Hf, K, La, Mg, Mn, Rb, Se, Si, Sm, Sr, Th, Ta, and Ti.
2. Elements that are volatilized on combustion and preferentially get adsorbed on the fly ash as flue gas cools down. These include As, Cd, Ga, Mo, Pb, Sb, Se and Zn.
3. Elements that remain almost entirely in the volatilised state tend to escape to the atmosphere as vapours. These are Hg, Cl and Br.

Fly ash is an alumino silicate glass consisting of the oxides of Si, Al, Fe and Ca with minor amounts of Mg, Na, K, Zn and S and various trace elements. The concentration associated with the ash may be either adsorbed on the surface of particle or incorporated into matrix (Natusch *et al.*, 1974). A mechanism that appears to be common for all ashes during their formation is the condensation of metal and metalloid vapours on refractory core materials. As the ash particles and gas stream exist from the combustion chamber and proceed upto the flue gas, this results in locally higher concentrations of many trace elements at the surface of ash particles and accounts for the generally higher concentration of these elements as particle size decreases (Markowski and Filbly, 1985). The association between trace elements and major elements/minerals may be an important factor in determining the leachate composition of water in contact with ashes.

It is recognised that the health hazards and environmental impacts from coal fired thermal power stations result from the mobilization of toxic elements from ash. The large amount of ash that accumulates at thermal power plants, its possible reuse and the dispersion and mobilization of toxic elements from it, requires greater attention (Palit et.al., 1991). Mobilization of various elements from the ash into the environment depends on climate, soils, indigenous vegetation and agricultural practices (Page et. al., 1979).

Leaching Aspects of Fly Ash

The leaching characteristics of fly ash are mainly controlled by factors such as its chemical composition, mineralogy and morphology. Fly ash from thermal power plants vary in chemical composition not only from plant to plant but also within the same plant. Chemically fly ash consists of Si, Al, Mg, Ca, K, Ti and Fe in greater proportion with many trace elements as V, Mn, Cr, Cu, Ni, As, Pb, Cd and smaller quantity of various potential toxic elements. Chemical composition study shows mostly the presence of four major elements *viz.* aluminium, silicon, iron and calcium in the fly ash. Other such as potassium, magnesium, barium, cobalt, cadmium, zinc, molybdenum, lead etc. are present in traces. Though in the traces, compared to the original coal, most of the elements are enriched in the fly ash, giving birth to the growing environmental concerns in the disposal and utilisation environment due to release of trace/heavy elements metals.

Crystalline phases in fly ash as major constituent is quartz and others such as mullite, magnetite, haemetite etc. are also present as minor constituents. The mineralogy of fly ash refers to both amorphous and crystalline phases or other mineral fractions. The leaching behaviour of fly ash can be related to their mineralogical characteristics. Most of the toxic elements reside on major phases and are easily available during leaching.

Morphologically, particles of fly ash may be irregularly shaped particles, solid spheres and cenospheres. Morphology of fly ash similarly, controls the leaching of toxic elements/heavy metals. Morphology of fly ash is related to the conditions of combustion and composition of inorganic materials in coal. Particle morphology is one of the most important morphological aspects of fly ash. In this particle size particle distribution also holds its importance. Finer particles mean more surface area and so more concentration of condensing elements. In fact, specific surface area of fly ash is determined by both particle size and particle morrphology.

The leaching of fly ash is a time dependent phenomenon. The initial leaching of the fly ash can be characterized by the surface hydrolysis and the dissolution of reactive phases formed under high temperature combustion. A close examination of the leaching studies show a rapid early dissolution followed by a later, slower release of the elements. The water-soluble fraction of a combustion residue may reflect the early dissolution process in the natural environment. In fact, the early dissolution mainly involves the soluble salts or the oxides on the particle surface of the fly ash. So, the dominant features of the initial dissolution stage are a high dissolution rate and the solution chemistry being controlled by buffering components of the fly ash. At this stage of high dissolution rate, release of salts or heavy metals associated with surface phases occur. The long term leaching of the fly ash occurs primarily in the alluminosilicate glass and some acid-soluble magnetic spinel phases. These phases usually cosntitute the matrix of the fly ash. The dominant leaching features of the matrix phases are a low dissolution rate and phase alteration over a along period of time.

Present study on environmental characterisation is in continuation with earlier studies, to evaluate leaching of trace elements from coal ashes from a few Thermal Power Stations situated in eastern India are presented in this paper (Gurdeep Singh and Sanjay Kr, 1996, 1999, 2000; Gurdeep Singh and Kumari Vibha, 1999). The possible water contamination is also envisaged through the leachate analysis from ash pond disposal sites in real life situation.

Experimental Methodology

Sampling

Fly ash samples from Thermal Power Stations were collected on five different days from each of the power plant over a week and a final homogenised sample was prepared while appropriately mixing various portions. Leachate samples were periodically collected twice a month from the final discharge of the ash pond which receives fly ash (80 per cent) and bottom (20 per cent) in slurry form from the plant.

Leaching Chemistry

Short-term leaching (shake) tests and open column percolation leaching experiments were carried out on the fly ash samples to ascertain its leachate chemistry as briefly described below.

Strong Acid Digest Test

This is short term leaching study and is carried out in the presence of strong acids to provided the available concentration levels of trace/heavy elements in the samples.

For the purpose of the experiment 0.5g fly ash sample is taken in a conical flask and to it is added 10ml nitric acid and 2ml perchloric acid. Mixture is then heated till dryness on hot plate. The conical flask is covered during the process of heating by funnel. The dried residue in the conical flask is then boiled with 2ml HNO_3 and then filtered. This is repeated with distilled water, warmed up and filtered till no residue is left in the conical flask. The filter paper in the silica dish is kept in the muffle furnace and heated to 850°C. Silica dish is then allowed to cool with residue in it and residue is collected for further analysis. The filtered solution is stored in polypropylene beaker for further analysis (elemental).

ASTM Shake Test

This is the rigorous and short term leaching study. This test is run for twenty four hours and deionized water is used as the leaching medium. Shake tests can only be useful to a limited extent, and with variability in results in identifying the elements most likely to leach out of a material and to estimate the equilibrium constants for some of the reactions that takes place during the test. However, shake tests are poor indicators of the conditions that might be expected in the field.

In this test 80gm fly ash sample is taken and put in the measuring bottle and to it is added two litre of distilled water. The bottle with the sample is then taken in the rotary agitator and rotated for twenty-four hours and this way sample with water is agitated for thorough mixing to result leaching. The extraction is performed in triplicate on each fly ash sample and the three replicates are then mixed to get the composite leachate sample. The leachate so obtained is filtered using Whatmann No. 42 filter paper. The filtrate so collected is then stored in the polypropylene bottle for further analysis. Once the potentiometric analysis is over, few drops of 6N nitric acid are added to the leachate collected in polypropylene bottle to avoid contamination and the bottle stored to be used for further elemental analysis of the leachate by AAS.

30-days Shake Test

This test is similar to 24-hr shake test and is carried over a period of 30 days and is intended to indicate the solubility of elements that may reach equilibrium with the surrounding water more easily. This test is required by some regulatory agencies of some of the States in the USA.

Toxicity Characteristics Leachate Procedure

The toxicity characteristic leachate procedure (TCLP) requires the use of an extraction fluid made of buffered acidic medium to run the test. For this the selection of the extraction fluid is made prior to conducting the test. Once the extraction fluid to be used in the test is determined, 40g fly ash sample is taken and then extraction fluid equal to twenty times the amount of sample taken is added in the zero head extractor under pressure. The system is tightly closed and then placed in an end-over-end rotary shaker for 18 hours, rotating at 30±2 rpm at a room temperature of about 25°C. Leachate after the said period of shaking is pressure filtered using 0.7 micron pore size filter paper.

Modified Synthetic Leachate Procedure (SLP)

This test is a modified SLP rather than a standard SLP. Here, unlike the standard SLP test which make use of a mineralized synthetic leaching medium prepared from deionized water, water from an actual field site (containing ions and impurities similar to those found in groundwater in an area of interest) is used. This test aims to detect the ion exchange reactions that can only be observed in the field like ion bearing water. This test is run for 24 hours.

ASTM Column

In the ASTM column procedure, one pore volume of distilled water is forced through a packed column of fly ash each day in a saturated upflow mode. Leaching in this column is conducted under a nitrogen atmosphere and thus present leaching in an oxygen poor environment. In the field, this type of leaching would occur below the water table where there is low concentration of oxygen. One pore volume corresponds to the void space between the material grains in the test column. The rate of percolation of the leaching medium regardless of the hydraulic conductivity of the material is controlled by applying variable nitrogen pressure. The test is run for 16 days and leachate collected after 1, 2, 4, 8, 16 days of leaching.

Open Percolation Column Experiments

In these experiments, deionized water is percolated through a packed column of fly ash in the presence of oxygen at a rate which depends on the natural permeability of the material. The open columns for leaching experiments were made of PVC pipe four inches in diameter and two feet in length. The column setup involved packing the coal ash material at optimum moisture and density conditions as determined by the Proctor test (ASTM Standards, 1990). The fly ash material was packed into the column in two inch lifts with a 2" x 2" wooden rod, about 4 feet long. Each packed layer was scarified, by lightly scraping the top of the packed layer with a long thin rod to ensure proper interlocking of the material. The top six inches of the column was left unpacked to allow for the addition and maintenance of the leaching medium. About 200 ml of leaching medium (de-ionized water) was added at the top of the column once every alternate day to maintain sufficient supply of water to the packed coal ash material. The top end of the column was exposed to the atmosphere and the bottom end was connected to quarter inch tubing. The columns discharged the leachates through this tubing into the 250 ml polypropylene beakers. The leachates were collected in these beakers and analysed.

Elemental Analysis of Leachates

The leachate samples were filtered and acidified with 2 ml of nitric acid and then preserved in polypropylene sampling bottles. The samples were kept in a refrigerator until further analysis. Sodium and potassium were determined using Systronics flame photometer. Concentration levels of trace elements were evaluated using Atomic Absorption Spectrophotometer (AAS) GBC-902. Working/standards solutions were prepared according to instructions given in the operation manual of the GBC-902 AAS.

Optimized operating conditions such as lamp current, wave length, slit width, sensitivity, flame type etc. as specified in the manual, were used for analysis of a particular element.

Merck-AAS standards were used for standardisation and calibration of AAS. Three standards and a blank of the concerned element were used to cover the range 0.1-0.8 Abs. The calibration was performed by using the blank solution to zero the instrument. The standards were then analysed with the lowest concentrations first and the blank was run between standards to ensure that baseline (zero point) has not changed. Samples were then analysed and their absorbance recorded. The calibration was performed in the concentration mode in which the concentration of sample was recorded.

Results and Discussion

Comparative evaluation of short-term leaching (shake) tests is presented in Table 16.1. Acid digest data provides the available concentration levels of Trace elements in fly ash whereas in comparison to this shake tests resulted significantly lower concentration levels particularly in 24 hr. ASTM, 30 Days and SLP shake tests. TCLP leachate data however, gives rise to significantly higher levels as this involved leaching in slightly acidic buffered conditions and as such does not reflect the actual behaviour. It may be emphasised that these shake test presented the accelerated leaching because of 1:20 solid to liquid (leachant) ratio. Nevertheless these short term (shake) leaching tests provide an immediate and rapid indication of leachable concentration levels of trace elements from fly ash.

Analysis of twenty two elements were carried out from each of the leachate samples collected from open column experiments and the observations are summarised in Tables 16.2 and 16.3 for fly ash, pond ash and actual ash pond leachates, respectively. It is noticed from the observations that the concentration of thirteen elements, namely, chromium, nickel, cobalt, cadmium, selenium, aluminum, silver, arsenic, boron, barium, vanadium, antimony and molybdenum were below the detection limit (.001 mg/l) in the entire study period. Among the other nine elements only calcium and magnesium were observed in the leachates throughout the study period while the concentration of other elements showed a decreasing trend to below detection limit (.001 mg/l). In the leachates from actual ash ponds, lead and manganese were found absent but iron, calcium, magnesium, sodium, potassium, copper and zinc were present throughout the study period.

A comparison of the concentration levels observed in the leachates of fly ash, pond ash and also leachates from actual ash pond disposal site with the permissible limits as per IS:2490, is presented in Tables 16.2 and 16.3 which indicates that the concentration levels of all the elements during the entire study period were either below detection limits (BDL) or below the permissible limits.

Table 16.1: Comparative Leachate Analysis Results of Shake Tests for Fly Ash.

Parameter	*Acid digest*	*TCLP*	*24-hr*	*30-D*	*SLP*	*SLP Blank*
pH	—	4.29	6.22	6.26	7.08	7.06
Conductivity	—	3.56	0.096	0.099	0.075	0.085
TDS	—	1.78	48	63	51	58
Iron	82.41	0.089	0.045	0.05	0.029	0.038
Lead	BDL	BDL	BDL	BDL	BDL	BDL
Magnesium	7.579	7.512	3.15	4.50	2.4	2.8
Calcium	304.00	304.13	3.37	5.12	63	70
Copper	0.094	0.215	BDL	BDL	BDL	BDL
Zinc	0.276	2.140	0.020	0.025	0.180	0.185
Manganese	0.638	0.314	0.031	0.030	0.021	0.028
Sodium	54.60	1452	39.80	41.10	6	10
Potassium	7.60	8.10	2.80	5.36	3	4
Chromium	0.860	0.803	BDL	BDL	BDL	BDL
Nickel	0.118	0.112	BDL	BDL	BDL	BDL
Cobalt	BDL	BDL	BDL	BDL	BDL	BDL
Cadmium	BDL	BDL	BDL	BDL	BDL	BDL
Selenium	BDL	BDL	BDL	BDL	BDL	BDL
Aluminium	BDL	BDL	BDL	BDL	BDL	BDL
Silver	BDL	BDL	BDL	BDL	BDL	BDL
Arsenic	BDL	BDL	BDL	BDL	BDL	BDL
Boron	BDL	BDL	BDL	BDL	BDL	BDL
Barium	BDL	BDL	BDL	BDL	BDL	BDL
Vanadium	BDL	BDL	BDL	BDL	BDL	BDL
Antimony	BDL	BDL	BDL	BDL	BDL	BDL
Molybdenum	BDL	BDL	BDL	BDL	BDL	BDL
Mercury	BDL	BDL	BDL	BDL	BDL	BDL

BDL: Below Detectable Limit; Concentration of Elements in ppm; TDS in ppm; Conductivity in mmhos/cm.

It can be inferred that no significant leaching occurs and toxicity is manageable with respect to trace elements both in the ash pond disposal site as well as in the open column leaching experiments. Further, analysis results of leachates from open column percolation experiments resemble closely with those of actual ash pond leachates. The physical set up of the open columns more closely resembles with because the flow of the leaching medium is influenced by gravity alone and the solid to liquid ratio is more close to the field situation. Hence, open column leaching experiments may be used in predicting the long term leaching behaviour that can be observed in the field. Fly ash leachates as generated from open percolation column leaching experiments and those from ash pond disposal site closely resemble and as such do not pose any significant environmental impacts in the disposal system. Overall, fly ash would not seem to pose any environmental problem during its utilization and/or disposal.

Table 16.2: Summary of the Leachate Analysis of Fly Ash from Thermal Power Stations # 1.

Parameter	*Open Percolation Column Experiments Leachates*			*Ash Pond Leachate*	*(IS: 2490, 1981)*
	Samples				
	FA#A	*FA#B*	*PA*		*Inland Surface Water*
pH	5.97-10.51	5.82-9.10	5.86-9.03	6.95-8.26	5.5-9.0
Conductivity	0.042-0.750	0.037-0.820	0.052 0.920	543-796	–
TDS	21-375	19-410	30-460	272-400	2100
Iron	BDL-0.740	BDL-1.220	BDL-1.369	0.89-1.983	–
Lead	BDL-0.420	BDL-0.396	BDL-0.490	0.121-0.462	0.1
Magnesium	BDL-15.53	0.039-38.00	0.065-44.00	17-29	–
Calcium	1.00-103.92	0.265-189.20	0.798-102-20	21-58	–
Copper	BDL-0.190	BDL-0.068	BDL-0.090	0.023-0.055	3
Zinc	BDL-0.380	BDL-0.372	BDL-1.529	0.295-1.763	5
Manganese	0.009-0.057	0.010-0.105	0.007-0.076	0.027-0.089	–
Sodium	3-56	3-49	3-47	19-43	–
Potassium	2-42	2-36	2-33	7-51	–
Chromium	BDL	BDL	BDL	BDL	2
Nickel	BDL	BDL	BDL	BDL	3
Cobalt	BDL	BDL	BDL	BDL	–
Cadmium	BDL	BDL	BDL	BDL	2
Selenium	BDL	BDL	BDL	BDL	0.05
Aluminium	BDL	BDL	BDL	BDL	–
Silver	BDL	BDL	BDL	BDL	–
Arsenic	BDL	BDL	BDL	BDL	0.2
Boron	BDL	BDL	BDL	BDL	2
Barium	BDL	BDL	BDL	BDL	–
Vanadium	BDL	BDL	BDL	BDL	–
Antimony	BDL	BDL	BDL	BDL	–
Molybdenum	BDL	BDL	BDL	BDL	–
Mercury	BDL	BDL	BDL	BDL	0.01

BDL: Below Detectable Limit; Concentration of Elements in ppm; TDS in ppm; Conductivity in mmhos/cm.

Table 16.3: Summary of the Leachate Analysis of Fly Ash from Thermal Power Stations # 2

Parameter	*Open Percolation Column Experiments Leachates*			*Ash Pond Leachate*	*(IS: 2490, 1981)*
	Samples				
	FA#1	*FA#2*	*PA*		*Inland Surface Water*
pH	4.98-9.92	4.38-8.90	4.71-8.92	7.2-8.58	5.5-9.0
Conductivity	0.060-0.962	0.070-0.848	0.036-0.973	645-892	–
TDS	30-481	35-424	30-487	320-445	2100
Iron	BDL-3.850	BDL-3.120	BDL-3.120	1.02-2.941	–
Lead	BDL-0.098	BDL-0.080	BDL-0.249	–	0.1
Magnesium	BDL-37.9	BDL-36.4	BDL-21.0	10-19	–
Calcium	1-87.6	2-72.2	2.12-48.0	18-46	–
Copper	BDL-0.094	BDL-0.088	BDL-0.052	0.011-0.047	3
Zinc	BDL-1.082	BDL-1.100	BDL-1.290	0.93-1.015	5
Manganese	BDL-0.099	BDL-0.092	BDL-0.069	–	–
Sodium	BDL-48	BDL-23	BDL-82	5-10	–
Potassium	BDL28	BDL-36	BDL-18	8-18	–
Chromium	BDL	BDL	BDL	BDL	2
Nickel	BDL	BDL	BDL	BDL	3
Cobalt	BDL	BDL	BDL	BDL	–
Cadmium	BDL	BDL	BDL	BDL	2
Selenium	BDL	BDL	BDL	BDL	0.05
Aluminium	BDL	BDL	BDL	BDL	–
Silver	BDL	BDL	BDL	BDL	–
Arsenic	BDL	BDL	BDL	BDL	0.2
Boron	BDL	BDL	BDL	BDL	2
Barium	BDL	BDL	BDL	BDL	–
Vanadium	BDL	BDL	BDL	BDL	–
Antimony	BDL	BDL	BDL	BDL	–
Molybdenum	BDL	BDL	BDL	BDL	–
Mercury	BDL	BDL	BDL	BDL	0.01

BDL: Below Detectable Limit; Concentration of Elements in ppm; TDS in ppm; Conductivity in mmhos/cm.

Concluding Remarks

On the basis of the study of the leaching of trace elements from coal ashes, following conclusions can be drawn:

1. In the study period of about three(3) years there was practically no leaching of thirteen elements namely, chromium, nickel, cobalt, cadmium, selenium, aluminum, silver, arsenic, boron, barium, vanadium, antimony and molybdenum from all the ash samples.
2. Out of the nine elements found in the leachates only calcium and magnesium were found to be leaching in the entire period. The leaching of other seven elements namely, iron, lead, copper, zinc, manganese, sodium and potassium was intermittent. The leaching of sodium and potassium practically stopped due to first flash phenomenon after 35 and 40 days, respectively. It is emphasised long-term leaching results should be considered to arrive at the environmental screening of such materials.
3. The concentration of the elements in the leachates was invariably well below the permissible limits for discharge of effluents as per IS:2490 and also for drinking water as per IS:10500.

Overall, the fly ash samples from various Thermal Power Stations evaluated in this study were found to be environmentally benign and can be engineered for their bulk utilization particularly for mined out areas reclamation and for soil amendment for good vegetation.

The Centre of Mining Environment at ISM Dhanbad is currently engaged in evolving low technology high volume field demonstration to show that fly ash particularly fly ash can be disposed and utilized as fill material in an environmentally acceptable way in reclamation of abandoned mines (Bradley C. Paul, Gurdeep Singh *et al.*, 1995, 1998). Fly ash has been successfully used as backfill in material in reclamation of mined out (goaf) area and at the top of the surface Helipad is set up, at Jamadoba Tata Steel Mining Area. This has attracted a lots of public attention as a result of aesthetic and scenic value provided in the Eco-Park thus resulted. Similarly a considerable quantity of fly ash has also been utilised at Ghantotand OB Dumps and Damoda worked out opencast mine sites. At these sites trace elements leaching even after three years of monitoring does not seem to pose any environmental problem. With these encouraging results cooperative arrangements are being made by power utilities and mining authorities for utilisation of fly ash in reclamation of various mined out areas in SECL, MCL, NCL, BCCL, SCCL etc.

Use of fly ash as backfill material for reclamation of mined out sites provide benefits such as easy availability, cheaper to transport because empty coal carriers returning from the power plant can "back haul" it to the mine site. From the standpoint of the power plant, this is essentially a waste material which requires large costs of handling and a disposal to comply with environmental regulations. From the environmental point, this waste material will go back to the same place where it was mined and use of this material serves as extra benefit to power plants. Studies are also in progress to use fly ash for agriculture development.

References

1. Annual Book of ASTM Standards, vol. 11, 10, 1990.
2. Annual Book of ASTM Standards, Vol. 04.01, 1991.

3. Atomic Absorption Spectrophotometer Operating Manual, AAS, GBC-902, Australia. 1990.
4. Bahor, M.P., Mclaren, R.J., Niece and Pedersen, H.C., Coal Ash Disposal Manual, Second Edition, Electric Power Research Institute, Final Report (CS-2049), 1981.
5. Bradley C. Paul, and Gurdeep Singh, Environmental evaluation of the feasibility of disposal and utilization of coal combustion residues in abandoned mine sites. Proceedings, First World Mining Environment Congress. Dec. 11-14, 1995. N. Delhi. Oxford and IBH Publ. Co. Pvt. Ltd. New Delhi. pp.1015-1030, 1995.
6. Bradley C. Paul, Gurdeep Singh and Chaturvedula, S. 1995. Use of FGD by products to conrol subsidence from underground mines: groundwater impacts. Proceedings International Ash Utilization Symposium Lexington, Ky, USA, October 23-25, 1995.
7. Bradley C. Paul, Gurdeep Singh, Steven Esling, Chaturvedula and Pandal, H. The impact of scrubber sludge on groundwater at an abandoned mine site. Environmental Monitoring and Assessment, 50:1-13, 1998.
8. Carloon, C. L. and D. C. Adriano, Environmental Impacts of Coal Combustion Residues, Journal of Environmental quality 22 pp227-247, 1993.
9. Gurdeep Singh and Kumar Sanjay, Environmental evaluation of coal ash from Chandrapura Thermal Power Stationof Damodar Valley Corp. Indian J. Environmental Protection. 18(12) 884-888, 1999.
10. Gurdeep Singh and Kumar Sanjay, Environmental evaluation on leaching of trace elements from coal ashes: a case study of Chandrapura Thermal Power Station. Journal of Environmental Studies and Policy 2(2): 135-142, 2000.
11. Gurdeep Singh and Kumari Vibha 1999. Environmental assessment of fly ash in its disposal environment at FCI, Ltd. Sindri, Poll. Res. 18(3) 339-343.
12. Gurdeep Singh and Sanjay Kumar Gambhir, Environmental evaluation of fly ash in its disposal environment. Proceedings, International Symposium on Coal - Science Technology, Industry Business and Environment, Nov. 18-19, 1996. editors. Kotur S. Narsimhan and Samir Sen. Allied Publishers Ltd. New Delhi. pp.547-556, 1996.
13. India Energy Forum, Power India Year Book-2005-06.
14. Jain, R. K., Environmental Assessment of Coal Combustion By-products of Burnpur Thermal Power Station, M. Tech. Thesis Submitted to Indian School of Mines, Dhanbad, 1998.
15. Kumari Vibha, Environmental Assessment of Fly Ash in its Disposal Environment at F.C.I, Sindri, Jharkhand (India), M. Tech. Thesis, Submitted to Indian School of Mines, Dhanbad, 1998.
16. Markowski, G.R. and Filbly, R., Trace element concentration as a function of particle size in fly ash from a pulverised coal utility boiler. Environmental Science and Technology, vol. 19, pp. 796-800, 1985.
17. Ministry of Coal, Annual Report, 2005.

18. Ministry of Petroleum and Natural Gas, Annual Report, 2005.

19. Ministry of Power, Annual Report, 2005.

20. Natusch, D.F.S., Wallace, J.R. and Evans, C.A., Toxic trace elements: preferential concentration in respirable particles. Science, Vol. 183, no. 4121, pp. 203-204, 1974.

21. Page, A.L., Elseewi, A.A. and Straughan, I.R., Physical and chemical properties of fly ash from coal fired power plants, Res. Review, vol. 71, pp. 83-120, 1979.

22. Palit, A., Gopal, R., Dube, S.K. and Mondal, P.K. Characterisation and utilisation of coal ash in the context of Super Thermal Power Stations. Proceedings International Conference on Environmental Impact of Coal Utilisation, IIT, Bombay, pp. 154-155, 1991.

23. Sampath, R. "Fly Ash -its pollution and potential for utilisation", Seminar on Fly Ash Utilization, New Delhi, 26-27 March, 1996.

24. Singh A. P., Environmental Evaluation of Coal Combustion Residues of Jamadoba Fluidised Bed Combustion Plant of TISCO with Special Emphasis on Stabilisation of Soil, M. Tech. thesis submitted to Indian School of Mines, Dhanbad, 2000.

25. Trehan. A., Krishnamurthy, R., Kumar. A. "NTPC'S experience in ash utilization" Seminar on Fly Ash Utilization, New Delhi, 26-27 March, 1996.

26. http://www.cea.nic.in

27. www.energymanagertraining.com/energy_overview_COAL.htm

28. www.teriin.org

Chapter 17

Management of Air Pollutions in Paint Shops: A Case Study

M.Z. Hasan, Manju Agarwal and M.M. Agarwal

Global Environmental and Agriculture Centre
HIG–79, Sector–E, Aliganj, Lucknow

ABSTRACT

Paints are chemical compounds which give a particular colour to the surface where it is applied. Paints are used both for better look and protection of surface of appliances and walls. In an industry the area where painting is done is known as paint shop. Paints are actually smooth blend of toxic chemicals. Paints contain many volatile organic compounds (VOCs) which may cause several aliments and general discomfort. VOCs can be removed using solvent recovery, solvent abatement and bio-filtration technique. An air pollution control system using incineration with scrubber and cyclone was studied and observed to be effective in controlling air pollution.

Keywords: *Paint, VOC, Toxic, Emissions, Bio-filtration, Abatement, Incinerator, Cyclone, Stack.*

Introduction

Paint has become an integral part of our lives. Paints are used both for better look and protection of surface of walls, doors, body of appliances and over all surface of a machine. In a house everybody likes to see fresh, colored soothing wall rather than tired looking walls. We prefer to keep good looking house appliances. In the selection of an appliance, its colour plays an important role. We select our vehicles on the basis of their coloure also.

Every factory has an unit for painting. The end process of manufacture of home appliance automobiles and even railways wagons bogies and engines, is painting unit. This unit is called a paint shop. Paint shops use several types of paint and apply them on the surface of the finished products most often by spray painting. This practice gives rise in the builtup of high level of various components ofpaint in the work environment. Though the workers engaged in the painting process use mask and aprons, yet accidental exposure can not be completely avoided. Paint shops utilize air pollutions control system to reduce air pollution level in the work environment.

Paints

Paints are chemical compounds which give a particular colour to the surface where these are applied. The colored surface gives soothing feeling to the eye and at the same time restricts the deterioration of the surface where it is applied. In fact a fine layer of paint is formed on the surface of the body of the product. The product thus finds a good look. The colour is given seductive environment friendly names like, Natural green, 'Red Rose', Prime Yellow or 'Sunset pink'. These colors give false sense of eco-purity. Paints are actually smooth blend of toxic chemicals. They generally include the colour carried by a resin and or binder, a solvent to help the paint application, and a dryer. There are paints based on vinyl and acrylic compounds. These paints also comprise polymers. Several paints contain formaldehyde, trace metals, thinners and foamers.

Toxic Effect of Paints

Paints contain many volatile organic compounds (VOCs) that outgas and escape into the air after they are applied on the surface. These VOCs present in the paint cause headaches, nausea, achey bones and general discomfort. The paint also triggers allergies and asthmatic reactions irritate skin, eyes and air ways and exherts increased stress on vital organs such as the heart. The World Health Organization (WHO) has reported a 20 per cent - 40 per cent increased risk of certain types of carcinoma, in particular lung cancer for those who come in to regular contact with paint. Recently researchers pointed to wards the added possibility of neurological damage.

Volatile Organic Compounds in Paints

The US Environment Protection Agency (USEPA) puts paints on its list of environmental hazardous. The main hazard from paints is from volatile organic compounds (VOCs). VOCs include chemicals such as benzene, formaldehyde, toluene, xylene etc. Some of these chemicals have been shown to cause cancer or neurological problems. Paint solvents are of particular concern to people with heart conditions because methanol and certain other compounds stress the heart. Those who bear contact lenses suffer from irritation lenses absorb strong vapours and hold them against eyes.

Emissions of VOC

In a painted building when new paints dries, the level of indoor VOC shots up 1000 folds of the out door levels of VOC. In newly developed residential areas paint emissions from houses and other structures in the region gives off more VOC vapours than its emissions from gas stations and refineries. The reasons why paints give off so many is because of the VOCs evaporate into the air easily and because of the large surface areas typically covered by paint. Paints are usually applied in occupied in occupied spaces and, therefore, people are more likely to be exposed to the vapours from paint surfaces. The VOC emissions from paints can continue even six months after application. The main sources of these VOCs are gloss paint and wood strain varnishes in newly constructed homes. The largest emissions of VOCs occur mainly from the solvents in paints.

Different Types of Paints

Paint with excessive level of VOCs poses grave health problems. In view of this many types of paint have been made which have reduced toxic effects. Some of these paints are described.

Low–VOC Paints

It is preferable to use paint with low VOCs to reduce toxic aspects of the paint. Convention oil based paints contain about 50 per cent petrochemical by weight. VOCs are found in these paints at about 430 parts per gallon. By contrast most water based latex paints contain about 5–15 per cent petrochemicals, but they still may contain harmful solvents. In order to be considering low VOC paint, the VOC levels have to be less than 100 parts per gallon. Low–VOC interior wall paints are generally to flat, eggshell, and semi glass finishes in white and soft pastels. Brighter shinier colors require the use of protection means that gives off VOCs. Low VOC on paint label also means that manufactures has not exceeded a certain level of cancer causing chemicals. USEPA has already prescribed limit for level of VOCs. The claim of manufacturers that the paint is free from VOCs is never correct. In fact all paints contain VOCs to some extent.

Other Kind of Paints

Latex paint are also available in low biocide varieties which means they are about 95 per cent free of the preservatives and fungicides of ten added to water based paints to fend off melden and mold oil paints do not need VOCs added to them sense they are naturally toxic to mold and milden.

There are all natural paints available which do not contain any petrochemical products, but they may contain naturally occurring VOCs such as citrus based solvent d, l–limonene, turpentine, tung oil or pine resins. These natural VOCs cause reactions such as watery eyes or respiratory problems in people sensitive to these chemicals. Most natural paints use solvent from citrus or other plant oils. They also contain plant resins, finely ground minerals and earth pigments. These paints are costly, take longer duration to dry and require more coats than latex paint.

The least toxic of all paints are milk paints and white washes. Milk paints are made with milk proteins earth pigments and clay. These paints are free of solvent. The areas painted with these paints, however, may need to be painted more often. White washed contains lime paste, water and salt. These paints are preferably used on plaster, cement or stucco walls.

VOC Control Process

The use of large amount of solvents for coating, painting and finishing purposes, poses a grave health hazardous problem. Generally the workers are provided masks and aprons but the work environment levels of VOC need to be under control. VOCs are controlled using either solvent recovery or solvent abatement techniques.

Solvent Recovery Techniques

The techniques utilize separation, condensation and adsorption. Separation is based on different process like osmosis, gas permeation, gel permeation, dialysis and electro dialysis. Separation systems for their relative simplicity are an interesting solution at least for low streams. Condensation is appropriate with boiling paint about 400C and concentration above 5000ppm. Absorption is a gas stream scrubbing that takes place in side tower as to optimize the mass exchange between gas and liquid phases. Liquid may be atomized, misted or vaporized. After the scrubbing, liquid is treated and solvent recovered. In adsorption system gas stream is brought in to contact with activated carbon or other solids that are able to physically attract and hold onto their pours surface the compound present in the gas stream.

Abatement Process

Thermal oxidation is performed in an incinerator operating at high temperature (1500°C and above). High temperature combustion causes undesired nitrogen oxides production. Abatement is also done by catalytic oxidation at relatively low temperature of 800°C. Catalysts are poisoned by gases and require periodic replacement and proper disposal

Biofiltration

In a biofilter, the waste gas is forced to rise though a layer of packing material colonized by microbial communities, in up and down flow configuration. As a contaminated gas stream passes through the reactor, target pollutants are transported into the bio film where they are utilized as carbon or energy source by the micro-organisms. Under aerobic conditions, they are able to oxidize the compound concerned into eater and carbon dioxide.

A Case Study

A study was undertaken in a paint shop for air pollution control. The paint shop was at the end of a continuous manufacturing process of big appliances. The chamber was closed and appliances were continuously entering the shop hanging on rolling wheels with chain. The appliances were painted by spray painting. Workers were provided with safety gagets like masks and aprons. Arrangement for fitting filters on two walls roof and bottom surface were made. Powerful exhaust fans were attached with each filter. Arrangement for flushing fresh air was also made. Filters were periodically removed and send to air pollution control system. The system comprises an incinerator, a wet scrubber, a cyclone and a stack. The exhaust of incinerator was allowed to enter to the scrubber where water was trickling from the top. The out let of treated emissions from scrubber entered the cyclone and finally it was released to the atmosphere through the stack. A bye pass was also provided from scrubber to stack. Sampling port was provided on the stack for monitoring.

Filters were brought from paint shop and fed to incinerator. The power was switch on and filters were incinerated for an optimum time previously determined by repeated experiments. SPM, CO, NOX and SOX were monitored in the stack. The ash from incinerated chamber and liquid waste from scrubber were collected and analysed. Stack emission quality for five cycles is given in Table 17.1.

Table 17.1: Stack Emission Quality (ppm).

Cycle No.	*SPM*	*CO*	*NO_x*	*SO_x*
1	76	95	22	24
2	62	90	24	20
3	72	100	25	26
4	78	95	24	26
5	68	90	20	20

Observations and Conclusion

Stack emission data showed that the system was capable of reducing the air pollution considerably. The characteristics of ash and liquid waste (Tables 17.2 and 17.3) suggested that ash should be disposed using practice specified for hazardous waste. The liquid waste is to be given primary treatment for the removal of heavy metals and then mixed with sewage for further treatment.

Table 17.2: Characteristics.

Batch	*Cd*	*Cr*	*Cu*	*Ni*	*Heavy Metal*	
					Pb	*Zn*
1	0.5	1.0	5	0.8	2.0	5.0
2	0.4	1.2	8	0.6	2.0	8.0
3	0.5	1.0	7	0.8	2.5	7.5
4	0.4	1.0	6	0.7	3.0	5.0
5	0.5	1.0	8	0.8	2.5	6.5

Table 17.3: Physico-chemical Characteristics of Liquid Waste,

Parameter	*Batch*				
	1	*2*	*3*	*4*	*5*
pH	6.9	6.7	6.8	7.0	6.9
TDS	505	490	480	510	520
Alkalinity as $CaCO_3$	220	230	240	210	240
Chloride as Cl	45	50	42	46	42
SO_4^{--}	48	46	46	52	50
Cd	0.005	0.005	0.004	0.005	0.005
Cr	0.01	0.01	0.02	0.01	0.01
Cu	0.5	0.4	0.5	0.5	0.5
Pb	0.05	0.05	0.04	0.05	0.05
Ni	0.04	0.04	0.05	0.04	0.05
Zn	2.0	1.0	1.0	1.5	1.0

All values are in mg/L except pH.

References

1. YuC, crump D: A review of the emission of VOCs from polymeric materials used in buildings. Building and Environment 33(6), 357–374 (1998).
2. USEPA: Evaluation of low–VOC latex paints, inside IAQ: EPA indoor Air Quality Research update, EPA/600/N–98/003 Fall/Winter, (1998).
3. Anon: The right paint, Mother Earth News, June/July, 80–1, (1999).
4. Ghosal, A.K. and Manjare S.D. selection of appropriate adsorption technique for recovery of VOCs: an analysis, Jour of loss prevention in the Process Industries 15,413–421 (2002).

Chapter 18

Water Availability and Quality Assessment using Hydrological Model for Mirik Lake, Darjeeling

A.K. Biswas[1], Pankaj Roy[2], Asis Majumder[2], Amal Kanti Mishra[2] and T.K. Ghosh[1]*

[1]CSIR-National Environmental Engineering Research Institute, Nagpur

[2]Jadavpur University, Kolkata

ABSTRACT

Mirik lake is situated in the Kurseong sub-division of Darjeeling district of West Bengal spread over an area of 0.1618 km^2 having water area of 0.005732 km^2 (or about 3.5 per cent of the land area). It is located at an attitude of 1767 m above MSL in approximately midway between Siliguri and Darjeeling road. The sources of pollution of Mirik Lake may be classified by point and non-point source. The total 72 kg/day of pollution load drained from surface with carrying municipal/industrial sewage for the entire Mirik Lake. In the present study, the software package Hydrologic Engineering Centre, Hydrologic Modeling System and Neuro Dimention 5 on Artificial Neural Network (ANN) have been used for calibration, simulation, validation, and optimization of runoff for Mirik Lake, Darjeeling. Also, Due to paucity of discharge data, the ANN model was used for determination of historical discharge data. Daily rainfall and stream flow data of ten years (January 1998 to March 2006) for the Mirik Lake has been used for the calibration and validation in the HEC-HMS model. The calibration error, expressed as a per cent difference of simulated value with respect to observed value, was found to lie within the range of 0.1 to 10 per cent for the lake. The present study is an attempt to assess the water availability and quality control which have immense importance in domestic and agricultural scenario as well as eco-tourism in north eastern India with the help of hydrological model. The results depict that the per capita surface water availability for the population dependent on the Mirik Lake has gone down from the present scenario of 45 m^3/capita/year in 2006 to 25 m^3/capita/year during the years 2010 to 2020, by keeping the growth of projected population as 2 per cent per decade as depicted in UN (1998). The study reavealed that most of the time quality of lake water exceeds the CPCB, class 'B' standards, thus require the adoption of techno-economically viable technology to revive the Mirik Lake.

Keywords: *HEC-HMS, ANN, Water availability, Water quality, Validation, Calibration.*

Introduction

Poor land management practices are causing accelerated sedimentation of water bodies. As natural sinks with longer residence times than rivers, the lakes, which are relatively more fragile ecologically than the rivers, and susceptible to degradation from such threats. Unless these threats could be controlled and/or reduced through better management practices, which also call for appropriate investment decisions, humanity may altogether lose access to these readily available sources with better water quality. The rising water consumption has also resulted increase in the release of wastewater which in one way and other is contaminating the surface sources. Many of the freshwater sources, *e.g.* lakes, reservoirs, ponds, streams and rivers, etc. are threatened more and more by the unabated discharge of the polluted water from various sources including agricultural washout. Therefore, lakes need special attention as they are quite often the catalysts in the development of the region by supplying water for variety of uses. However, many lakes are reported to have undergone quantitative and qualitative degradation in the last few decades. Since behaviour of the lake ecosystem depends to a large extent on its hydrological regime, understanding the hydrological regime of the lakes is very significant to develop strategies for their conservation, management and rejuvenation. Generally eutrophication is the main cause of lake degradation. High nitrate and phosphate content in water leading to extreme growth of producers, causing degradation of water resources is referred to as eutrophication. Transparency of water will be destroyed by eutrophication and the lake water becomes turbid.

This chapter deals with the water availability and Quality Prediction using Hydrological model namely HEC-HMS and ANN model for Mirik lake, Darjeeling and suggested remedial measures for quality and quantity assurance of lake water for public use.

Study Area

Mirik Lake (26°54'N and 83°13'E) is situated in the Kurseong sub-division of Darjeeling district in West Bengal. It is located at an altitude of 1767 m above Median Sea Level (MSL) in approximately midway between Siliguri and Darjeeling on the road.

The development of Mirik is mainly oriented around the Lake at Mirik. The topography of the area is such that the whole area is sloping towards the Lake and as a result the wastewater and the surface run-off reach to the Lake. The Mirik Lake is the principal source of drinking water after conventional treatment and disinfection for Mirik town. The western bank of the Lake is having only outlet through a gap between the hillocks. For the residents, besides natural spring water, the water of the Lake is also the main source of water supply. The Mirik town is not having any sewerage systems, therefore, raw sewage along with agricultural runoffs are directly reaching the Mirik Lake and polluting it. So, the lake and its surrounding areas are under acute pressure due to non-regulated anthropogenic activities. As a result the lake water fails meeting water quality criteria for class 'B' (for outdoor bathing) and class 'C' (for drinking water source after conventional treatment and disinfection) as recommended by CPCB. The major identified threats are as follows:

- Discharge of wastewater on the lake through 39 points
- Non-point source of pollution: washing, bathing, disposal of solid wastes, cleaning of vehicles, storm water, etc
- Pollutants often lead to anoxic condition of lake bottom, causing excess release of nutrients (nitrogen, phosphorus) and greenhouse gases (CH_4, CO_2)
- Sedimentation leading to remarkable decrease in water depth
- Odour, fish kill and possible proliferation of mosquito breeding
- Possible source of pathogens causing cholera, typhoid, scabies etc.
- Encroachment of the lake area by constructing hotel
- Indiscriminate use of lake surroundings by picnic parties from neighbouring areas/districts

Materials and Methods

Data Acquisition: Wastewater and Water

Wastewater sampling were conducted during the study period October 2005 to September 2007 from the selected wastewater channels, out of 39 channels joining the lake, having flow rate more than 5 m^3hr^{-1} and samples were analysed as per Standard Method (20th Ed)for the major physico-chemical and bacteriological analysis, *viz.* Dissolved Oxygen (DO), Biochemical Oxygen Demand (BOD), Chemical Oxygen Demand (COD), Sulphate (as SO_4), Total Phosphate (as PO_4), Chloride (as Cl), Total Coliform (TC) and Fecal Coliform(FC). Sampling points are shown in Figure 18.1 and the waste characterization are summarised in Table 18.1.

Subsequently, water samples were also collected from the eight surface points of lake for physico-chemical, bacteriological analysis as per Standard Methods. Sampling points are shown in Figure 18.2 and the characteristics of lake water is summarized in Table 18.2.

Study revealed that waste volume joining the lake waterbody varies between $15m^3.hr^{-1}$ to 680 $m^3.hr^{-1}$ at different months from the major wastewater channels

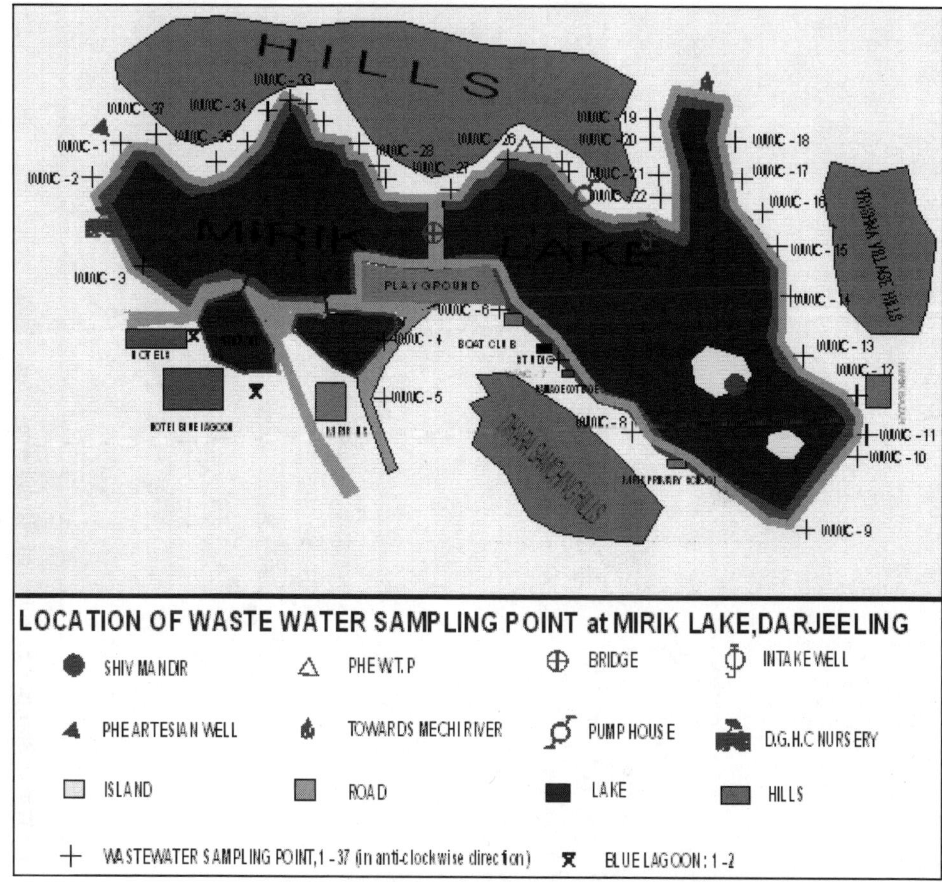

Figure 18.1: Sampling Points Location of Wastewater in Mirik Lake.

joining the lake water, thus the water availability fluctuates along with surface run-off to the Mirik lake.

Regarding quality of lake water, it appears that most of the time, BOD values exceeds the limits of 'B' category of CPCB standards throughout the study periods.

Hydrological Model for Prediction of Water Availability and Quality for Lake

The software package Hydrologic Engineering Centre, Hydrologic Modeling System (Technical Reference Manual 2004 and user's 2000) and Neuro Dimention 5 on Artificial Neural Network (ANN) have been used for calibration, simulation, validation, and optimization of runoff for Mirik Lake, Darjeeling. The HEC-HMS developed by Hydrologic Engineering Centre (HEC) of US Army cop of Engineers, is a well-known hydrologic model. However, ANN model is well established method for describing physical process occurring in the aquatic environment. The development in information technology over the last decade has provided

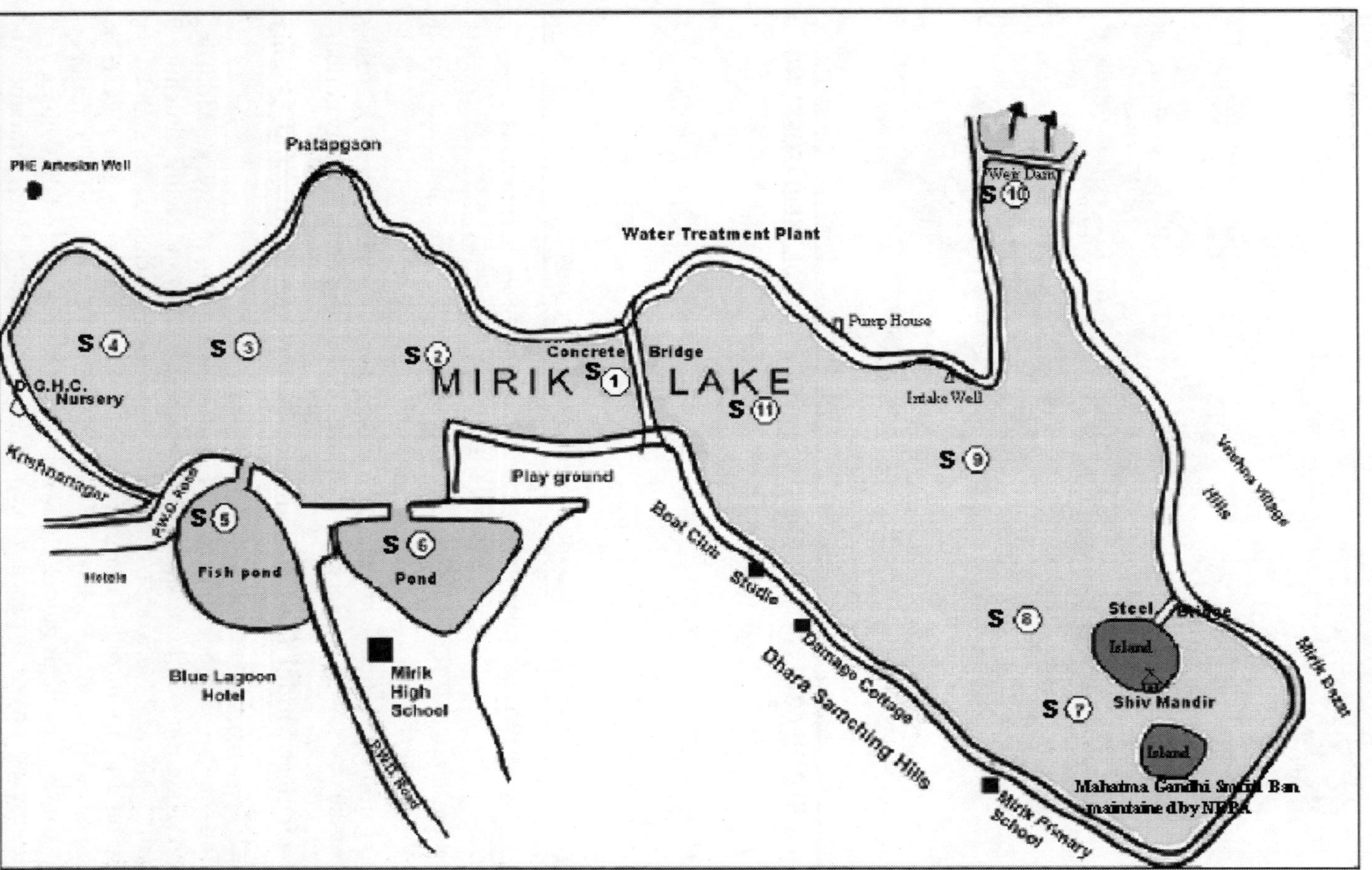

Figure 18.2: Location of Surface Water Sampling Points at Mirik Lake, Darjeeling.

Table 18.1: Seasonal Variation in Physico-Chemical and Bacteriological Characteristics of Wastewater Joining Mirik lake, Darjeeling (Range values)

Parameter	*Sampling Points*															
	WWC–3				*WWC BL–1*				*WWC–4*				*WWC–9*			
	PM	*W*	*S*	*M*	*PM*	*W*	*S*	*M*	*PM*	*W*	*S*	*M*	*PM*	*W*	*S*	*M*
Flow	35–115	5–62	3–26	37–295	32–195	3–72	95–122	94–170	8–54	5–28	7–15	16–48	11–32	2–9	2–10	10–95
pH	6–6.5	6.5–7.1	6.5–7.2	5.8–7	6.0–6.5	6.0–7.5	6.5–7.4	6.2–6.5	6.0–6.5	5.5–6.5	6.5–7.0	6.0–6.2	6.0–6.5	6.0–7.5	6.0–6.5	6.5–6.7
TSS	15–48	21–68	7–36	12–90	6–22	5–10	8–11	9–20	13–20	7–72	12–31	17–122	10–22	5–20	8–24	19–78
DO	2–6	0–3.6	1–3	2–4	2–8	3.4–6.3	2–7	2–5	2.5–4.5	1.7–5.5	3–4	2–5	1–4	0–1.5	1–3	2–3
BOD_3^{27}	18–40	36–143	16–35	5–13	5–7	7–20	1–10	1–6	5–12	13–96	20–28	6–36	28–40	30–148	4–20	7–124
COD	8–15	47–212	31–74	12–26	8–9	18–26	4–18	4–7	8–19	19–118	23–54	15–19	14–50	46–206	11–32	26–42
Cl	8–20	12–53	30–38	7–9	14–20	20–26	18–24	4–6	10–16	12–20	10–28	5–11	10–64	36–112	8–22	18–65
TC x 10^3	37–60	2.65–290	7–25	6–12	3.0–4.1	0.11–0.3	0.14–0.22	0.15–0.37	2–2.5	0.4–90	1.3–4	5–11	80–100	15–240	18–35	153–220
FC x 10^3	7–12	1.56–220	5–15.8	1–0.5	4–1	0.05–0.1	0.065–0.115	0.1–0.2	1–1.2	0.28–75	0.5–3	1–5	20–21	1.4–196	2.5–3.5	100–510

NB: All value are expressed in mg/l; except pH, TC and FC in organisms/100ml, and Flow in m^3/hr.

M: Monsoon. (June–August); PM: Post Monsoon. (Sept–Nov); W: Winter (December–February); S: Summer (March–May)

Wastewater Channels: WWC-3, WWCBL-1, WWC-4, WWC-9.

Table 18.2: Seasonal Variation in Physico-Chemical and Bacteriological Characteristics of Water in Mirik Lake, Darjeeling (Range values)

Parameter	*Sampling Points*															
	S–1				*S–2*				*S–4*				*S–5*			
	PM	*W*	*S*	*M*	*PM*	*W*	*S*	*M*	*PM*	*W*	*S*	*M*	*PM*	*W*	*S*	*M*
pH	6.5–7.5	6.5–7.4	6.8–7.5	6.8–7.3	6.5–7.5	6.5–6.8	6.8–7.5	7.0–7.3	6.0–6.5	6.5–6.5	6.5–7.2	6.0–7.1	6.5–6.5	6.0–6.8	6.5–6.8	6.4–7.1
Temp.	11.0–21	5.5–10.0	12.0–23.0	18–23.5	11.5–22	5.0–9.5	12.5–24	18–24	10.2–17	4.0–8.5	12.5–23	17–21	10.2–15	5.0–9	12–22	17–21
TSS	5–9	8–11	ND	ND	6–11	12–18	ND	ND	6–15	11–66	ND	ND	7–9	12–46	ND	ND
DO	7.0–9.5	7.0–7.0	8.0–9.1	8.0–9.9	7.2–9.8	7.3–10.3	7.0–8.9	7.5–8.4	5.0–8.1	7.2–8.9	7.0–8.1	5.8–7.8	3.8–6.6	6.0–7.5	6.0–6.5	2.3–6.6
BOD_3^{27}		3.0–5.0	4.0–6.0	2.0–5.0	3.0–6.0	6.0–8.0	3.0–6.0	3.0–4.0	4.0–5.0	4.0–6.0	2.0–3.0	1.0–6.0	1.0–3.0	3.0–5.0	2.0–5.0	1.0–2.0
COD	11.6–20	1–8	6.0–16	45–53	11–12	9–16	12–18	15–23	15–15.5	12.0–39.0	8.0–14.0	10–29	ND	11.0–15.0	ND	ND
Cl	11–13	10–16	11	9–19	7–15	11–14	9.0–16	9–19	5–10	13–32	6.0–10	5–14	18–12	16–37	17	12–17
TC	ND	ND	ND	ND	45–220	50–130	65–85	55–80	35–130	50–280	35–90	35–120	20	135	35	15–35
FC	ND	ND	ND	ND	20–55	30–60	44	30–40	15–45	10–16	20–22	4–20	8	36	11	8–20
PO_4	–0.016	ND	ND	0.039–0.312	0.008	ND	ND	0.001–0.277	0.041	0.047–0.139	ND	ND	0.042	ND	ND	0.015–0.528
SO_4	7.69	ND	3.54	ND	ND	ND	ND	ND	ND	2.62–3.85	4.31–9.54	1.23	5.69	ND	ND	ND

Contd...

Table 18.2–*Contd...*

Parameter	*Sampling Points*															
	S–6				*S–7*				*S–10*				*S–11*			
	PM	*W*	*S*	*M*	*PM*	*W*	*S*	*M*	*PM*	*W*	*S*	*M*	*PM*	*W*	*S*	*M*
pH	6.0–6.5	6.0–7.0	6.5–7.5	5.5–6.8	6.5–7.5	6.5–7.6	6.8–7.8	6.5–7.0	6.5–7.5	6.5–7.6	7.8–7.8	6.5–7.4	6.5–8.5	7.0–7.9	7.5–7.8	6.0–7.5
Temp	12–21	5.5–10	12.5–23	19–24	13.5–23	6.5–10	10–22	20–23.5	18–23	5.5–10	12.5–21	19–21	12–22	5.5–9.5	12–23	18–22
TSS	6–10	50–65	ND	ND	3–15	22–57	18.6–19	9–48	10–15	4–22	9–12	5–24	5–11	ND	ND	4–12
DO	7.5–8.2	8.6–10.4	8.0–8.1	5.3–9.1	8.2–9.9	7.4–10.2	9	7.5–8.6	7.7–10.6	6.6–9.6	9.4	5.2–106	8.2–10.7	8.4–10.5	9.3	7.0–10.1
BOD_3^{27}	3.0–5.0	4.0–8.0	1.0–5.0	3.0–5.0	5.0–5.0	5.0–6.0	4.0–7.0	2.0–6.0	4.0–6.0	4.0–8.0	3.0–8.0	3.0–6.0	5.0–6.0	5.0–8.0	2.0–6.0	3.0–5.0
COD	7.7–23	12.0–33	6.0–12.0	4.0–10.0	11.6–61.4	7.1–26.3	8.0	8.0–23.0	7.7–11.6	4.0–23.0	8–11	15–30	7–18	ND	ND	7.0–15.0
Cl	7–12	13–32	6.0–14	5–10	6–11	12–19	8–10	8–19	6–11	10–16	11–13	7–20	6–11	10–15	7–9	8–17
TC	30–90	12–16	35–240	30–65	25–420	52–240	150	20–210	10–150	40–220	70–75	90–140	20–170	ND	ND	120–170
FC	5–55	1–60	20–28	12–30	7–20	22–80	75	4–30	5–30	10–26	36	10–65	15–25	ND	ND	35–60
PO_4	0.016	0.081–0.185	ND	0.019–0.513	0.073	0.079–0.176	ND	0.05–0.55	0.06	0.056–0.118	ND	BDL 0.258	0.085	0.056–0.118	ND	0.245–0.458
SO_4	6.46–6.92	5.85–9.08	5.7	0.92	4.92–5.38	2.92–6.92	5.4	0.77	2.31–3.08	2.0–3.38	3.08	0.31	ND	ND	ND	0.46

NB: All value are expressed in mg/l; except pH, TC and FC in organisms/100ml, and Flow in m^3/hr.

M: Monsoon. (June–August); PM: Post Monsoon. (Sept–Nov); W: Winter (December–February); S: Summer (March–May)

Sampling points: S–1, S–2, S–4, S–5, S–6, S–7, S–10, S–11.

opportunities of extended computational ability together with improved data manipulation, storage and retrieval. The use of the models require input of daily rainfall, soil cover, land use pattern, discharge data etc.

Data towards daily rainfall, temp, relative humidity, wind speed etc. for a period of nine years (January 1998 to March 2006) for Mirik Lake, Darjeeling were collected from Tea Estate, Thurbu, Mirik. Evapo-traspiration has been calculated following Penman equation by using temperature, relative humidity, wind speed etc. Horton's equation has been used for determining the infiltration rate. The package used separate modules to represent each component of the runoff process.

Also, due to paucity of discharge data, the ANN model was used for determination of historical discharge data. Daily rainfall and stream flow data of ten years (January 1998 to March 2006) for the Mirik Lake has been used for the calibration and validation in the HEC-HMS model as shown in Figure 18.3. The calibration error, expressed as a per cent difference of simulated value with respect to observed value, was found to lie within the range of 0.1 to 10 per cent for the lake.

The future water availability of Mirik Lake in Darjeeling has been evaluated using the projected daily precipitation and temperature data for the period of 2008-2020, which has been generated based on ANN model, with the help of HEC-HMS model by considering fixed CO_2 as 1990's level which represents the key forces, such as, economic growth, population, land use pattern and soil cover. The land use pattern of the area, as shown in Figure 18.4, highlights that there are forest land, several other water bodies, agricultural area, solid waste disposal site and rural settlement area.

The neuro dimension software was used to development of the model and to output the optimized solution for the said area. The data *i.e.* rainfall and temperature of January 1998 to March 2006 was used to develop the model by considering all the losses. The model should an average MSE near to 0.2 per cent. The data generated for the same period were optimized using the HEC-HMS model by keeping the same land use pattern and land cover as in future also considered all losses which has been used by different formulae.

HEC-HMS Package

In this present research work, the software Package HEC-HMS (Technical Reference Manual, 2000 and User's Manual, 2001) has been used for the simulation of runoff hydrograph for four important eastern river basins in India. The HEC-HMS developed by Hydrologic Engineering Centre (HEC) of US Army Corps of Engineers is a well-known hydrologic model specially designed for the simulation of flood events in river basin. The package uses separate modules to represent each component of the runoff process. The precipitation losses are found for each computation time interval and is subtracted from the Mean Areal Precipitation (MAP) depth for that interval. The remaining depth is referred to as precipitation excess. This depth is considered uniformly distributed over a watershed area, so it represents a volume of runoff. In HEC-HMS, the transformation of rainfall excess to streamflow is accomplished by the unit hydrograph procedure (Flood estimation techniques, 1988-

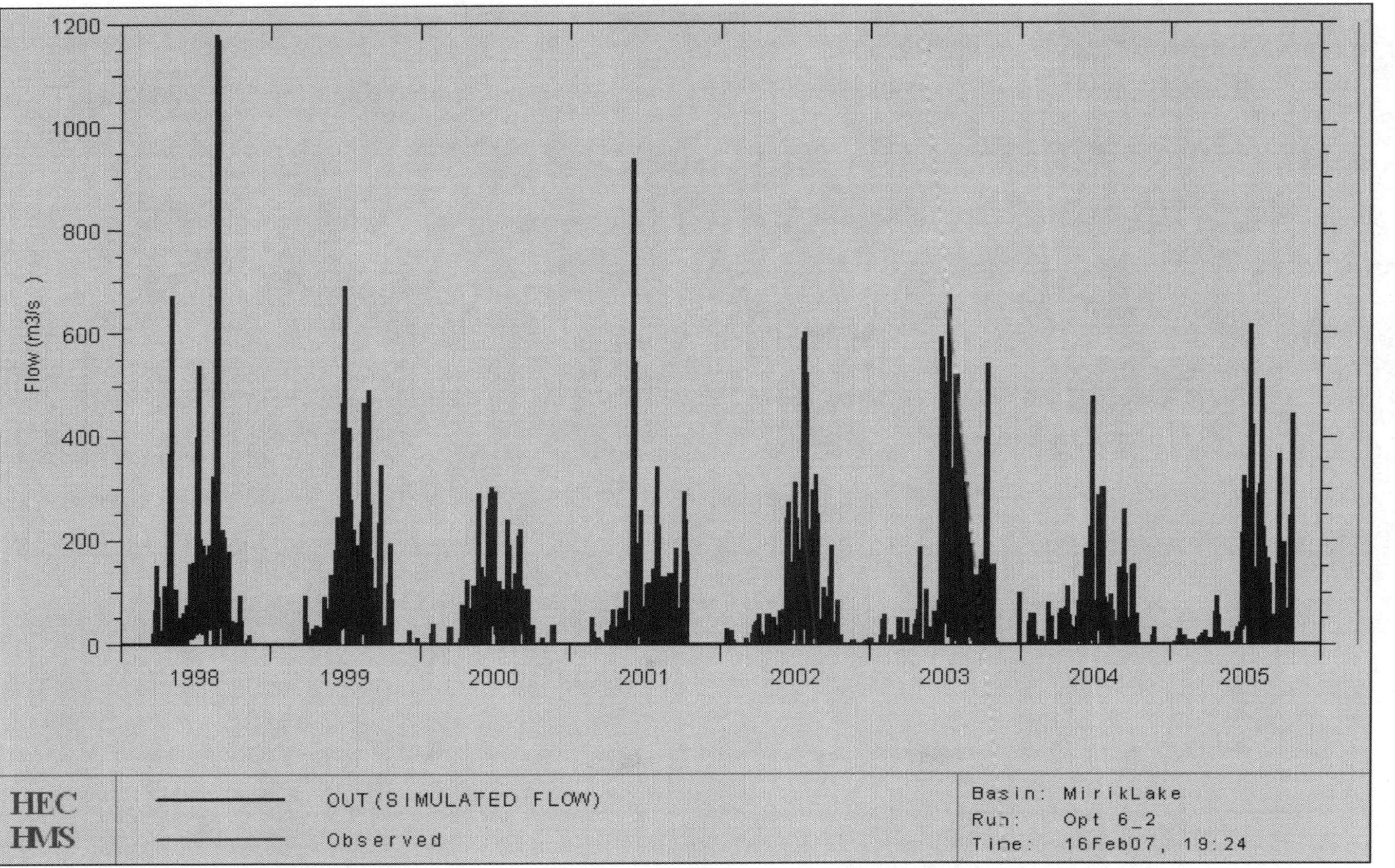

Figure 18.3: Calibrated and Validated Streamflow Hydrograph of Mirik Lake by HEC-HMS.

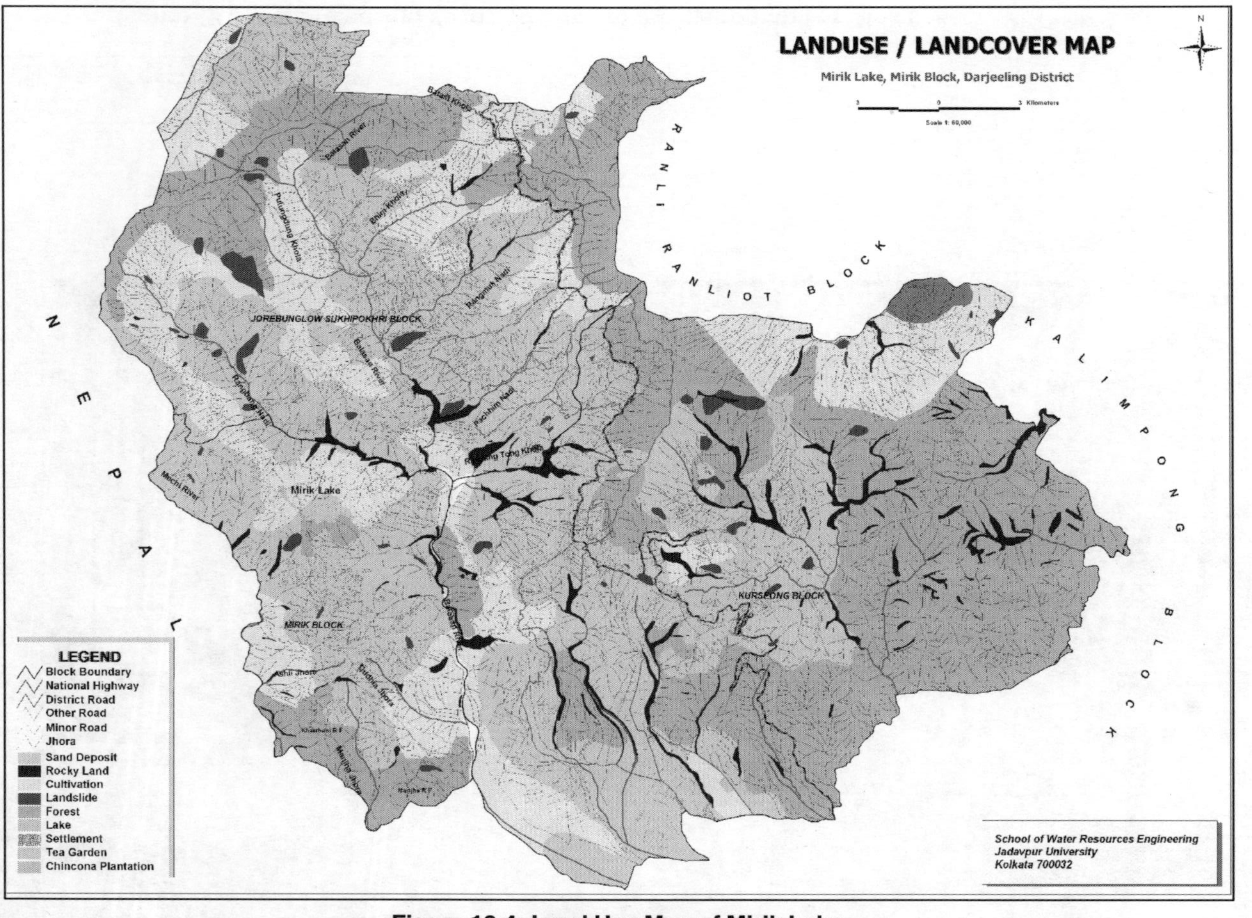

Figure 18.4: Land Use Map of Mirik Lake.

89 and Soil Conservation Service, 1972). There are six unit hydrograph methods available in HEC-HMS, namely user-specified unit hydrograph, Clark's, Snyder's, SCS, Mod Clark and Kinematics Wave methods (Bhaskar et.al., 1997; Lowing et. al., 1981; Maidment, 1992; Sherman, 1932 and Singh, 1988). However, Snyder's method is selected for the present study for the estimation of runoff. The Snyder method does not define a complete unit hydrograph, so the HEC-HMS programme utilizes the Clark method in a trial-and-error procedure to complete the hydrograph. Initial Clark Parameters are derived from the given Snyder parameters, Tp and Cp, and these are used to compute a Clark unit hydrograph. A new set of Snyder parameters computed from this hydrograph is compared with the given Snyder parameters, and this process is continued for twenty iterations or until the difference between the computed and the given values is less than 1 per cent. The basic governing equation is discussed in Technical Reference Manual, 2000.

In HEC-HMS the models of channel flow computes a downstream hydrograph, given an upstream hydrograph as a boundary condition. There are five routing methods available in HEC-HMS, namely Lag, Muskingum, Modified Puls, Kinematic Wave and Muskingum Cunge. However, the Muskingum method is selected for the present study for the estimation of the runoff on account of its capability to accommodate more complicated looped relationship of storage vs. outflow. In HEC-HMS, the model of channel flow computes a downstream hydrograph, given an upstream hydrograph as a boundary condition. The basic governing equation is discussed in Technical Reference Manual, 2000 delineated below.

The package HEC-HMS have many methodological options for calculation the surface runoff volumes. On of the methods used for estimating runoff volume is the initial and constant loss methods.

Artificial Neural Networks

An Artificial Neural Network is nothing but a collection of interconnected processing elements (PEs). The connection strengths, also called the network weights, can be adopted such that the network's output matches a desired response although; it is a data based process model.

Figure 18.5 depicts a typical multilayer perception, which has been used in this study, which resembles a black box model, where a set of data like x_1, x_2, x_3, etc. are fed directly to the network through the input layer and subsequently produces expected result 'y' in the output layer.

In the present study, two input and one output parameters for rainfall runoff modeling and one input and 9 output for water quality modelling were chosen or configured and with the help of Neuro Dimension 5 and an ANN software was prepared by multilayer perception algorithm was used to identify the streamflow data for the period of 1998 to 2006 and water quality data for the period of 2008 to 2020 of Mirik Lake at outlet point as well as prediction for the same. The daily rainfall data with meteorological parameters collected from January 1998 to March 2006 and the historical average water quality parameters from August 2005 to September, 2007 as depicted in Table 18.3 determined by the School of Water Resources Engineering,

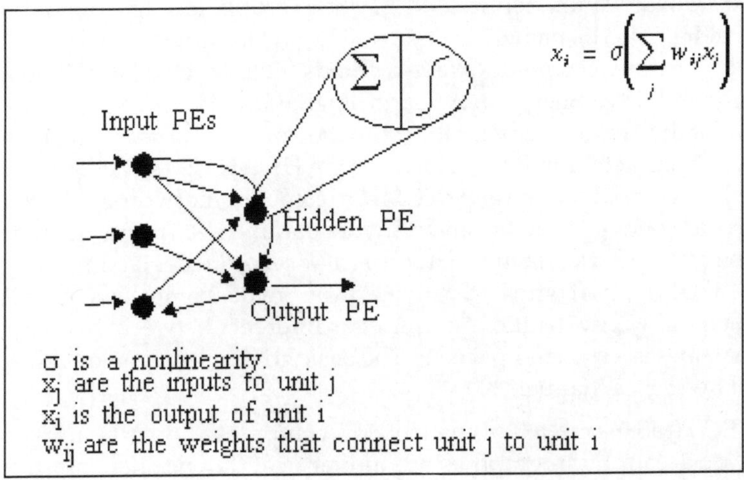

Figure 18.5: Typical Architecture of Single Hidden Layer Neural Network.

Jadavpur University and National Environmental Engineering Research Institute, Nagpur were used as model data set for training (70 per cent), Cross Validation (15 per cent) and testing (15 per cent). In the training procedure, the available data set was used to find or optimal weights of the network where the least Mean Square Error (MSE) indicates the completion of the training. In the testing procedure, the model was compared with the output data *i.e.* streamflow and water quality parameters which was found from the single simulation. The comparison highlights a MSE that helps the user for deciding the continuation of the training.

Table 18.3: Average Water Quality Parameter 2007.

Season	*DO*	*BOD*	*COD*	*Chloride*	*Sul-phate*	*Phos-phate*	*pH*	*FC*	*TC*
Summer	8.06	5	20.47	10.68	6.970	ND	7.1	35	94
Monsoon	7.48	4	11.10	12.37	0.736	0.061	6.7	26	84
Post Monsoon	7.95	4	16.63	10.43	5.270	0.046	6.8	25	111
Winter	8.31	6	22.90	15.99	4.580	0.054	6.8	33	117

NB: All values are expressed a mg/L except pH and TC and FC in organism/100 ml.

ND: Not Detected.

Model Creation, Training, Testing and Prediction with the Help of Historical Data

The model was created and trained in the following way:

Stream Flow Data

The available historical rainfall along with date data was arranged in such a way so that the hydrographs at outlet in Mirik catchment were shown in the first two columns. The same for streamflow for Mirik Lake has been shown in the remaining 1 column of an MS Excel spreadsheet. The Neuro solution or Neuro dimension for Excel software was used to identify the input and output columns. Here, the first 2 columns as input and the remaining column of the spreadsheet were identified as output. Then the same software was used to create the model. After the model was created the same software was used for training the model. The model was trained for 25 times. In each of the training the simulation was applied for either 10 or 12 times or more precisely Run. In each run the model has 1000 epoch training. In short the model was trained for nearly 100 times or EPOCH.

Water Quality Data

The available historical water quality in seasonally along with date data was arranged in such a way so that the water quality at outlet in different seasons in Mirik catchment was shown in the first one column. The same for water quality parameters like DO, BOD_5, COD, Chloride, Sulphate (SO_4), Phosphate (PO_4), pH, Faecal Coliform(FC) and Total Coliform(TC) for Mirik Lake have been shown in the remaining 9 columns of an MS Excel spreadsheet. The Neurosolution or Neuro dimension for Excel software was used to identify the input and output columns. Here, the first 1 column as input and the remaining 9 columns of the spreadsheet were identified as output. Then the same software was used to create the model. It is an Elman networks extend the multilayer perceptron with context units, which are processing elements (PEs) that remember past activity. Context units provide the network with the ability to extract temporal information from the data. In the Elman network, the activity of the first hidden PEs are copied to the context units, while the Jordan network copies the output of the network. Networks which feed the input and the last hidden layer to the context units are also available.

Results and Discussion

Predicted Stream Flow Analysis

The predicted surface water hydrographs for the Mirik catchment for the period 2008–2020 against the historical hydrographs for the periods 1998-2006 are found out by using the model and only one hydrograph for the year 2020 is depicted in Figure 18.6. The hydrographs for the second decade depict changes in seasonal pattern. The hydrograph also displays smaller peaks in the months of April and June. However, values of streamflow in July, August and September were much lower than those in May. The reduced peak in all the discharge hydrographs may be attributed to low intensity of rainfall per day. A high flow in November was also observed in the Mirik catchment. The higher values of peak flow in the first decade as compared to the second decade (experiencing highest annual rainfall in the projected scenario) may be attributed to the occurrence of more intense one-day spells of precipitation. The run-off volumes for the periods 2008-2020 are 0.427152, 0.42068, 0.388320, 0.3236, 0.255644, 0.239464, 0.236228, 0.221666, 0.200632, 189306, 0.16989, 0.153717 and 0.119732 MCM/D respectively. The highest value of peak flow (1.7393 m^3/s) in the year 2020 as compared to the lowest value (0.032215 m^3/s) in 2010,

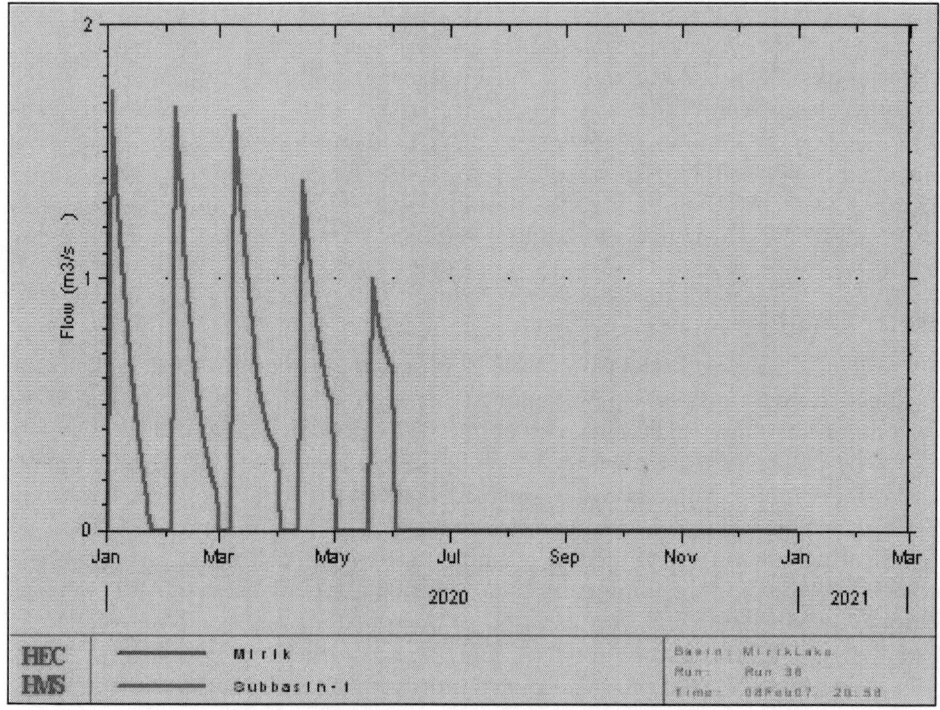

Figure 18.6: Projected Stream Flow for Mirik Catchment for the Period 2020.

experiencing highest annual rainfall in the projected years, may be attributed to the likelihood of the occurrence of flood. The per capita surface water availability for the population dependent on the Mirik Lake has gone down from the present scenario of 45 m^3/capita/year in 2006 to 44, 43, 39, 33, 26, 24, 23, 22, 19, 18, 16, 15 and 11 m^3/ capita/year respectively during the years 2010 to 2020, by keeping the growth of projected population as 2 per cent per decade as depicted in UN (1998).

Predicted Water Quality Analysis

The predicted average values of Physico-chemical and Bacteriological characteristics of Mirik Lake basin of eleven locations is given in Tables 18.4.

Central Pollution Control Board (CPCB) has classified surface water as per water quality criteria and designated best-use. Using the same standard the Mirik Lake water can be classified as class 'B' shown in Table 18.5. However if the lake water could be protected from organic pollution through implementation of environmental management plan in a short term measure, the lake could be upgraded to class 'A' in near future. Intensively use of Mirik lake by surrounding people would cause one of the major threats to lake in future and the results of water quality predicted by ANN for the period of 2010 to 2020 may be deteriorated as Class 'C'.

Table 18.4: Average Water Quality Parameters for the Predicted Years (2008–2020).

Season	*DO*	*BOD*	*COD*	*Chloride*	*Sul-phate*	*Phos-phate*	*pH*	*FC*	*TC*
Summer	7.73	2.0	18.76	6.0	4.050	ND	9.0	22	54
Monsoon	5.34	2.0	11.12	17.0	4.210	0.117	8.0	46	172
Post Monsoon	4.81	2.0	6.92	8.0	0.860	0.080	9.0	53	143
Winter	9.60	13.0	16.00	15.0	0.789	0.420	9.0	19	106
Average	6.90	5.0	8.50	12.0	2.470	0.206	8.8	35	118

NB: All values are expressed a mg/L except pH and TC and FC in organism/100 ml.

ND: Not Done.

Table 18.5: Designated Best Use Classification of Inland Surface Water (CPCB Standards).

Class	*Designated Best Use*	*Criteria*
A	Drinking water source without conventional treatment but after disinfection	pH: 6.5 to 8.5 Dissolved Oxygen: 6 mg/l or more Biochemical Oxygen Demand: 2 mg/l or less Total Coliform: 50 MPN/100 ml
B	Outdoor bathing (Organised)	pH: 6.5 to 8.5 Dissolved Oxygen: 5 mg/l or more Biochemical Oxygen Demand: 3 mg/l or less Total Coliform: 500 MPN/100 ml
C	Drinking water source with conventional treatment followed by disinfection	pH: 6.5 to 8.5 Dissolved Oxygen: 4 mg/l or more Biochemical Oxygen Demand: 3 mg/l or less Total Coliform: 5000 MPN/100 ml
D	Propagation of wild life and fisheries	pH: 6.5 to 8.5 Dissolved Oxygen: 4 mg/l or more Free Ammonia: 12 mg/l
E	Irrigation, industrial cooling and controlled waste disposal	pH: 6.5 to 8.5 Electrical Conductivity: 2250 μ mhos/cm Sodium Absorption Ratio: 26 Boron: 2 mg/l

Conclusion(s)

Considering the rate of predicted wastewater discharge in the lake and alteration in water quality of the lake, it appears that the state of the lake would be in acute danger in forthcoming years. In view of this, there is an urgent need to adopt techno-economically viable technology to revive the Mirik Lake *viz.* reed bed technology, waste stabilisation pond etc. Since there is continuous anthropogenic pressure on the lakes with reference to wastewater discharge, deforestation, disposal of solid waste, siltation, encroachment, washing/cleaning etc., certain long term measures need to be adopted for revival of the lake.

References

1. Biswas, M (2008). Participatory Management of Eco-System Services:A Study of Wetland in West Bengal, Ph.D. thesis, awarded from School of Water Resources Engineering, JAdavpur University, Kolkata
2. HEC-HMS, Version 2.2.2, 2000. Hydrologic Simulation Package, Technical Reference Manual, US Army Corps of Engineers, Hydrologic Engineering Centre
3. HEC-HMS, Version 2.2.2, 2001. Hydrologic Simulation Package, user's Manual, US Army Corps of Engineers, Hydrologic Engineering Centre.
4. Sherman, L.K., 1932. Streamflow from Rainfall by the Unit hydrograph Method, Eng. News Rec., Vol.108, pp. 501-505.
5. Shiklomanov, I. A. (1998) *World Water Resources—A New Appraisal and Assessment for the 21st Century*. UNESCO, Paris, France.
6. Singh, V.P., 1988. Hydrologic Systems Rainfall-Runoff Modeling, Vol.1, Chapter 10, Prentice Hall, New Jersey.
7. Soil Conservation Service, 1972. National Engineering Handbook, Section 4, Hydrology, SCS, US Department of Agriculture, Washington, D.C.
8. NeuroDimensions, (2005).NeuroSolutions, www.nd.com,
9. Neuro DimensionTM for Excel v.5 Manual, Publisher-Neurosolution
10. Sankhua, R N, "Neural Network in Hydrological Modelling a case study of river Brahmaputra" IIT (Kharagpur).
11. Lowing, M.J and Mein, R.G., 1981. Flood Event Modeling; A Study of two Methods, Water Resources Bulletin, 17(4), pp.559-606
12. Maidment, D.R. (Edited) 1992. Handbook of Hydrology, Chapter 9, McGraw Hill, New York.
13. United Nations (UN), 1998. World Population Prospect: The 1998 Revision, Vol.I, Sales No. E99XIII.9.
14. Revival of Mirik lake, Darjeeling for Designated Best Use, 2008; under National Lake Conservation Programme by National Environmental Engineering Research Institute, Kolkata and School of Water Resources Engineering, Jadavpur University, Kolkata.
15. Standard Methods for Examination of Water and Wastewater (1998). 20th edition, American Public Health Association/American Water Works Association/ Water Environment Federation, Washington DC, USA.

Chapter 19

Spatial Assessment of Coastal Aquifer Groundwater Quality due to Anthropogenic Activities: Puri City, India

Ritesh Vijay[1], P.K. Mohapatra[2], P.R. Pujari[1] and R.A. Sohony[1]

[1]CSIR-National Environmental Engineering Research Institute, Nehru Marg, Nagpur

[2]Orissa Water Supply and Sewerage Board, Water Works Road, Puri

ABSTRACT

Puri city is situated on shoreline of the Bay of Bengal and groundwater is the only source available to meet water supply of the city. Due to anthropogenic activities and absence of sewerage system, groundwater quality is deteriorating though open discharge of domestic waste, open defecation, septic tanks, soak pits, contaminated water pools and salt water intrusion by continuous withdrawal of groundwater. Groundwater samples were collected during summer season from the water fields which provide water supply, hand pumps, open wells, open water bodies and production wells in the city. Groundwater table data were also collected to know fluctuation of water level in the city. The objective of the paper is to assess the groundwater level and quality spatially using geospatial technologies and evaluated with Indian drinking water standard for domestic use. Further measures are suggested based on the field observations and quality analysis to safeguard the groundwater resources and check the groundwater contamination.

Keywords: *Groundwater, Water table, Contamination, GIS.*

Introduction

Groundwater is one of the most valuable natural resources, which supports human health, economic development and ecological diversity. Most of the Earth's liquid freshwater is found, not in lakes and rivers, but is stored underground in aquifers and constitute nearly 95 per cent of freshwater of planet. Overexploitation and unabated pollution of this vital resource is threatening our ecosystems and even the life of future generations (Jha *et al.*, 2007). Assessing the quality of groundwater is important to ensure sustainable safe use of these resources. However, describing the overall water quality condition is difficult due to the spatial variability of multiple contaminants and the wide range of indicators (Babiker *et al.*, 2007). Groundwater protection is a long-term key issue which needs regulation, management and planning, and should be a subject of public information. (Nieto *et. al.*, 2005; De Leo *et al.*, 2002).

The groundwater monitoring and assessment is the task that involves the combination of multiple factors which includes land use land cover, climatic change, human impact, saltwater intrusion into the coastal aquifer system as well as economical and administrative issues. In contrast with surface water pollution, sub-surface pollution is difficult to detect, is even more difficult to control, and may persist for years, decades, or even centuries (Todd, 1980). Study of aquifer vulnerability to pollution are becoming more in demand because on the one hand groundwater represents the main source of drinking water, and on the other hand high concentrations of human/economic activities, *e.g.* industrial, agricultural, and household represent real or potential sources of groundwater contamination (Rahman, A., 2008; Lambrakis, 1997). The management of the water supply systems that use groundwater from aquifers prone to saline water contamination is a complex task (Julio *et al., 2006)*.

The Grographic Information System (GIS) technique has provided an efficient tool for assessing and analyzing the vulnerability to groundwater pollution (Srivastava and Bhattacharya, 2006). The movement, area and the distributional problem of the contamination can be addressed through this technique. The unbiased data representation capability of the geospatial technology made the whole thing possible with the reliable field data as its input. In the present study an attempt has been made to assess spatial variation of water level and quality in a costal aquifer. In this regards the identification of potential source of groundwater pollution and analysis of biological and chemical constituents are measured to monitor and asses the groundwater resources.

Study Area

Puri is situated on the shoreline of the Bay of Bengal lies between 19° 47′ to 19° 50′ North latitudes and 85° 48′ to 85° 52′ East longitudes (Topsheet 74 E/13). The general topography of the town slopes smoothly towards south-eastern side though there are several local slopes in all directions due to the presence of sand mounds and dunes. The ground elevation varies from sea level to 18.8 m above msl (mean sea level) within municipal boundary. The climate of Puri is warm–humid with the maximum and minimum temperature are 37.5°C and 27.0°C in the summer while 28.2°C and 15.2°C in winter respectively. The mean annual rainfall is about 1520mm from the southwest monsoon (NEERI, 2000). The population of the city is 1,42,028 as per 2001 census records. Since the city is on shoreline and nearby rivers is estuarine, surface water is not suitable for water supply. The city receives water supply (20.44 MLD) only from groundwater sources by two sweet water zones namely Chakratirtha and Balia Panda water fields situated 2-3 km away from the seashore (NEERI, 2008) as shown in Figure 19.1.

Methodology

The methodology developed for systematic flow of the work plan for assessment of groundwater quality. The whole process includes two phases. In the first phase, primary data was collected and analyzed while in second phase thematic database was prepared and field data was transfered to the GIS environment for analysis.

Phase-I: Data Collection

In this phase, well inventory data such as the water table depth, lithologs and groundwater samples were collected form the production wells, hand pumps, open wells and open water bodies. Most of the production wells situated in Chaktratirtha and Balia Panda water fields are the major source of water supply to city. Samples were collected during summer (June-2007) as presented in Table 19.1.

Table 191: Summary of Sampling Locations and Sources.

Location	*Source*	*Summer*
Chakratirtha Water Fields	Production Wells	10
Balia Panda Water Fields	Production Wells	6
Puri City	Hand Pumps	27
	Production Wells	2
	Open Wells	2
	Open water body	6
Total samples		53

Samples were preserved and analyzed as per standard method described in APHA (1995), NEERI (1986), and Chemistry for Environmental Engineering (Sawyer *et al.,* 1994) for physico-chemical and bacteriological parameters. Groundwater samples were analyzed for ten physico-chemical (pH, Total Dissolved Solids, Total

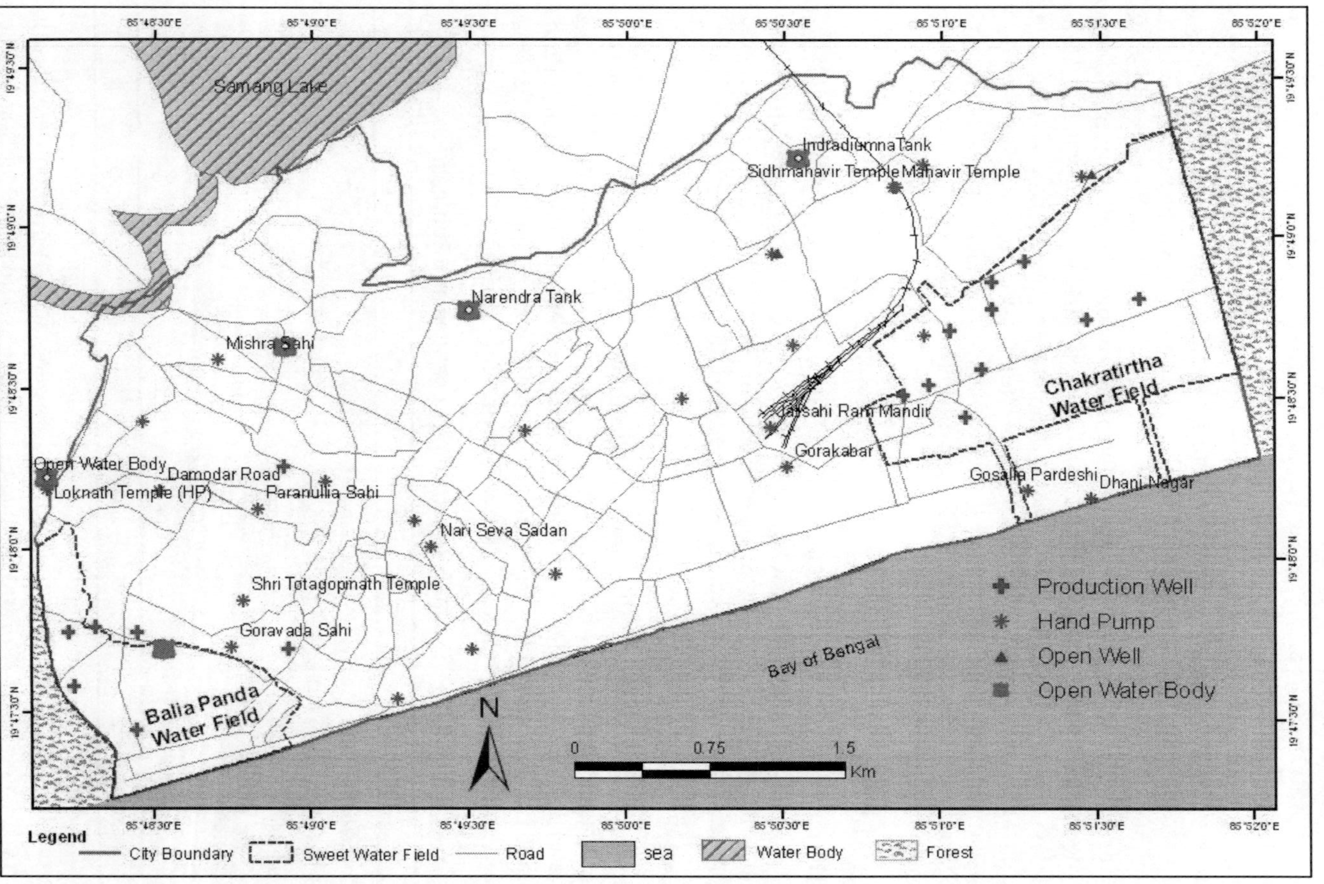

Figure 19.1: Details of Study Area and Sampling Locations.

Hardness, Alkalinity, Calcium, Magnesium, Sulphate, Nitrate, Chloride and Fluoride), two bacteriological (Total Coliform - TC and Fecal Coliform - FC) and 1 metal (iron) of individual sample.

Phase-II: GIS Based Analysis

The permanent spatial feature such as city boundary, sweet water zones and other basic features are demarked with the help of Survey topo sheet and related data source in GIS environment (ArcGIS 9.3). The locations of samples were transferred on GIS based on global positioning system (GPS) readings. The entire thematic layers follow the same spatial reference system (UTM/WGS 84) for a better accuracy and alignment during the analysis. Groundwater quality parameters based on laboratory analysis and water table information (msl) were attributed to each sampling location. The water table contours were generated by the interpolation process to know fluctuation in the city. The concentration of water quality parameters which deviates form it's desirable and permissible limit are over laid on groundwater table to show its level of contamination. Different quality assessment aspects of the aquifer system were shown by their distinct symbology for the better visual perception.

Results and Discussion

Groundwater levels were measured during summer sampling and transferred to msl to estimate the fluctuation in water table. Figure 19.2 presents the fluctuation in groundwater level in the city area. Based on the field observations, water table depletes more in water fields around 3 m compare to city area around 1.85 m due to continuous withdrawal of groundwater for city water supply. The minimum groundwater levels was observed at Production well PC7 of Balaia Panda water field due to continuous water supply and maximum was observed 3.892 m at Production well around Narendra tank.

Water table depth is one of the most important factors because it determines the thickness of material through which water infiltrates and reaches saturated zone. The spatial distribution of groundwater level was carried out to know the trend of the water table as shown in Figure 19.3. It reveals that water level varies form 0–5 m above msl. The area under water level range 0 -1 m is around 3.63 sq. k.m and covers 21.2 per cent of whole city due to more withdrawal of groundwater during summer period.

The groundwater quality based on physico-chemical and metal analysis is presented in Table 19.2. Individual parameter is presented based on average and standard deviation of the samples in specific areas. Groundwater analysis for various parameters including total dissolved solids, hardness, alkalinity, carbonate, chloride, sulphate, fluoride and nitrate are more in city area compared to the water fields. Except nitrate most of the parameters are within desirable and permissible limit as per Indian drinking water standard (IS:10500, 1991). The concentration of nitrate was high at 10 locations in groundwater and mainly due to human activities like domestic effluent, human and animal excreta, septic tanks, soak pits and no sewerage system. The spatial distribution of nitrate in groundwater along with water level contour is presented in Figure 19.3. The permissible limit of nitrate in drinking water

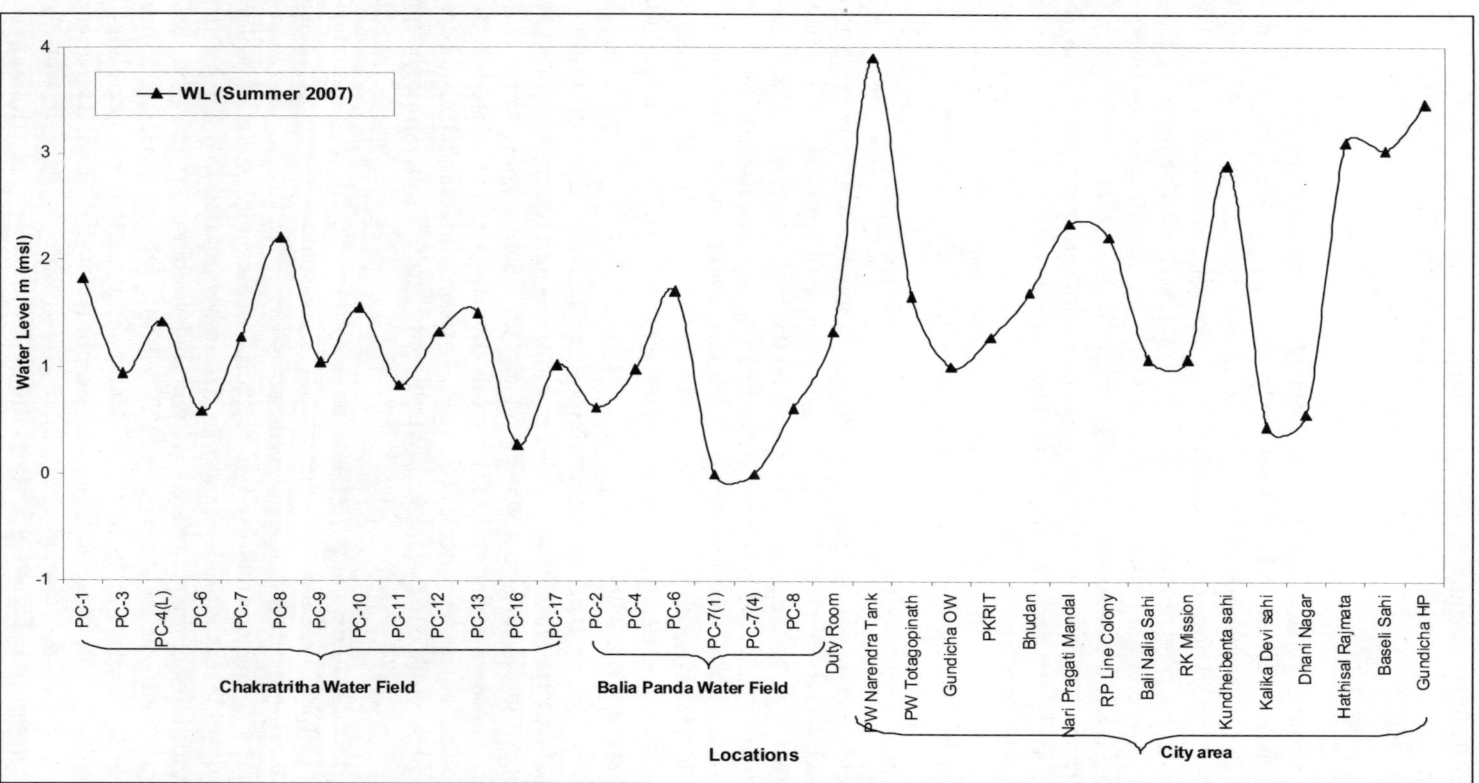

Figure 19.2: Variation in Groundwater Level.

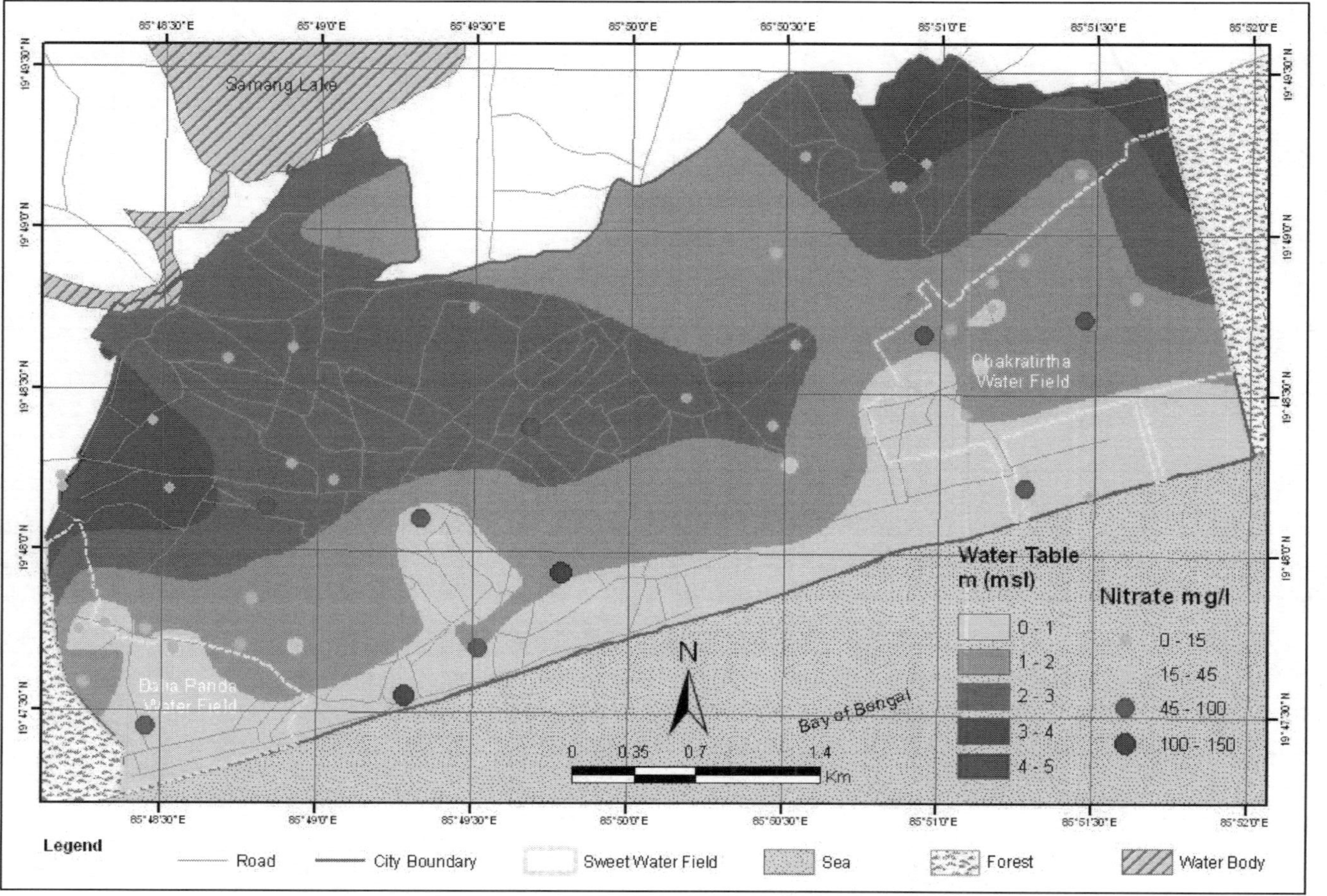

Figure 19.3: Distribution of Nitrate and Groundwater LeveL

Table 19.2: Statistics of Groundwater Quality in Specific Areas.

Parameter	*Chakratirtha Water Field*	*Balia Panda Water Field*	*City Production Wells*	*City Hand Pumps*	*City Open Wells/ Water Bodies*	*Sample Higher than Permissible Limit (Total)*	*Indian Drinking Water Standard*
						No. and per cent	*Limit*
pH	7.6±0.1	7.1±0.2	7.1±0.4	7.5±0.3	7.3±0.1	0	6.5 - 8.5
TDS	215±103	198±104	339±131	505±419	518.5±67	1(1.88)	500-2000
Total Hardness	114±42	101±45	128±11	162±124	157±21	0	300 - 600
Alkalinity	90±45	93±60	132±28	151±69	186±110	0	200 - 600
Ca	15±4	16±7	22±7	15±9	19±6	0	75 - 200
Mg	18±9	15±10	17±7	30±26	27±15	0	30 - 100
SO_4	31±16	17±11	31±18	52±61	53±40	0	200 - 400
NO_3	16±16	14±23	14±10	31±41	16±12	10(18.86)	45
Cl	51±17	49±8	75±3	147±215	124±51	1(1.88)	250 - 1000
F	0.09±0.02	0.16±0.09	0.11±0.01	0.43±0.54	0.72±1.31	3(5.66)	1.0 - 1.5
Fe	0.79±0.63	0.62±0.43	1.24±0.2	1.52±0.86	1.67±0.56	30(56.6)	0.3 - 1.0

as per Indian standard is 45 mg/l, beyond this it may cause methaemoglobinemia or blue baby disease in infants, disorders of alimentary canal, respiratory, nervous system and carcinogenic in adults (Basappa Reddy 2003; Shankar *et al.*, 2007). The level of contamination varies where water level in the range 0- 3 m and severely contaminated in the range 0-2 m. Nearer the coast where water level touches the ground surface that leads to possible contamination than the other part of the city.

Among heavy metals Iron is found as most critical metal in the groundwater. Most of the samples collected from water fields during summer are within desirable and permissible limits while samples form hand pumps are beyond permissible limit as demarcated in Figures 19.4 and 19.5. Iron is found beyond the permissible limit in 57 per cent groundwater samples as per Indian drinking water standard. Iron bearing groundwater is often noticeably orange in color, causing discoloration of laundry, and has an unpleasant taste as observed which is apparent in drinking and food preparation.

The presence of bacteria in groundwater of specific areas in the city including water fields is summarized in Table 19.3. Figures 19.6 and 19.7 represent the spatial distribution of FC and TC in groundwater respectively. Based on bacteriological analysis, FC and TC were found even in both the water fields which are the only source of water supply. Open wells and open water bodies in the city were highly contaminated in terms of harmful bacteria. This reveals the presence of harmful bacteria in shallow as well as deeper aquifers due to percolation of municipal sewage, practice of open defecation and poor sanitary conditions. Even some of the production wells in both the water fields are not free from bacteriological contamination which indicates movement towards sweet water zones due to pumping. The bacteriological contamination always leads to the spreading of water borne diseases causing like cholera and severe digestive system disorder problems.

Table 19.3: Bacteriological details in Specific Areas (Number and per cent of sample).

Parameter	*Chakratirtha Water Field*	*Balia Panda Water Field*	*City Production Wells*	*City Hand Pumps*	*City Open Wells/Water Bodies*	*Indian Drinking Water Standard*	*Total Positive Sample*
TC	4 (7.54)	1 (1.79)	nil	3 (5.66)	7 (13.2)	Nill	15 (28.19)
FC	1 (1.88)	1 (1.88)	nil	1 (1.88)	6 (11.32)	Nil	9 (16.96)

Conclusion and Suggestion

Since city is situated along the coast and groundwater is only source to meet water supply from Chakratirtha and Balia Panda water fields, there is a dynamic balance of fresh and salt water interface. Both the water fields contain fine to medium sand, which acts as a slow sand filter to maintain the quality of groundwater and being substantially large in size which facilitate the recharge of groundwater significantly. Based on field observations, various open water bodies, tanks and ponds in the city contribute recharge as well as contaminants in the groundwater.

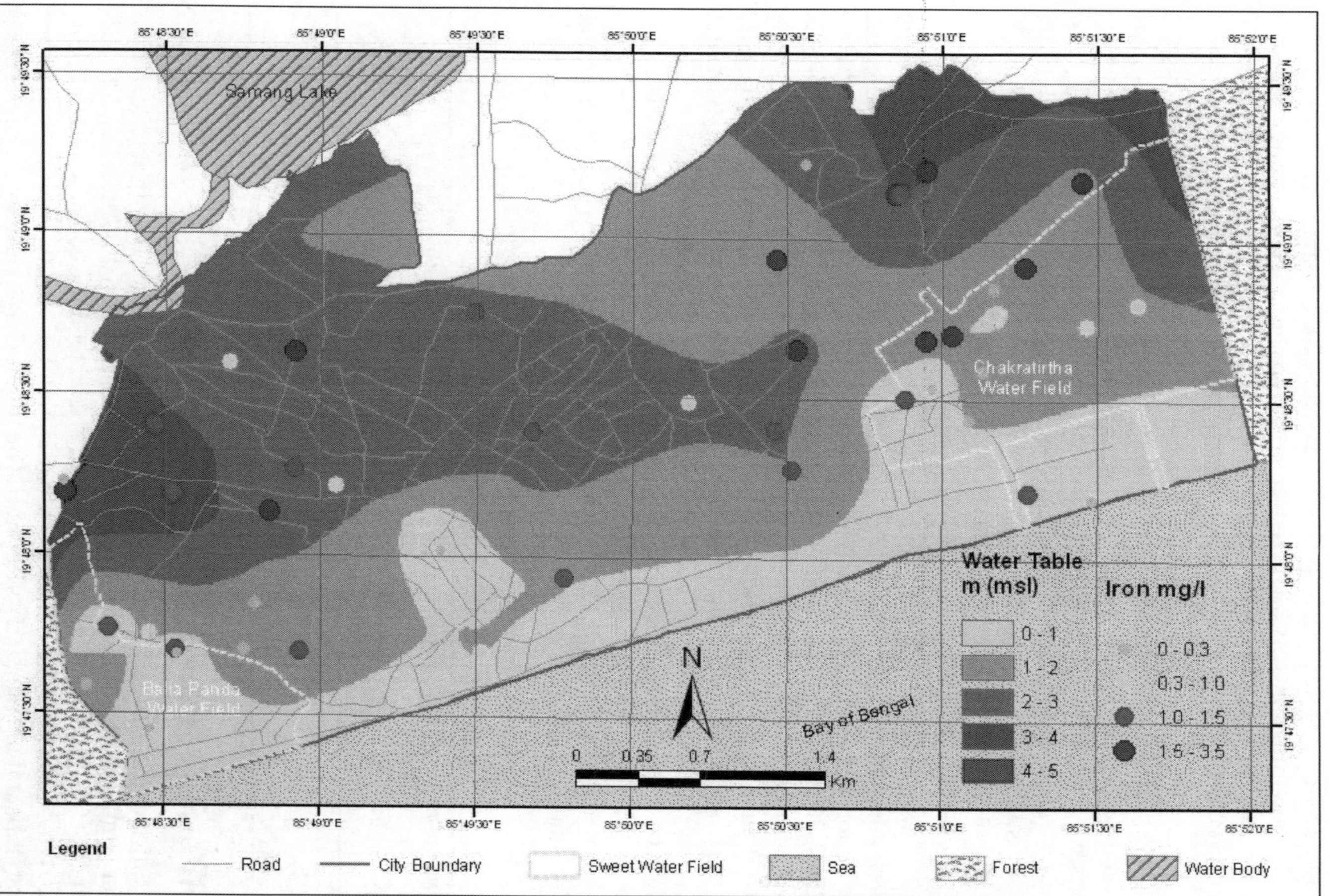

Figure 19.4: Distribution of Iron and Groundwater Level.

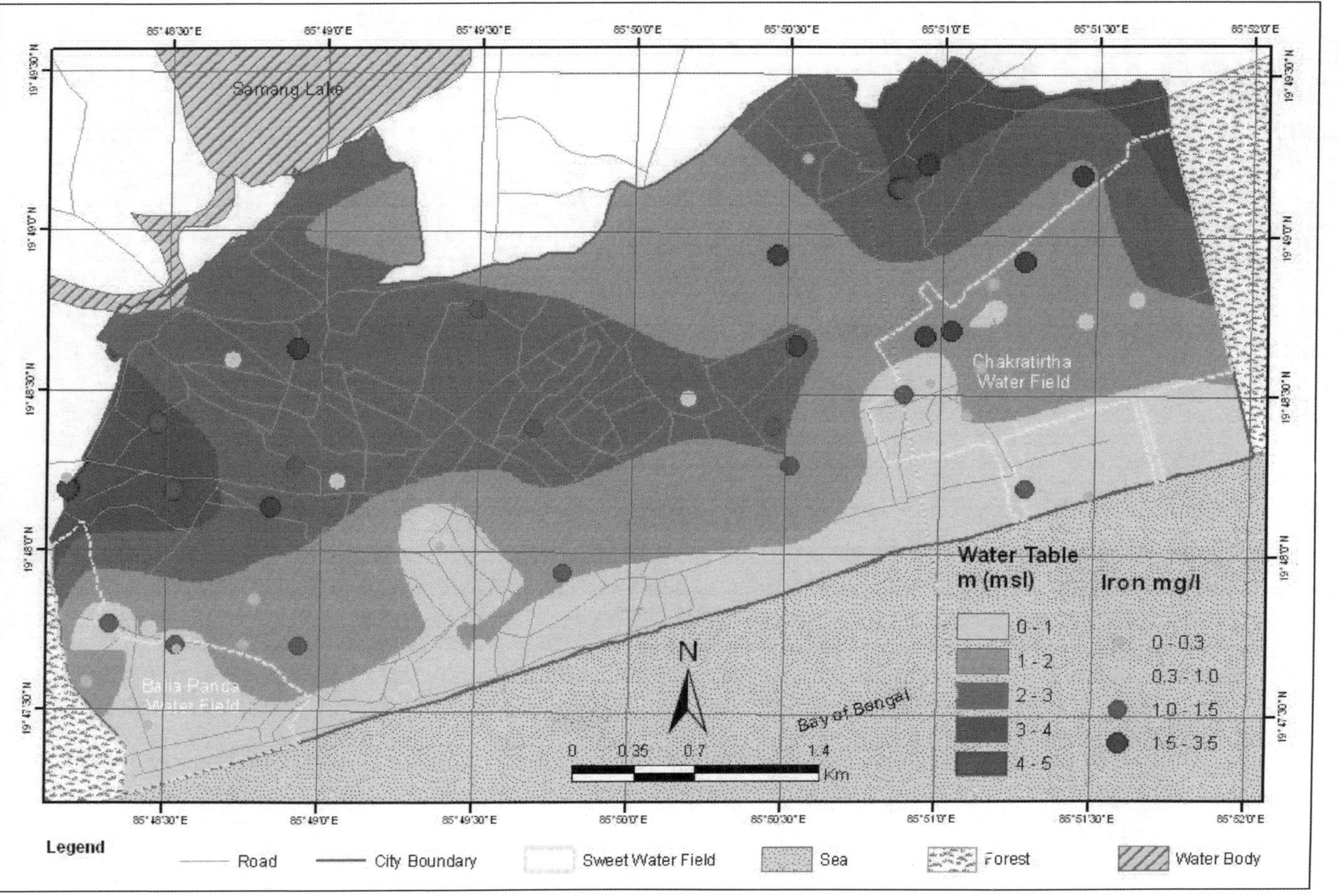

Figure 19.5: Distribution of Iron and Groundwater Level.

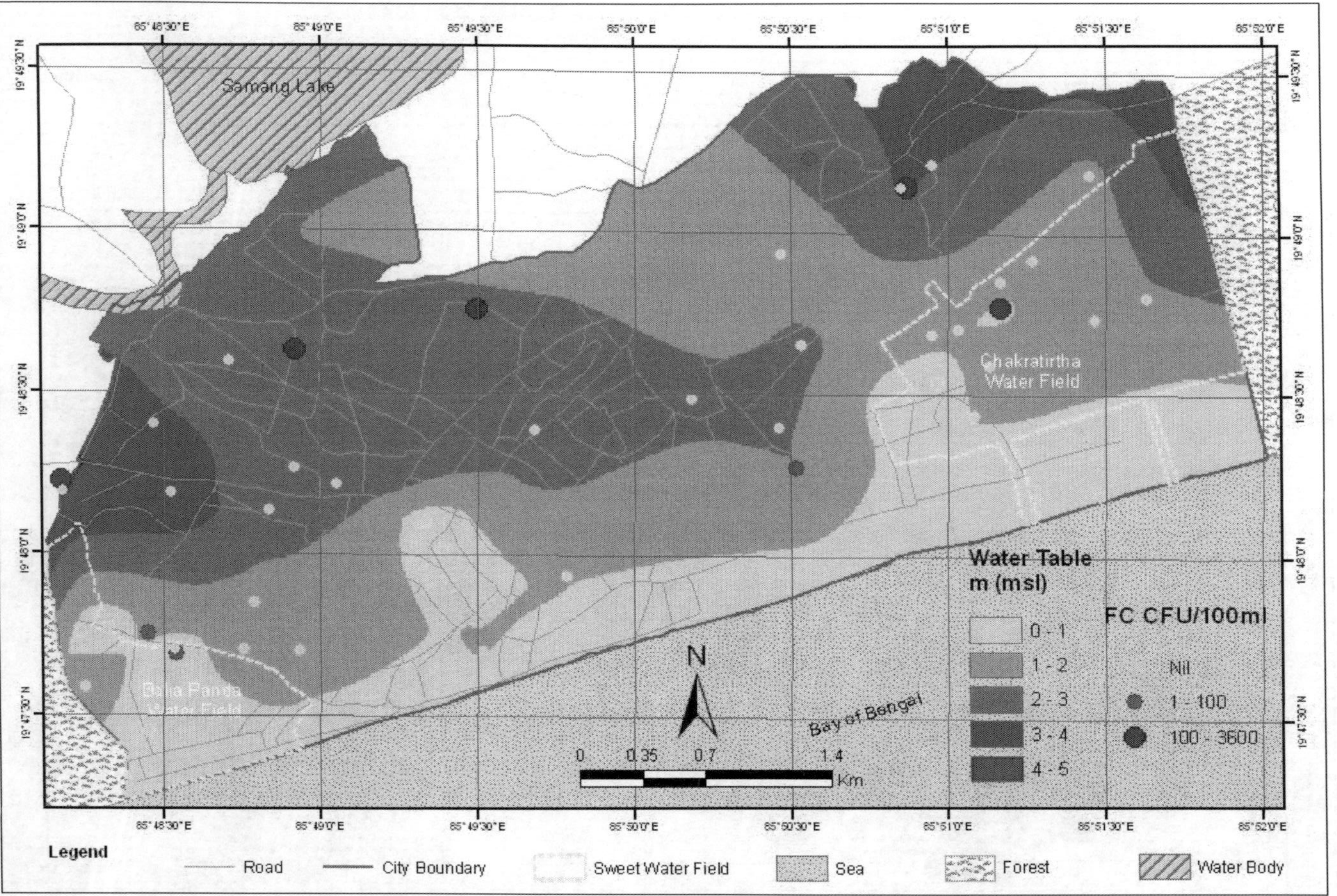

Figure 19.6: Distribution of FC and Groundwater Level.

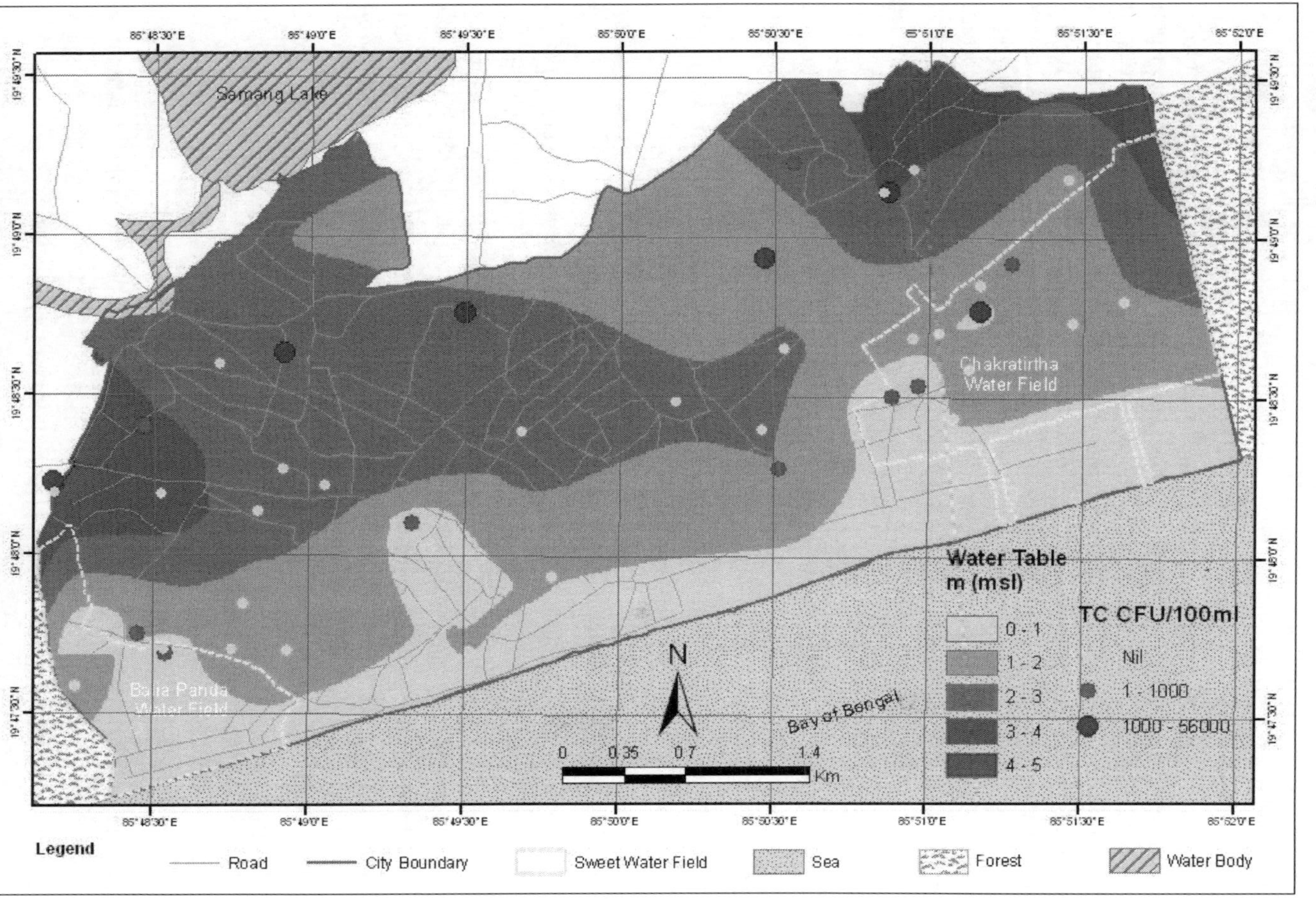

Figure 19.7: Distribution of TC and Groundwater Level.

The groundwater level in city is not varied much therefore shallow water table depth is governing factor for contamination. Even, contaminant move in the vicinity of the water fields due to continuous pumping.

Based on groundwater quality and spatial assessment contaminants are more in city area compared to water fields. Although contaminants are moving towards the water fields as observed due to continuous withdrawal, require protection from the man made activities in terms of slums, discharge of domestic waste, open defecation, bathing, cleaning of vehicles etc. Those activities should be stopped immediately to safe guard the groundwater resources. Since the sewerage system of city is under construction and treatment of sewage will also help in improving the groundwater quality. Bacteriological results reveal that presence of harmful bacteria in groundwater as well as in water fields require chlorinated water supply to city. Further steps should be taken to stop practicing of soak pits, open defecation, contaminated water pools and unorganized waste dumping to check groundwater contamination.

References

1. APHA (1995). Standard Methods for the examination of water and wastewater. American Public Health Association, American Water Works Association, Water Pollution Control, Federation, NY, Washington DC.
2. ArcGIS 9.3, (2004). What New in ArcGIS 9, Environmental System Research Institute, Jeremiah Lindemann, Krista Page, Redlands, CA, USA.
3. Babiker, S.I., Mohamed. A. A and Hiyama, T. (2007). Assessing groundwater quality using GIS Water Resource Manage, 21, 699-715.
4. Basappa Reddy, M. (2003). *Status of groundwater quality in Bangalore and its Environs*. Technical Report, Department of Mines and Geology, Bangalore.
5. De Loe, R., Di Giantomasso, S., and Kreutzwiser, R. D. (2002). Local capacity for groundwater protection in Ontario.Environmental Management, 29(2), 217–233
6. Jha, M. K., Chowdhury, A., Chowdary, V.M., and Peiffer, S. (2007). Groundwater management and development by integrated remote sensing and geographic information systems: prospects and constraints. Water Resource Management, 21, 427-467.
7. Julio, F., Ferreira da Silva and Naim Haie. (2007). Optimal Locations of Groundwater Extractions in Coastal Aquifers. Water resource management, 21, 1299–1311.
8. IS:10500 (1991). Drinking Water Standard, Manad Bhaban, New Delhi
9. Labrakis, N. J. (1998). The impact of human activities in the Malia coastal area (Crete)
10. NEERI (1986). Laboratory manual on water analysis, Nagpur.
11. NEERI (2000). Report, Design of Sewerage, Sewage Treatment and Utilization Systems at Jagannath Puri, Orissa.

12. NEERI (2008). Study of groundwater aquifer system at Jagannath Puri and its protection from contamination, Orissa.

13. Nieto, P., Custodio, E., and and Manzano, M. (2005). Baseline groundwater quality: a European approach, Environmental Science and Policy, 8, 399-409.

14. on groundwater quality, Environmental Geology, 36(1–2), 87-92

15. Rahman, A. (2008). A GIS based DRASTIC model for assessing groundwater vulnerability in shallow aquifer in Aligarh, India, Applied Geography, 28, 32 - 53

16. Shankar, B. S.; Balasubramanya, N. and Maruthesha, Reddy M. T. (2008). Impact of industrialization on groundwater quality–a case study of Peenya industrial area, Bangalore, India, Environmental Monitoring and Assessment 142:263–268.

17. Srivastava, P. K., and Bhattacharya, A. K. (2006). Groundwater assessment through an integrated approach using remote sensing, GIS and resistivity techniques: a case study from a hard rock terrain. International journal of Remote Sensing, 27(20), 4599–4620

18. Todd, D. K. (1980). Groundwater Hydrology (2nd Ed.), New York, NY: Wiley.

Chapter 20

Need for Environmental Policy and Legal Issues at National Level

D.T. Devale

Maharashtra Pollution Control Board, Mumbai

Tiwari Committee appointed by the Department of Science and Technology, Govt. of India for the recommendations of the Legislative Measures and Administrative Machinery for the protection of environment submitted its report on15th September, 1981. After review of a number of Environmental Legislations observed that

1. Many of the existing Environmental Laws are outdated.
2. They lack statements of explicit policy objectives.
3. They are mutually inconsistent
4. They lack adequate provisions for helping the implementing machinery.
5. There is no procedure for reviewing the efficacy of the Laws.

The committee suggested certain recommendations for making provisions for making it more effective such as,

(a) Comprehensive review and reformation of some Central and State Acts like Insecticides Act, 1968; The Water (Prevention and Control of Pollution) Act, 1974; Indian Forests Act, 1927 etc.;

(b) New Legislation for the areas not covered by the present Laws (*e.g.* Toxic Substances); and

(c) The introduction of "The Environment Protection" in the concurrent list of the seventh schedule.

The Parliament had taken number of steps after receipt of the report, such as, a comprehensive review was taken of the existing specific Environmental Laws and reformation done on the lines of the Environment (Protection) Act, 1986, by incorporating drastic amendments in the Legal Control and Remedies under the provisions of the Water (Prevention and Control of Pollution) Act, 1974 and the Air (Prevention and Control of Pollution) Act, 1981. The Water (Prevention and Control of Pollution) Act, 1974 has been made more stringent by introducing higher cess rates for non-complied industries and local bodies as well as covering almost all the industries, operations and processes consuming water more than 10 M^3 as well as reducing the rebate from the 70 per cent in the initial Act upto that of 25 per cent of compliance of the consent conditions meeting all the environmental standards and consumption of the water to the prescribed Authority with by it. This was to ensure conservation of precious nature resource-water.

Besides that grant of consent/s to new industries, operations or processes and new discharges or outlets have been made two stages *i.e.* (i) Consent to Establish and (2) Consent to Operate. This has facilitated to secure installation of the pollution control devices prior to grant of Consent to Operate itself, thereby, prevent, control or abatement of pollution at source itself.

Further Quasi-Judicial powers are given to the Boards to regulate the activities of the polluting industries by issuance of appropriate directions including the directions of closure or prohibition or regulation of polluting activities.

The Parliament has passed a general and comprehensive Legislation after catastrophic event of the Bhopal Tragedy to fill up the gaps in the existing specific Environmental Legislations covering more sources of environmental pollutions, including solid, liquid and gaseous substances responsible for the pollution of the environment for the first time, such as, handling of hazardous substances, bio-medical wastes, plastic wastes, municipal solid waste, lead acid batteries, ozone depleting substances etc. The concept of the Environment Impact Assessment, Public Hearing and Environmental Clearance has been introduced to examine most polluting activities before taking any effective steps.

However, in spite of above review and reformation of various Environmental Laws, it is observed that in the legal matters, we continue to follow blindly on inherited tradition legislating laws after Laws in a set pattern without ever inquiring, whether such Laws are the eight instruments for the purpose. For example, what is the rational behind making violations of the Environmental Laws, a matter of criminal offence ? when it is clear that the penalties imposed after lapse of a considerable time and effect have no relation to the injustice done. There is no jurisprudential thinking, but we continue the tradition left by the British.

Therefore, Maharashtra Pollution Control Board had taken a review of it's legal actions in 1995 and observed that filing of prosecutions against defaulting industries, operations or processes is not an effective remedy. It was noticed that the penal provisions remained as threats for a while and fade away after some years, the Board has therefore resorted to preventive measures provided under Section 33A/31A for both under the Water and Air Acts respectively. Today, preventive measures are playing very important role in implementation of both the Acts. The Board had taken effective action under Section 33A and 31A of the above Acts by issuing directions including upgradation of existing pollution control devices, adopting cleaner technologies and prescribing most stringent conditions to decrease pollution load by restraining the discharges into the stream or river and direct disposal on land for irrigation/gardening purposes including re-circulation of the water in the process itself. The results of such directions are found to be encouraging. The Board could have secure about 150 crores investment from the industries in the State of Maharashtra for the installation of most modern technologies for the treatment and disposal or emission etc. The Board has also introduced a system of bank guarantee for ensuring the compliance of the above directions and in case of non-compliances, the bank guarantees are forfeited and credited to the accounts of the Board.

As far as the remedies available to the individual, organizations or public at large under the above Acts, they can file prosecution after giving 60 days notice but in practice, nobody has resorted to this remedy taking into consideration very technical nature of offences, non-availability of technical reports and data as well as vexatious nature of the proceedings in the Courts of Law. Therefore, the individual, organization or public at large have resorted to public interest litigations, wherein our Hon'ble High Court and Supreme Court have played an important role in achieving the desired results by issuing the noteworthy Directions/Judgments/Principles. The Board has shifted it's legal actions from filing of prosecutions to that of initiating Quasi-Judicial proceeding of issuance of the directions after 1995-96.

Tiwari Committee further observed in it's report that till then there had not been a single in-depth study evaluating the factors, which would help the law to achieve the national objections mentioned in the National Environmental Policy.

The present National Policies for the environmental management are contained in the National Forest Policy, 1988, the National Conservation Strategy and Policy Statement on the Environmental and Development, 1992 and the Policy Statement on Abatement of Pollution, 1992.

Besides that some sect-oral policies such as the Agricultural Policy, 2000 and the National Population Policy, 2000 and the National Water Policy, 2002 also contributed towards environmental management. All these policies recognize the need for sustainable development. Our National Environmental Policy, seeks to extend coverage and fill in the gaps that still exist in the light of the present knowledge and accumulated experience. It does not displace, but builds on the earlier policies.

Though the National Forest, 1988 formulated for the conversation and development of forests/green belts, our specific Environmental Laws like the Water and Air Acts, Environment (Protection) Act, 1986 could not make requisite provisions for maintaining green belts or not disturbing forest areas. Day by day, forest coverage in India alongwith the endangered species depleting.

Our Agricultural Policy could not discourage the use of chemical fertilizers and thereby, the fertility of huge acres of land has been considerably affected.

The National Population Policy of India could not promote voluntary measures for controlling population growth, we are now nearby Chinese population because it has been successfully controlled it's population growth. National Water Policy, 2002 also could not achieve it's objections.

Out of 22 Municipal Corporations, 8 Corporations are discharging their sewage without treatment and remaining are having inadequate capacity of STP to treat total generation of sewage from their areas and thereby, causing pollution of the river, creek and other natural resources. Similarly, the status of the sewage treatment at Municipal Councils is the worst than that of the Corporations. They are causing discharge of untreated sewage into the natural water bodies and thereby, almost all the rivers in the State of Maharashtra have become polluted. The problem of drinking water at the downstream of Panchganga river, which is the source of drinking water for the downstream villages become more acute in the summer season, giving rise to a number of agitations and unrest in the said area.

However, when we go through our Environmental Legislations, we find that there is no integrated approach to achieve the objections in the National Policy. Thereby, we could not introduce environmental concerns in all development projects. The proximate drivers of environmental degradation are population, inappropriate technologies, consumption choices and poverty leading to changes in relations between the people and eco-systems as well as development activities such as intensive agriculture, polluting industries and unplanned urbanization. It is increasingly evident that the poor environmental quality has adversely affected human health.

It is said that the law relating to the environment is derived from the two principal sources, namely (i) common law developed by the Courts through Judicial Precedents; and (ii)the statutory law with regulations or bye-laws made thereunder. For example, a person can sue for nuisance when an act endangers his life, health, property and comfort or obstructs his enjoyment of the rights common to all people. It is an act or illegal omission, which causes any common injury, danger or annoyance to the public who dwell, or occupy property in the vicinity or person, who have occasion to exercise public rights.

With the development of science and technology and with the ever-increasing world population, tremendous changes brought in the human environment. It, therefore, become necessary to regulate human behavior and social transactions with new laws, designed to suit the changing conditions and values. A new branch of the law, called Environmental Law, grew at this stage in order to manage and face the myriad challenges of such a system.

Environmental Law is of the recent origin. Environmental Laws, therefore, relates to the management of the environment and strategies for tacking the problems affecting the environment.

Environmental Law involves conservation of natural resources for their better use by the present day society as well as by future generations and it also governs the inter-relationship between natural resources and human beings and also extends to relationship between natural resources and other living creatures.

In this paper, an humble attempt is made to highlight some of the factors responsible for the journey of specific Environmental Laws and Legal Policy followed for the various Environmental Protection Laws and major problems of implementation in our Legal Policy as under:-

Development of Environmental Jurisprudence through our High Courts and Supreme Court of India

Firstly, the principles developed in the International Conventions and Treaties were brought to be incorporated in our Municipal Laws as a part of the Customary International Law.

Secondly, our High Courts and Supreme Court of India taken cognizance of the non-implementation of the various Environmental Laws and the Public Interest Litigations has become a growing mechanism in the field of environmental protection in India. In India, a class action against public nuisance can be brought under Section 91 of the Code of Civil Procedure and Section 133 of the Code of Criminal Procedure. Our High Courts and Supreme Court of India invoked jurisdiction under Article 226 and Article 32 of the constitution respectively and have made significant contributions to the development of the Public Interest Litigations (PIL) through environmental cases.

In order to formulate legal principles for environmental protection, the Indian Judiciary rejected inadequate and outdated doctrines by relaxing rule of locus-standi to ensure environmental justice through the Public Interest Litigations and by evolving an indigenous Environmental Jurisprudence by rejecting exceptions to the traditional common law doctrine of strict liability in Shriram Fertilizer case of hazardous and inherently dangerous industry.

India plays an important role in several significant international initiatives concerned with the environment. It is a party to key multi-lateral arrangements and recognizes the interdependence among and trans-boundary character of several environmental problems. National Environmental Policy is also intended to be statement of India's commitment to make a positive contribution to international efforts. In fact, our Air (Prevention and Control of Pollution) Act, 1981, the Environment (Protection) Act, 1986 are passed under the Legislative subject, "the decisions taken at the United Nations Conference on Human Environment held at Stockholm in June, 1972".

Our National Commitment to a clean environment, mandated in the Constitution, in Article 48A and 51A (g), strengthened by the Judicial Interpretation of Article 21 of the Constitution of India.

The Constitution (Forty-second Amendment) Act, 1976, inserted in the Constitution for the first time specific provisions to, "protect and improve" the environment. These provisions are duty oriented. One of the directive principles required that State shall endeavor "to protect and improve the environment" and "to safeguard the forests and wild life of the country"

The Amendment Act also introduced a new Part: IV-A dealing with the fundamental duties lays down the duty of every citizen of India to protect and improve the natural environment including forests, lakes, rivers and wild life and to have compassion for living creatures under Article 51A(g). This particular amendment (42nd) of the Constitution transformed the Democratic Republic of India into a "Socialist" one and accordingly, the changes were effected to suit this wavelength."

The International Concepts like Global Warming, adoption of cleaner technologies, sustainable development, precautionary and polluters pay principle, Public Trust etc., which have been incorporated in various Judgment and Orders passed by the Hon'ble High Courts and Hon'ble Supreme Court of India have not so far been incorporated in the various Environmental Laws. The result can be summerised as under:

1. End of pipe treatment technologies have been adopted by the chemical and allied industries for several years. This approach first allows, in some cases, the production of waste and wasteful use of resources in manufacturing and then, consumes further resources to convert the waste generated into different potentially usable/recoverable material or otherwise to other forms without reducing the net contributions to the environment. As experienced in the industrialized countries, such an approach is a short term remedy and can not lead to sustainable industrial development in the long term.
2. Our existing Environmental Laws could not make effective provision to deal with 'Global Warming' *i.e.* the increase in average temperature of the Earth's near surface air and oceans in recent decades and its projected continuation. Increasing global temperature will cause sea levels to rise and expected to increase the intensity of extreme weather events. Other effects include changes in agriculture yields, trade routes, glacier retreat, species extinctions and increases in the ranges of disease vectors. Necessary policies will have to be formulated to contain the effects of global warming by incorporating necessary provisions in our Enviornmental Laws.
3. Though, the Public Interest Litigation has been an effective remedy in the last 20 years and a number of land mark orders have been passed by our High Courts and Hon'ble Supreme Court of India, the Judgment of the U.S. Court in the Bhopal Case has been criticized by the Indian Jurists as ignoring the affirmation of public interest by the sovereign state of India, the Mehta dictum (authoritive opinion) has been criticized by the Jurists outside India as 'imposition of liability by Judicial dictate (to give orders to) regardless of the general principles of law, which would 'undermine the creditability of India Courts and lead to accusations of bias from foreign investors.
4. In Union Carbide Corporation v/s Union of India (1989 3 SCC 38) the Supreme Court of India stated that securing immediate relief to the victims was found necessary to enter into compromise with Union Carbide Corporation.

It is said that the Supreme Court seems to have deliberately missed an opportunity to develop new principles in relation to multinational corporations operations with inherently dangerous technologies in the developing countries.

It has been observed that the basic jurisprudential theory that governs the Environmental Laws is, a Policing the Society Theory. Whereunder, the legislators as well as administrators assumed that their task is to act as vigilant policeman, who detects crime and bring the culprits to the courts. In the courts, the matters are tried on the basis of whatever evidence recorded and not on the basis of the non-compliance brought to the notice of the courts.

Solutions Suggested

1. The offences coming under the Environmental Laws can be redefined from being one of criminal liability to that of being a liability in tort. Litigation concerning tortuous liability would shift the problems from being one in criminal law to that in civil law, which are more simpler than criminal laws.
2. Under Tort Law, the compensation to be paid by the defaulting parties can be rationalized on an economic basis, directly linking it to the degree of harm done.
3. If, compensation is to be made equitable in Tort, it becomes necessary that the concerned Boards and the Departments of Environment be given the right to make the required assessment.
4. The non-violators should be provided with sufficient incentives to comply with the laws, by allowing sufficient tax deductions for those, who do comply with the laws. This will link it up with the civil assessment of violators.

The Boards will have to plan and suggest remedial measures for various sources of the pollution. The function of the Board as such has become multi-disciplinary after entrusting to it the implementation of various Environmental Laws including the Environment (Protection) Act, 1986 and various Notifications issued thereunder. Therefore, the question is not one how to solve the pollution problem, but how to solve the problem of determining priorities by the Government.

Chapter 21

Mathematical Modeling of Infectious Disease: A Case Study

Manju Agarwal, M.M. Agarwal and Samina Hasan

Global Environmental and Agriculture Centre,
HIG–79, Sector–E, Aliganj, Lucknow

ABSTRACT

In this chapter, a Predator-Prey (SEI) epidemic model is investigated. This model describes the transmission of avian influenza among birds and humans. Bird population is considered as the prey population and human population is assumed to be the predator population, thus we arrive at an eco-epidemiological model in which bird population acts as a reservoir of disease and become infected by acquiring that disease. Predator or the human population is assumed to contract disease when they come into contact with infected birds population. Mathematical analysis of the model regarding stability and persistence has been performed. We identified the parameters which influence the persistence of all population and noted that the basic reproduction ratio R_0 plays an important role in shaping the dynamics of the system. Since, the predator feeding is not instantaneous; we have considered time delay due to gestation. We have also incorporated the effect of diffusion due to migration of different individuals of the population. Also some numerical simulation for supporting the theoretical is given.

Keywords: *Predator-prey, Susceptible, Exposed, Infected, Impermanence, Delay, Diffusion, Infectious.*

Introduction

The use of mathematical models for understanding the dynamics of infectious diseases has been in practice for more than a century. In recent times this has grown in popularity and it is now often confused with epidemiology of infectious diseases. Mathematical modelings of infectious diseases use equation systems to represent elements of the dynamics of infectious processes, involving agent, host, and environment. As any model, mathematical models are useful for it is a powerful instrument for analysis, to predict situations, and even for evaluating the potential capacity of certain interventions to change the likelihood of new cases. The difficulty in transforming the complex situations involved in the process of transmission of many infectious agents into mathematical models is a limitation to their use in many situations.

In mathematical models of infectious diseases population is divided into three states: susceptibles (S) who can acquire infection; infectives (I) who can transmit infection to susceptibles; and removals (R) who are immune or dead as a consequence of infection. Such models of endemic infection are therefore known as the SIR model. The population in recovered subclass may either become immune or again become susceptible after recovery depending upon the disease under consideration. However, there are many diseases which incubate inside the host for a period of time before the hosts becomes infectious in latent period. The resulting models are of SEIR or SEIRS types which are more realistic than the former SIR and SIRS type models. A key concept in mathematical modeling of infectious diseases is the basic reproduction number (originally named as basic reproductive ratio or rate) (R0) meaning "the average number of secondary infections produced when one infected individual is introduced into a host population in which every host is susceptible" (in the case of microparasitic infections) or "the average number of offspring (or female offspring in the case of dioecious parasites) produced throughout the reproductive lifespan of a mature parasite that survive to maturity in the absence of density dependent constraints to population growth" (in the case of macroparasitic infections). There are three basic potential situations related to the R0 value. If R0 = 1, the number of cases is stable and indicates that the disease is endemic, if R0>1 the number of cases is increasing and it will eventually become an epidemic; and if R0<1 this means that the number of cases is decreasing and epidemic would not spread in the population.

Infectious diseases constituted the most serious health issue in the world until the beginning of the 20th century when chronic degenerative diseases began to dominate this scenario in developed countries. McNeill [1] analyses the importance of infectious diseases in the history of humanity and concludes that the role of infectious diseases in the set of factors that defined the course of the historical evolution of human civilization has been underestimated, and considers that this role was as important as that of economic and military determinants.

Hippocrates [2] established the existence of links between the occurrence of diseases and the environment in which populations lived. The idea that some diseases could be transmitted between people (or be contagious) is also old and became the basis for the institution of preventive actions even before the existence and importance of microbian agents had been scientifically proved. Fracastoro [3] wrote the first theorisation of the concept of the transmission of diseases between people through minute particles. However, it was not scientifically shown before the middle of the 19th century, when Snow's [4] indicated the water used by the population as a possible source of transmission of cholera. However, it was only at the end of the 19th century that the idea of contagion finally overthrew its opponent, the miasmatic theory, in the intellectual battles nurtured by new scientific advances that truly reflected different visions and philosophical perspectives on the world, on society, and on its diseases. The advances in microbiology that took place during that century, particularly the works of Koch and Pasteur, [5]confirmed the role of live microscopic agents as an immediate, fundamental cause of a wide range of morbid entities, thereby provoking profound changes in the understanding of the causes of infectious diseases and the consequent strategies for their prevention.

The terminology and concepts used today in epidemiology of infectious diseases (EID) evolve from a complex set of scientific fields that studied their agents, their causes and determinants, the dynamics of transmission and diffusion of these agents, and their means of prevention being. Bearing in mind that in its early phases of development EID constituted epidemiology itself, many of the concepts used in epidemiology are derived from this time. More recently, an inverse phenomenon has taken place in which terms used in EID have originated from other areas of epidemiology or other disciplines.

In this paper, a Predator-Prey (SEI) epidemic model is investigated. This model describes the transmission of avian influenza among birds and humans. Mathematical analysis of the model regarding stability and persistence has been performed. We identified the parameters which influence the persistence of all population and noted that the basic reproduction ratio R_0 plays an important role in shaping the dynamics of the system. Since, the predator feeding is not instantaneous; we have considered time delay due to gestation. We have also incorporated the effect of diffusion due to migration of different individuals of the population.

The Mathematical Model

In the present study, our model consists of two populations: (1) the avian population, whose total density is denoted by $N(t)$, (2) the human population whose population density is denoted by $Y(t)$.The avian population is divided into three groups namely, susceptible S(t), exposed $E(t)$, and infectious $I(t)$.Therefore, at any time t the total density of the avian population is

$$N(t) = S(t) + E(t) + I(t).$$

We make the following assumptions in our model:

1. We have considered avian population as the prey population and human population as the predator population. In the absence of disease, the prey population grows according to the logistic law with intrinsic growth rate $r(r>0)$ and carrying capacity $K(K>0)$. We assume that infectious avian population suffers an extra disease related death with constant rate α. The natural death rate is assumed to be constant d for all hosts and it is assumed that prey population does not recover or become immune.
2. We assume that disease spread among the avian population through direct contact. It is assumed that after getting infection a host stays in latent period

before becoming infectious The incidence term of the disease is of bilinear mass action form βIS where β is the probability per unit time of transmitting infection between individuals. μ is the transfer rate among the corresponding classes. $1/\mu$ is the mean latent period during which prey population remains in exposed class.

3. We assume that disease spread among the human population when they come into contact with infectious avian population either by eating undercooked flesh of infectious birds or by touching the surface contaminated by avian influenza virus. Natural death rate of human population is assumed to be constant, $\nu > 0$.
4. The interaction coefficient of human and avian population is p (>0). q, (0<q<1) is the conversion rate of susceptible human population.

Based on the following assumptions our model takes the following form:

$$\frac{dS}{dt} = rS\left(1 - \frac{S+E+I}{K}\right) - \beta IS - pSY,$$

$$\frac{dE}{dt} = \beta IS - (d+\mu)E - pEY, \tag{1}$$

$$\frac{dI}{dt} = \mu E - (d+\alpha)I - pIY,$$

$$\frac{dY}{dt} = -\nu Y + qpSY + qpEY - qpIY,$$

with initial conditions $S(0) = S_0 > 0$; $E(0) = E_0 > 0$; $I(0) = I_0 > 0$; $Y(0) = Y_0 > 0$.

Our analysis utilizes the techniques discussed in Freedman and Waltman [6]. In particular, we use a lemma of G. J. Butler and R. P. McGhee given in [8], which roughly states that if the omega limit set of bounded semi orbit contain an isolated equilibrium P, then either P is the omega limit set, or there are points (different from *P*) of the omega limit set in both the stable and unstable manifolds of *P*. We define a vector function of t, x(t) with positive components, to exhibit persistence if for each component $x_i(t)$, $\liminf x_i(t) > 0$. We mean that each solution vector has this property, for a system of differential equation to be persistent [6]. Biologically, persistence means the survival of all populations in future time. Mathematically, persistence means that the strictly positive solution does not have omega limit points on the boundary of the non negative cone. We say that a system persists uniformly whenever each component persists uniformly [7].

Basic reproduction ratio is found which is denoted by R_0 and is defined as the average number of secondary infections in a completely susceptible population when one infective enters the population. When $R_0 < 1$, each infected individual produces, on average, less than one new infected individual, and we therefore predict that the infection will be cleared from the population or the microparasite will be cleared from the individual. If $R_0 > 1$, the pathogen is able to invade the susceptible population. This threshold behavior is the most important and useful aspect of the R_0 concept. In an endemic infection, we can determine which control measures, and at what magnitude, would be most effective in reducing R_0 below one, providing important guidance for public health initiative.

The Mathematical Model with Delay

There exists a discrete time delay due to gestation of the predator in real life situation and, therefore time delay is to be considered to make the model more realistic. If time delay τ is considered in our model we arrive at the following set of differential equations:

$$\frac{dS}{dt} = rS\left(1 - \frac{S+E+I}{K}\right) - \beta IS - pSY,$$

$$\frac{dE}{dt} = \beta IS - (d+\mu)E - pEY,$$

$$\frac{dI}{dt} = \mu E - (d+\alpha)I - pIY, \tag{2}$$

$$\frac{dY}{dt} = -\nu Y + qpS(t-\tau)Y(t-\tau) +$$

$$qpE(t-\tau)Y(t-\tau) - qpI(t-\tau)Y(t-\tau), (2)$$

With the initial conditions,

$$S(t) = S_0 > 0; E(t) = E_0 > 0; I(t) = I_0 > 0; Y(t) = Y_0 > 0,$$

Wher, $S_0(t), E_0(t), I_0(t)$ and $Y_0(t)$ are given nonnegative functions on $-\tau < t < 0$.

The Delayed Model with Diffusion

Since there is continuous movement of different individuals between the compartments so diffusion plays an important role in mathematical modeling. In this section we have considered the effect of diffusion in the delayed

SEI model. On applying diffusion in the delayed model we get the following set of differential equation:

$$\frac{\partial S}{\partial t} = rS\left(1 - \frac{S+E+I}{K}\right) - \beta IS - pSY + D_S\frac{\partial^2 S}{\partial x^2},$$

$$\frac{\partial E}{\partial t} = \beta IS - (d+\mu)E - pEY + D_E\frac{\partial^2 E}{\partial x^2},$$

$$\frac{\partial I}{\partial t} = \mu E - (d+\alpha)I - pIY + D_I \frac{\partial^2 I}{\partial x^2},$$

$$\frac{\partial Y}{\partial t} = -vY + qpS(t-\tau)Y(t-\tau) +$$

$$qpE(t-\tau)Y(t-\tau) - qpI(t-\tau)Y(t-\tau) + D_Y \frac{\partial^2 Y}{\partial x^2},$$

Where, D_S, D_E, D_I and D_Y are the diffusion coefficient of the susceptible, exposed, infectious prey and predator respectively.

Discussion

Now a days, various infectious diseases are spreading all over the world. Even though vaccines are available still they cause suffering and mortality all over the world. An epidemiological model uses a microscopic description to predict the microscopic behavior of disease spread through a population.

In this paper, we have considered an eco- epidemiological model namely SEI model. Since, the predator feeding is not instantaneous; we have considered time delay due to gestation. We have also incorporated the effect of diffusion due to migration of different individuals of the population. The system is analyzed for its equilibria and their stability. The main results of our investigation can be summarized as follows:

- ☆ The analysis in section (3) revealed that the death rate of predator plays an important role in shaping the dynamics of the system. The parameter R_0 determines whether the disease persists or not. The parameter R_0 is the basic reproduction ratio [1]. It represents the average number of secondary infections from a single host in a totally susceptible population. It is observed that if $R_0 > 1$ and the death rate of predator v remains below a certain threshold value, the infection does not tend to zero. We have derived criteria for impermanence. It is found that the system (1) is impermanent if either death rate of predator v exceeds its net energy intake rate at the equilibrium density or the equilibrium density of the susceptible prey remains below a certain threshold value.
- ☆ In section (4), we have modified the original model (1) by considering the delay effect due to gestation of the predator. It is found that for large τ the existence and stability properties of the axial and boundary equilibria remain unchanged.
- ☆ In section (5), the effect of diffusion of different populations on the delayed system is studied. It is observed that stability of the system depends on the diffusivity and the wave number of perturbation (p_1). A high value of diffusivity of the infected and exposed prey population (D_I) and (D_E)

respectively and low diffusivity of predator population (D_Y) along with high death rate of predator (ν) ensures the stability of the disease free equilibrium.

References

1. McNeill W H. Plagues and peoples. New York: Doubleday, 1977.
2. Hippocrates. Airs water and places. Cambridge: Harvard University Press, 1948.
3. Fracastoro G. Il contagio, la malattie contagiose e la lora cura. Firenze: Leo S Olschki Editore, 1950.
4. Snow J. On cholera. New York: Commonwealth Fund, 1936.
5. H.I.Freedman and P.Waltman, Persistence in a model of three competitive populations, Math. Biosci. 73:89-101(1985).
6. V.Hutson and G.T. Vickers, A criterion for permanent co-existence of species with an application to a two prey one predator system; Math.Biosci.63 (1983),253-269.
7. G.J.Butler, H.I.Freedman and P.Waltman, Uniformly persistent system, Proc.Am.Math.Soc.96 (1986), 425-430.

Chapter 22

Biodesulfurization of Diesel in Internal Loop AirLift Reactor

Somnath Nandi

Department of Petroleum and Petrochemical Engineering, MAEER's Maharashtra Institute of Technology, Paud Road, Kothrud, Pune – 411 038, Mh.

ABSTRACT

Hydrodesulfurized diesel with sulfur content 200 ppm (mg/dm^3) is used for biodesulfurization study using *Rhodococcus* sp., so that resultant diesel is almost free of sulfur and compatible with US and European standards. An internal loop AirLift reactor of overall volume 23.675 dm^3 was indigenously designed. Hydrodynamic and mass transfer studies were conducted previously with air-water systems with the designed set-up. We had also developed generalized mathematical models of the system which were cross validated with experimental findings (Nandi et. al., 2008). In the present paper we propose a detailed modeling and simulation study to demonstrate efficacy of AirLift reactor for biodesulfurization. Monod type growth kinetics was considered and the intrinsic kinetic parameters were taken from Mukhopadhyaya et. al., 2006.

Keywords: *Air lift reactor, Biodesulfurization, Modeling, Simulation, Environment friendly process.*

Introduction

Environmental regulation norms are becoming stringent day by day forcing refineries to take several corrective measures and change and/or improve on their standard processes in order to make the processes mode green and the saleable product within the specified norms. Sulfur is considered to be an inevitable element present from crude till the finished refined products. Conventionally hydrodesulfurization is carried in presence of Cobalt, Nickel and Molybdenum catalyst at moderate temperature (above 400 ^{0}C) and high pressure (4–10 MPa) using hydrogen as a reactant, which reacts with sulfur present as sulfide or disulfide or thiophene or mono or disubstituted thiophene in the refined diesel (Nandi and Deotare, 2007). But it is very difficult to remove polyaromatic sulfur compounds like alkylated di-benzothiophene, naphthothiophenes etc. So in order to reduce sulfur to an ultra level [10 ppm (mg/dm^3) or below] to meet global specifications, we need to introduce some other process. Microbial biodesulfurization is a promising area of research, as several biocatalysts has been identified which can transform sulfur compounds efficiently which are otherwise difficult to remove in conventional hydrodesulfurization process (Maghsoudi *et al.*, 2001; Chang *et al.*, 2001). Very recently Mukhopadhyaya *et al.* (2006) has performed and experimental and modeling study on biodesulfurization of hydrotreated diesel in a trickle bed reactor using bacterial strain *Rhodococcus* sp.

Airlift bio-reactors are becoming popular due to their simple constructions and the less shear force imposed on the living cells and microbes and low cost of operation and low maintenance is used now-a-days by many pharmaceutical industries and in wastewater treatment plants as well. Mixing in airlift bioreactors is usually imperfect and mathematical models for them cannot be described by either perfect mixing (continuous stirred tank reactors, CSTR) or by plug flow behavior (plug flow reactors, PFR) (Znad *et al.*, 2004). In the present investigation, a mathematical model based on tanks-in-series with back flow has been developed for a previously designed internal loop airlift reactor or volume 23.675 dm^3 and the same has been simulated for biodesulfurization of diesel. To best of our knowledge and belief, this is the first time we are proposing a rigorous phenomenological model based on fist principles and simulate it for biodesulfurization in Airlift reactor.

Modeling of Bio-desulfurization in ALR

Airlift reactor (ALR) is modified version of bubble column, which is divided into regions of gas-liquid up flow called riser section and the region containing gas liquid downflow, termed as downcomer. The internal loop airlift reactor have two concentric tubes where inner one acts as riser and annular section performs duty of the downcomer region. Mixing in ALR is generally imperfect. From mixing point of view, ALR can be sub-divided into four main sections: (i) the bottom where gas distributor is placed and medium is recycled from the downcomer, (ii) the riser where gas liquid flow upwards, (iii) the disengaging section at top of the assembly where some portion of gas is separated from the gas–liquid mixture, and (iv) the downcomer section where gas-liquid flow downwards. Each section has separate mixing characteristics so they are to be modeled separately. Two major modeling approaches are available in literature namely: axial dispersion model (ADM) and tanks in series model (TSM). Out of these two methodologies, TSM is more realistic and advantageous compared to ADM (Kanai *et al.*, 1996). So we have considered tanks in series approach for modeling of desulfurization of diesel in airlift reactor.

A modification of TSM can be done considering back mixing parameter *m*, as the ratio of back flow rate to net forward liquid flow rate. Flow in ALR is considered as flow through a series of equal sized well-mixed stirred tanks. The bottom section ($i = 1$) is treated as well-mixed stage. The riser and top sections ($i = 2, 3, .., A$) are described as tanks-in-series with back flow. As the flow in downcomer ($i = A + 1, A + 2, .., B$) is relatively well defined, the back flow in the downcomer is neglected. The model developed is schematically represented in Figure 22.1. At top section, most of the gas bubbles disengages and a small quantity is only entrained by liquid recirculation through downcomer. So the flow through downcomer is almost single phase and relatively well defined and hence backmixing effect (m) is neglected in this section. The riser, including top and bottom sections are divided into A hypothetical well mixed stages. Consequently, mixing in downcomer is represented by (B–A) stages without backflow. The stages in the riser are numbered upwards and that of downcomer are numbered downwards. The oxygen concentrations in gas phase is assumed to be uniform and is equal to the equilibrium dissolved oxygen concentration (C_O^*), was assumed constant throughout the airlift bioreactor and its value is 0.00651 g/dm^3 (*i.e.*, value of C_O at t = 0). At bottom section, the gas feed and recycled flow from downcomer are mixed well. The temperature is considered to be kept constant and hence no energy balance is taken into account. As the concentration of microbes or the reactants are negligibly small compared to overall volume of the reactor assembly, the effect of enthalpy change, if any, during desulfurization reaction is almost insignificant.

Material Balances in different Sections of ALR

The tanks in series model with back flow provides simultaneous 1st order ODEs which are material balances of the micro-organisms, substrate and dissolved oxygen for hypothetical well-mixed tanks or stages (Figure 22.1). The unsteady state material balances of components can be written as follows:

Bottom Section (i = 1)

For micro-organisms and substrate (C = X, and S)

$$\frac{dC_1}{dt} = \frac{Q}{V_b\left(1-\varepsilon_r\right)}C_b + \frac{bQ}{V_b\left(1-\varepsilon_r\right)}C_2 - \frac{\left(1+b\right)Q}{V_b\left(1-\varepsilon_r\right)}C_1 + r_{C,1} \tag{1}$$

For the dissolved oxygen (O)

$$\frac{dO_1}{dt} = \frac{Q}{V_b\left(1-\varepsilon_r\right)}O_b + \frac{bQ}{V_b\left(1-\varepsilon_r\right)}O_2 - \frac{\left(1+b\right)Q}{V_b\left(1-\varepsilon_r\right)}O_1 + \left(k_la\right)_r\left(O_1^* - O_1\right) + r_{O,1} \tag{2}$$

Riser Section (i = 2, 3, .., A–1)

For micro-organisms and substrate (C = X, and S)

$$\frac{dC_i}{dt} = \frac{m\,Q}{\frac{V_r}{(A-2)}\left(1-\varepsilon_r\right)}\left[C_{i+1} - C_i\right] + \frac{(1+m)\,Q}{\frac{V_r}{(A-2)}\left(1-\varepsilon_r\right)}\left[C_{i-1} - C_i\right] + r_{C,i} \tag{3}$$

For the dissolved oxygen (O)

$$\frac{dO_i}{dt}=\frac{m\,Q}{\frac{V_r}{(A-2)}(1-\varepsilon_r)}\left[O_{i+1}-O_i\right]+\frac{(1+m)\,Q}{\frac{V_r}{(A-2)}(1-\varepsilon_r)}\left[O_{i-1}-O_i\right]+(k_l a)_r\left(O_i^*-O_i\right)+r_{O,i} \tag{4}$$

Disengaging Section (i = A)

For micro-organisms and substrate (C = X, and S)

$$\frac{dC_M}{dt}=\frac{Q\,(1+m)}{V_t\,(1-\varepsilon_r)}\left[C_{A-1}-C_A\right]+r_{C,A} \tag{5}$$

For the dissolved oxygen (O)

$$\frac{dO_1}{dt}=\frac{Q\,(1+m)}{V_t\,(1-\varepsilon_r)}\left[O_{A-1}-O_A\right]+(k_l a)_d\left(O_A^*-O_A\right)+r_{O,A} \tag{6}$$

Downcomer Section (i = A+1, A+2, .., B)

For micro-organisms and substrate (C = X, and S)

$$\frac{dC_i}{dt}=\frac{Q}{\frac{V_d}{(B-A)}(1-\varepsilon_d)}\left[C_{i-1}-C_i\right]+r_{C,i} \tag{7}$$

For the dissolved oxygen (O)

$$\frac{dO_i}{dt}=\frac{Q}{\frac{V_d}{(B-A)}(1-\varepsilon_d)}\left[O_{i-1}-O_i\right]+(k_l a)_d\left(O_i^*-O_i\right)+r_{O,i} \tag{8}$$

Kinetic Model

Assumptions are (i) feed is sterile, (ii) internal mass transfer resistance is negligible and hence effectiveness factor is unity, (iii) organo-sulfur compounds are only growth limiting substrates, (iv) microbial growth follows Monod type kinetics, (v) microbial reaction occurs only at outer surface of the bio-film.

$$r_{X,i}=\frac{dX_i}{dt}=\frac{\mu_{\max}\,XS}{(K_S+S)} \tag{9}$$

$$r_{S,i}=\frac{dS_i}{dt}=-\frac{\mu_{\max}\,XS}{Y(K_S+S)} \tag{10}$$

$$r_{O,i}=\frac{dO_i}{dt}=-\delta\frac{dX_i}{dt}-\phi X_i \tag{11}$$

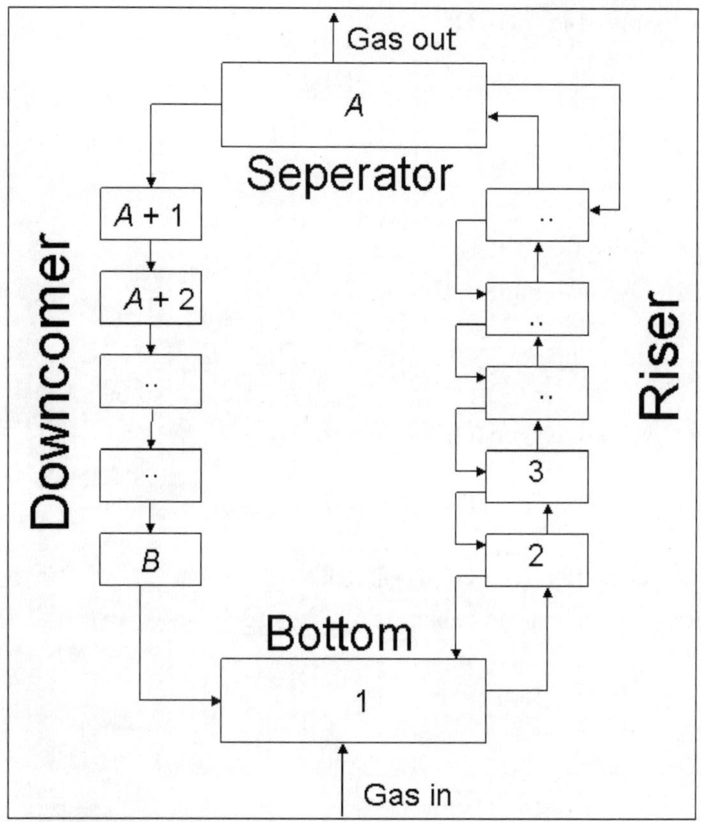

Figure 22.1: Tanks in Series Model for Airlift Reactor.

where δ is parameter in the Luedeking-Pirt-like equation for oxygen uptake g oxygen/g of biomass and ϕ is another parameter for *g* oxygen/(*g* biomass. *h* oxygen uptake). μ_{max} is the maximum growth rate = 0.096 h^{-1}, K_s is saturation constant = 71 mg/dm^3, and *Y* is yield coefficient = 0.2 g biomass consumed/g of substrate consumed (Mukhapadhyaya *et al.*, 2006).

Experimental Part

The schematic diagram of the internal loop airlift reactor is shown in Figure 22.2. The reactor was constructed of a 0.18 m inner diameter borosilicate glass column with 2 m in height. The draft tube was of inner diameter 0.1 m and 1.5 m in height also made up of borosilicate glass with wall thickness of 0.002 m. Therefore cross-sectional area of the draft tube acting as riser is 0.0254 m^2. The draft tube was placed at 0.15 m above the bottom of the reactor. The gas sparger at the bottom has a diameter of 0.06 m containing perforations of diameter 1 mm with a triangular pitch of 7 mm. A series of experiments were performed by varying the flow rate of gas monitored through calibrated rotameter with help of precisely controlled valve in the range of 0.1–0.7 lit/s. The superficial gas velocity obtained was in the range of 0.025–0.09 m/s. Air

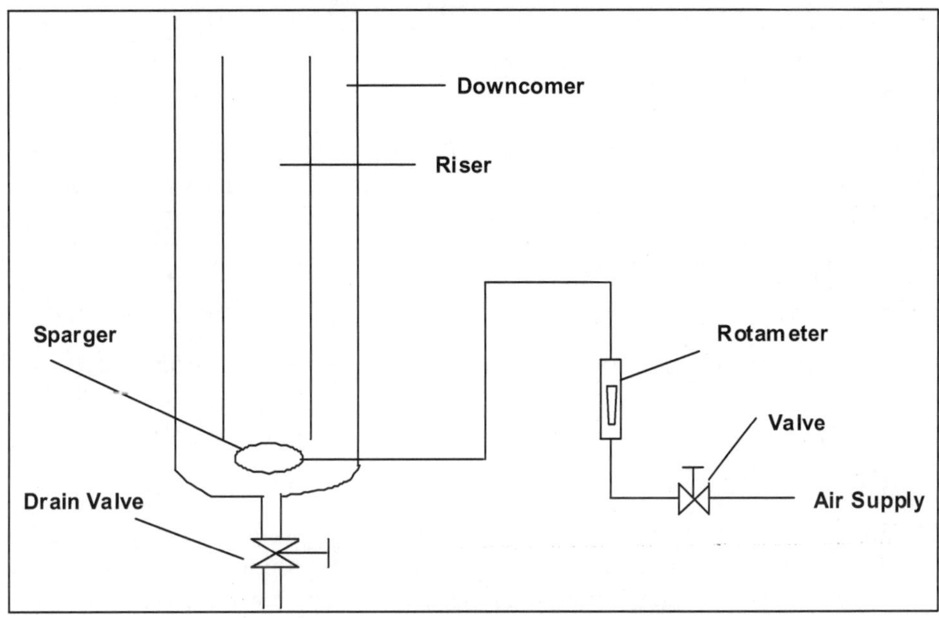

Figure 22.2: Schematic of the Experimental Set-up of ALR.

and water were used as gas and liquid phases. Individual gas hold ups of riser and downcomer sections were measured by standard procedure (Chisti, 1989) with help of U-tube manometer filled with iodine mixed carbon-tetra-chloride as manometric fluid (with density of 1.58 g/cm^3). The liquid circulation velocity in the airlift system was measured with help of a deep fried almond seed tracer of dimension $0.02 \times 0.01 \times 0.001$ m^3 having density of 1030 kg/m^3, with the help of stop watch to measure time taken to travel between two marked points at 1 m distance apart. Several such readings were taken for a single set and the data obtained were tested statistically to judge the accurate circulation velocity. The details of the experimentation procedure and results can be obtained from Nandi *et al.*, 2008. All the design parameters and operating conditions are provided in Table 22.1.

Results and Discussion

Determination of Model Parameters

The growth kinetic parameters like μ_{max}, K_s and Y were taken from Mukhopadhyaya et. al., 2006 (values are reported in kinetic modeling section). The other parameters like d and f do depends on the dimension and type of reactor. These two parameters were fitted for our ALR system with help of previously obtained experimental data of Mukhapadhyaya *et al.*, 2006, and the values obtained are: δ = 0.00164 (—) and ϕ = 0.00614 (h^{-1}).

Estimation of Number of Stages for Riser and Downcomer

Downcomer operates as a plug flow reactor. Without any significant back-mixing (*i.e.*, $m = 0$). Now relationship between ADM and TSM is given by (Znad *et al.*, 2004)

$$\frac{1}{Pe}=\frac{m+\left(\frac{1}{2}\right)}{B} \tag{12}$$

where, *Pe* is Peclet number and *B* denotes number of stages present in airlift reactor. Now, if back mixing is absent completely ($m = 0$), equivalent number of tanks (*B*) is given by $Pe/2$.

$$Pe=\frac{V_{ld}\,h_D}{D_{ax,d}}=\frac{0.27\times 1.77}{0.025}=19.116 \tag{13}$$

where, h_D is height of dispersion and $D_{ax,}$ d is axial dispersion coefficient = 0.025 m$^{2/}$ s.

$$h_D=\frac{h_L}{1-\varepsilon_g} \tag{14}$$

So, we get, the value of $B = 10$. Using same procedure, we can obtain $A = 12$. So the riser section is sub-divided into 12 equal parts and that of downcomer is divided into 10 parts. Bottom section, riser and top section is completely mixed and hence the value of back mix parameter *m* is considered to be 1 for each of them.

Simulation of Sulfur Removal

We have done extensive experimentations with the internal loop airlift assembly earlier (Nandi et. al., 2008) and have estimated all the important parameters of the system like overall gas hold up (e_g), individual hold ups for riser (e_r) and downcomer (e_d), liquid circulation velocity in riser (V_{lr}) and downcomer (V_{ld}) as well as overall-mass-transfer coefficients in riser $(k_la)_r$ and downcomer $(k_la)_d$ as a function of superficial gas velocity (U_{gr}). Theoretical models were also developed for prediction of all of these parameters with variation of superficial gas velocities. In the present study, we have extensively utilized the theoretical understanding and practical knowledge gathered beforehand.

We have simulated the bio-desulfurization process in the internal loop airlift reactor with gas free liquid column height of 1.7 m., with constant flow rate of air maintained at 5.5 m^3/hr. The reactor is operating in a semi-batch mode with the hydro-desulfurized diesel fed to the ALR in batch mode and passing air continuously.

All the mass balance equations along with the kinetic equation are solved simultaneously using stiff equation solver routine of MATLAB (Beers, 2007). The initial conditions considered for biomass (X_B), organosulfur compounds (*S*) and dissolved oxygen concentrations are (*O*) as shown below:

At time, $t = 0$, $X_{B0} = 0$, $S_0 = 500$ mg/dm^3 (ppm) and $O_0 = 0.00651$ g/dm^3

We have considered inlet organo-sulfur concentration of diesel as 500 ppm as it is very difficult to reduce sulfur concentration below this level by conventional hydrodesulfurization (Nandi et. al., 2007). So in order to achieve new US and European norms to reduce sulfur below 10 ppm or further, we should hydro-desulfurize diesel first in trickle bed reactor conventionally using Co/Mo/Ni catalysts and subsequently feed the output to the airlift reactor for further biodesulfurization in presence of some suitable Microbes. We have used *Rhodoccus* sp. in this study as the suitable microbial starin as it is demonstrated to be quite effective and the kinetic parameters are available in literature (Mukhopadhyaya *et al.*, 2006).

Table 22.1: Design and Operating Parameters of ALR.

Parameters	*Value*
Overall Volume, V (dm^3)	23.675
Volume of Riser V_r (dm^3)	11.781
Volume of Downcomer V_d (dm^3)	6.805
Volume of Bottom Section V_b (dm^3)	3.817
Volume of Top Section V_t (dm^3)	1.272
Vol. flow Rate of Gas Q_g (m^3/s)	$1.524´10^{-3}$
Holdup of Riser ε_r (-)	0.06
Holdup of Downcomer ε_d (-)	0.034
O2 trf. Coeff. of Riser $(k_la)_r$ (s^{-1})	0.027
O2 trf. Coeff. Downcomer $(k_la)_d$ (s^{-1})	0.022
Velocity of liquid in riser, V_{lr} (m/s)	0.32
Velocity of liq. downcomer, V_{ld} (m/s)	0.27

The Figure 22.3 represents simulated results of biomass growth with passage of time and it is clear that, biomass concentration increases with time and reaches saturation around 120 hr. From Figure 22.4 it is quite evident that concentration of organo-sulfur compounds diminishes with passage of time, due to microbial action. It can also be noted that concentration of organo-sulfur compounds reaches to almost zero level around 120 hr when biomass concentration also reaches saturation. So we can conclude that using the internal loop airlift assembly with the specified design and operating parameters, we can able to remove almost entire quantity of sulfur compounds present

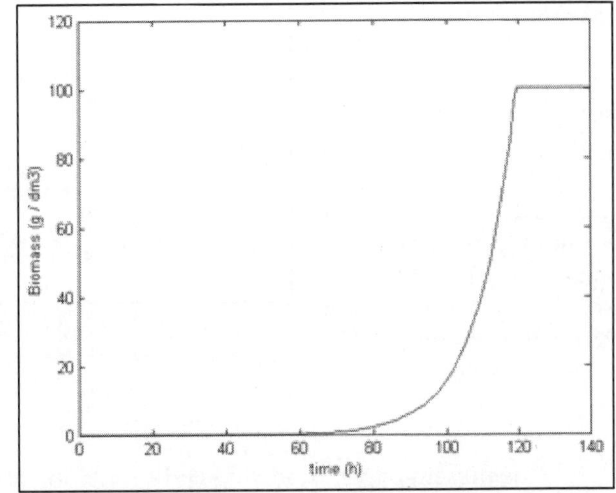

Figure 22.3: Effect of Time on Biomass Concentration for Internal Loop Airlift Reactor Operating in Semi-Batch Mode. Biomass concentration increases with time and stabilizes around 120 h operation.

in the previously hydrotreated diesel in about 120 hr. So airlift bioreactor can be employed for deep and ultra desulfurization. As no other chemicals are employed except air and water, the route is green and environment friendly one.

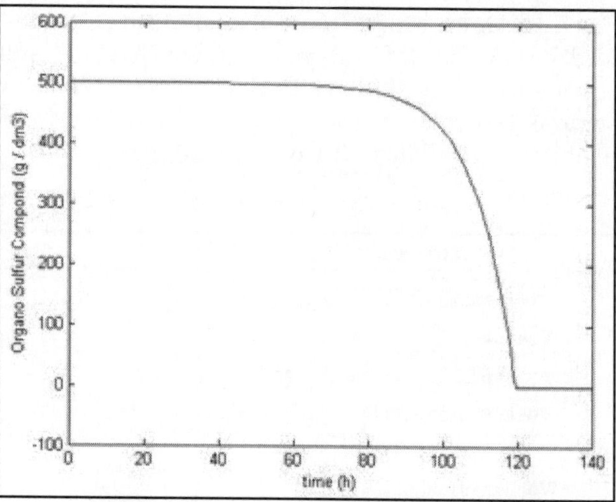

Figure 22.4: Simulated Results of Organo-sulfur Compounds Present in Diesel as a Function of Time for Semi-batch Operation of ALR.

The profile of dissolved oxygen concentration during the study is also plotted in Figure 22.5. We can see that dissolved oxygen quantity in the bio-culture is decreasing with time, this is due to the fact that oxygen is consumed for bio-reaction (desulfurization) as well as required by the microbes for their survival and growth. As time increases, oxygen concentration decreases marginally from earlier value of 6.51×10^{-3} g/dm^3 (which is the maximum solubility of oxygen in water at room temperature) and stabilizes around 2.0×10^{-4} g/dm^3 at a time of 120 hr. So even at the end of the desulfurization, the fermentation broth is not completely devoid of oxygen.

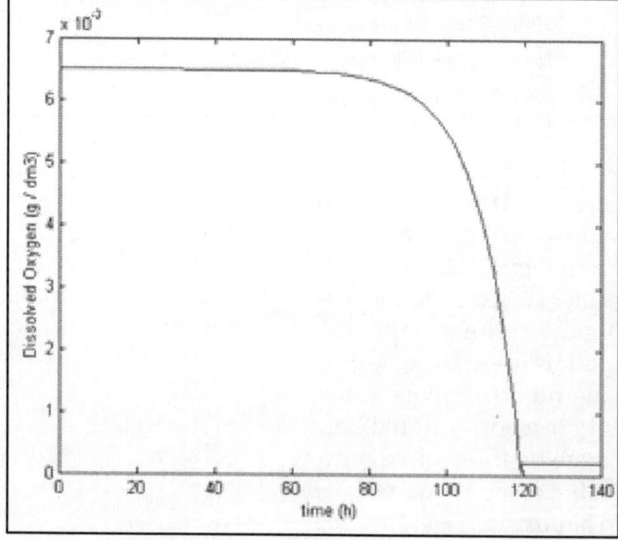

Figure 22.5: Profile of Dissolved Oxygen for Semi-batch Operation of Internal Loop Airlift Reactor.

Conclusion

In the present paper, modeling and simulation of internal loop airlift reactor is done using tanks in series model (TSM) under unsteady state conditions for bio-desulfurization of previously hydrodesulfurized diesel oil. The reactor shows relatively uniform axial dissolved oxygen concentration varying form small band of 6.51×10^{-3} g/dm^3 to 2.0×10^{-4} g/dm^3 at a time of 120 hr (Figure 22.5).

It has been clearly demonstrated that internal loop airlift bioreactor can be used quite effectively to reduce organo-sulfur concentration from 500 ppm to almost nil in about 120 hr. So we can conclude that, the designed airlift assembly can efficiently be employed for ultra and deep desulfurization operation in semibatch mode. The obtained product is well within the stringent specification of US and Europen norms. The process discussed here is a green and environment friendly one. Here we have done only simulation study with the bacterial strain *Rhodococcus* sp. The procedure is sufficiently generalized one and we can extend the same with other suitable microbes as well. Further experimental study is essential to accommodate effect of change of initial liquid column height in the airlift reactor and mass transfer effects on the overall performance of the process.

Nomenclature

A: No. of stages in riser section

B: No. of stages in downcomer section

C_X: Concentration of biomass (g/dm^3)

C_S: Concentration of substrate/organo-sulfur (g/dm^3)

D_{ax}: Axial dispersion coefficient (m^2/s)

h_D: Height of dispersion (m)

K_S: Saturation constant (mg/dm^3)

$k_l a$: mass transfer coefficient for oxygen (per s)

m: Factor for back-mixing

O: Concentration of oxygen in aqueous (g/dm^3)

Pe: Peclet number

Q: Flow rate (m^3/s)

O^*: Eqm. dissolved oxygen concentration (g/dm^3)

S: Substrate, organo-sulfur compounds present

U_{gr}: Superficial gas velocity in riser (m/s)

V: Volume (dm^3)

V_{ld}: Linear liquid velocity in downcomer (m/s)

V_{lr}: Linear liquid velocity in riser (m/s)

X: Biomass

Greek Symbols

ε: Gas hold up (–)

μ_{max}: Maximum growth rate (per h)

δ: Luedeking–Pirt–like kinetic parameter (–)

ϕ: Parameter for kinetics (per h)

Subscript

b: Bottom section, where gas is sparged in airlift

d: Downcomer section of Airlift assembly

r: Riser section of ALR

t: Top section or disengaging section of ALR

References

1. Beers, K. J., *Numerical Methods for Chemical Engineering: Applications in MATLAB*, Cambridge University Press, New York (2007).
2. Chang, J. H, Kim, Y. J., Lee, B. H., Cho, K. S., Rya, H. W., Chang, Y. K., Chang, H. N., "Production of a Desulfurization Biocatalyst by 2-stage Fermentation and its Application for the Treatment of Model Diesel Oil", *Biotechnol.*, 17, 878–880 (2001).
3. Chisti, Y., "Airlift Bioreactors", Elsevier, London (1989).
4. Kanai, T., Uzumaki, T., Kawase, Y., "Simulation of AirLift Bioreactors: Steady State Performance of Continuous Culture Processes", *Comp. and Chem. Eng.* 20, 1089–1099 (1996).
5. Maghsoudi, S., Vossught, M., Kheirolomoom Tanaka, A. E., and Katho, S. "Biodesulfurization of Hydrocarbons and Diesel Fuels by Rhodococcus strain P32C1." *Biochem. Eng. J.*, 8, 151 - 156 (2001).
6. Mukhopadhyaya, M., Chowdhury, R., and Bhattacharya, P. "Biodesulfurization of Hydrodesulfurized Diesel in Trickle Bed Reactor–Experiments and Modeling", *Jour. Sci. and Ind. Res.*, 65, 432–436 (2006).
7. Nandi, S. and Deotare, K. N., "Simulating an Industrial Scale Trickle Bed Reactor for Hydro-desulfurization of Diesel", pp. 169–175 in *Modeling and Simulation (Vol. I)*, Edited by S.R.K. Prasad, R. Prabhakar, E. Ramaswamy and R. Sarvanan, Allied Publishers Pvt. Ltd., Chennai (2007).
8. Nandi, S., Narayan, A., Jha, M. K., and Jaju, S., "Hydrodynamic Study of Air Lift Reactor", oral presentation at CHEMCON 2008 at Punjab University, December 2008, available in CD-ROM (Ref: TP 63) (2008).
9. Znad, H., Bales, V., Markos, J. and Kawase, Y. "Modeling and Simulation of Airlift Bioreactors", *Biochem. Eng. J.*, 21, 73–81 (2004).

Chapter 23

Biofilms and Biosurfactants: An Environmental Perspective

Sakshi P. Korde[1], *Smita S. Bhuyan*[2], *Urmila P. Jagtap*[1], *Sharvari V. Gaidhani*[1], *Karishma R. Pardesi*[2] *and Balu A. Chopade*[1]

[1]*Institute of Bioinformatics and Biotechnology,*
[2]*Department of Microbiology,*
University of Pune, Pune – 411 007, Mh.

ABSTRACT

Biofilms are three dimensional structures with high level of differentiation. Biofilms can be beneficial while some are detrimental. Exopolysaccharides of biofilms provide protection against desiccation and presence of antimicrobial agents; additionally they act as reservoir of nutrients and ions. Biofilms find applications in wastewater cleanup by means of decomposition of organic matter. However, biofouling is a major cause of concern for variety of industrial and shipping equipments. A variety of medical implants are subject to colonisation by biofilm bacteria; biofilm formation thereby leading to antibiotic treatment failures. These detrimental biofilms can be cleaned up with the natural cleansers- "Biosurfactants". A large number of these surface active agents act antagonistically against detrimental biofilms and neutrailse their negative effects. Antiadhesive activity complemented with antimicrobial activity exhibited by most of biosurfactants, can be used as an infection preventive strategy. Few of them help in bacterial surface motility, while still others enhance adhesion of beneficial biofilms. All these properties make biosurfactants ideal environmental tools for eradication of detrimental biofilms. Thus with the aid of biosurfactants to keep harmul biofilms at bay, and search for newer beneficial biofilm formers and enhancement of existing ones will go a long way in a cleaner and safer environment.

Architecture of Biofilms

Although micro-organisms are often considered as simple life forms, they are capable of complex differentiation and behaviours[1]. Micro-organisms form complex matrix enclosed communities which express particular genes in a pattern and are capable of producing chemicals like violacein which cannot be produced by planktonic forms of the microbes. Violacein is produced as a targeted chemical defence against eukaryotic predation by *U. australis* biofilms, *Microbulbifer* sp. and *Pseudoalteromonas ulvaev*[2]. These complex structures are called Biofilms. Microbial cells are held together by a thick mesh of exopolysaccharides (EPS), also known as glycocalyx[3]. It is made up of high molecular weight polysaccharides fabricated from cells to the surfaces forming an organised mesh. It is known that EPS forms water network channels in biofilms[4]. The role of EPS in differentiation of bioflm is supported by Danese *et al.*[5] who reported that EPS and colonic acid are essential for formation of complex biofilms in E. coli K-12. Exopolysaccharides play important role by not only providing protection against harsh environmental conditions such as desiccation and presence of antimicrobial agents but also acts as a reservoir of nutrients and ions[6].

During biofilm formation the cells undergo various changes not evident in individual cells. Initially, biofilms were considered as mere mass of polysaccharides embedded with single cells. The discovery of the well defined water channels in the interiors of biofilms opened new dimensions in the understanding of biofilm structure. There are three proposed mechanisms for congregation of cells on substratum to form biofilm. First is by redistribution of cells by surface motility. Studies by O'Toole and Kolter[7] on Pseudomonas aeruginosa mutants suggested that type IV pili-mediated twitching motility plays a role in surface aggregation. Twitching motility is a mode of locomotion used by *P. aeruginosa* in which type IV polar pili are believed to extend and retract, propelling bacteria across a surface. Second is by binary fission of the attached cells where the cells grow outward from the central point to form the colony[8]. A third mechanism of aggregation is recruitment of cells from the bulk fluid to the developing biofilm[9]. The contribution of each mechanism in biofilm formation depends on the type of the organism and the production of several substances by micro-organisms.

The biofilm formation involves four basic stages. The free living bacteria are attracted to the substratum by certain surface properties like availability of adsorbed organic substrates *i.e.* conditioned surfaces[10]. After the first stage of weak attachment the aggregated microbes start producing the exopolysaccharides. As the EPS mass increases attachment of microbial colony to the surface becomes irreversible. This triggers the process of biofilm maturation. The final stage of biofilm differentiation is formation of an intricate channel network in the biofilm structure. As the complex EPS polymer channels spread over substratum, redistribution of the attached bacterial cells takes place. Also striking difference between the gene expression in planktonic cells and biofilm has been shown by Sauer *et al*[11]. The difference between expression of almost 50 per cent of the proteome of planktonic cells and biofilm of *P. aeruginosa* was reported.

Anabolic and respiratory activities of *S. aureus* biofilm are stratified; the region of the biofilm proximal to the oxygen source or the nutrient source shows highest metabolic activity, whereas biomass at the substratum surface is inactive.[12] However, only 10 per cent of the cells in biofilm are not viable. The cells below the metabolically active cells respire anaerobically as the oxygen availability is less. These findings suggest that the biofilm consists of four distinct types of cells, those growing aerobically, those growing fermentatively, dead and the dormant ones. Thus this inactive biomass harboured in the *S. aureus* biofilm may explain the persistent temperament of the biofilms in nature[13].

To establish a well coordinated colony, interaction between neighbouring cells is necessary. Hence cells undergo transition from planktonic state to community based existence where the EPS matrix holds them together. It is proved that biofilm formation in Gram negative bacteria is initiated by strong communication signals between microbial cells. These signals are produced by accumulation of specific substances till their threshold concentrations are reached. This phenomenon is called as Quorum Sensing[14, 15, 16]. Biofilm formation is triggered by various environmental signals. Factors like nutrient availability, temperature, pH, osmolarity and oxygen availability affect the initiation of biofilm formation. These external factors (Quorum sensing in Gram negative bacteria) activate targeted genes required for biofilm formation.

Benefits of Biofilms

One of the best examples of successful, application of biofilms is in the area of wastewater cleanup. Biofilms are used as biobarriers to control the spread of contamination of algal plumes[17]. Decomposition of organic matter in plants and animals is achieved by micro-organisms; Decay happens (partly) as the micro-organisms feed on the tissues of the dead matter. With this understanding, perhaps one could engineer a system that uses decomposer micro-organisms (in the form of a biofilm) to process wastewater and sewage: Wastewater cleanup in trickling filters is achieved by micro-organisms in biofilm which degrade the harmful organic material from these wastewaters[18].

Interestingly, scientists and water treatment authorities have discovered that waters processed by biofilm communities are biologically more stable than water filtered by conventional means of treatment systems. This just means that there is likelihood of less micro-organism contamination in water that has passed through a biofilm-based filter than would be in waters that have been processed by alternative treatment systems. This implies that biofilm treated water typically has lower disinfectant demand (*e.g.*, use of chlorine) and disinfection by-product formation (*e.g.*, the taste and smell of chlorine) than water treated by other means, if the water is rich in organic carbon content[18].

Phototrophic biofilms are effectively used in the treatment of wastewaters. Their net oxygen production is high which facilitates the degradation of organic matter. Phototrophic biofilms build biomass from carbon dioxide. Cyanobacterial biofilms accumulate nitrogen in form of ammonia, nitrates and urea, implying that this biomass could be used as biofertilizer. Phototrophic biofilms are implemented in oil degradation, heavy metal removal and aquaculture by virtue of reducing the concentrations to nitrogenous compounds[19]. Pseudomonads associated with the plant root rhizosphere act as growth promoting factor and also as biocontrol agents[20].

Biofilms are used in developing membrane bioreactors. The moving-bed-biofilm reactor with a submerged membrane is employed as biomass separation reactor[21]. The most recent design of bioreactor is the double-deck aerated biofilm reactor with integrated designs of sludge hopper, thickener, and velocity-zone concept for particle settlement. An aerated moving-bed biofilm reactor and a flat-sheet membrane module for a biofilm membrane bioreactor process is used to overcome the problem of membrane clogging in bioreactor designs [22]. Biofilm bioreactors are used as powerhouse for bioconversion of industrially important substrates. A fungal biofilm of *Tyromyces sambuceus* was used for continous bioconversion of emulsified castor oil to decalactone [23]. Generally biofilms are considered as derogatory to the surface but it is reported that the formation of biofilm by iron reducing bacteria reduces the corrosion of steel. The iron reducing bacteria reduce Fe^{+3} to Fe^{+2} by scavenging the oxygen in water channels of biofilms. Thus the consumption of the oxygen in vicinity of the steel surface signifiacantly reduces the corrosion [24].

Derogatory Biofilms

The most common problem is the formation of a slimy layer of biofilm on the internal surfaces of the water pipelines. The biofilms which are irreversibly attached to the surface can withstand the hydrodynamic shear force of flowing water, additionally presence of uneven surface on the inner sides of the pipeline increases frictional resistance to water flow. The reducing cross sectional area of the pipelines decreases the net amount of fluid transport. Also if the biofilm forming bacteria have pathogenic traits, the risk of water contamination cannot be neglected.

Biofilm formation is widespread in industrial equipments which use sea water as a coolant. Thick coat of the EPS of biofilm impedes the effective heat transfer from surface to the coolant causing high energy loss. Also bacterial colonies release certain biocides in the coolant fluid which adversely affect marine environment [25]. Biofilm formation on sensor surfaces, sighting equipment etc is very common in humid environments thereby affecting functionality of these devices.

The moment the surface is submerged in the water body, the process of biofouling is initiated. Organic salts and matter adheres to the surfaces, conditioning the surface for microbial adherence. This is seen on the ship surfaces which are perpetually submerged in sea water. Marine bacteria readily form strong layers of biofilm on these surfaces. Metabolic activities of the colonising bacteria start process of corrosion of the metal surface. This significantly affects the life span of the vessel. The rough surface of the vessel bottom increases the friction against the water flow and reduces the efficiency of transportation. Moreover the EPS mesh formation enhances the attachment of other sea organisms like barnacles and sponges[26]. Zardus *et al.*, found that presence of biofilm coat on the submerged surface significantly increased adhesion strength in the ascidian *Phallusia nigra*, the polychaete tubeworm *Hydroides elegans*, and the barnacle *Balanus amphitrite* at one or more developmental stages[27].

In the marine environment, most natural surfaces are also susceptible to bacterial colonization and formation of microbial biofilms, an event that precedes biofouling. Biofilms can cause damage to certain life forms such as marine sea grass and algae [28]. In this scenario, the microbial biofilm induces the settlement and metamorphosis of macrofoulers that can block light access to the plants photosynthetic cells on the fronds. The microbial films also attract predatory grazers searching for food. While researchers have been working towards anti-fouling materials to coat manmade objects, some marine plants have evolved a mechanism to inhibit the growth of microbial biofilms by producing compounds that inhibit the initial act of biofilm formation. The most prominant example is the Australian red alga, *Delisea pulchra*, which produces a group of water insoluble secondary metabolites that inhibit the density-dependent cell-cell communication pathways that controls biofilm formation [29, 30].

Microbially Induced Concrete Corrosion (MICC) caused by the sulphur reducing bacterial biofilm is a prime concern in cities. The manholes provide favourable condition for formation of strong and persistent biofilms. The microbial colonies reduce the available sulphur dioxide producing sulphuric acid. The acid attacks the basal concrete layer leading to the corrosion of the construction [31]. Using reclaimed wastewater for crop irrigation is a practical alternative to discharge of wastewater treatment plant effluents into surface waters. However, biofouling has been identified as a major contributor to emitter clogging in drip irrigation systems distributing reclaimed wastewater. Not much is known about the biofilms formed in the emitter device but studies regarding the mature stages of biofilm and effect of fluid flow on the biofilm architecture would help us prevent the clogging problem [32]. Due to polyelectrolyte nature of EPS, biofilms can act as electron sink at cathode and as conductors, interrupting the electronic signals of the computer chips, a deleterious effect causing malfunctioning of the computers[4].

Biofilms in Plant Pathogenesis

Biofilms on different substrates have unique properties, reflecting prevailing conditions at those sites. Plant surfaces provide a safe and nutrient rich surface for biofilm formation. Biofilms are formed on phylloshpere, phylloplane, vascular bundles, seeds and rhizosphere. Attachment is required for biofilm formation, and bacteria interact with plant tissues through adhesins including polysaccharides and surface proteins, with initial contact often mediated by active motility. Recognition between lectins and their cognate carbohydrates is a common means of specificity. Biofilm development and resulting intimate interactions with plants often require cell-cell communication between colonizing bacteria. *Xylella fastidiosa* is deposited as biofilm into the xylem of plants by sap-feeding leafhoppers, and can induce Pierce's disease of grapevine and citrus variegated chlorosis[33]. Marques[34] and colleagues showed *Clavibacter michiganensis* as the causative agent for the ring rot in potato. The involvement of cellulose in bacterial biofilms has long been known for *Acetobacter xylinum*, which forms a biofilm at air-water interfaces and on the surface of decaying fruits[35]. *Agrobacterium tumefaciens* produces cellulose fibrils which enhance the root surface adherence of bacteria. Matrix formation also plays role in infection of the bacteria [36].

Biofilms and Infections

Biofilm formation is basically a transition from free living to sessile form. But why would microbes prefer this form of colonisation? Earlier studies clearly reflect the advantages conferred by biofilm formation on the survival strategies of micro-organisms. Most noted advantage is that, biofilms are exceptionally resistant to antibacterial or antifungal agents. Also the invasive competitor or predator microbes are kept at bay by production of certain chemical defence by coordinated colonies of microbes. Hence most persistent infections are due the formation of strong and recalcitrant biofilms. Surface attachment is an initiating step in pathogenesis of most of the organisms. Kuroki *et al.*[37] have reported the direct correlation between infectious capacity and biofilm formation of *Bacillus cereus* and *Bacillus thuringiensis*. The most common biofilms are that of *Staphylococcus epidermidis* and *Staphylococcus aureus* [38]. Chronic Pseudomonas aeruginosa infections in the lungs of cystic fibrosis patients are also due to biofilm formation. Respiratory infections caused by nontypeable *Haemophilus influenzae* (NTHi) are a major medical problem. Evidence suggests that the ability to form biofilms on mucosal surfaces may play a role in NTHi pathogenesis [39]. The pneumococci form biofilms *in vivo* and this process may be intertwined with the formation of neutrophil extracellular traps. These findings provide new insights into the potential causes of antibiotic treatment failures and bacterial persistence in chronic pneumococcal otitis media [40].

Medical implants, such as contact lenses, central venous catheters, needleless connectors, endotracheal tubes, intrauterine devices, mechanical heart valves, pacemakers, peritoneal dialysis catheters, prosthetic joints, tympanostomy tubes, urinary catheters, voice prostheses, prosthetic heart valves and joint replacements can be colonised by micro-organisms that form an adherent biofilm on the surface of the device. Most common isolates obtained from infected devices are Gram positive *Enterococcus faecalis, Staphylococcus aureus, Staphylococcus epidermidis*, and *Streptococcus* viridans; and the gram-negative *Escherichia coli, Klebsiella pneumoniae, Proteus mirabilis,* and *Pseudomonas aeruginosa* [41]. Inhabitation by biofilms leads to blockage, loss of function and reduced life span in implants. Such colonised implants must be immediately replaced. The development of surface-attached biofilm bacterial communities is considered an important source of nosocomial infections. The Bacillus strains found in hospitals cause severe bacteremia through catheters infection. The location of biofilm depends on the microenvironment available to the microbes and the period of infection. Prolonged infection is characterised by formation of biofilm on internal surfaces of the catheters also leading to the infection in blood stream of patients [41]. *Candida* sp. initially forms single species biofilms and in the later phase develops a multi species biofilm. This imparts high level of drug resistance to fungal biofilms. The medical devices in use which are predisposable to biofilm formation of Candida spp. and subsequent infection include vascular catheters, joint prostheses, dialysis access devices, cardiac devices like prosthetic valves, pacemakers, cardioverter defibrillators, ventricular assist devices, intrauterine devices, central nervous system devices such as ventriculo-peritoneal shunts, urinary catheters implants[42]. Dental plaque is the most natural biofilm formed. The dental plaque is mainly formed by anaerobic microbes along with layers of aerobic, microaerophilc and facultative microbes. All the dental caries and periodontitis (gum diseases) are caused by changes in physiology and behaviour of dental plaques [43].

Biosurfactants as a Solution for Detrimental Biofilms

A conditioned surface is the most basic requirement for initiation of biofilm. In the natural environment, favourable niche is provided by adsorption of organic salts on surface whereas in medical implants the host serum proteins and nutrients help in adherence to the substratum. It is a well known fact that many marine bacteria produce surface active agents. These secondary metabolites are known as biosurfactants/bioemulsifiers. Biosurfactants possess a unique ability to reduce interfacial tension between two liquid phases and form emulsions. This property is attributed to the amhiphathic nature of these molecules. The presence of hydrophobic and hydrophilic moiety enables the molecules to accumulate at the liquid–liquid interphase. The hydrophobic moiety is oriented towards the hydrophobic liquid. The emulsion formation is characterised by formation of micelles. Micelles are spherical structures consisting of hydrophobic core and hydrophilic exterior. The miniscule droplets of immiscible liquid are contained in the micelle structure which facilitates the emulsion formation. Biosurfactants are either secreted extracellularly by the organisms or are membrane bound.

Why do Microbes Produce Biosurfactants?

The natural role of biosurfactants is to facilitate adhesion of microbial cells to the substrate molecule. Once the soluble carbon source is exhausted the bacteria consume insoluble substrates as hydrocarbons. The membrane bound biosurfactants play an important role in adhesion of cells to the oil droplet and detachment from the consumed droplet. Biosurfactants are secreted form microemulsions so that, the oil droplet can be ingested via phagocytosis by the micro-organism[44]. Most of the reported biosurfactants demonstrate antimicrobial and antifungal properties. Hence production of biosurfactants is a defence mechanism against the predator as well as the competitor organisms. Surfactin, a potent biosurfactant and antibiotic is produced by *B. subtilis* is effective against most multidrug resistant organisms like *Enterococcus faecalis* (11 strains), *Staphylococcus aureus* (6 strains), *Pseudomonas aeruginosa* (7 strains) and *Escherichia coli* CI 18 (1 strain) but it is most effective against *E. faecalis* [45]. Many secreted biosurfactants can chelate toxic ions and reduce the environmental toxicity thereby ensuring survival of the producing micro-organism. Todd *et al.*, [46] in 2000 have reported that rhamnolipid reduces cadmium toxicity in soil by forming a complex with the metal ions.

Biosurfactants are known to change the surface hydrophobicity of the cells and play role in cell attachment to substratum. Surfactants affect the development of flagella and swarming motility suggesting effect on the attachment capability of bacteria [47]. Biofilm formation and organism dispersal leading to spread of infection by P. aeruginosa are dependent on motility. Swimming *i.e.* motion on semisolid medium requires flagella[48] and swarming *i.e.* motility on wet surface requires quorum sensing. As mentioned earlier biofilm formation initiates with quorum sensing suggesting that swarming is must for biofilm formation. Quorum sensing is a major regulatory phenomenon in nitrate metabolism (a prime source for rhamnolipid production), biofilm formation and motility. Nitrate metabolism and nitrogen sensing are directly related to *P. aeruginosa* pathogenesis [49].Biosurfactant like rhamnolipids are essential in motility through swarming as they reduce surface tension of aqueous phase on wet surface. This reduces the energy required for swarming and facilitates biofilm formation [50]. All the organisms compete with each other for environmental resources. The biosurfactant produced by the biofilm forming bacteria is also known to inhibit the biofilm formation of other microbes. As seen in case of *P. aeruginosa*, the rhamnolipid biosurfactant produced as an extracellular metabolite plays pivotal role in inhibiting the biofilm formation of *Staphylococcus epidermidis*[51]. Above data suggests the potential of biosurfactants in inhibiting the biofilm formation.

Several strategies can be employed to remove detrimental biofilms, depending on its characteristics.

These include:

- ✰ Mechanical cleaning of biofilms.
- ✰ Inhibition of biofilm growth by omission of essential nutrients required for growth environment.
- ✰ Use of specific antimicrobial agents.
- ✰ Inhibiting microbial attachment to a surface and
- ✰ Promoting direct biomass detachment from the substratum [52].

Mechanical cleaning of the fouling can be very expensive and not always feasible. Though removal of the nutrients from the medium is a very attractive method but controlling the medium composition is not always possible as in case of medical implants.

Hence, use of antimicrobial compounds like vancomycin against bacterial pathogens and flucanazole against fungal pathogens is in vogue. In polymicrobial biofilms the fungal culture can modulate the action of antibacterial agent and the bacterial counterpart can affect the antifungal agents by production of slime in the biofilm matrix thus increasing the barrier in drug penetration [53].Thus prevention of initial attachment to the substrate is the most appealing option.

Most widely used method of inhibition of biofouling is application of toxic chemicals like tributyltin (TBT) on surfaces. Unfortunately, slow degradation of these toxic chemicals causes lethal effects on marine organisms. Organotin moieties in these antifouling paints degrades slowly and causes Imposex (development of male characteristics in female genitalia) in dog-whelks, defective shell growth in oysters and debilitation of immunological defences in fishes [54]. Hence an environment friendly method of biofouling prevention must be devised. The numerous advantages of biosurfactants, such as mild production conditions, lower toxicity, higher biodegradability and environmental compatibility make it a lucrative option for treatment of marine biofilms. These attributes make biosurfactants a suitable alternative to replace toxic chemicals. *Bacillus* and *Pseudomonas* spp. were predominant among the antifouling agents producing organisms. The secondary metabolites produced by these strains like pyolipic acid is reported to be efficient antifouling agent and is also reported to show surfactant like properties [55]. These natural compounds can be used to manufacture antifouling paints [56]. Studies on Pseudomonas fluorescens surfactant proved potential of biosurfactant as corrosion preventing agent for stainless steel [57].

Biosurfactants find dual applications as antimicrobial and anti-adhesive agents in prevention of biofilm formation. Biosurfactant forms a conditioning layer on the surface, changing its physio-chemical properties like wettability. This either results in increase in adherence of certain organisms depending on the cell surface hydrophobicity (*e.g. Pseudomonas aeruginosa*) or inhibition of adherence in organisms like Candida. Surfactin, the most studied biosurfactant can inhibit growth of preformed biofilm causing dispersal of colonies and inhibits *Salmonella enterica* adhesion to polyvinyl chloride (PVC) sheets without inhibiting the growth. Surfactin and rhamnolipids are capable of preventing *Escherichia coli* and *Proteus mirabilis* biofilm attachment[58]. Rhamnolipid-preconditioned Poly-tetra-fluoro ethylene (PTFE) surface was able to reduce *Listeria monocytogenes* attachment[59]. Also presence of fats and fatty acids affected the biofilm formation by *Listeria monocytogenes* in ready to eat meat processing environments[60]. Further work demonstrated that the prior adsorption of P. fluorescens surfactant in stainless steel also favoured the bactericidal effect of disinfectants[61]. It was found that conditioning of silicone rubber with rhamnolipids resulted in reduction of *Streptococcus salivarius* and *Candida tropicalis* adhesion rates by 66 per cent. Minimal planktonic cell attachment to the surface was reported after conditioning[62] Rodrigues et al., 2006c L.R. Rodrigues, I.M. Banat, H.C. van der Mei, J.A. Teixeira and R. Oliveira, Interference in adhesion of bacteria and yeasts isolated from explanted voice prostheses to silicone rubber by rhamnolipid biosurfactants, Journal of Applied Microbiology (2006) (3), pp. 470–480. Full Text via CrossRef | View Record in Scopus Cited By in Scopus (7).

Role of Quorum sensing in the programmed cell disruption of *Pseudomonas aeruginosa* biofilms was proposed by Schooling S. R. *et al.*[63] in 2004. When medium supplemented with rhamnolipid which is required for efficient quorum sensing, was used to grow the *P. aeruginosa* cultures, biofilm formation was completely inhibited. Biosurfactants from two *Bacillus* spp. namely *B. subtilis* and *B. licheniformis* are very efficient in inhibiting pathogenic bacteria. It is observed that the surface adhesion is greatly reduced on precoating of biosurfactant to the surface. Also introduction of biosurfactants prior to inoculation proved to be effective. Adhesion of *Candida albicans* and *Candida tropicalis* strains isolated from naturally colonized voice prostheses was inhibited by *Streptococcus thermophilus* B biosurfactant. The biosurfactant molecules are produced in response to the competition between bacteria and yeast for surface attachment. Biosurfactant forms a conditioning layer and interferes with yeast adhesion on the surface. *S. thermophilus* restricts the yeast cell attachment to the surface of the silicone rubber. The crude biosurfactants released are mixtures of various components, with a glycolipid-like component being the most surface active[64].The preadsorbed layer of biosurfactant is effective in inhibition of biofilm formation. Recent studies in medical settings showed that adhesion of uropathogenic micro-organisms could be inhibited by a biosurfactant produced by a *Lactobacillus* strain.[64,65]. Initially it was observed that consumption of dairy products prevented contamination of implants. This is due to a biosurfactant produced by *Lactobacillus* sp. present in food products. The biosurfactant produced by *Lactobacillus* was able to inhibit the adhesion of Enterococcus faecalis by 99 per cent [66,65]. Lactococcus lactis and Streptococcus thermophilus produce biosurfactants which decrease the amount of bacteria in a multi-species biofilm on voice prosthesis[67]. The authors had postulated presence of O antigen (surface lipopeptide) as surface active agent. The lipopeptide biosurfactants consists of two fractions. It is proved that fraction belonging to fengycin-like family is active as anti-adhesive agent against pathogenic cultures mainly *Escherichia coli* and *Staphylococcus aureus*[68]. The disruption of *Bordetella bronchiseptica* biofilms by rhamnolipids was demonstrated by Irie and O'Toole[69]. Antiadhesive activity complemented with the antimicrobial activity exhibited by most of biosurfactants, can be utilised to devise a very proficient infection preventive strategy. A device to prevent the biofilm formation on catheters and other implants has been developed using bisurfactants. Surfactants used for practice of the present invention include biosurfactants and surfactant polymers. Preferred biosurfactants include surlactin, surfactin, visconsin, rhamnolipids, glycolipids, lipopeptides, depsipeptides, phospholipids, substituted fatty acids, and lipopolysaccharides. And preferred surfactant polymers include surfactant polymers having a poly (vinyl amine) backbone, having hydrophilic poly(ethylene oxide) sides chains and hydrophobic hexanal side chains.

The device comprises of,

1 A reservoir, which includes a polymer matrix portion,
2. An antimicrobial agent disposed within the polymer matrix portion; and
3. A surfactant region disposed over the reservoir at an outer surface of the device.

The device was successfully implanted within the body of a patient for a period of at least three months which is a considerably longer lifespan for a medical implant[70].

The bacteria residing in oral cavity are also capable of releasing biosurfactants. These glycolipid type of biosurfactants from two *Streptococcus mitis* strains (BA and BMS) found in oral cavity significantly reduced adhesion to glass surface by Streptococcus mutans NS with or without salivary coating. This suggests, that biosurfactants play a fundamental role in preventing oral biofilms[71]. These studies prove the huge potential of biosurfactant as anti adhesive agents. The different types of biosurfactant molecules produced by variety of micro-organisms can be specifically used against detrimental biofilms. Due to increasing reports on understanding of formation (see review by Gaidhani *et al.*, 2008[72]) and regulation of biofilms, the scope of application of biosurfactant in various aspects of inhibition of biofilms is widened.

Conclusion(s)

Our understanding of the complex structural and functional intricacies of biofilms has improved over the last decade. The benefits of biofilms are explored and innovative technologies have been established. The role of biofilms in biofouling and pathogenesis is well understood. Studies on biosurfactant properties have encouraged the researchers to view these versatile molecules as an appropriate solution to detrimental effects of biofilms. Cost economic production of biosurfactants and their actual implementation on field remains to be investigated.

References

1. O'Toole.G *et al.* (2000). Ann. Rev. Microbiol. 54: 49–79.
2. Matz C *et al.* (2008), PLoS ONE 3(7): e2744.
3. Hughes K.A *et al.* (2002). J.Appl Microbiol, 85(3): 583–590.
4. C. Potera, (1996). Science. 273: 1795-1797.
5. Danese PN, *et al.* (2000). J. Bacteriol.182: 3593–96
6. Allison, D. G. (1993). Microbiol Eur, 16±19.
7. O'Toole G A. *et al.* (1998). Mol. Microbiol. 30: 295–304.
8. Heydorn A. *et al.* (2000). Microbiol. 146: 2395–407.
9. Tolker-Nielson T *et al.* (2000). J. Bacteriol. 182: 6482–89.
10. W.G. Characklis and K.E. Cooksey, (1983). Adv Appl. Microbiol. 29: 93-138.
11. Sauer K *et al.* (2002). J Bacteriol, 184: 1140–1154.
12. De Beer D *et al.* (1994). Biotechnol. Bioeng. 43: 1131–38.
13. Rani S A *et al.* (2007). J. Bacteriol.189(11): .4223–4233
14. Davies D G. *et al.* (1998). Science, 280: 295–298.
15. Parsek M R. *et al.* (2000).
16. Sauer K *et al.* (2000). J. Bacteriol.184: 1140–54.
17. B. Chen and Y. Li, Proc 1999 Conference on Hazardous Waste Research.
18. Wilderer P. A. *et al.* (2000). Water, Air, and Soil Pollut, 123: 147–158.
19. Roeselers *et al.* (2008). J Appl Phycol, 20(3): 227–235.
20. Lugtenberg B J J *et al.* (2001). Annu Rev Phytopathol, 39: 461-490.
21. Leiknes T.and Odegaard H, (2007). Desalination, 202,(1-3),135-143.
22. Phattaranawik J. and Leiknes T., Environ.Energy Engg,(2008).
23. Pakula R and Freeman A. (1995). Biotechnol. Bioengg.
24. Dubiel M. *et al.* (2002). App environ microbiol,68(3): 1440–1445.
25. Eguía E *et al.* (2008). Inter Biodeterior and Biodegrad, 62(2),79-87.
26. Nogata Y *et al.* (2003). Biofouling, 19: 193–196.

27. Zardus J D. *et al.* (2008). Biol. Bull. 214: 91–98.
28. Dworjanyn, S.A. *et al.* (1999). Biol. 133, 727^736.
29. De Nys R, *et al.* (1993), Tetrahedron. 49: 11213-20.
30. Gram L. *et al.* (1996). Appl Environ Microbiol, 62(11): 4284–4287.
31. Satoh H, *et al.* (2009). Water Res.
32. Dazhuang Y. *et al.* (2009). J Environ Sci, 21(6): 834-841.
33. Purcell A H. *et al.* (1996), Ann Rev Phytopathol, 34: 131-151.
34. Marques L L R. *et al.* (2003). Phytopathology, 93: S57.
35. Cannon R. E. and Anderson, S. M. (1991), Crit. Rev. Microbiol. 17: 435-447.
36. Matthysse A G. *et al.* (2005). MPMI,18(9): 1002–1010.
37. Kuroki R *et al.* (2009). Intern Med.;48(10): 791-6.
38. Huebner J. and Goldmann, D. A. (1999). Ann Rev Med 50, 223–236.
39. Izano EA and Kaplan JB. (2009). Microbial Pathol., 46(4): 207-13.
40. Reid SD *et al.* (2009). J Infect Dis.,15: 199(6): 774-5.
41. Donlan R M. (2001). Emerging Infect Diseases,7(2).
42. Kojic EM and Darouiche RO. (2004). Clin Microbiol Rev; 17: 255–267.
43. Sissons, C. H. (1997), Adv. Dent. Res. 11: 110–126.
44. Ron.E and Rosenberg.E, (2001)Environ microbiol,3(4). 229-236
45. Fernandes, P. A. *et al.* (2007). Braz. J. Microbiol, 38, 704–709.
46. Todd R. S *et al.* (2000). Appl Environ Microbiol, 66(10): 4585-4588.
47. Splendiani A *et al.* (2006). Biotechnol Bioengg, 94: 15–23.
48. Harshey, R. M. (2003). Ann Rev Microbiol, 57: 249-273.
49. Nadine E. Van Alst *et al.* (2007). Infect. Immun, 75(8): 3780-3790.
50. Koch, A. K. *et al.* (1991). J Bacteriol, 173: 4212-4219.
51. Qin Z. *et al.* (2009). Microbiol,155, 2148-2156.
52. Stewart, P. S. *et al.* (2000)Process Anal Appl,373–405.
53. Adam. B *et al.* (2002). J. Med. Microbiol. 51, 344–349
54. Yebra D M. *et al.* (2004). Antifouling Technol, Prog Org Coat 50: 75–104.
55. Vandyke M I *et al.* (1993). Can J Microbiol,39: 1071–1078.
56. Burgess J G. *et al.* (2003). Biofouling, 19: 197–205.
57. O'Toole.G *et al.*, 2005.
58. Mireles JR 2nd *et al.* (2001). J Bacteriol 183: 5848–5854.
59. Meylheuc T. *et al.* (2006). Inter J Food Microbiol,109(1-2): 71-78.
60. Somers E B. and Lee wong A C., Biofilms, Ann report 1999,

61. Meylheuc T. *et al.* (2003). Biofilms. Med, 17–19.
62. Rodrigues L *et al.*, 2006.
63. Schooling S. R. *et al.* (2004). Biofilms, 1: 2: 91-99.
64. Busscher HJ, (1997). Appl. Environ. Microbiol, 63(10): 3810-3817.
65. Velraeds M. M *et al.* (1996). Appl Environ Microbiol 62: 1958–1963.
66. Velraeds MM *et al.* (2000). World J Urol,18: 422–426.
67. Rodrigues L *et al.* (2004). Appl Environ Microbiol, 70: 4408–4410.
68. Rivardo.F *et al.* (2009). Appl Microb Cell Physiol.
69. Irie, Y. *et al.* (2005). FEMS Microbiol. Lett. 250: 237-243.
70. Bucay-couto and Weenna, Li, Jamie (Jianmin),(2005). US Patent 6921390
71. *Van Hoogmoed. C et al. (1997). Appl Environ Microbiol, 63(10): 3810–3817.*
72. Gaidhani, S. *et al.* (2008).Proc Indo–Ialtalian Conference, 903–938.

Chapter 24

Bioremediation of High Nitrate Containing Nitro-Explosive Wastewater using Yeast Cultures

P.P. Kanekar[1], S.S. Sarnaik[1], N.P. Gujarathi[1], S.P. Kanekar[1], V.V. Phalke[1], P.N. Shede[1], M.R. Kedargol[1], K.F. Reardon[2], T. Soman[3] and V.D. Raut[3]

[1]*Microbial Sciences Division, Agharkar Research Institute, G.G.Agarkar Road, Pune*

[2]*Colorado State University, Fort Collins, CO 80523-1370, Colorado, USA*

[3]*High Energy Materials Research Laboratory (HEMRL), DRDO, Sutarwadi, Pune*

ABSTRACT

A large number of nitro compounds are manufactured for application in various fields like agriculture, defense, automobile, textile, etc. Their residues are subsequently released into the environment. To remediate sites contaminated with such toxic chemo pollutants, use of biological entities is assuming importance. Bioremediation technology makes use of versatile micro-organisms that degrade the toxic compounds to non-toxic products. The present chapter describes bioremediation of high nitrate and acetate containing wastewater using two yeast cultures namely *Williopsis* sp. and *Pichia* sp. isolated from garden soil by enrichment and bating technique. Fixed film bioreactors with insertion of baffles were developed separately for both the cultures using brick pieces as supporting media and operated in batch mode for a period of one week. The culture of *Williopsis* sp. showed better removal of both nitrate and acetate *i.e.* 72.6 and 95.94 per cent res. from the wastewater as compared to that of *Pichia* sp. 51.05 and 90.35 per cent res. The insertion of baffles in the reactor was found to be beneficial for *Pichia* sp. in removing acetate from the wastewater. Thus the fixed film bioreactors could be developed using yeast cultures for bioremediation of wastewaters containing high concentration of nitrate as well as acetate.

Keywords: *Nitrate, Acetate, Pichia sp., Williopsis sp., Fixed film bioreactor, Bioremediation.*

Introduction

Nitro compounds are produced by incomplete combustion of fossil fuel or nitration reactions and are used as chemical feedstock for synthesis of explosives pesticides, herbicides, dyes, pharmaceuticals, etc. The indiscriminate use of nitro compounds in the past due to wide applications has resulted in inexorable environmental pollution. Among these nitro aromatics are recognized as recalcitrant and given Hazardous Rating-3. Environmental contamination by nitro compounds is associated principally with industrial chemicals such as nitro-explosives, dyes, polyurethane foams, herbicides, insecticides, solvents etc. Higher intake of nitro compounds can cause methaemoglobinaemia in human beings and are harmful to the liver and central nervous system in rats, mice, and rabbits.

Although several conventional clean up methods currently used for the removal of nitro compounds, none has proved to be sustainable. Recently, remediation by biological systems has attracted worldwide attention to decontaminate nitro compound polluted sources. The incredible versatility inherited in microbes has rendered these compounds as a part of the biogeochemical cycle. Several microbes catalyze mineralization and/or non-specific transformation of nitro-aromatics either by aerobic or anaerobic processes.

Since nitro-explosives form one of the major group of nitro compounds, their production and use for military operations have resulted in their dissemination into the environment, where their presence in waterways and soil poses an ecological and health hazards. Therefore removal of nitrates from such nitro-explosive contaminated sites becomes essential. Some of the bioremediation methods used for removing nitrate from nitro-explosives wastewater/soil are reported (White *et al.*1996; Rodgers and Bunce 2001; Kanekar *et al.*, 2003, 2009; McCormick *et al.*, 1984; Kaplan 1993; Kitts *et al.*, 1994; Toumi *et al.*, 1997; Cooper *et al.*, 1998; Strietelmeier *et al.*, 2006; Ahmad *et al.*, 2007, Kulkarni and Chaudhari, 2007). The present chapter describes bioremediation of high nitrate and acetate containing wastewater generated during production of nitro-explosive using yeast cultures in a fixed film bioreactor.

Materials and Methods

Characterization of the Wastewater

Collection and Chemical Characterization of Wastewater Samples

The high nitrate containing wastewater under study was collected from High Energy Material Research Laboratory (HEMRL), Pune as and when required and preserved in the laboratory at ambient temperature for further studies. This wastewater was generated during production of recently developed nitro-explosive which contained high concentration of nitrates along with acetate. Hence it was characterized for the parameters like pH, concentration of nitrate as per Standard Methods (Greenberg *et al.*, 1992, Shivran *et al.*, 2006). Acetate concentration present in the wastewater was estimated using Gas Chromatograph (Perkin Elmer, USA) equipped with Flame Ionization Detector (FID). Nitrogen having flow rate of 28 ml/min was used as carrier gas and mixture of Hydrogen and air (1:10, v/v) was used for flame. Temperatures of oven, injector and detector were 140, 170 and 190°C respectively. Acetic acid (Sigma Chemicals, USA) was used as reference standard. The analysis of wastewater for each parameter was performed for six times.

Bioremediation of Wastewater by Yeast Cultures

Isolation and Identification of the Yeast Cultures Capable of Growing on Wastewater

Two yeast cultures were isolated from soil samples using enrichment technique as described by Kanekar *et al.* (2009) and identified on the basis of morphological and biochemical characteristics (Lodder 1970; Krutzman and Fell 2000).

Flask Culture Studies for Removal of Nitrate and Acetate

Initially, the tolerance of the isolates to high concentration of nitrate and acetate incorporated in Davis Mingioli's synthetic medium as source of nitrogen and carbon. Growth of these isolates in the suitably diluted wastewater was also studied. Ability of the cultures to remove nitrate and acetate from the wastewater was studied at different initial pH values of the wastewater and under shake and static culture in neutralized wastewater at ambient temperature (28± 2°C).

Development of Fixed Film Bioreactor

Two fixed film bioreactors were developed in 10 L rectangular containers using brick pieces as supporting medium for formation of film of the individual cultures separately. The rubber baffles were introduced in the bioreactors breadth wise in opposite direction across the supporting media at a distance of 12 cm from each end.

Five kg brick pieces were used as supporting media and placed randomly in the container. 96 h grown culture broth having cell density 1.65 ´ 10^7 cells/ml was mixed with equal volume *i.e.* 2.5 L of wastewater and recycled through brick bed for five days at ambient temperature (28± 2°C) for formation of film on brick pieces. The development of film of the organisms was checked by microscopic observation of the growth of the cultures on brick pieces. The bioreactors were operated for one week *i.e.* 168 hrs. and the performance of reactors was evaluated with respect to removal of nitrate, acetate and change in pH of the wastewater at 24 h interval. The analysis of the cell free supernatant was carried out after centrifugation of the samples at 14,000 rpm for 15 min at 20°C (Kubota 6930, Japan).

Results and Discussions

Chemical Characterization of the Wastewater

Chemical characterization of the wastewater under study is detailed in Table 24.1. pH of the wastewater is seen to be highly acidic *i.e.*2.38. Since production of

Table 24.1: Characterization of Wastewater under Study.

Parameters	*Concentration*	*MPCB Limits*
pH	2.38	6.5–7.5
Nitrate	3,34,000 mg/L	45 mg/L
Acetate	2,27,000 mg/L	–

nitro-explosive is carried out using acetic acid and acetic anhydride as solvents, it is quite obvious that pH of the wastewater is acidic in nature and the waste contains high concentration of acetate. The wastewater thus could be preserved easily at ambient temperature (28±2°C). Production of this nitro-explosive involves nitration process using concentrated nitric acid and ammonium nitrate as raw materials and hence the nitrate content of the wastewater is very high *i.e.* of the magnitude of 3,34,000 mg/L. The permissible limit of MPCB for nitrate is only 45 mg/L. The acetate content of the wastewater was 2,27,000 mg/L. There are no permissible limits specified by Maharashtra Pollution Control Board (MPCB) for acetate content but still the acetate content of the wastewater can be considered as very high. Nitrogen present in the form of nitrates may be acquiescent to degradation by the micro-organisms but becomes toxic when present in higher concentration. The concentration of nitrates in the wastewater under study is of the magnitude of 3,34,000 mg/L which is highly toxic and detrimental to living systems. Hence the removal of nitrates from such wastewater becomes very essential.

Bioremediation of Wastewater Containing Nitrate and Acetate by Yeast Cultures

Isolation and Identification of the Yeast Cultures Capable of Growing on Wastewater

Based on the morphological and biochemical tests performed, one isolate was identified as *Pichia* sp. and the other one as *Williopsis* sp. These cultures are deposited in MACS's Culture Collection of Micro-organisms (MCM) recognized by World Federation of Culture Collection (WFCC- Registration Number 561) with the accession number as MCM Y-3 and MCM Y- 4 respectively and preserved in glycerol stocks at -20°C.

The wastewater under study has highly acidic pH and contains high concentration of nitrate in the wastewater, therefore, it was necessary to neutralize the wastewater and then dilute it suitably so as to have nitrate concentration suitable for growth of the micro-organisms. Both the yeast isolates were found to tolerate nitrate and acetate at the concentration of 4 g per cent *i.e.* 40,000 mg/L and 2 g per cent *i.e.* 20,000 mg/L respectively when incorporated in Davis Mingioli's synthetic medium and utilize ammonium nitrate and sodium acetate as nitrogen and carbon source for their growth. The 1:10 (v/v) diluted wastewater had nitrate and acetate concentrations near the tolerance limit of these organisms *i.e.* nitrate as 33,400 mg/L and that of acetate as 22,700 mg/L respectively. Hence 1:10 diluted wastewater was used for further studies.

Flask Culture Studies for Removal of Nitrate by both the Yeast Cultures

Flaks culture studies *Williopsis* sp. showed nitrate removal to the extent of 45 per cent and reduction in acetate content as 79 per cent at initial pH of 7 of the wastewater after 72 h of incubation at ambient temperature (28± 2°C) under shake culture condition. In case of *Pichia* sp. nitrate removal was 49 per cent and acetate removal was 76 per cent under similar experimental conditions.

Performance of Fixed film Bioreactors using *Pichia* sp. and *Williopsis* sp.

The bioreactors were evaluated for change in pH of the wastewater and removal of nitrate from the wastewater. The results on the change in pH of the wastewater and the removal of nitrate and acetate from the bioreactors using *Pichia* sp. and *Williopsis* sp. are detailed in Figures 24.1 and 24.2 respectively. In general the bioreactor containing the culture of *Williopsis* sp. showed better removal of both nitrate and acetate (72.6 and 95.94 per cent res.) from the wastewater as compared to that of *Pichia* sp. (51.05 and 90.35 per cent res.) The change in pH was also considerable in the bioreactor inoculated with both the cultures. The change in pH was from initial 7.0 to 9.3 in case of *Williopsis* sp. and 5.91 to 8.04 in case of *Pichia* sp. Thus both the yeast cultures can be used for bioremediation of high nitrate/acetate containing wastewaters either individually or as consortium.

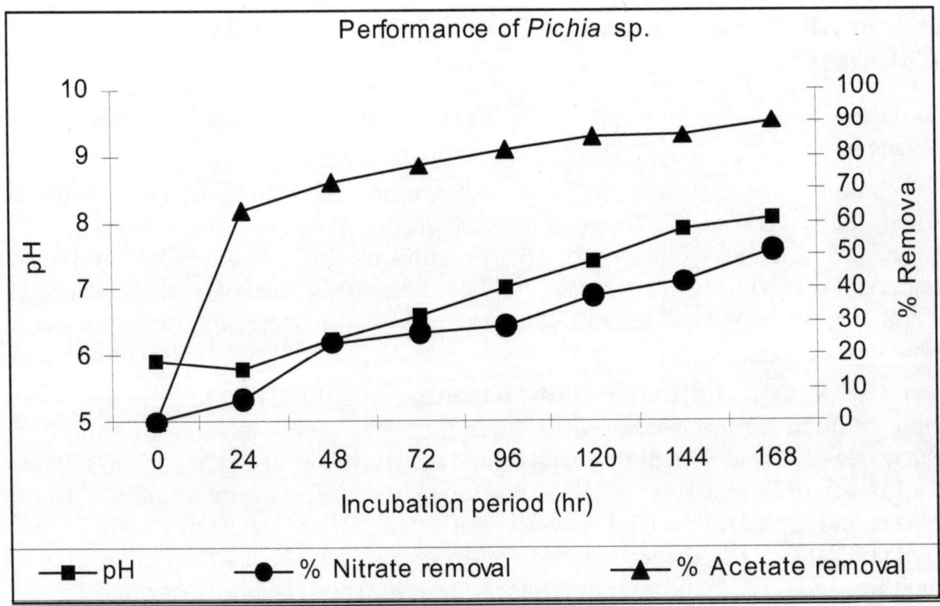

Figure 24.1: Performance of *Pichia* sp. with Respect to Removal of Nitrate, Acetate from the Wastewater (Initial concentration 33,400 and 22,700 mg/L res.) and Change in pH.

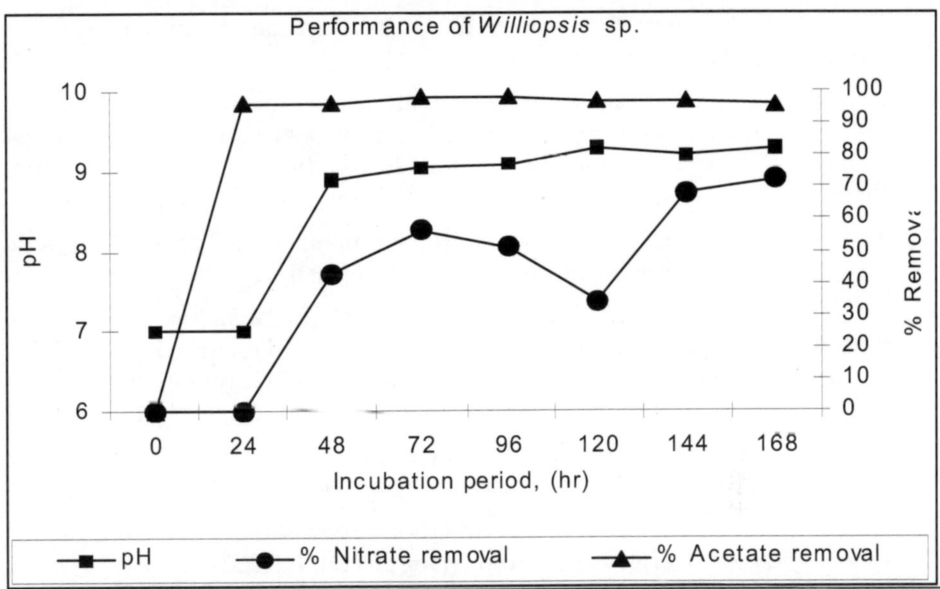

Figure 24.2: Performance of *Williopsis* sp. with Respect to Removal of Nitrate, Acetate from the Wastewater (Initial concentration 33,400 and 22,700 mg/L res.) and Change in pH.

Conclusion(s)

The present studies showed potential of soil isolates of *Pichia* sp. and *Williopsis* sp. in bioremediation of wastewater generated during production of nitro-explosive which was highly acidic in nature and contained high concentrations of nitrate and acetate. The culture of *Williopsis* sp. showed better removal of both nitrate and acetate from the wastewater as compared to that of *Pichia* sp. Thus both the yeast cultures can be used for bioremediation of high nitrate/acetate containing wastewaters either individually or as consortium.

References

1. Ahmad F, Schnitker SP, Newell CJ (2007). Remediation of RDX- and HMX-contaminated groundwater using organic mulch permeable reactive barriers. J Contam Hydrol, **90 (1-2):** 1–20
2. Cooper PW, Card ER, Autenrieth R (1998). 'Treatment of HMX and RDX Contamination' In Explosives Engineering, Wiley-VCH, New York
3. Cruickshank R. (Ed) (1965). Medical microbiology, The English Language Book Society, Eleventh edition.
4. Greenberg AE, Clesceri LS, Eaton AD (1992). *Standard methods for the examination of water and wastewater*. American Public Health Association, American Water Works Association, Water Environment Federation, Washington DC.

5. Kanekar PP, Dautpure PS, Sarnaik SS (2003). Biodegradation of nitro-explosives. Indian J Exp Biol, **41:** 991–1001

6. Kanekar SP, Kanekar PP, Sarnaik SS, Gujarathi NP, Shede PN, Kedargol MR, Reardon KF (2009). Bioremediation of nitro-explosive wastewater by an yeast isolate *Pichia sydowiorum* MCM Y-3 in fixed film bioreactor. J Ind Microbiol Biotechnol. 36: 253–260

7. Kaplan, D.L. (1993). Biotechnology and bioremediation for organic energetic compounds. Nova Science Publishers, Inc., New York

8. Kitts CL, Cunningham DP, Unkefer PJ (1994). Isolation of three hexahydro-1, 3, 5-trinitro-1, 3, 5-triazine-degrading species of the family Enterobacteriaceae from nitramine explosive-contaminated soil. Appl Environ Microbiol, **60 (12):** 4608–4611

9. Krutzman CP, Fell JW (eds). (2000). The Yeasts: A Taxonomic Study, Elsevier Science B. V. Netherlands

10. Kulkarni M, Chaudhari A (2007). Microbial remediation of nitro-aromatic compounds: An overview. J Environ. Mang. **85 (2):** 496-512

11. Lodder J. (1970). The Yeasts: A Taxonomic study, Second edition, North-Holland Publishing Company.

12. McCormick NG, Cornell JH, Kaplan AM (1984). The fate of hexahydro-1, 3,5-trinitro-1, 3,5-triazine (RDX) and related compounds in anaerobic denitrifying continuous culture systems using simulated wastewater. *Technical Report Natick/ TR-85/008, ADA149462*. US Army Natick Research and Development Center, MA.

13. Rodgers JD, Bunce NJ (2001). Treatment methods for the remediation of nitro-aromatic explosives. *Water Research,* **35 (9):** 2101 - 2111

14. Shivran HS, Kumar D, Singh RV (2006). Improvement of water quality through biological Denitrification. Environ Sci Engin, **48 (1):** 57-60.

15. Strietelmeier BA, Coyne PJ, Leonard PA, Miller WL, Brian JR (2006). Initial characterization of a highly contaminated high explosives outfall in preparation for in situ bioremediation. http://www.osti.gov/energycitations/servlets/purl/752372-4ucf9X

16. Toumi E, Coover MP, Stroo HF (1997). Bioremediation using composting or anaerobic treatment for ordnance-contaminated soils. Annals New York Aca Sci, **829:** 160–178.

Chapter 25

Bioremediation: A Preventive Solution for Pollution

***Urmila P. Jagtap*[1]*, Smita S. Bhuyan*[2]*, Sakshi P. Korde*[1]*, Karishma R. Pardesi*[2]*, Sharvari V. Gaidhani*[1] *and Balu A. Chopade*[1,2]**

[1]*Institute of Bioinformatics and Biotechnology, University of Pune, Pune*

[2]*Department of Microbiology, University of Pune, Pune*

ABSTRACT

The rapidly deteriorating environment due to excessive use and misuse of hydrocarbons is major cause of concern of this century. The terrestrial as well as marine life forms are severely affected. There is an urgent need for implementation of eco-friendly measures for a speedy and effective clean up. In the present scenario, bioremediation appears to be the most lucrative option for restoring the environments delicate balance. Remediation with micro-organisms, either aerobic or anaerobic is preferred over conventional methods. Biosurfactant producing micro-organisms can speed up the process and outcome. The present work aims to summarize the global pollution problem and management and remedial measures for reverting back to a cleaner environment.

Introduction

There is a global concern about water pollution due to release of hydrocarbons either by industries or accidental oil spills. Contaminated environments include freshwaters, seas, (lakes, rivers and groundwaters), including beaches (mainly containing sand) and inland soils (composed of mixtures of sand, silt and clay).

Commercial oil and gas explorations have been taking place in the seas for over 150 years. During the estimations carried out by Aghai Diba *et.al.* [1] in the year of 2000, each year an average of 60,000 metric tonnes of petroleum byproducts, 24,000 tonnes of sulfites, and 400,000 tonnes of chlorine are discharged into the Caspian Sea. The debris of mineral exploration, particularly oil extraction and pipeline construction, have contributed to the pollution of about 30,000 hectares of land.

Hydrocarbons are one of the major environmental pollutants and are found to be persistent in many polluted environmental niches. The problem of water pollution is very serious due to the fact that hydrocarbons are non-biodegradable and toxic thereby endangering marine life. Currently, chemical surfactants are majorly used for environmental cleanup. However, the synthetic surfactants themselves are non-degradable and have toxic effects which are contradictory to all current environmental policies. Biosurfactants, on the other hand are preferred over synthetic surfactants; because of their bioavailability, biocompatibility, eco-friendly nature, low production cost, structural diversity and biodegradative nature[2]. Another reason for the search for biological surfactants is the depletion of oil resources required for the production of synthetic surfactants. This makes use of microbial surfactants even more promising [3].

Today, there is a high market for the surfactant industry. The total quantity of surfactants produced during 1989 in United States and globally has been estimated to be around 7.6 X 10^9 and 15.5 X 10^9 lb, respectively[4]. Almost half of the surfactants produced (5.2 million tonnes in 1991) are made for washing and cleaning[5].The steady growth of the surfactant market achieved in the last years was hit by the current economic recession. The world market of surfactants reached US$23, 9 billion in 2008, roughly 2.5 per cent up from the previous year. The surfactant market value is expected to grow by 2.8 per cent by 2010 when overall chemical industry experiences slowdown[6]. The market is expected to see a recovery in two years in the forecast and the growth by 4 per cent is expected thereafter. Therefore, the tremendous environmental damage resulting due to the use of huge amount of synthetic surfactant can be avoided by using biosurfactant which works in an eco-friendly manner. Multi-biotech, a subsidiary of Geodyne Technology, has already commercialized biosurfactants for enhanced oil recovery applications.

The microbial biosurfactants have different ways of action. Micro-organisms with the ability to degrade crude oil are ubiquitously distributed in soil and marine environments[7]. The extracellular or cell bound biosurfactants function by promoting uptake of the substrate, the cell bound biosurfactants take up the pseudosolubilised substrate by facilitating attachment of the cells to the insoluble substrate[8]. In addition to altering surface tension, they can affect the movement of materials across aqueous interfaces and change the solubility of compounds in aqueous systems. Recently, biosurfactants are largely being used in oil industry (oil production, processing, and transportation, recovery of oil from polluted objects), hydrocarbon biodegradation, cleaning of tanks and soil bioremediation[9,10]. The extent of hydrocarbon biodegradation in contaminated soils is critically dependent upon three factors, namely the creation of optimal environmental conditions to stimulate biodegradative activity, the predominant petroleum hydrocarbon types in the contaminated matrix and the bioavailavility of the contaminants to micro-organisms[11]. Bioremediation is also being proposed and used as a treatment option at many hydrocarbon contaminated sites[12].

Mineral Oil Biodegradation

What is Biodegradation?

By definition, biodegradation is the chemical transformation of a substance caused by organisms or their enzymes. The primary biodegradation refers to the minimum transformation that alters the physical characteristics of a compound, leaving the molecule largely intact. The mineralization of the intermediary metabolites which are known to be more toxic than the original substrate in certain cases, ultimately results in the production of carbon dioxide, methane, water, mineral salts, and new microbial cellular constituents[13]. This achieves the complete process of biodegradation. By establishing conditions which maximize rates and extents of microbial growth, hydrocarbon access, and transformation, highly accelerated and bioreactor-based petroleum waste degradation processes have been implemented in business[14].

Mineral oils constitute a huge proportion of oil products which are used as lubricants in the industries worldwide. They occur as complex mixtures including hydrocarbons predominantly and in minor quantities sulfur, nitrogen, oxygen and metals. The petroleum industries, where the annual production reaches approximately billions of metric tonnes, have always been the hub of crude oil trading[15]. Mineral oils, which are a class of petroleum products, include white oils, lubricating oils, light fuel oils, residual fuel oils, as well as transformer and cable oils[16]. Though they are economically important, the waste lubricants in the perspective of environment, act as pollutants due to their non-biodegradable nature, exerting harmful effects on plants, animals and marine organisms. Some of them, *e.g.* greases and hydraulic fluids, possess the toxic characteristics, therefore if released, cause incurable damage to environment. Additionally, certain intermediate products of degradation can be carcinogenic to living forms[17,18]. Therefore, there is an urgent need for implementation of strategies for eco-friendly measures like bioremediation to be implemented to save the deteriorating environmental balance[19].

Aerobic Hydrocarbon Metabolism

The development of newer techniques and availability of useful infrastructure *e.g.* genetic engineering have significantly enhanced our understanding of hydrocarbon catabolism. Cellular and physiological adaptations of micro-organisms to the presence of hydrocarbons as well as the biochemical mechanisms involved in hydrocarbon acquisition and uptake is being studied extensively[20]. The use of genetically engineered microbes for bioremediation has also been considered[21].The main factor on which the rate of petroleum biodegradation depends is the molecular composition of hydrocarbon. Petroleum products contain thousands of individual hydrocarbons and related compounds. Their main components are usually sub-divided into saturates (n- and branched-chain alkanes or isoprenoids and cycloparaffins) and aromatics (mono-, di-, and polynuclear aromatic compounds containing alkyl side chains and/or fused cycloparaffin rings). Many hydroxy aromatic compounds can be toxic when present at elevated levels[22]. The degradation of aromatic compounds by micro-organisms is accomplished by diverse pathways. Condensed aromatic hydrocarbons are poorly degradable; and heterocyclic compounds, even those characterized by high degrees of condensation, undergo slight degradation. Due to complexity of oil products, biodegradation caused by mixed cultures is more effective than that by pure cultures.

Number of bacteria *viz. Burkholderia cepacia* sp., *P. putida* sp., *B. cepacia* sp., *Acinetobacter* sp., *P. aeruginosa* sp., *Nocardiodes* sp. *strain CF8, Rhodococcus* sp. etc. are able to degrade alkanes. When large number of strains were characterized, it appeard that the regulation of alkane degradation varies among bacteria[23]. Different micro-organisms differ in their ability to degrade hydrocarbons, which mainly depends on the complexity level of alkanes. These differences are due to repression of alkane degradation by citrate and the maintenance of repression during stationary phase. Most of these bacteria synthesize biosurfactant which has a major role in hydrocarbon biodegradation. The type of biosurfactant and its nature (cell membrane bound/ extracellular) also adds to the diversity. The environmental conditions have a direct impact on metabolism of microbes and thus; indirectly on enzymes like monooxygenases, reductases and hydroxylases with a resultant difference in the degree of degradation.

Romine et. al.[24] in 1999 sequenced the pNL1 (H"184 kb) plasmid of *Sphingomonas aromaticivorans* F199. This plasmid confers the ability to degrade toluene, xylene, salicylate, biphenyl, dibenzothiophene, flourene, and benzoate. More recently, a *Rhodococcus* sp. that metabolizes anthracene to 1,2-dihydroxyanthracene and then to either 3-(2-carboxyvinyl)naphthalene-2-carboxylic acid or 6,7-benzocoumarin has been observed by Dean-Ross *et al.*[25] in 2001. In both *Azoarcus* sp. strains EB1 and EbN1, ethylbenzene oxidizes (under denitrifying conditions) to produce phenylethanol followed by oxidation to acetophenone by the action of an enzyme $\alpha\beta\gamma$-Mo-Fe-S heterotrimer[14].

Monoxygenases:

Different micro-organisms produce the enzyme monooxygenase, which initiates the transformation of fairly persistent compounds like aliphatic, aromatic compounds and PHAs by the process known as monooxygenesis (MMO). MMO initiates transformation of Trichloroethylene (TCE) which is a common Dense Non-Aqueous Phase Liquid (DNAPL) found in contaminated soil and groundwater to TCE-epoxide by oxidizing the TCE with one oxygen molecule while incorporating the other oxygen into a water molecule[26].

Steps involved in degradation of TCE[26] (Figure 25.1):

1. Conversion of TCE to TCE-epoxide.
2. Extrusion of TCE-epoxide from the cell.
3. TCE-epoxide being very unstable is broken down spontaneously in the aqueous environment existing outside the cell into CO, formate, glyoxylate or dichloracetate.

3. Break down products are further metabolized by methanotrophs, which use them as an energy source and by heterotrophs in absence of methanotrophs. In methanotrophic consortium TCE is completely mineralized to CO_2 and converted into cell biomass.

Many strains employ either monoxygenases or both mono- and di-oxygenases for the metabolism of single-ring PAHs[27]. Two distinct monooxygenases, a Cu-containing monooxygenase and an integral-membrane, binuclear-iron monooxygenase have been described in *Nocardiodes* sp. strain CF8[28]. Cu-containing monooxygenase is commonly observed in the microbes degrading alkanes whereas, those which can degrade more than six carbons induce the binuclear-iron monooxygenase.

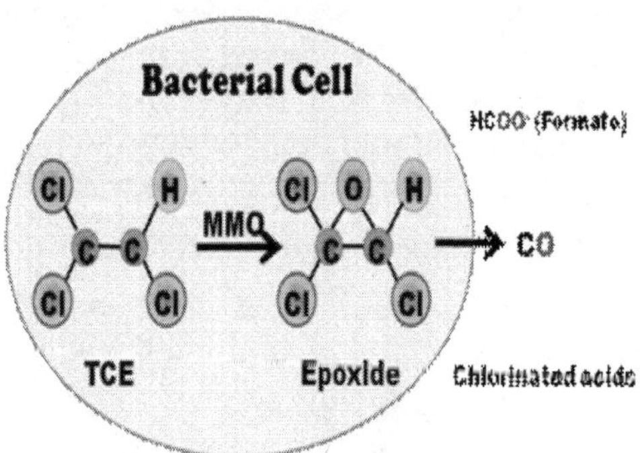

Figure 25.1: Mechanism of TCE Degradation.

Anaerobic Hydrocarbon Metabolism

A diverse set of bacteria including members of α-, β-, γ-, and Δ-subclasses of proteobacteria, posses biochemical and molecular mechanisms to accomplish anaerobic hydrocarbon metabolism. Studies on huge microbial consortia, enrichment cultures, and microcosms have illustrated that different hydrocarbons such as toluene[29], alkylbenzenes including *m*-, *o*-, and *p*-xylene and trimethylbenzenes[30], benzene[31], naphthalene and phenanthrene[32], methylnaphthalene and tetralin[32], $>C_6$ *n*-alkanes[33], branched alkanes[34], and hydrocarbon mixtures can be metabolized under anaerobic conditions.

Among these, toluene has been the most studied hydrocarbon with respect to enzymatic and molecular genetics in *Azoarcus* sp. strain T, *Thauera aromatica* strain K172, and *Thauera* sp. strain T1[35]. Two- and three-ring PAHs may also be metabolized under anaerobic conditions. In sulfate-reducing[36] and denitrifying[37] bacteria, naphthalene gets carboxylated to form 2-naphthoate. Mixed cultures are found to enhance biodegradation as compared to the pure cultures. In all the above mentioned hydrocarbons, benzene has remained exception as its mechanism for anaerobic biodegradation has not been elucidated till date. However, two strains of *Dechloromonas* namely RCB and JJ belonging to the β-proteobacteria class that mineralizes benzene with nitrate as the electron acceptor has been isolated recently[38]. All these reactions take place under Fe(III)-reducing, denitrifying, and sulfate-reducing conditions, by anoxygenic photosynthetic bacteria, or in presence of syntrophic consortia of proton-reducing and methanogenic bacteria. As oxygen is absent, other terminal acceptors like manganese oxides[39], soil humic acids and anthraquinone-2,6-disulfonate (a humic acid compound), and fumarate are used.

Bioremediation

In-situ bioremediation is a technology designed to control subsurface conditions by making naturally occurring bacteria thrive and grow where the contaminant is actually located[26]. The rate of biodegradation increases with an increase in the growth rate of degrading bacteria; provided the growth conditions are optimal. The contaminants can be potentially reduced into the final end products: CO_2, water and cell biomass. Bioremediation has been successfully carried out at petroleum contaminated sites for decades. The use of this technology; therefore; has been extended to more recalcitrant contaminants such as halogenated hydrocarbons.

In situ bioremediation offers several potential advantages over conventional technologies[26]:

Advantages

- Less disturbance:
  - i) Soil and groundwater do not have to be removed.
  - ii) No disposal of waste compounds.
  - iii) Smaller footprint for above ground component of treatment system.
- Safe: Minimum risk associated with hazardous waste transport and handling.
- Eco-friendly and Permanent: Complete restoration to basic constituents (release of waste to atmosphere doesn't occur.)
- Appealing: Technology utilizes natural occurring micro-organisms to degrade toxics into harmless products.
- Cheap: The long term costs and low maintenance dominate over the relatively high initial costs.

Disadvantages

- *High up front costs*: Determining the feasibility and parameters for system design requires detailed site characterization.
- *Communication*: It is difficult to convince the client, that up front sampling with detailed analysis are necessary and that the associated costs are justified.
- *Technical skills*: High level of training is necessary for technicians.

The advantages and disadvantages of different remediation technologies should always be considered because *in situ* bioremediation may not be applicable always or other technologies are more cost effective. Number of parameters like hydrogeology, transmissivity, temperature, water content, toxicity, competitive substrate, mineral nutrient bioavailability, gratuitous biodegradation, desorption rates, pH are essential in the process of "*in situ* bioremediation".

The Microbiological Engine

Numerous studies have recently been dedicated to the safety of global environment. Of special interest, are studies concerning the biological remediation of sea and land areas polluted with crude oil and its products. The microbial degradation of organic contaminants in terrestrial and aquatic ecosystems is main process in natural decontamination and enhancement of this phenomenon is the basis of bioremediation technologies. The importance of micro-organisms in the words of Eric Anderson and Rikke Andersen[26] is such that "The combined microbiological metabolic activity of the interacting suite of indigenous subsurface bacteria is the engine that drives the bioremediation process".

A large number of micro-organisms are found to be potent degraders of organic compounds. Many of them do so with the help of biosurfacrants/bioemulsifiers. Biosurfactants produced by a strain of *Pseudomonas aeruginosa* SB30 were used to remove oil from gravel in the Exxon Valdez oil spill in Alaska[40]. In another study, *Acinetobacter* sp. strain W-17 was found to enhance phenol biodegradation while *P. aeruginosa* could degrade the aliphatic hydrocarbon α, w-tetradecane and hexadecane potentially on rhamnolipid supplementation[19]. Some strains like *Rhodococcus* spp. DS-07, DS-21[41] and *Pseudomonas* spp. DS-09 and DS-22 are known to degrade masut, ethanol–benzene resins and petroleum oils at low temperatures (4-6^{0}C). These strains could be used to develop the consortia for bioremediation of oil-polluted sites under cold climatic conditions. Around 30 per cent increase in total oil recovery from underground sandstone by using trehalolipids from *Nocardia rhodochrous* sp. has been documented[42]. Due to the complexity of oil derivatives, biodegradation elaborated by mixed cultures is more effective than that by pure cultures.

Role of Biosurfactants

Amphiphatic compounds like biosurfactants reduce surface tension by accumulating at the interface of two surfaces/phases thus increasing the surface area of insoluble compounds, leading to increased bioavailability and subsequent biodegradation of the hydrocarbons[43]. Micro-organisms may also take up insoluble hydrocarbons by adhering to hydrocarbons at the water-hydrocarbon liquid or solid interface[44].

A significant problem in biodegradation of hydrocarbon contaminants is their extremely low solubility in aqueous phase. During the course of time the bioavailability of contaminants decreases by aging. They increase the solubility, mobility, bioavailability and subsequent biodegradation of hydrophobic compounds *i.e.* alkanes, aromatic compounds like chlorinated benzenes. Biosurfactants, being natural in origin have different and specific mode of action. They are not toxic, are relatively easy to synthesize and have wide range of applicability. Most of the organisms, growing on water immiscible substrates synthesize high- or low-molecular weight biosurfactants.

The oil industry is a rapidly emerging field where biosurfactants find application for cleanup[9,10]. *B. licheniformis* JF-2, produces a very effective biosurfactant under anaerobic conditions, (CMC 10 mgliter^{-1}; interfacial tension of saline against decane lowered to 10^{-3} dynes.cm^{-1}), and finds potential usefulness in in-situ microbially enhanced oil recovery[45]. Hayes *et al.* (1986)[46] demonstrated that when heavy crude oil, was treated with emulsan, its viscosity reduced from 200,000 to 100 cP[45], while chemical surfactant treatment proved ineffective. In yet another interesting application, a strain of *Rhodococcus* sp. which was able to degrade chlorobenzene substrates at temperatures as low as 0°C was identified[47]. This can help in the hydrocarbon contaminated sites at cold places.

In the perspective of soil bioremediation, the term bioavailability is used to describe the extent to which mass transfer limitations determine a reaction (*e.g.* biodegradation rate or uptake rate by a specified organism) or an effect (*e.g.* toxicity to a specified organism or biodegradability by an organism). The processes that are involved in the transfer of contaminants from soil to the catalyzing enzymes of the degrading micro-organisms determine the bioavailability of contaminants. According to Zhang and Miller[10], a three-way interaction between biosurfactant, substrate, and cells is very critical to achieve an enhanced biodegradation rate.

The process of bioremediation involves following steps:

1. Contact of the substrate with the degrading micro-organism.
2. Transport of the substrate to the aqueous phase by desorption, dissolution or other means
3. Reduction of interfacial tension between the two phases (surfactants/ emulsifiers)
4. Intake of oil droplet or its constituent breakdown products.

The efficiency of biosurfactants for the remediation of metal[48], phenanthrene[49], and polychlorinated biphenyl[50] contamination in soil has recently been shown. Biosurfactants are also useful in bioremediation of sites contaminated with toxic heavy metals like uranium, cadmium, and lead[48]. 56 per cent of the aliphatic and 73 per cent of the aromatic hydrocarbons were recovered from hydrocarbon-contaminated sandyloam soil by treatment with *P. aeruginosa* biosurfactant[51]. Bioremediation using immobilized cells has been widely investigated for the treatment of numerous toxic chemicals such as pentacholorophenol[52], 4-chlorophenol[53] and chlorobenzoates[54]. When the bioemulsifier alasan from *Acinetobacter radioresistens* KA53 was checked for PAH solubility, the degradation results showed 8-fold increase in the solubilities of phenanthrene, fluoranthene and pyrene, respectively[55]. Similar results on PAH solubilization have been obtained with rhamnolipids produced by *Pseudomonas aeruginosa* sp. and other *pseudomonads*[56]. In case of increasing the mass transfer of PAHs into the aqueous phase, Page *et al.*[57] found that the biosurfactant from *Rhodococcus* strain H13-A was up to 35-fold more effective than Tween 80.

There are certain key differences between bioremediation of metal and organic-contaminated soils. The most obvious difference is that unlike organics, metals cannot be biodegraded. In some cases metals can be transformed but transformation often only increases metal toxicity (*e.g.*, $Hg^{++} \rightarrow CH_3\text{-}Hg^{+}$). A second difference is that organics mostly occur as neutral molecules, while metals mostly exist as cationic species. Thus, since contaminant sorption depends on the chemical properties of both the soil and the contaminant, the choice of surfactant used for contaminant complexation is very important. Certain microbial products such as bacterial and algal exopolysaccharides have been shown to bind a variety of metals. Emulsan, produced by *Acinetobacter* RAG-1 was found to bind uranium (UO_2^{2+}). Similarly, a *Pseudomonas* exopolysaccharide bound upto 96 µg uranium/mg exopolymer[58]. *Klebsiella* exopolysaccharide bound comparable amounts of cadmium (11 µg/mg exopolymer) as well as copper (22 µg/mg exopolymer)[59]. A study on several marine *Pseudomonas* sp. exopolysaccharides showed complexation of copper, iron, lead, nickel, and zinc[60].

Biosurfactants need to fulfill certain prerequisites before they can be employed for the process of bioremediation. The foremost being ability to enhance the bioavailibilty of contaminants which otherwise would be difficult to attack (*e.g.* polycyclic aromatic hydrocarbons (PAHs), polychlorinated biphenyls (PCBs)). Second, the use of biosurfactants for bioremediation should be economically acceptable *i.e.* they must be available at any time in high quantities at cheaper rates and at the same time, they must be efficient at low concentrations. For enhanced oil recovery application, chemical surfactants cost in the range of $4 to 6 per kg (209b), whereas lignin-based sulfonates are 40 per cent cheaper; at these prices, they are economically viable only at a sale price of oil of $35 and 25 per barrel, respectively. This can be a major limiting hurdle in terms of higher costs of producing biosurfactants. Containment and intrinsic bioremediation is also worth considering, since it is seldom cost effective to clean contaminated sites down to virgin land quality, before redevelopment. Risk assessment based on health criteria is gaining more importance in handling of the tremendous challenge of prioritizing and targeting the remediation efforts. Remediation efficiency of xenobiotic pollutants by microbial cells remains a major challenge to microbial ecologists and process engineers[61]. Thus, combination of treatments, containment and intrinsic remediation will probably be one of the major challenges in the future remediation of contaminated soil and groundwater.

Biosurfactants thus have a wide range of applicability. However, it appears that biodegradation of mineral oils and hydrocarbons by biosurfactants is restricted by their high production costs. Current efforts are therefore directed towards the isolation and manipulation of micro-organisms exhibiting desired degradative capabilities with simultaneous production of potent biosurfactants. Interestingly, the 10- to 40-fold-lower CMC of biosurfactants compared with chemical surfactants has narrowed the actual gap in cost compared with their chemical counterparts.

Conclusion(s)

A detailed understanding of the hydrocarbon metabolism has been accomplished with enormous amount of research work done in the last century. Bioremediation with aerobic and anaerobic micro-organisms aided with biosurfactants is the most preferred method of choice for cleaner and safer environment. Bridging the gap between research and actual field application remains the larger problem at hand to work upon.

References

1. Aghai Diba *et.al.* 2003. Payv and Iran News.
2. Leahy, J. G., and R. R. Colwell. 1990. Microbiol. Rev. 54:305–315.
3. O. V. Turkovskaya *et.al.* 2001. Appl. Biochem. Microbiol., 37(1), pp. 71–75

4. Goswami, P. and H. D. Singh. 1991. Biotechnol. Bioeng. 37:1–11.
5. Hans-Joachim Daniel *et.al.* 1998. Biotechnol. Lett., 20(12):1153–1156
6. Book: World surfactant market. 2008. Publisher: Acmite Market Intelligence.
7. Venkateswaran, K. *et.al.* 1995. Microbiol. 41, 767.
8. Volkering, F. *et.al.* 1998. Biodegrad. 8: 410–417.
9. Ganitkevich *et.al.* 1988. Biotechnologiya; 4(5), pp. 575–583.
10. Zhang, G. and Miller, R.M. 1995. Appl. Environ. Microbiol. 61(6), pp. 2247–2251.
11. Facundo j. Márquez-rocha *et.al.* 2001. Water Air Soil Pollut. 128: 313–320.
12. Braddock *et.al.* 1997. Environ.Sci. Technol. 31(7), 2078.
13. US Army Corps of Engineers; 1999. US army Manual EM1110-2-1424
14. Jonathan D. *et.al.* 2003. Microbiol. Mol. Biol. Rev.; 67(4): 503–549.
15. Proskuryakova, V.A. and Drabkina, A.E. 1987. Chemistry of Mineral Oil and Gas: a University Text-book. Eds., Leningrad. Moscow: Khimiya.
16. Gary JH, H and werk GE. 2001. Petroleum Refining Technology and Economics." 4th Edition Marcel Dekker Inc. New York USA
17. Mil and E *et.al.* 1996. J. Chem. Tech. Biotechnol. 67: 227–236.
18. Emmanuel O. Aluyor and Mudiakeoghene Ori-jesu. 2009. African J. of Biotechnol. 8 (6), pp. 915-920, 20.
19. E. V. Pleshakova *et.al.* 2001. Appl. Biochem. and Microbiol., 37 (4), pp. 342–347.
20. De Bont, J. A. M. 1998. Trends Biotechnol. 16:493-499.
21. Garibsu, C., and I. Alkorta. 1999. J. Chem. Technol. Biotechnol. 74:599-606.
22. Mil and E *et.al.* 1996. Chem. Tech. Biotechnol. 67: 227–236.
23. Jonathan D. Van Hamme *et.al.* 2003. Microbiol. Mol. Biol.; 67(4): 503–549.
24. Romine *et.al.* 1999. J. Ind. Microbiol. Biotechnol. 23:303-313.
25. Dean-Ross, D. *et.al.* 2001. Lett. 204:205-211.
26. Eric Elmer Anderson and Rikke Granum Andersen; 1996; Groundwater pollut. primer, 1996.
27. Annweiler *et.al.* 2000. Appl. Environ. Microbiol. 66:5329-5333.
28. Hamamura *et.al.* 2001. Appl. Environ. Microbiol. 67:4992-4998.
29. Elshahed, M. S. and M. J. McInerney. 2001. FEMS Microbiol. Ecol. 35:163-169.
30. Chen, C. I., and R. T. Taylor. 1997. Appl. Microbiol. Biotechnol. 48:121-128.
31. Burland, S., and E. Edwards. 1999. Appl. Environ. Microbiol. 65:529-533.
32. Coates, J. D. *et.al.* 1996. Appl. Environ. Microbiol. 62:1099-1101.
33. Ehrenrich, P. *et.al.* 2000. Arch. Microbiol. 173:58-64.

34. Bregnard *et.al.* 1997. Appl. Environ. Microbiol. 63:2077-2081.
35. Biegert, T. *et.al.* 1996. Eur. J. Biochem. 238:661-668.
36. Galushko *et.al.* 1999. Environ. Microbiol. 1:415-420.
37. Rockne *et.al.* 2000. Appl. Environ. Microbiol. 66:1595-1601.
38. Coates *et.al.* 2001. Nature 411:1039-1043.
39. Langenhoff *et.al.* 1997. FEMS Microbiol. Ecol. 22:119-127.
40. L.M.Santa Anna *et.al.* 2002. Brazilian J. Chem. Engg. 19(02): 159–166.
41. Whyte *et.al.* 1998. Appl. Environ. Microbiol. 64:2578–2584.
42. Rapp *et.al.* 1977. DESCHEMA Monogr. Biotechnol. 81:177–185.
43. Bragg *et.al.* 1994. Nature 368:413–418.
44. Atlas, R. M. *et.al.* 1993. Oceanus. 36:71–81.
45. Jitendra D. Desai and Ibrahim M. Banat. 1997. Microbiol. Mol. Biol. Rev. 61(1): (47–64)
46. Hayes *et.al.* 1986. Microbial surfactants. Chemtech 239–243.
47. Peter Rapp and Lotte H. E. 2003. Gabriel-Ju¨rgens Microbiology, 149, 2879–2890
48. Miller, R. M. 1995. Environ. Health Perspect. 103:59–62.
49. Providenti *et.al.* 1995. FEMS Microbiol. Ecol. 17:15–26.
50. Van Dyke *et.al.* 1993. J. Ind. Microbiol. 11:163–170.
51. Scheibenbogen *et.al.* 1994. J. Chem. Technol. Biotechnol. 59:53–59.
52. O'Reilly T, Crawford L *et.al.* 1989. Appl. Environ. Microbiol. 55: 2115–2118.
53. Balfanz J and Rehm J 1991. Appl. Microbiol. Biotechnol. 35: 662–668.
54. Sahasrabudhe R *et.al.* 1988. Biotechnol. Bioeng. 31: 889–893.
55. Navon-Venezia, S. *et.al.* 1995. Appl. Environ. Microbiol. 61, 3240–3244.
56. Mulligan *et.al.* 2001a. Engg. Geol. 60, 193–207.
57. Page *et.al.*; 1999 Environ. Engg. Science 16, 465–474.
58. Marques AM *et.al.* 1990. Appl. Microbiol. Biotech. 34:429-431.
59. Bitton G *et.al.* 1978. Microbial. Ecol. 4:119-125.
60. Corpe WA *et.al.* 1975. Dev. Ind. Microbiol. 16:249-255.
61. Usama Beshay *et.al.* 2002. Biotechnol.Lett. 24: 1295–1297.

Chapter 26

Wastewater Management in a Steel Plant: A Case Study

R.A. Pandey and A.N. Vaidya

CSIR-National Environmental Engineering Research Institute, Nehru Marg, Nagpur

ABSTRACT

An integrated steel plant is a collection of various small component plants, which carry out different unit operation involved in steel making. These plants include Raw Material Handling Plant (RMHP), Sintering Plant (SP), Blast Furnace (BF), Coal Handling Plant (CHP), Coke Oven Plant (COP), Coke Chemical division (CCD), Refractory Material Plant (RMP), Captive Power Plants (CPP), Medium Pressure Boiler (MPB), Steel Melting Shop (SMS), Rolling Mills (RM), and Fertilizer Plant (FP). These plants generate wastewaters of different characteristics in varying quantities. Wastewater management in these sub units has a direct impact on the overall wastewater management in an integrated steel plant. The paper outlines wastewater management practices adopted in these plants in relation to the overall wastewater management in steel plant.

Keywords: *Steel plant, Wastewater, Sludge, Treatment, Disposal.*

Introduction

Steel plant under investigation has a capacity to produce 1.8 million tones of ingot steel per year. There are four blast furnaces, three each with capacity of 1,000 tonnes per day of hot metal (molten pig iron), and the fourth with a capacity of 1,500 tonnes per day. The steel plant has five Linz-Donawitz (LD) converters each with production capacity of 50 to 60 tonnes of steel, and four Open hearth Furnaces each with capacity of 80 tonnes of steel. Molten iron from blast furnaces is converted to steel in these converters by bringing down carbon, manganese, sulphur, phosphorus and silicon contents in the iron to the desired level. During this process, pure oxygen is blown into the LD converters. Oxygen for the LD converters is produced in the tonnage oxygen plant, which has a production capacity of 16,000 Nm^3/hour of 99.05 per cent pure oxygen. The molten steel thus produced in tapped into ladles and teemed into ingot moulds.

These steel ingots are heated to a rolling temperature of 1300°C in soaking pits and the soaked ingots are rolled into slabs in the Blooming and Slabbing Mill. The cold slabs are reheated to the rolling temperatures of 1250°C, and after removal of scales, are fed to the Plate Mill and the Hot Strip Mill. The Plate Mill has a capacity to produce 280,000 tonnes per year of finished wide and heavy plates of 7 mm to 63 mm thickness. The Hot Strip Mill has the production capacity of 1.1 million tones of hot-rolled coils annually, and it converts slabs into strips ranging from 1.6 mm to 10 mm in thickness.

Electrical Sheet Mill has a capacity to produce 50,000 tonnes of electrical sheet annually. The sheets, sheared at the Hot Strip Mill, are reheated and then rolled in the Roughing Mill and packed rolled in the Finishing Mill. The Pipe Plant is designed for producing large diameter Electric Resistance Welded (ERW) pipes.

A Sintering Plant with production capacity of 6,000 tonnes/day of self-fluxing sinter is installed for steel manufacture. The plant utilizes iron ore and limestone fines produced during mechanical mining and other wastes generated in the plant, *e.g.* mill scale and coke breeze.

The steel plant under investigation also has several related production units, *viz.* Coke Ovens and By-product Recovery Plants, Hot Rolling Mill, Cold Rolling Mill, and a Fertilizer Plant for the production of ammonium nitrate. Cold Rolling Mills have pickling lines, cold rolling units, tandem mill and annealing line. The cold reduced coils are further processed depending upon the desired product.

Amongst the service sections, the Captive Power Plant (CPP) and Medium Pressure Boiler for the two major sections. There are two CPPs identified as CPP-1 and CPP-II. CPP-I consists of four coal fired and two gas fired boilers of 30 MW capacity each while, CPP-II consists of two boilers of 60 MW capacity each firing both fuel oil and coal.

Existing Wastewater Management Systems in the Steel Plant and its Performance

The different units in the steel plant under investigation have their own wastewater treatment systems. The details of these treatment systems and their performance are summarized hereunder.

Biological Oxidation and Detoxification (BOD) for Coke Oven and Coal Chemical Division

The existing BOD plant for the treatment of wastewater generated from coke oven and coal chemical sections constitutes an equalization tank, dissolved air flotation unit and three stage biological treatment units comprising of two activated sludge units, a trickling filter unit and sludge drying beds. The schematics of the BOD plant are presented in Figure 26.1 while the performance of the systems is presented in Tables 26.1 and 26.2. Tar catch pits are maintained properly. No spillage was observed around the pits. All units of the BOD plant were in operation and maintained properly. House keeping was satisfactory. Raw wastewater distribution system of the trickling filter plant was functioning properly. Biomass was observed on the packing matrix of the trickling filter. The excess sludge generated at the BOD plant was being recycled to the aeration tank as observed during the inspection period. Records of nutrient/chemical inputs for the BOD plant are not maintained.

Pickling Liquor Neutralization in Cold Rolling Mills (CRMs)

The system for neutralization of pickle liquor generated at CRM comprises neutralization tank with agitation and aeration facilities, settling pits, and sludge drying beds. The schematics of the neutralization system are presented in Figure 26.2 while the performance of the system is presented in Table 26.3. Flow measuring devices for monitoring the flow of influents into the treatment system have not been installed. House keeping was satisfactory. Recycle of spent acid with concomitant recovery of ferrous sulphate was observed during the inspection period.

Hot Rolling Mills (HRM)

Oil reclamation system at HRM consists of settling tank fitted with oil skimmer, sand filter for removal of traces of oil and back wash facilities. The schematics of the oil reclamation system are presented in Figure 26.3 while the performance of the system is presented in Table 26.4. Flow measuring devices have not been installed of monitoring the flow of influents into the treatment system. Oil reclamation system was being operated properly during the inspection period. Backwash water generated from the sand filter was being recycled in the process after oil reclamation. Discharge from the HRM section to the outfall No. 7 was not observed. Reclaimed oil was stored in barrels and was being lifted by the chemical/soap manufacturers during the inspection period. House keeping was satisfactory.

Silicon Steel Mills (SSM)

The existing effluent treatment system at silicon steel mill includes oil de-emulsifier with oil skimmer, neutralization tank fitted with agitators and aerators, clariflocculator and coke filter bed. The schematics of effluent treatment system are presented in Figure 26.4 while the performance of the system is presented in Table 26.5. Flow measuring devices have not been installed for monitoring the flow of influents into the treatment system. Oil skimmer was functioning during the inspection period. Proper dosing of chemicals to the treatment system was not installed. Sludge generated from the silicon effluent treatment plant was being disposed off on sludge drying beds. House keeping was satisfactory.

Blast Furnace

The existing treatment facility for the wastewater generated from gas cleaning unit at blast furnace includes clarifier and vacuum sludge filters. The schematics of

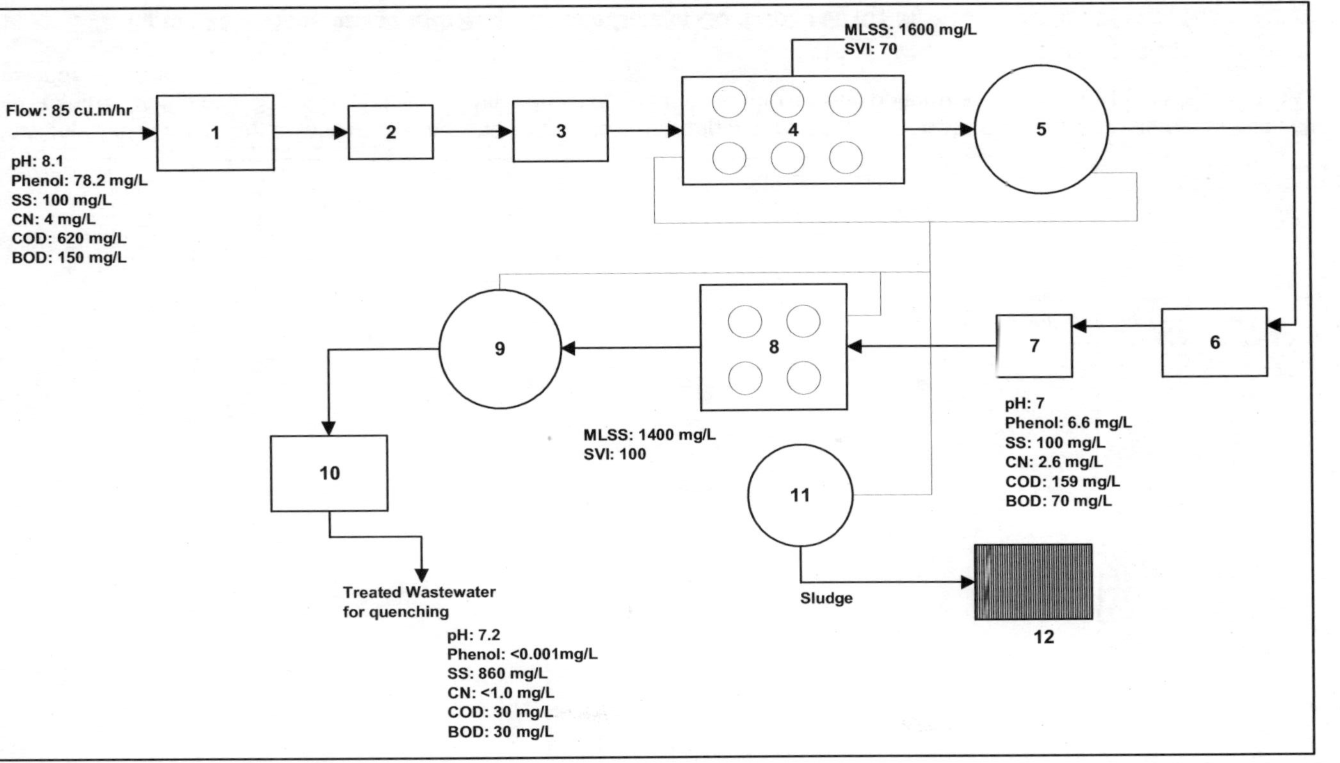

1: Equilisation tank (14m dia x 6.1m SWD); 2: Influent sump (4m dia x 2.75m SWD); 3: DAF cum dosing pump house (10m x 8m); 4: Aeration tank-I (36.5m x 24.5m x 3.5m SWD Aerators 6 x 40 hp); 5: Clarifier-I (16mdia x 1.8m media depth); 6: Trickling filter (36m dia x 1.8m media depth); 7: Trickling filter sump (6m x 5m x 1.5m SWD; 8: Aeration Tank - II (32m x 32m x 3.5m SWD Aerators 4 x 50 hp); 9: Clarifier-II (17m dia x 3.0m SWD); 10: Treated wastewater sump (6.5m dia x 3.05m SWD); 11: Thickner (70m dia x 4.0m SWD); 12: Sludge Drying Bed.

Figure 26.1: Schematics with Performance Data of Biological Oxidation and Detoxification Plant.

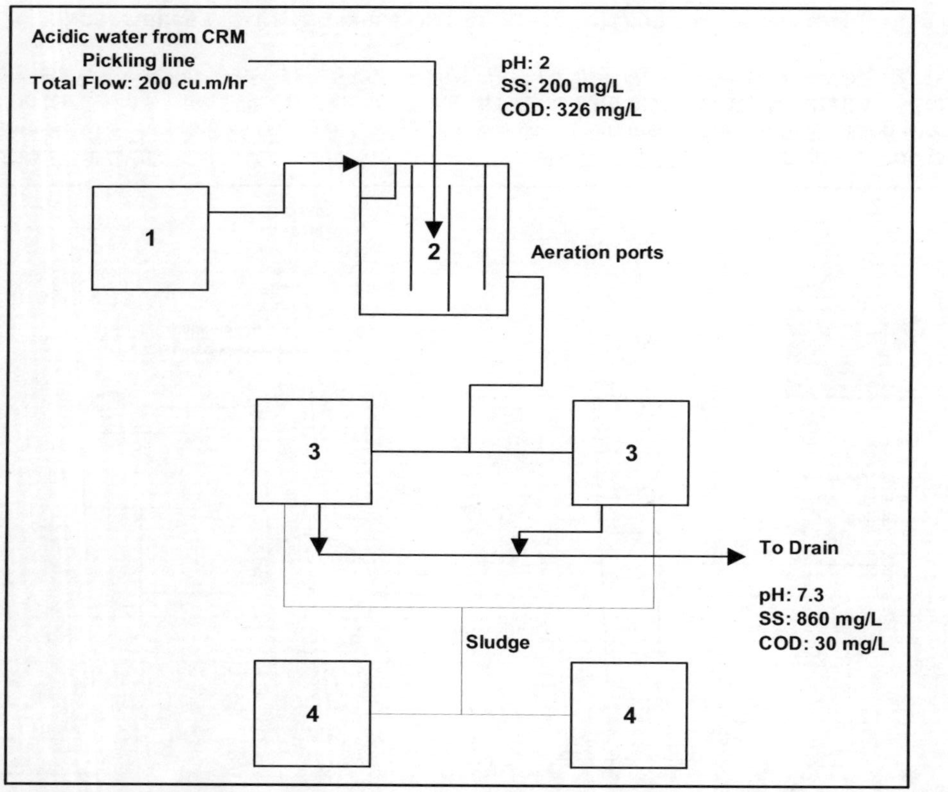

1: Solution sump (5m x 4m x 1m); 2: Mixing channel (zig zag type) (total length 47m; breadth of each channel 0.5m; depth of each channel 1.25m; total volume 29.375 m^3); 3: Settling pits (9m x 9m x 6.5m depth, 2 Nos. with conical bottom); 4: Sludge drying beds (capacity to hold 140 cu.m sludge volume.

Figure 26.2: Schematics with Performance Data of Cold Roll Mill (CRM) Neutralization Plant.

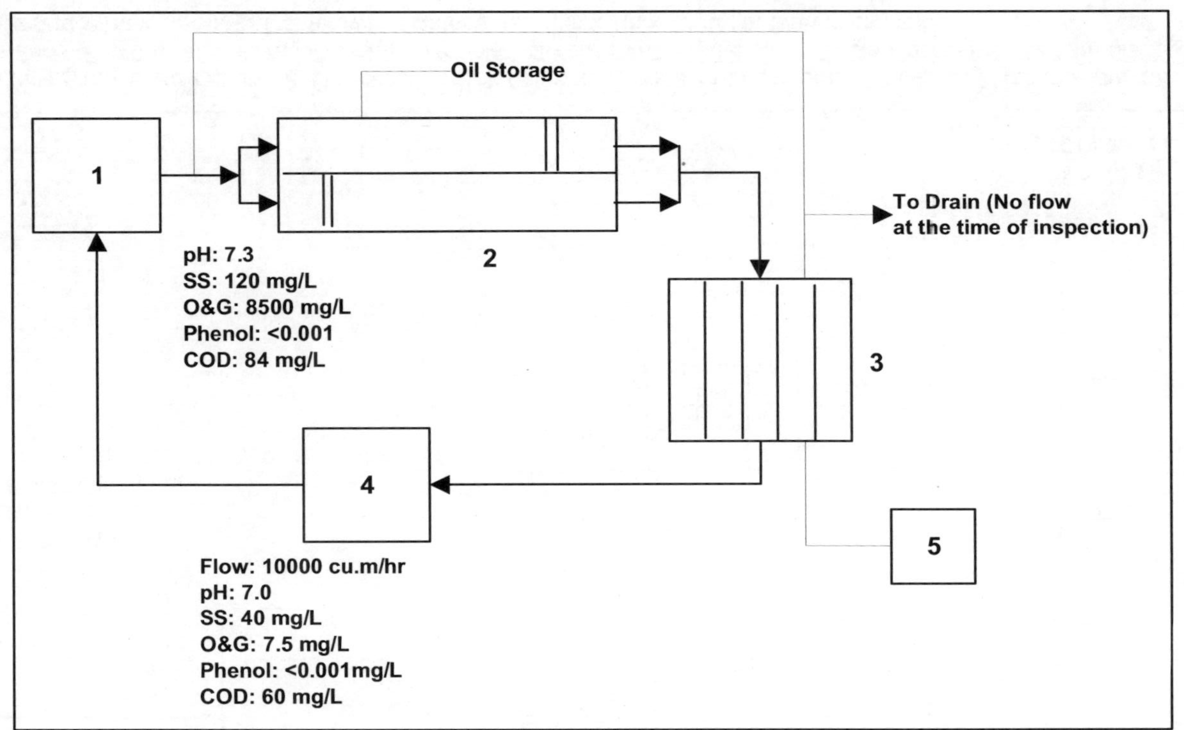

1: HRM; 2: Settling tank with skimmer (4 Nos. 55 x 8m x 3.5 m); 3: Sand filter beds (2 chambers, 25m x 3m each); 4: Sump; 5: Backwash water.

Figure 26.3: Schematics with Performance Data of Rolling Mill Pump House Oil Reclamation Plant.

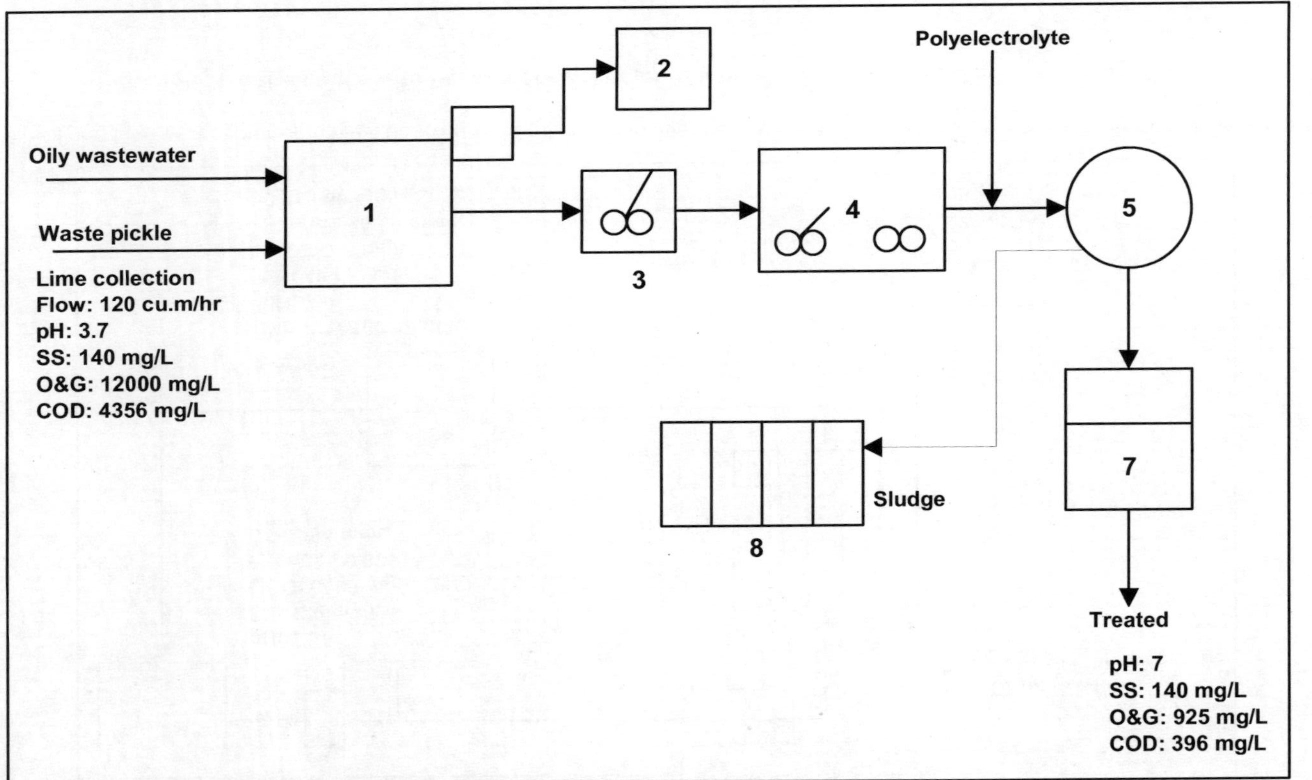

1: Equalisation tank (7m x 6.4m x 4.2m); 2: Oil collection tank (2.9m x 2.9m x 3.7m) agitator rpm 30-40; 3: Flash mix tank (9.8m x 4.52m x 3.77m) agitator 2nos. 30-40 rpm; 4: Aeration tank (4.87m dia x 2.74m SWD, center well); 5: Clariflocculator (12.19m dia x 3.65m SWD); 6: Coke filter bed (2m x 1m x 1.5m); 7: Treated wastewater sump (2m x 1m x 1.5m); 8: Sludge drying bed

Figure 26.4 Schematics with Performance Data of Effluent Treatment Plant at Silicon Steel Mill (SSM).

Table 26.1: Performance Evaluation of BOD Plant at the Steel Plant (Initial samples).

Parameter	*Influent (Magnitude)*	*Effluent Magnitude*		
		Clarifier-I	*Trickling filter*	*Clarifier-II*
Temperature (°C)	40	32	30	30
pH	8.4	7.7	7.4	7.2
Total dissolved solids	640	1100	1200	1050
Total suspended solids	80	80	100	150
Dissolved oxygen		1.2	2.8	3.0
Chloride as Cl	312.8	380.8	265.2	190.4
Oil and Greese	120	42	10	8
Phenol	72.2	82.11	ND	ND
COD	594	217.8	178.2	160
BOD_5, 20°C	180	–	95	80
Sulphide as S^{--}	6.9	ND	ND	ND
Thiosulfate as $S_2O_3^{--}$	86.8	5.6	ND	ND
Sulphate as SO_4^{--}	42.5	180	177.5	107.5
Thiocyanate as SCN^-	68	78	0.98	0.62
Nitrite as NO_2-N	ND	ND	ND	ND
Nitrate as NO_3-N	42	63	66	66
Ammonia Nitrogen as -N	115	88	80	75
Total Nitrogen as–N	180	160	150	120
Cyanide as CN^-	4	4	1.6	1.0
Dissolved phosphate PO_4^{--}	2.77	109.7	36.4	28.8
Heavy metals				
- Cadmium	0.01	0.01	0.01	0.01
- Chromium	0.09	0.18	0.22	0.16
- Copper	0.17	0.11	0.18	0.06
- Iron	10.5	1.93	4.10	1.01
- Manganese	0.06	0.18	0.34	0.10
- Nickel	ND	ND	0.01	0.01
- Lead	0.03	0.01	0.03	ND
- Zinc	0.41	0.29	0.28	0.12
Flow m^3/hr	90			

All the parameters were analyzed as per APHA, AWWA and WEF (1995).

All the values are expressed in mg/L except pH, temperature and flow.

Table 26.2: Performance Evaluation of BOD Plant at the Steel Plant (Composited samples every one hour for a period of six hours).

Parameter	*Influent (Magnitude)*	*Effluent Magnitude*	
		Trickling filter	*Clarifier-II*
Temperature (°C)	41	32	30.5
pH	8.1	7.0	6.9
Total dissolved solids	630	740	680
Total suspended solids	100	100	120
Dissolved oxygen		2.8	3.0
Chloride as Cl^-	367.2	190.4	142.8
Oil and Greese	80	12	6
Phenol	78.2	6.6	ND
COD	620	159.4	139.6
BOD_5, 20°C	150	70	40
Sulphide as S^{--}	6.4	ND	ND
Thiosulfate as $S_2O_3^{--}$	67.2	ND	ND
Sulphate as SO_4^{--}	38.7	83.7	72.5
Thiocyanate as SCN^-	76	3.53	1.49
Nitrite as NO_2-N	ND	ND	ND
Nitrate as NO_3-N	38	45	63
Ammonia Nitrogen as -N	712	80	55
Total Nitrogen as–N	180	175	160
Cyanide as CN^-	3.8	2.6	1.8
Dissolved phosphate PO_4^{--}	5.55	50	12.7
Heavy metals			
- Cadmium	0.01	0.01	0.01
- Chromium	0.18	0.26	0.19
- Copper	0.05	0.09	0.01
- Iron	1.09	0.98	0.74
- Manganese	0.12	0.14	0.14
- Nickel	0.01	ND	ND
- Lead	ND	0.01	0.01
- Zinc	0.26	0.19	0.15
Flow m^3/hr	90		

All the parameters were analyzed as per APHA, AWWA and WEF (1995).

All the values are expressed in mg/L except pH, temperature and flow.

Table 26.3: Performance Evaluation of Treatment Plant for CRM Neutralization at the Steel Plant.

Parameter	*Influent (Magnitude)*	*Effluent (Magnitude)*
Temperature (°C)	34	32
pH	1.92	7.3
Total dissolved solids	10500	540
Total suspended solids	200	860
Dissolved oxygen	—	3.5
Chloride as Cl^-	2184	2000
Oil and Greese	ND	ND
Phenol	ND	ND
COD	326	29.7
BOD_5, 20°C	60.0	30.23
Sulphide as S^{--}	ND	ND
Thiosulfate as $S_2O_3^{--}$	14	ND
Sulphate as SO_4^{--}	1900	830
Thiocyanate as SCN^-	ND	ND
Nitrite as NO_2-N	–	–
Nitrate as NO_3-N	28	3
Ammonia Nitrogen as -N	78	3
Total Nitrogen as–N	90	35
Cyanide as CN^-	ND	ND
Dissolved phosphate PO_4^{--}	18.8	ND
Heavy metals		
- Cadmium	0.009	0.001
- Chromium	0.570	0.050
- Copper	0.326	0.028
- Iron	–	3.77
- Manganese	–	1.139
- Nickel	0.156	ND
- Lead	0.147	0.024
- Zinc	0.482	0.332
Flow m^3/hr	200	

All the parameters were analyzed as per APHA, AWWA and WEF (1995).

All the values are expressed in mg/L except pH, temperature and flow.

Table 26.4: Performance Evaluation of Rolling Mill Pump House at the Steel Plant.

Parameter	*Influent (Magnitude)*	*Effluent (Magnitude)*
Temperature (°C)	31	30
pH	7.3	7.0
Total dissolved solids	500	1200
Total suspended solids	120	40
Dissolved oxygen	–	3.8
Chloride as Cl^-	600	580
Oil and Greese	8500	7.5
Phenol	ND	ND
COD	84	60
BOD_5, 20°C	15	8
Sulphide as S^{--}	50	25
Thiosulfate as $S_2O_3^{--}$	0.6	26
Sulphate as SO_4^{--}	300	250
Thiocyanate as SCN^-	ND	ND
Nitrite as NO_2-N	ND	ND
Nitrate as NO_3-N	4.5	3.0
Ammonia Nitrogen as -N	15	14
Total Nitrogen as–N	30	25
Cyanide as CN^-	ND	ND
Dissolved phosphate PO_4^{--}	2.4	0.80
Heavy metals		
- Cadmium	0.20	0.15
- Chromium	0.60	0.18
- Copper	0.05	0.01
- Iron	3.10	0.80
- Manganese	1.80	0.30
- Nickel	ND	ND
- Lead	ND	ND
- Zinc	1.20	0.6
Flow m^3/hr	10000	

All the parameters were analyzed as per APHA, AWWA and WEF (1995).

All the values are expressed in mg/L except pH, temperature and flow.

Table 26.5: Performance Evaluation of Effluent Treatment Plant of Silicon Steel Mill at the Steel Plant.

Parameter	*Initial Samples*		*Composited Samples*	
	Influent	*Effluent*	*Influent*	*Effluent*
Temperature (°C)	33	32	33.5	31.5
pH	3.9	7.9	3.7	7.0
Total dissolved solids	8340	5000	8000	4600
Total suspended solids	146	110	140	140
Dissolved oxygen	–	–	–	–
Chloride as Cl^-	300	278	340	306
Oil and Greese	10000	78	12000	925
Phenol	ND	ND	ND	ND
COD	5302	459	4356	396
BOD_5, 20°C	–	–	–	–
Sulphide as S^{--}	ND	ND	ND	ND
Thiosulfate as $S_2O_3^{--}$	ND	ND	ND	ND
Sulphate as SO_4^{--}	12.5	11.2	–	ND
Thiocyanate as SCN^-	ND	ND	0.7	0.6
Nitrite as NO_2-N	ND	ND	ND	ND
Nitrate as NO_3-N	ND	ND	ND	ND
Ammonia Nitrogen as -N	ND	ND	ND	ND
Total Nitrogen as–N	ND	ND	ND	ND
Cyanide as CN^-	ND	ND	ND	ND
Dissolved phosphate P	ND	ND	ND	ND
Heavy metals				
- Cadmium	0.01	ND	0.01	0.01
- Chromium	0.28	0.22	0.18	0.06
- Copper	0.21	ND	0.34	0.01
- Iron	120	0.28	–	0.02
- Manganese	0.21	0.03	–	0.13
- Nickel	0.16	ND	0.16	ND
- Lead	0.08	0.00	0.02	ND
- Zinc	0.44	0.11	0.84	0.06
Flow m^3/hr	120			

All the parameters were analyzed as per APHA, AWWA and WEF (1995).

All the values are expressed in mg/L except pH, temperature and flow.

the treatment unit are presented in Figure 26.5 while the performance of the system is presented in Table 26.6. Flow measuring devices have not been installed for monitoring the flow of influents into the treatment system. Complete recycle of the clarifier water was operative during the inspection period. Vacuum filter system for the sludge dewatering was functioning. Sludge generated in the treatment system was being lifted for reuse in ore blending section.

LD4 and LD5 units of SMS

The treatment system for the management of wastewater generated at the LD4 and LD5 units constitutes classifier, clarifloculator and vacuum sludge filter. The schematics of the treatment unit are presented if Figure 26.6 while the performance of the system is presented in Table 26.7. The sludge generated at ETP is sent to OB and BP for recycle. House keeping was satisfactory.

Fertilizer Plant

Steel plant under investigation has a ammonium nitrate manufacturing plant which utilize by products *viz.* nitrogen and hydrogen, which are produced in the steel plant. The nitrogen and hydrogen are converted into ammonium nitrate. The schematics of treatment plant for the wastewater generated from the fertilizer plant are presented in Figure 26.7.

Captive Power Plant (CPP)

Captive power plant is operated using coal, furnace oil and mixed gases generated from coke oven batteries and blast furnaces. The existing flyash disposal system at CPP-I includes ash slurry preparation unit and ash settling ponds. The schemcatics for the ash disposal system are presented in Figure 26.8 while the performance of the system is presented in Table 26.8. Ash slurry generated from captive power plant was being discharged into the existing common ash pond. Flow measuring devices have not been installed for monitoring the flow of influents into the treatment system. Polyelectrolyte addition for treating ash slurry generated from captive power plant was observed during the inspection period.

Characteristics of chemical sludges generated from BOD, CRM, SSM, blast furnace and LD4 and LD5 treatment plants at the steel plant under investigation are presented in Table 26.9.

Lagoon

A lagoon has been constructed for common effluent treatment unit. The network of different wastewater outfalls along with the lagoon is presented in Figure 26.9. Flow measuring devices have not been installed for monitoring the flow of influent into the lagoon. A fishpond has been constructed on the advice of the regulatory agency for using lagoon water for pisciculture. The overflow of lagoon meets the river traversing a distance of about 200 meters.

Performance data of the existing Lagoon for the treatment of combined wastewater are presented in Table 26.10. The treated effluent from the lagoon does

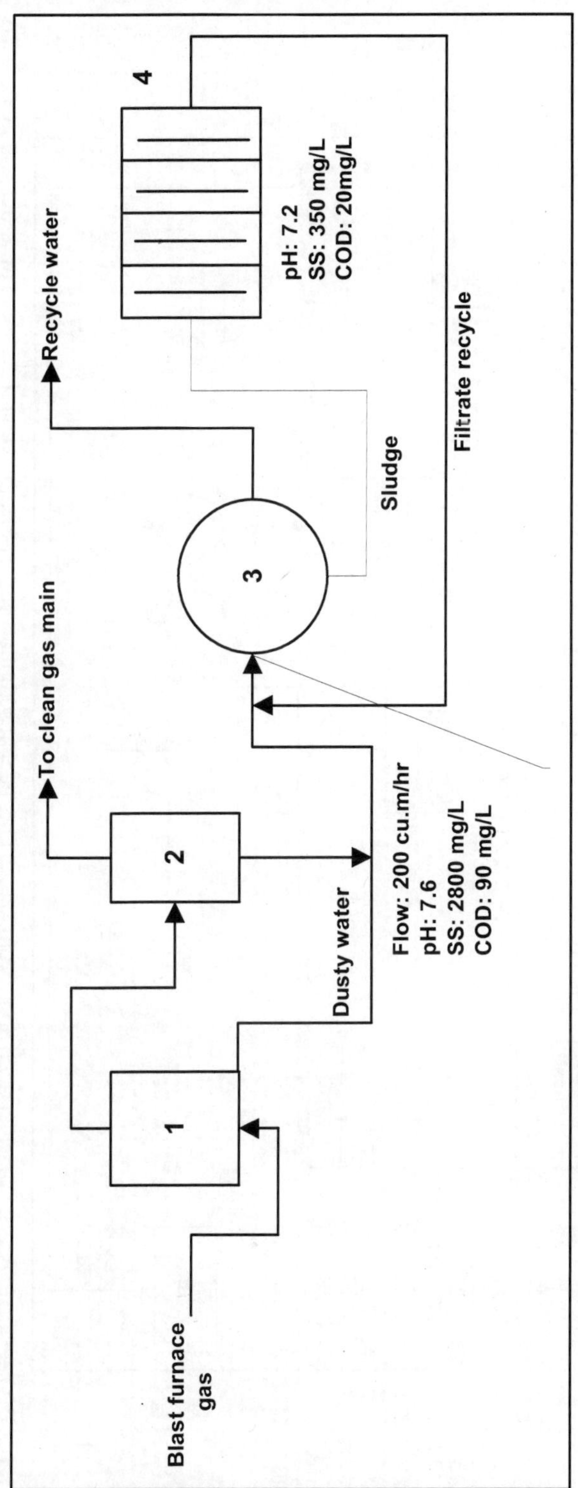

1: Washer; 2: Electrostatic Precipitator; 3: Clarifier (2Nos. 42 media x 3m SWD); 4: Vacuum filter

Figure 26.5: Schematics with Performance Data of Effluent Treatment Plant at Blast Furnace Plant.

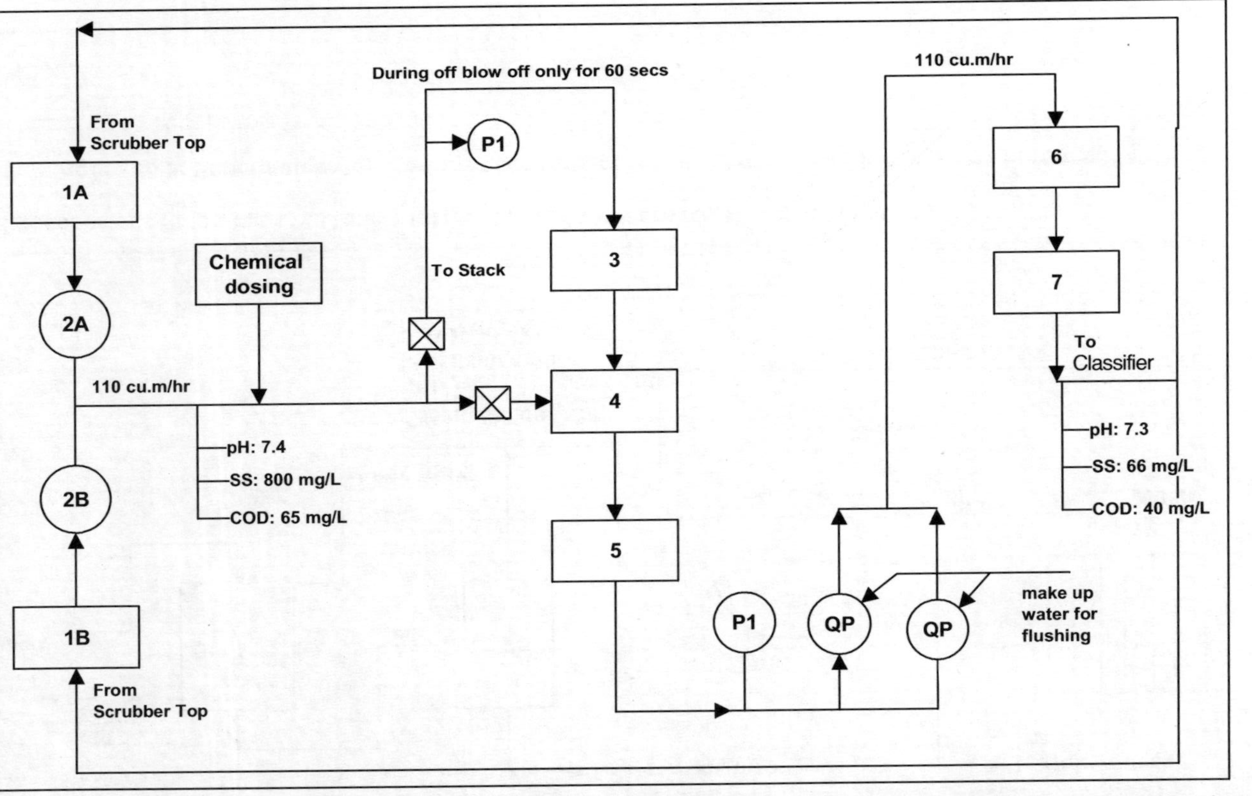

1: A, 1B: Clasifier; 2: A, 2B: Classifier each about 10m. dia; 3: Gas duct at scrubber top; 4: Movable venturi; 5: Sump at scrubber bottom — To retain 110 cu.m/hr of water; 6: Quencher venturi; 7: Scrubber top part; QP: Quencher pump — capacity 110 m^3/hr; P1: Pumps for circulation to different streams

Figure 26.6: Schematics with Performance Data of Effluent Treatment Plant at LD4 and LD5 units.

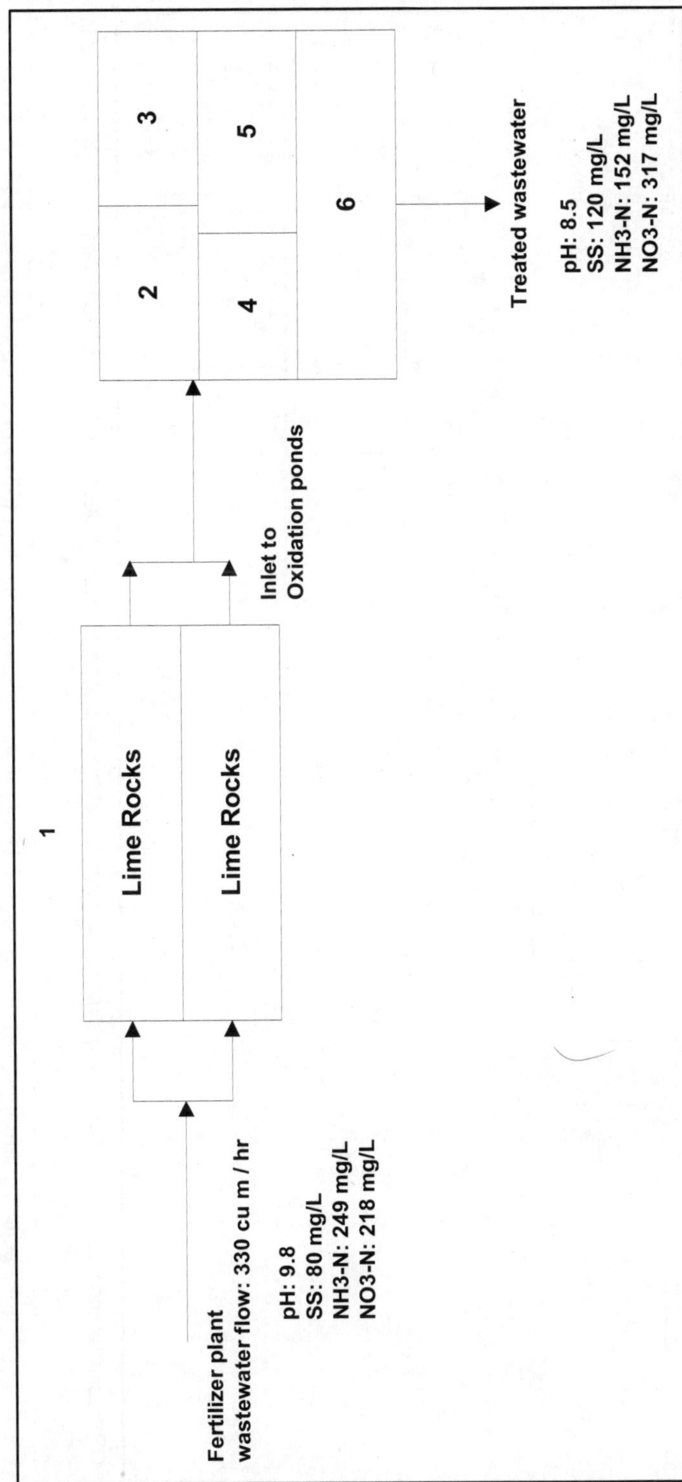

1: Neutralization Pit (40m x 10m x 1.0m); 2: Oxidation Pond-I (9000 cu.m. capacity); 3: Oxidation Pond-II (9000 cu.m. capacity); 4: Oxidation Pond-III (9000 cu.m. capacity); 5: Oxidation Pond-IV (9000 cu.m. capacity); 6: Oxidation Pond-V (9000 cu.m. capacity)

Figure 26.7: Schematics with Performance Data of Fertilizer Wastewater Treatment Plant.

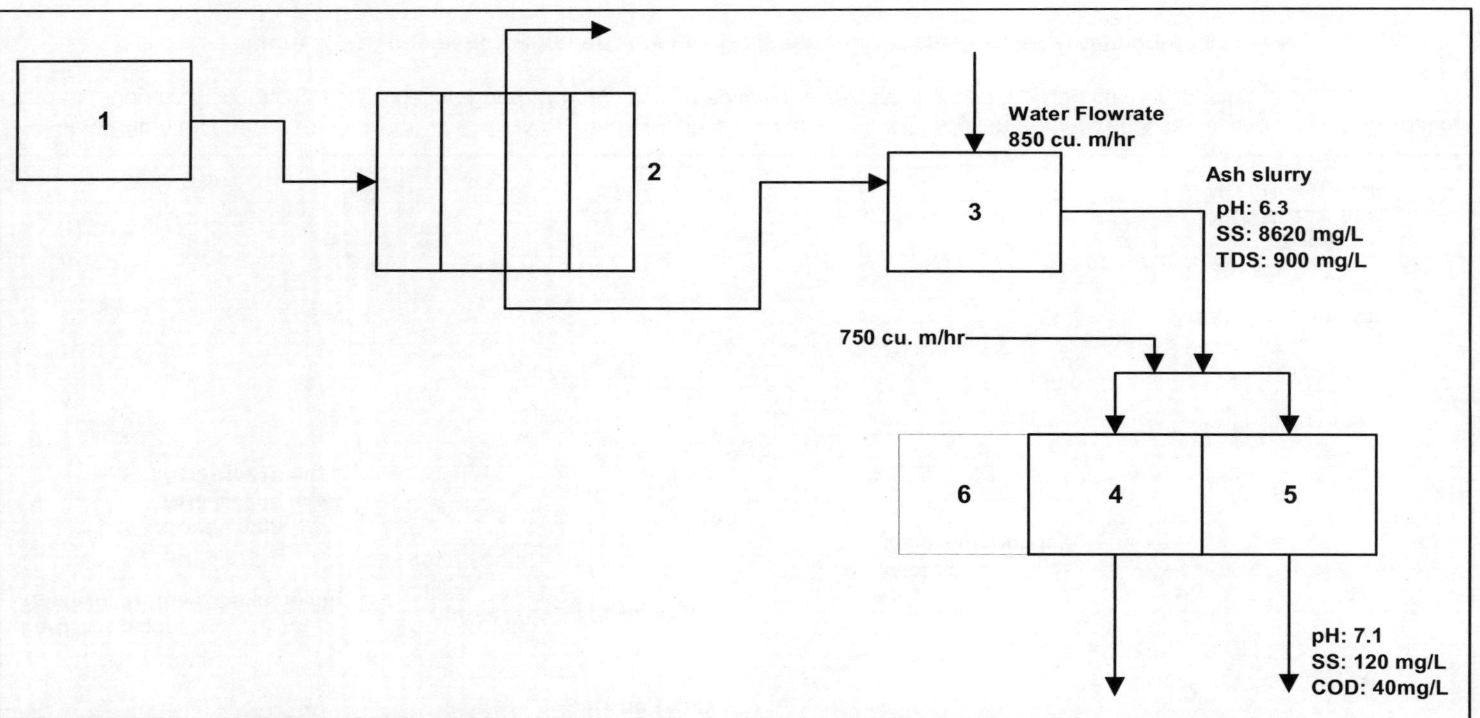

1: CPP boiler section; 2: Electrostatic Precipitator; 3: Ash slurry making tank; 4: Ash pond - A (29.5 Hct. area); 5: Ash pond - B (15.6 Hct. area); 6: Ash pond - C (22.5 Hct.area)

Figure 26.8: Schematics with Performance Data for Ash Disposal System for Captive Power Plant (CPP).

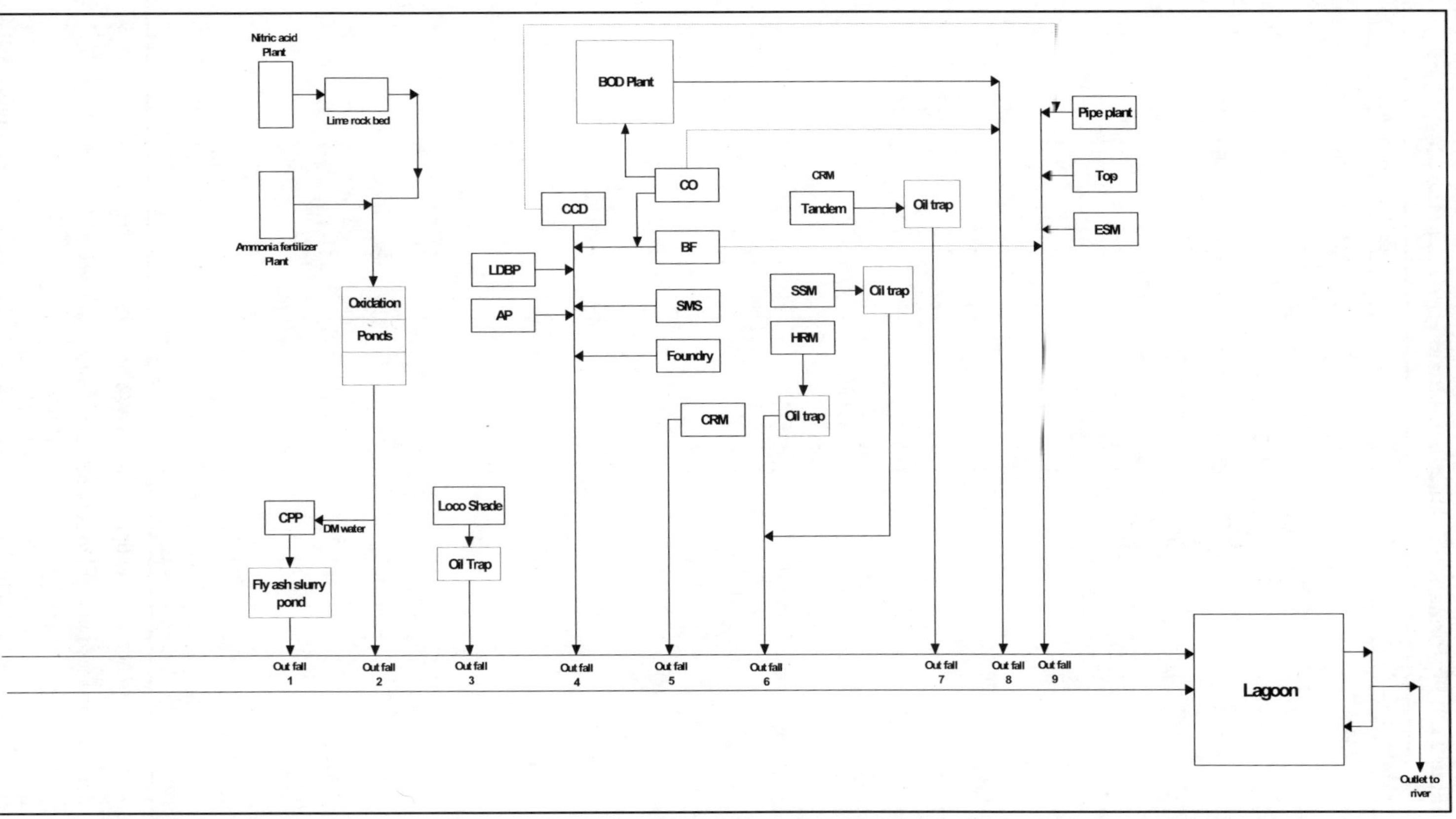

Figure 26.9: Collection of Treated Wastewaters from different Units of the Manufacturing at Steel Plant.

Table 26.6: Performance Evaluation of Blast Furnace at the Steel Plant

Parameter	*Influent (Magnitude)*	*Effluent (Magnitude)*
Temperature (°C)	34	32
pH	7.6	7.2
Total dissolved solids	1300	1160
Total suspended solids	2800	350
Dissolved oxygen	–	–
Chloride as Cl^-	900	816
Oil and Greese	20	ND
Phenol	ND	ND
COD	90	19.8
BOD_5, 20°C	ND	–
Sulphide as S^{--}	ND	ND
Thiosulfate as $S_2O_3^{--}$	ND	ND
Sulphate as SO_4^{--}	ND	ND
Thiocyanate as SCN^-	0.8	0.4
Nitrite as NO_2-N	ND	–
Nitrate as NO_3-N	5.0	3.3
Ammonia Nitrogen as -N	15	12
Total Nitrogen as–N	20	18
Cyanide as CN^-	ND	ND
Dissolved phosphate PO_4^{--}	2.0	1.11
Heavy metals		
- Cadmium	0.01	0.01
- Chromium	0.25	0.19
- Copper	ND	ND
- Iron	15	2.2
- Manganese	1.9	1.28
- Nickel	ND	ND
- Lead	0.03	0.01
- Zinc	1.2	0.01
Flow m^3/hr	200	

All the parameters were analyzed as per APHA, AWWA and WEF (1995).

All the values are expressed in mg/L except pH, temperature and flow.

Table 26.7: Performance Evaluation of Effluent Treatment Plant LD4 and LD5 at the Steel Plant.

Parameter	*Influent (Magnitude)*	*Effluent (Magnitude)*
Temperature (°C)	32	31
pH	7.4	7.3
Total dissolved solids	1298	1120
Total suspended solids	800	66
Dissolved oxygen	–	–
Chloride as Cl^-	780	780
Oil and Greese	5	ND
Phenol	ND	ND
COD	65	40
BOD_5, 20°C	18	12
Sulphide as S^{--}	ND	ND
Thiosulfate as $S_2O_3^{--}$	ND	ND
Sulphate as SO_4^{--}	ND	ND
Thiocyanate as SCN^-	0.5	0.1
Nitrite as NO_2-N	ND	ND
Nitrate as NO_3-N	5.0	7.0
Ammonia Nitrogen as -N	12.0	8.0
Total Nitrogen as–N	18.0	16.0
Cyanide as CN^-	ND	ND
Dissolved phosphate PO_4^{--}	1.0	0.8
Heavy metals		
- Cadmium	ND	ND
- Chromium	0.01	0.02
- Copper	ND	ND
- Iron	12	14
- Manganese	0.80	0.60
- Nickel	ND	ND
- Lead	0.02	0.01
- Zinc	1.1	1.6
Flow m^3/hr	110	

All the parameters were analyzed as per APHA, AWWA and WEF (1995).

All the values are expressed in mg/L except pH, temperature and flow.

Table 26.8: Performance Evaluation of Treatment Plant for CPP at the Steel Plant.

Parameter	*Influent (Magnitude)*	*Effluent (Magnitude)*
Temperature (°C)	32	30
pH	6.3	7.1
Total dissolved solids	900	1300
Total suspended solids	8620	120
Dissolved oxygen	–	3.0
Chloride as Cl^-	40.8	34.0
Oil and Greese	12	5
Phenol	ND	ND
COD	70	40
BOD_5, 20°C	ND	ND
Sulphide as S^{--}	ND	ND
Thiosulfate as $S_2O_3^{--}$	ND	ND
Sulphate as SO_4^{--}	40	45
Thiocyanate as SCN^-	0.14	ND
Nitrite as NO_2-N	ND	ND
Nitrate as NO_3-N	2.5	1.5
Ammonia Nitrogen as -N	10	8
Total Nitrogen as–N	16	12
Cyanide as CN^-	ND	ND
Dissolved phosphate PO_4^{--}	3.0	2.5
Heavy metals		
- Cadmium	0.01	ND
- Chromium	0.06	0.12
- Copper	0.02	0.03
- Iron	0.45	0.28
- Manganese	0.12	0.01
- Nickel	ND	ND
- Lead	ND	0.04
- Zinc	0.08	0.06
Flow m^3/hr	550	

All the parameters were analyzed as per APHA, AWWA and WEF (1995).

All the values are expressed in mg/L except pH, temperature and flow.

Table 26.9: Characteristics of Chemical Sludge Generated from BOD, CRM, SSM, Blast Furnace and LD4 and LD5 Treatment Plants at the Steel Plant.

Parameters	*BOD Plant Chemical Sludge (Magnitude)*	*CRM Plant Chemical Sludge (Magnitude)*	*SSM Plant Chemical Sludge (Magnitude)*	*Blast Furnace Plant Chemical Sludge (Magnitude)*	*LD4 and LD5 Plants Chemical Sludge (Magnitude)*
A. Physical characteristics					
Appearance	Blaskish	Greenish	Gray Black	Gray	Black
Odour	Tarry smell	Odourless	Odourless	Odourless	Odourless
pH	7.1	8.5	8.5	8	7.2
Proximate Characteristics (%) (On wet Basis)					
Moisture	62	50	48	20	42
Volatile matter	18	6	14	22	16
Ash	8	40	36	54	38
Fixed carbon	12	4	2	4	4
B. Chemical characteristics					
Elemental composition (%)					
Carbon	38.5	ND	12	20.5	0.5
Nitrogen	9	7	5.5	5.5	3
Sulphur	0.8	12	ND	ND	ND
Heavy Metals (mg/kg)					
Lead	ND	ND	10	27.2	81.8
Aluminium	ND	ND	ND	ND	250
Chromium	ND	ND	ND	ND	ND
Iron	ND	200	150	580	500
Nickel	200	25.2	25.2	3.4	ND
Silica	10	11.2	44	80	78
Zinc	15.6	11.2	32.8	159.4	119.4
Copper	2	32.8	ND	11.4	6.4
Manganese	9	15.8	302.8	300	260.2
Cadmium	ND	1.2	1.2	9	0.4
C. Sludge category as per MEF guidelines*	Waste category No. 3 and 12	Waste category No. 3 and 12	Waste category No. 3 and 12	Waste category No. 3 and 12	Waste category No. 3 and 12

All the parameters were analyzed as per APHA, AWWA and WEF (1995).

Table 26.10: Performance Evaluation of Lagoon at Steel Plant.

Parameter	*Initial Samples*		*Composited Samples*	
	Influent	*Effluent*	*Influent*	*Effluent*
Temperature (°C)	33.5	32	33	31.5
pH	7.1	6.8	7.4	7.4
Total dissolved solids	380	500	460	480
Total suspended solids	88	56	100	88
Dissolved oxygen	3.0	–	–	–
Chloride as Cl^-	135	122.4	95.2	88.4
Oil and Greese	15	8	12	5
Phenol	ND	ND	ND	ND
COD	39.6	19.8	56.8	19.4
BOD_5, 20°C	15.0	10.0	30.0	10.0
Sulphide as S^{--}	6.4	ND	3.2	ND
Thiosulfate as $S_2O_3^{--}$	ND	ND	ND	ND
Sulphate as SO_4^{--}	76.2	81.2	60.0	67.5
Thiocyanate as SCN^-	ND	ND	ND	ND
Nitrite as NO_2-N	ND	ND	ND	ND
Nitrate as NO_3-N	16	18	18	12
Ammonia Nitrogen as -N	22	28	30	17
Total Nitrogen as–N	50	40	62	45
Cyanide as CN^-	ND	ND	ND	ND
Dissolved phosphate P	2.87	2.77	2.49	1.44
Heavy metals				
- Cadmium	ND	ND	ND	0.01
- Chromium	0.07	0.05	0.25	0.23
- Copper	0.01	ND	0.02	0.01
- Iron	0.55	0.33	0.57	0.26
- Manganese	0.12	0.07	0.09	0.16
- Nickel	ND	ND	ND	ND
- Lead	ND	ND	0.01	0.01
- Zinc	0.09	0.15	0.31	0.07
Flow m^3/hr	5400			

All the parameters were analyzed as per APHA, AWWA and WEF (1995).

All the values are expressed in mg/L except pH, temperature and flow.

not conform to the standards prescribed by the regulatory agencies with reference to nitrate nitrogen for discharge into inland surface water.

Discussion

The flow measuring devices are essential unit for controlling the wastewater to different treatment systems. However, the flow measuring devices have not been installed for monitoring and control of flow of influent in to the pickling neutralization unit, hot rolling mills, oil reclamation unit, and effluent treatment system of silicon plant and wastewater treatment unit of the blast furnace. This results into malfunctioning of the treatment system due to uncontrolled discharge of the effluent. The treated effluent from the BOD plant, silicon steel mill, blast furnace and captive power plant do not conform to the standards prescribed by the regulatory agencies with reference to suspended solid.

In the BOD plant the suspended solids, biochemical oxygen demand (BOD_5), nitrate as NO_3-N, ammonia nitrogen as - N, cyanide as CN^-, dissolved phosphate as P was higher than the prescribed limit of regulatory agency as presented in Table 26.2.This may be attributed to the poor maintenance of biomass in the reactor since the biomass concentration in the reactor was in the range of 1400 to 1600 mg/L (Pandey et. al 1991, Pandey and Kumaran 1991, Metcalf and Eddy 1996) Further the Volatile suspended solids (VSS) of the Mixed liquor suspended solid (MLSS) was only in the range of 400 to 500 mg/L. Thus most of the biomass was inert as it was developed by cattle dung. The use of potential micro-organisms having capability of biodegrading the constituents of coke oven wastewater may improve the performance of the system. The Sludge Volume Index (SVI) of biomass is 100, which indicate poor settling properties of the biomass present in the aerobic unit of the BOD plant. This has created hurdle for built-up of biomass in the aerobic units in-spite of keeping 50 per cent recirculation of the settled sludge from clarifier-I and II of the systems to the respective aerobic units. The BOD of the treated effluent was found to be in the range of 80–100 mg/L. This indicates that the biomass present in the aerobic systems do not constitute the communities of micro-organisms, which can easily utilize the different pollutants of the wastewater.

The treated wastewater from the neutralization plant at cold rolling mills (CRMs) contains average suspended solids of 860 mg/L which is far high than the prescribed limit of 100 mg/L as per regulatory agency. The settling pits have adequate capacity for providing sufficient time for settling the neutralized wastewater.

In the Hot rolling mills (HRM) of the steel plant, there is only settling tank with oil skimmer and the oil skimmer in the present system was working efficiently and could remove oil and grease more than 99 per cent. The oil after skimming is sent to the storage tank for reuse while the effluent from oil skimmer is pass through sand filter beds which finally remove the oil and grease and the clear water is recycled in the process.

The treatment unit at Silicon steel mills (SSM) in the investigated steel plant indicates the treated effluent with a average suspended solids of 140 mg/L, oil and grease 78 mg/L with a COD value of 460 mg/L. This indicates that the treatment unit is not working properly. The coke filter bed was observed to be choked which results in improper filtration of the effluent emerging from the clariflocculator of the treatment unit. The coke bed requires proper washing at different time intervals, which will enable proper filtration of the effluent generated from clariflocculator.

The treatment plant at blast furnace constitutes electro static precipitator for cleaning the gases emerging from blast furnace after washing. The dusty water is passed through a clarifier and the settled sludge is routed through vacuum filter. The treated water from clarifier is recycled in the process and the filtrate of the vacuum filter is routed through clarifier in-order to remove the fine suspended particles. The treated effluent from the effluent treatment plant at blast furnace generates the treated effluent containing suspended solids of 350 mg/L, which is higher than the prescribed limits for the discharge. An additional clarifier is needed in-order to provide better settling of the suspended solids present in the wastewater from blast furnace section

The suspended solid in the treated effluent from the Captive power plant (CPP) is higher than the prescribed limit of the discharge. This may be attributed to inadequate capacity of settling ponds for the effluent generated from the CPP. This requires construction of additional settling pond for effective removal of suspended solids from the effluent of CPP.

All the effluent generated from treatment units falls in the lagoon and the treated combined effluent from the lagoon conforms to the prescribed limits of the regulatory agencies.

Thus it is a necessary to monitor the efficiency of all the wastewater treatment units installed at individual unit operation of the steel plant and combined effluent can be treated through lagoon. The overflow of lagoon water can be recycle and reuse as process water.

References

1. American Public Health Association (APHA), 1995. Standard Methods for the Examination of Water and Wastewater. (19th ed). APHA, AWWA and WEF, Washington, D.C. http://cpcb.nic.in/
2. Kumar, S., Kumar, R., Bandopadhyay, A., 2006. Innovative methodologies for the utilisation of wastes from metallurgical and allied industries. Resources, Conservation and Recycling, Available online.
3. Metcalf and Eddy, 1996. Wastewater Engineering, Treatment, Disposal, Reuse. (3rd ed.). Tata McGraw-Hill, Inc. India, pp. 370.
4. R.A. Pandey, N.M. Parahad, P. Kumaran, 1991. Biological Treatment of LTC Wastewater–A case study, *Water Research*, 25, 1555–1564.
5. R.A. Pandey, P. Kumaran, 1991. Treatment of coal carbonization wastewater - biological method, *Ind. J. Environ. Protect*. 11, 94–104.

Chapter 27

Microwave Oven Technology for Management of Waste Products of Benzimidazole Synthesis

***B.S. Kuchekar*[1], *A.R. Chabukswar*[1], *S.C. Jagdale*[1], *P.D. Lokhande*[2], *V.J. Kadam*[1] and *D.J. Desai*[1]**

[1]*MAEER's Maharashtra Institute of Pharmacy, MIT Campus, Paud road, Kothrud, Pune*

[2]*Department of Chemistry, University of Pune, Ganeshkhind Road, Pune*

ABSTRACT

Organic synthesis has wide scope of applications in drugs, paints, textiles, dyes and has many allied industrial applications. Many drug molecules and intermediates have been synthesized by using organic synthesis. Existing methods of synthesis involves use various solvents, chemicals and extreme conditions for the synthesis. The hazardous chemicals and waste products have posed a big challenge. Many approaches have been proposed for the elimination of waste products of organic synthesis. In the present work attempts have been made to synthesize the derivatives of the benzimidazoles by using a Microwave technology. In this technique microwaves have been utilized for synthesis of benzimidazoles. In the conventional methods longer duration, cost and treatment of by products are the major drawbacks. Hence the Microwave technology has been developed for the synthesis of benzimidazole derivatives. The process takes short time for the completion and offers good yield of the derivatives of benzimidazole over conventional method and found to be environment friendly.

***Keywords**: Microwave, Organic synthesis, Drugs, Benzimidazole, Environment.*

Introduction

Organic synthesis has been one of the most useful technology foe its applications in diverse fields. The synthesis of drugs and drug molecules is possible by the use of organic synthesis. Drugs like paracetamol, Isoniazide, Rifampicin, many antibiotics, anti-diabetics, anti-viral, anti-cancer and many life saving drugs have been synthesized by the organic synthesis. Conventional methods of synthesis involves use of multi-step synthesis of complex molecules with many hazardous chemicals, extreme heating, longer duration and tedious processes. The by products of the chemical reactions again creates a big problem of their disposal and even is of great environmental concern. Many new techniques and process alterations have been carried out to solve the problems associated with the conventional synthesis.Recently microwave technology for the synthesis of many drugs and drug intermediates have offer a great alternative over the conventional methods.

In the present work we have synthesized 6 derivatives of benzimidazoles by using microwave technology. Our Approach aim is to develop a scientifically based framework for greener preparation of these materials in a manner that renders the materials less mobile in the environment and reduces or eliminates the use and generation of hazardous substances and waste products. Benzimidazoles and their derivatives have been found to exert significant pharmacological activities. They have been found to possess anti-bacterial and anti-fungal activities.

Results and Discussion

Synthesized compounds were analyzed by Melting Points, Infra-red Spectroscopy and reactions were monitored by using Thin Layer Chromatography (TLC) Table 27.1. The values were found to be matched with literature values for all the compounds and the method has been found to be suitable for the synthesis of derivatives with improved yield and waste products have not been formed rendering the method more suitable for the clean environment.

Table 27.1: Physical and Sectoral Properties of the Synthesized Compounds.

Sl.No.	*Derivative*	*M.P. (°C)*	*% Yield*	*IR (KBr) ν (cm^{-1}) Values*
1.	Compound IV A	200	65	764 (Ar-CH st.), 1670 (C=N st), 2410(C-S-C st.), 3310 (Ar-NH_2)
2.	Compound IV B	210	75	810 (C-Cl), 1510 (Ar-CH st), 1680 (C=N st), 2410 (C-S-C st), 3410 (Ar-NH_2)
3.	Compound IV C	196	88	789 (C-Cl), 1510 (Ar-CH st), 1710 (C=N st), 2410 (C-S-C st), 3410 (Ar-NH_2)
4.	Compound IV D	218	68	786 (C-Cl), 1510 (Ar-CH st), 1723 (C=N st), 2410 (C-S-C st), 3410 (Ar-NH_2)
5.	Compound IV E	190	65	886 (Ar-CH st), 1410 (Ar-NO_2 st), 1680 (C=N st), 2410 (C-S-C st), 3420 (Ar-NH_2)
6.	Compound IV F	184	78	890 (Ar-CH st), 1410 (Ph-C-Ost), 1680 (C=Nst), 1810 (C=O), 2410 (C-S-C st), 3410 (AR-NH_2)

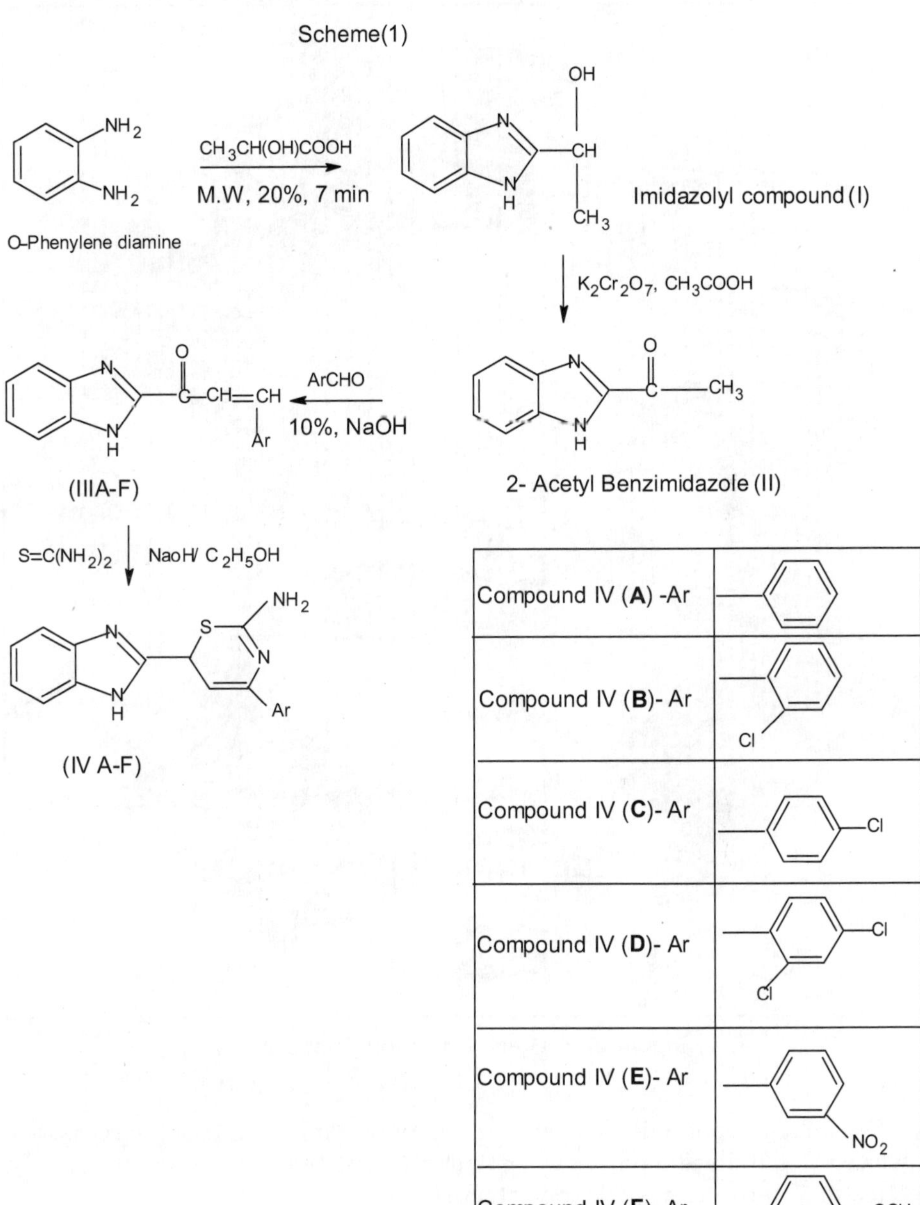

Microwave assisted synthesis means synthesis of molecules by using microwaves. Microwaves, a nonionising radiation incapable of breaking bonds, are a form of energy and not a heat and are manifested as heat through their interaction with medium or material where in they can be reflected (metals), transmitted (good insulators) or absorbed.

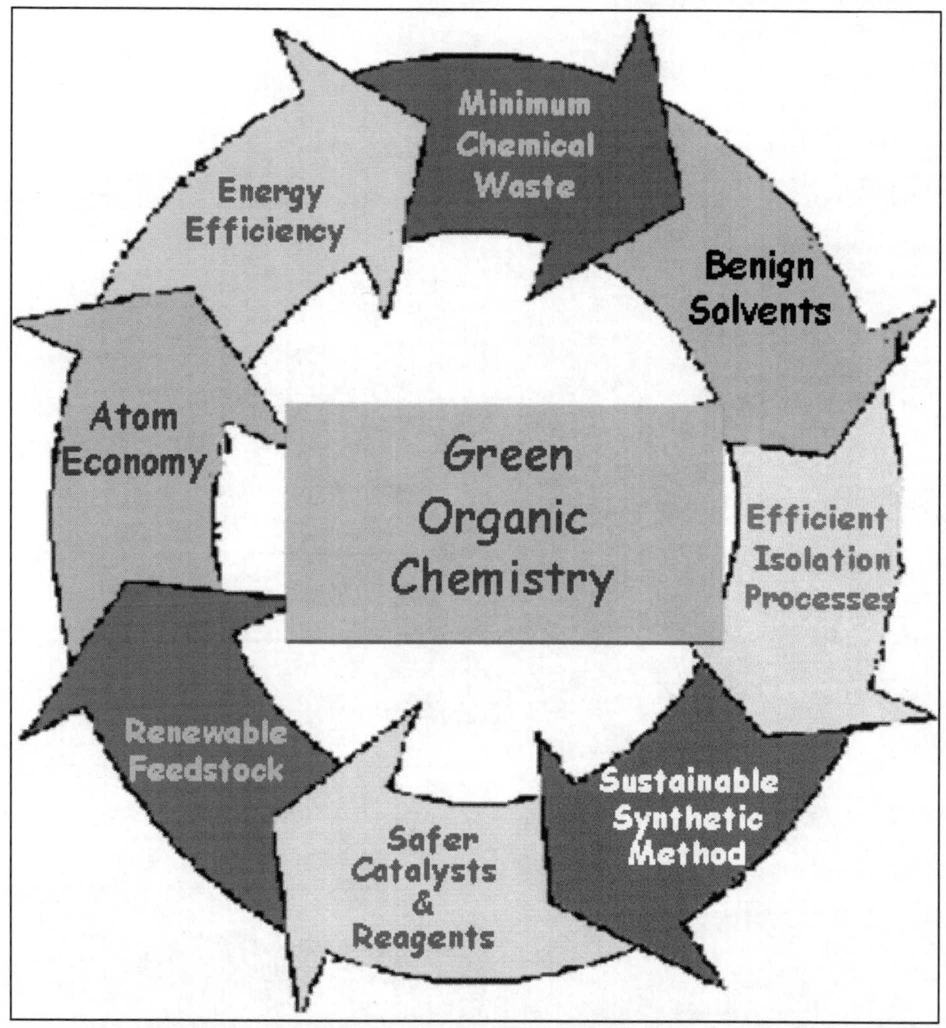

Figure 27.1: Green Chemistry Process.

Organic reaction under solvent free conditions are advantageous because of enhanced selectivity and efficiency, ease of manipulation, and more importantly, because toxic and often volatile solvents are avoided.

Reactions conducted through microwaves are cleaner and more environmentally friendly than conventional heating methods. Microwaves heat the compound directly; therefore usage of solvent in the chemical reaction can be reduced or eliminated.

The use of microwaves has also reduced the amount of purification required for the end products of chemical reactions, involving toxic reagents. In most of reactions, products are obtained which do not need further purification.The conventional synthesis can be performed in less time, with improved yield and more purified products can be obtained thus offering a technology which is more environment friendly and involves less waste products which is of great value in the present situation.

Figure 27.1 represents how the chemical reactions can be worked for Green organic chemistry which minimizes the burden on the environment and the waste product management.

Procedure

Synthesis of 1-(H-Benzo (d) Imidazole-2-yl) ethanol (Compound -1)

1(1h-benzo(d) imidazol-2-yl) ethanol was prepared by the reaction of 0.1 mole of o-Phenylene diamine with 0.1 mole of lactic acid. The mixture was refluxed with condenser in synthetic microwave oven in the intensity of 20 per cent (160 watts) for 6 mins. The mixture was cooled and added 10 per cent NaOH solution till basic. Then the precipitate was filtered, washed with cold water, dried and recrystallised with hot water.

Synthesis of 2-Acetyl Benzimidazole (Compound-2)

To a solution of the compound -1 (10m.mol) in aqueous acetic acid (5 per cent v/v), added at room temperature the solution of potassium dichromate (10m.mol)in aqueous acetic acid (5 per cent v/v) and the mixture was stirred for 12 min. The separated product was filtered, washed with water, dried and recrystallised from ethanol.

Synthesis of Chalcon Derivatives of Benzimidazole (Compound-3A-F)

To a solution of compound -2 (0.01mol) in aqueous NaOH (10 per cent) the respective aldehyde (0.01mol) in ethanol was added. The reaction mixture was stirred for 30mins. The separated solid was filtered and washed with water, dried and recrystallised from ethanol.

Synthesis of Thiazine Derivatives (Compound 4A-F)

A mixture of compound -3 (0.02mol) and thiourea (0.02mol) were dissolved in ethanolic NaOH (10ml), stirred for 3 hrs using a magnetic stirrer and then poured into 400ml of cold water with continuous stirring. This was kept in a cold condition in refrigerator for 12 hrs. the precipitate obtained was filtered, washed and recrystallised from ethanol.

Conclusion

Present method has been found to be useful for the synthesis of benzimidazole derivatives with improved yield, environment friendly and minimized waste by products. Thus method can be applied in the synthesis of the drugs and drug intermediates with less burden on the waste product management in organic reactions for synthesis of useful compounds.

References

1. Hayden F. G, Gwalthey J. M, Antimicrobial agent and chemotherapy, 1982, 21: 892-897.
2. Pandey V. K, Punita Garg, J. C. Katigar, Indian drugs, 1986, 11: 605-608.
3. Suroorahmed, Asgarmahmoodnandan, Indian Journal of Heterocyclic Chemistry, 1997, 7: 55-58.

Chapter 28

Red Mud Utilization: A Focused Review

Suchita B. Rai[1], M.J. Chaddha[1], Biswanath Sen[1], Kailas L. Wasewar[2] and Jyoti Mukhopadhyay[1]

[1]Jawaharlal Nehru Aluminium Research Development and Design Centre, Nagpur

[2]VNIT, Nagpur

ABSTRACT

World wide alumina is produced from bauxite by Bayer Process which generates highly alkaline solid waste (pH 12 - 13.5) known as Red Mud. Around 70 million tonnes of red mud per annum is generated world wide of which ~ 4 million tonne of red mud is generated annually in India. Disposal of such a huge quantity of alkaline solid waste is creating a big environmental issue leading to its alternative usage/utilization methods. Several routes of using red mud in effective ways have been tried and extensive research is still going on to find the most suitable and economic way of utilizing the same. Red Mud consists of Fe_2O_3 (35-55 wt per cent), Al_2O_3 (16-22 wt per cent), SiO_2 (4-16 wt per cent), Na_2O (3-6 wt per cent), TiO_2 (2-19 wt per cent), CaO (0.8-4.5 wt per cent) and its composition itself shows some ways of utilizing red mud to alleviate its disposal problem. Its alumina and silica content is quite encouraging towards the utilization of red mud for building materials. High iron content has also shows scope of metal/value added product recovery from the same. Many special adsorbent/compounds have been developed from red mud for pollution control purpose also. But these applications are not always economically viable because of source and surrounding conditions. So environmentally safe disposal has also been given importance; *i.e.* neutralization of red mud for safer disposal as well as amendment of red mud to support re-vegetation/vegetation cover in the plant and mine surrounding areas. In this paper, all different directions of red mud utilization will be discussed.

Keywords: *Red mud, Alkaline solid waste, Utilization.*

Introduction

Aluminium is the third largest element in the earth's crust and its ores are abundantly available. The only commercial route for aluminium production involves the extraction of alumina (Al_2O_3) from the bauxite ores (Bayer process) followed by the molten salt electrolysis of alumina dissolved in a cryolite bath (Hall- Heroult process). Over 90 per cent of the world's alumina is produced by the Bayer process and the solid waste generated during the processing of bauxite by this process is known as Red mud. It is also known as 'bauxite tailings' or 'Bauxite Residue'.

Characteristics of Red Mud

The chemical and mineralogical compositions of red mud vary widely depending on the bauxite source and the process parameters. It is a highly complex material containing six major oxides *viz.* Fe_2O_3 (35-55 wt per cent), Al_2O_3 (16-22 wt per cent), SiO_2 (4-16 wt per cent), Na_2O (3-6 wt per cent), TiO_2 (2-19 wt per cent), CaO (0.8-4.5 wt per cent). It also contains several minor/trace amount of numerous other elements as oxides of V, P, Ga, Cr, Zr, Mg, Sr, Ba, Li, K, Pb, Cu, Mn, Ni, Zn, Co, rare earth etc. Red mud is composed of as many as 15-21 mineral phases. In addition to all those mineral phases present in the bauxite processed, red mud also contains a number of other phases which are formed during the digestion process. Mineral phases such as gibbsite, bohemite, diaspore, hematite, magnetite, goethite, alumogoethite, rutile and anatase are present in the bauxite to be processed. The phases formed in the process include sodalites (sodium aluminosilicates, calcium aluminosilicates, cancrinite, sodium titanate, calcium titanate etc. It is also disposed as slurry having a high solid concentration of 30-60 per cent with a pulp density ranging between 1.1-1.6 gm/cm^3. It is extremely fine sized (over 60 per cent less than 10 microns) and have inherently poor settling properties[1]. In addition, red muds are also thixotropic and possess poor soil-mechanical properties. A huge amount of 70 million tones of red mud is generated around the world per annum, which is highly alkaline (pH 12 to 13.5) and creates environmental problems like caustic seepage in groundwater, deterioration of associated ground fertility, disposal space scarcity and many more.

Utilisation of Red Mud

Effective utilization of red mud is undoubtedly the best way to fight the situation. But high alkalinity of red mud is the main hurdle in its utilization path. Utilization of red mud in building materials production has got a promising chance due to presence of alumina, silica and aluminosilicate compounds which play important role in building materials property development. Though neutralization/soda fixation is achieved in case of sintered/fired bricks, presence of soda (high pH) creates problem in the case of stabilized blocks and unfired bricks as in those cases soda leaches out and contaminates surrounding water bodies. Soil amendment of red mud to neutralize it and reduce other harmful ions present in it to make it suitable for vegetation or vegetation support cover is also a good approach. But special treatment of such a huge volume of solid waste is very slow process. Direct neutralization of red mud followed by disposal is also a way which has been studied by many people around the world. Some bio-remediation routes have also been studied in Australia in which some microbial species which can survive in high pH environment were allowed to grow by consuming some organic substrate (*e.g.* source of glucose, sucrose, lactose etc; provided externally). During this growth period they generate some weak organic acid and some times CO_2 also which in turn neutralizes red mud. In some specific cases where Al_2O_3 or TiO_2 quantity is appreciable in red mud, recovery of the same has been tried. Specially treated red mud has also been tried as adsorbent in water pollution control purposes. Appreciable quantity of iron (35 per cent - 55 per cent) present in red mud makes a sense that successful removal/recovery of iron from red mud will lead to substantial reduction in red mud volume *i.e.* less disposal problem. Several methods have been tried so far *e.g.* magnetic separation for iron beneficiation and reducing red mud volume to be disposed off, acid leaching and liquid-liquid extraction etc. But partial or complete reduction of iron by different reductant has improved the quantity and quality of iron present in treated red mud which in turn made iron recovery from red mud much more feasible.

In this paper, different scopes of red mud utilization such as construction and building materials, metal value recovery, soil amendment and vegetation, pollution control have been reviewed.

Construction and Building Materials

Making red mud bricks may provide a partial answer to the alumina industry's waste disposal problem. Red mud has been used to make bricks for inexpensive housing by researchers from the Jamaica Bauxite Institute and the University of Toronto[2]. Recovery of iron from red mud was studied with direct reduction roasting process followed by magnetic separation, and then building materials were prepared from alumosilicate residues[3]. Brick specimens were prepared with alumosilicate residues and hydrated lime. Mean compressive strength of specimens was 24.10 MPa. It was indicated that main mineral phase transformed from nepheline ($NaAlSiO_4$) in alumosilicate residues to gehlenite ($Ca_2Al_2SiO_7$) in brick specimens through X-ray diffraction (XRD) technology. Also, presence of alumina, iron oxide and silica makes red mud a suitable ingredient for the preparation of suitable cements[4]. Different types of cements were made using admixtures of red mud, fly ash, bauxite, gypsum and lime. The strengths were found to be comparable to ordinary Portland Cement (OPC).

Bricks were made from red mud using clay and water fixing agent[5].

A method have been invented for making bricks, particularly light-weight construction bricks, from red mud in which strongly water-fixing organic or inorganic substances are added to the filter-wet red mud before processing in order to influence its consistency [6]. Lime stabilised red mud bricks have also been prepared.

Red mud can also be used in ceramic industry. Red mud was investigated for use in making of ceramic glazes such as porcelain, vitreous (sanitary ware glazes), tile and electroporcelain glazes in the ceramic industry[7]. Specific mixtures of red mud, fly ash and spent pot liner were used to prepare glass–ceramic products which showed excellent properties and aesthetic appearance for possible applications as decorative tiles in the building industry[8]. Use of red mud as plasma- spray coatings on aluminium has also been reported[9].

Metal Value Recovery

Metals such as iron, titania and alumina can be recovered from red mud depending upon their quantity present in the red mud. An extensive study on the possibility of magnetic separation of red mud was carried out on red mud generated by Bayer process from Fria Deposit (Guinea) and reported that ~ 85 per cent of the iron present in red mud was recovered at 0.06 Tesla magnetic intensity[10]. In another work Turkish red mud was mixed with dolomite and coke to make pellet and sintered (1100^{0}C) followed by smelting (1500^{0}C) to produce pig iron[11]. Studies were conducted on Jamaican red mud to recover undigested alumina by ash-sintering method followed by carbothermonic reduction of iron for metallization. Scope of magnetic separation and smelting of reduced material for pig iron production is also reported. If smelted, acid leaching of titania from smelter slag is also studied [12]. Unfortunately no study on metal value recovery from red mud (metallurgical industry waste) has been reported so far in India.

Soil Amendment and Vegetation

Bulk utilisation of red mud can be thought of by making its use in landfill and land reclamation. Bauxite mines or other quarries can be filled up by red mud after neutralization. For neutralization, lime, gypsum, seawater or other materials having similar neutralisation properties are mixed with red mud. Establishment of vegetation on the residues is an attractive desirable method of suppressing dust generation. However, there are constraints in achieving vegetation growth on the residues due to the inherent high pH and sodium levels. Use of gypsum and thermally dried sewage sludge as amendments for establishing the clover species Trifolium pratense on a red mud/process sand mix have been studied[13]. Gypsum amendment improved chemical conditions of the substrate and increased plant yield. Potential application of bauxite residue in soil/sediment remediation, soil/sediment stabilisation has studied. Wherever the agricultural land soil is acidic, red mud mixed with manures/composts to neutralise the soil. It was seen that under certain conditions the properties of bauxite residue potentially can allow its use for improving soil conditions [14].

Pollution Control

The most interesting applications of red mud are however in the environmental field. After adequate neutralisation red mud can be used for the remediation of contaminated sites and treatment of contaminated liquid waste. With this principal aim, several studies were carried out on the metal trapping and acid neutralisation capacities of red mud. The possibility to reuse treated red mud for contaminated waters and soils have been evaluated[15]. Acid mine drainage (AMD) is an environmental problem produced when sulphides come in contact with an oxidant (bacteria) and water, producing acid generation and metals leaching. One solution proposed is to use red mud which is very alkaline, to neutralize oxidized acidic tailings[16]. Plain soil was treated with red mud neutralized with either waste gypsum from the phosphate industry or ferrous sulfate from the titanium dioxide industry to investigate the reduction in the leaching of phosphorus (P) applied as superphosphate fertilizer from a very sandy Swan Coastal region[17].

In recent years, a great deal of concern has been expressed with regard to global climate change and its link to growing atmospheric concentrations of carbon dioxide. In order to decrease the impact of anthropogenic CO_2 on global climate, several strategies are under development that will potentially remove CO_2 from the atmosphere or decrease CO_2 emission. Addition of CO_2 to this caustic material will not only result in carbonate mineral formation but will also serve to decrease the pH of the residue, making disposal more environmentally sound. Moreover, bauxite residue (pH 13.5) can serve as an effective source of caustic material to treat oil and gas field wastewater brines (pH 3 to 5) for mineral carbonation. The use of bauxite residue/brine to capture and store CO_2 will serve not only to help mitigate the impact of anthropogenic CO_2 on global warming but will also help neutralize caustic industrial waste for safe storage and reuse [18]. Also the Department of Agriculture, Western Australia has been working with Alcoa World Alumina Australia Ltd. for investigating the potential to use bauxite residues as soil amendments for the poor, acidic, sandy soils of the Swan Coastal plain [19].

Studies have also been carried out to use red mud as adsorbent for cleaning of industrial gases and as synthetic coagulants.

Conclusion

As can be seen, there are several applications of red mud based on the physical and chemical characteristics. A considerable research has been carried out on the utilisation aspect of red mud. But, presently there is no large scale utilisation of red mud reported anywhere in the world. Many of the process developed are not commercially implemented because of the high cost involved in the processing of red mud. In all the alumina plants present in India, red mud is disposed off in red mud lakes or ponds and hence a cost effective process is to be developed where bulk and effective utilisation of red mud can be carried out to reduce its environmental hazards.

References

1. Chaddha M. J., S. Rai and R. N. Goyal, Characteristics of red mud of Indian alumina Plants and their possible utilisation, ENVICON 2007, 41-44.
2. Peter Newton, Making Bricks with Red Mud in Jamaica, IDRC: Resources: Books: Reports: Vol.21, No 2.
3. Wanchao Liu, Jiakuan Yang and Bo XiaRef, Application of Bayer red mud for iron recovery and building material production from alumosilicate residues, Available online 8 April, 2008.
4. Singh Maneesh, S. N. Upadhayay, P. M. Prasad, Preparation of special cements from red mud, Waste Management, Vol 16, Issue 8, 1996, Pages 665- 670.
5. Method for producing bricks from red mud, Document Type and Number: United States Patent 3886244
6. H H. Nakamura *et al.*, "Use of Bauxite Wastes for Lightweight Building Products," Ceramic Bulletin, March, 1971, pp. 248-250.

7. Paramguru K., P. C. Rath, V. N. Misra, Trends in red mud utilization, Mineral processing and Extractive Metallurgy Review: 26: 1-29, 2005.

8. Nimje M. T., R. J. Sharma, G. S. Sengar, Conversion of Aluminium Industrial wastes (Fly- ash, red mud) into glass- ceramic products, ENVICON 2007, 41-44, 132-135.

9. Satpathy Alok *et al.,* Red mud: A potential coating material, ENVICON 2007, 139-141.

10. Fofana Mohamad, *et al.,* Treatment of red mud from alumina production by high intensity magnetic separation, Magnetic and Electrical Separation. 1995. Vol.6, page 243-251.

11. Erça Erol; Reat Apak. Furnace smelting and extractive metallurgy of red mud: Recovery of TiO_2, Al_2O_3 and pig iron. Journal of Chemical Technology and Biotechnology. 26 Mar 1997. Volume 70, Issue 3, Pages 241 - 246.

12. Mishra B; A Staley; D Kirkpatrick. Recovery of value-added products from red mud, Minerals and Metallurgical Processing. 2002. vol. 19, No. 2, page 87-94.

13. Courtney R.; Timpson J.P.; Grennan E., Growth of Trifolium pratense in Red Mud Amended With Process Sand, Gypsum and Thermally Dried Sewage Sludge,: International Journal of Surface Mining, Reclamation and Environment, Volume 17, Number 4, December 2003, pp. 227-233(7)

14. Varnanas S. P., D. Boufounas and D. Fafoutis, Proceedings of the 9th intrenational conference on environmental science and technology, Rhodes island, Greece, 1-3 Sept. 2005.

15. Claudia Brunori, Carlo Creminsini *et al.,* reuse of a treated red mud bauxite waste: studies on environmental compatibility, Journal of hazardous materials (B117 92005)55-63.

16. M. Paradis, J. Duchesne, A. Lamontagne, and D. Isabel, Using red mud bauxite for the neutralization of acid mine tailings: a column leaching test, Canadian Geotechnical Journal 43(11): 1167–1179 (2006).

17. S Vlahos, KJ Summers, DT Bell and RJ Gilkes, Reducing phosphorus leaching from sandy soils with red mud bauxite processing residues, *Australian Journal of Soil Research* 27(4) 651–662.

18. Yee Soong; *et al.,* CO_2 Sequestration Via Brine/bauxite Residue Mixture National Energy Technology Laboratory, U.S. Department of Energy, Pittsburgh, PA 15236-0940,

19. Summers, R. N., Rivers, M.R. and Clarke, M. F., the use of bauxite residue to control diffuse phosphorous pollution in Western Australia- A win–win win outcome, proceedings of the 6th International alumna quality workshop 2002.

Chapter 29

Atmospheric Brown Cloud (ABC), Climate Change and Mitigation Measures

Tapan Chakrabarti

CSIR-NEERI, Nagpur

Introduction

Climate Change and its Causes

Factors that can shape climate are often called climate forcing which include

- Variations in solar radiation
- Deviations in the Earth's orbit
- Changes in greenhouse gas concentrations.

There are a variety of climate change feedbacks that can either amplify or diminish the initial forcing. Some parts of the climate system, such as the oceans and ice caps, respond slowly in reaction to climate forcing because of their large mass. Therefore, the climate system can take centuries or longer to fully respond to new external forcings.

Greenhouse Gases and Climate Change

Combustion of coal, oil and natural gas, and to a lesser extent, deforestation, land-cover change, and emissions of halocarbons and other greenhouse gases (GHGs), are rapidly increasing the atmospheric concentrations of climate warming gases.

GHGs include:

- ✰ Carbon Dioxide
- ✰ Methane
- ✰ NO
- ✰ Chloro-fluorocarbon
- ✰ Ozone

Natural changes in climate take place over many years, this allows animals and plants to adapt to new environments. Scientists have discovered that the earth's temperature is increasing too rapidly and that's due to the amount of carbon dioxide that is released into the atmosphere.

The warming of approximately 0.1-0.2°C per decade that has resulted is very likely the primary cause of the increasing loss of snow cover and Arctic sea ice,.of more frequent occurrence of heavy precipitation, of rising sea level, and of shifts in the natural ranges of plants and animals.

The global average temperature is already approximately 0.8°C above its pre-industrial level, and present atmospheric levels of GHGs will contribute to further warming of 0.5–1°C as equilibrium is re-established.

The rate of warming will further increase if emission controls are not put in place. Present estimates suggest that limiting the increase in global average surface temperature to no more than 2 to 2.5°C above its 1750 value of 15°C will be required to avoid the most catastrophic, but certainly not all, consequences of climate change. Accomplishing this will require reducing emissions sharply by 2050 and to near zero by 2100.

Climate change is causing the retreat of glaciers. Melting glaciers pose two main concerns: (i) a threat to water security; and (ii) an increase in the vulnerability to natural disasters. The Hindu Kush–Himalayan region has been severely affected by regular occurrences of Glacial Lake Outburst Floods (GLOF) in recent years. These areas are resource rich and have a dense population of around 150 million people, often clustered along steep river valleys, making them highly vulnerable to such disasters.

The Hindu Kush–Himalayas are the water source for 7 of the great rivers of Asia — Ganges, Indus, Brahmaputra, Salween, Mekong, Yangtze and Yellow. They also harbor some of the world's richest ecosystems in terms of biodiversity, and many protected areas lie in the proximity of potentially dangerous glacial lakes. Recent studies by the Atmospheric Brown Cloud (ABC) Science Team found that the atmospheric brown clouds (ABCs) over South Asia may speed up Himalayan glacier melting, in that they contribute as much as the recent increase in anthropogenic greenhouse gases to regional lower atmospheric warming trends. Results from the ABC high altitude observatory indicate presence of significantly higher concentrations of ABCs (including black carbon) in the Himalayan region.

India and Climate Change

At present, India is rated as the 6th largest contributor of CO_2 emissions and China the 2nd. However, our per capita CO_2 of 0.93t per annum is well below the world average of 3.87t per annum.

India is highly vulnerable to climate change as its economy is heavily reliant on climate sensitive sectors like agriculture and forestry. The vast low-lying and densely populated coastline is susceptible to rise in sea level.

Atmospheric Brown Cloud (ABC)

Atmospheric (Asian) Brown Cloud (ABC) is an emerging regional climate change issue. Recent studies show that ABC and its interaction with build-up of greenhouse gases significantly affect the regional climate, glacial melting, water budgets, agriculture and public health.

The important, but inadequately understood role that ABC can play in regional and global climate change necessitates investigation on human influence on earth's climate.

Soot in brown clouds over the Indian Ocean are absorbing solar radiation and warming the lower atmosphere in the region. As a result of the warmer temperatures, Himalayan glaciers are melting, affecting rivers that supply water to people living downstream.

The brown haze is caused by air pollution, mainly the sub-micron size aerosol particles, emitted from a wide range of anthropogenic and natural sources such as automobile exhaust, flue gas emission from thermal power plants and industrial combustion process.

Atmospheric brown clouds are mostly the result of biomass burning and fossil fuel consumption. They consist of a mixture of light-absorbing and light-scattering aerosols and, therefore, contribute to atmospheric solar heating and surface cooling. The sum of the two climate forcing terms—the net aerosol forcing effect—is thought to be negative and may have masked as much as half of the global warming attributed to the recent rapid rise in greenhouse gases.

The cloud contains tiny soot particles from the burning of fossil fuels, crop residues burned by farmers, and cow dung and firewood burned by household cooks in practices unchanged for centuries. It also contains industrial gases issued by the manufacturing plants of the world's fastest growing economies, and soil dust from plains rapidly turning into desert through drought, water diversion and aggressive farming (Figure 29.1).

UNEP's Initiatives in ABC Management

When effects of ABCs are considered, the effect of global warming (due to greenhouse gases) on glacier retreat and snow pack melting will be more pronounced, i.e, the melting will be accelerated significantly. There is a need to quantify the effects of ABCs on glacier melt and the potential impacts on the availability of water resources in Asia.

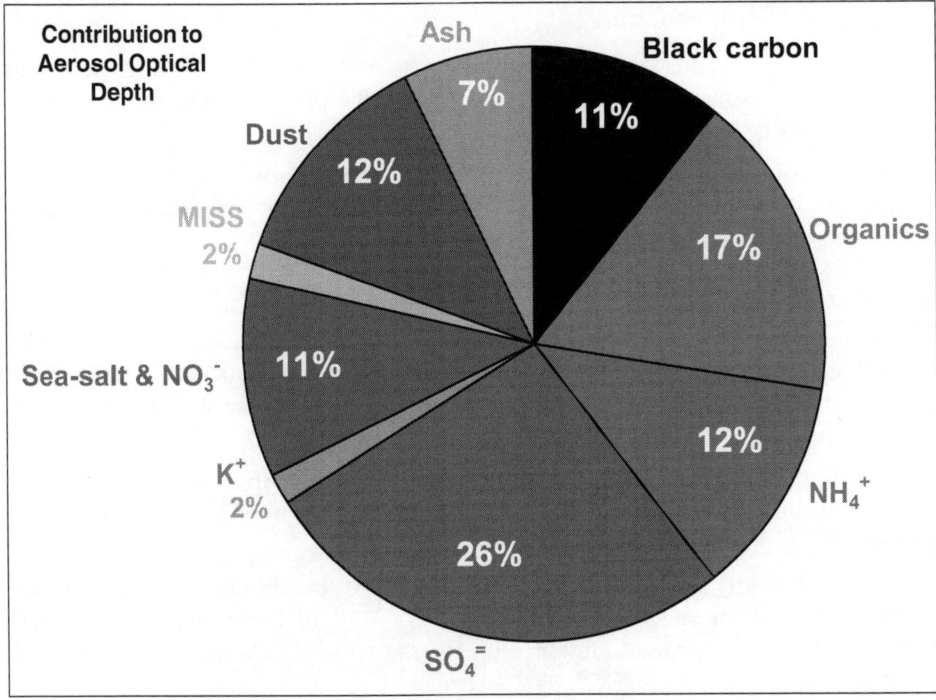

Figure 29.1: Brown Cloud Chemical Composition.

To study this emerging environmental issue in Asia, UNEP commissioned Project Atmospheric Brown Cloud (ABC) in 2002 in collaboration with a team of distinguished scientists from Asia, Europe and USA. Project ABC focuses on observation, impact assessment, and awareness and mitigation, accompanied by capacity building.

Climate Change Initiatives in India

Ministry of Environment and Forests, India

The Ministry of Environment and Forests, Government of India has launched a website on climate change which provides easy access to information related to climate change science, impacts, and international negotiations. The site also has a directory of organizations working on climate change related issues in India, information about AIJ and GEF projects in the country, and events being organized in the run-up to COP-8.

Centre for Ecological Sciences, India

The Centre for Ecological Sciences was established at the Indian Institute of Science in 1982 and carrying out several active research programmes in Basic and Applied Ecology.

Centre for Science and Environment (CSE)

CSE is creating awareness about climate change.

TERI, New Delhi, India

TERI is monitoring carbon black in Himalayan Glaciers and trying to link its findings to climate change. Observatories set up at Kanpur and Kharagpur under Project ABC provide detailed information on the nature, variability and seasonal fluctuations of ABC. India has established a centre for climate change research to provide data for modeling and monitoring climate change in the country.

The centre will be based at the Indian Institute of Tropical Meteorology in Pune and run by India's Ministry of Earth Sciences.

The centre will network with other national and international scientific institutes and universities that measure greenhouse gases and aerosols and monitor glaciers, temperature change and rainfall patterns.

The eight missions identified in the National Action Plan on Climate Change are:

- ✰ National Solar Mission
- ✰ National Mission for Enhanced Energy Efficiency
- ✰ National Mission on Sustainable Habitat
- ✰ National Water Mission
- ✰ National Mission for Sustaining the Himalayan Ecosystem
- ✰ National Mission for a Green India
- ✰ National Mission for Sustainable Agriculture
- ✰ National Mission on Strategic Knowledge for Climate Change.

Control of pollution from the Indian transport sector, identified as the major contributor to air pollution, calls for formulation of long-term firm auto-fuel policy and its effective implementation with cooperation of all sectors of society.

The first Indian emission regulations were idle emission limits, which became effective in 1989. These idle emission regulations were soon replaced by mass emission limits for both gasoline (1991) and diesel (1992) vehicles, which were gradually tightened during the 1990's. Since the year 2000, India started adopting European emission and fuel regulations for four-wheeled light-duty and for heavy-duty vehicles. India's own emission regulations still apply to two- and three-wheeled vehicles. On October 6, 2003, the National Auto Fuel Policy has been announced, which envisages a phased program for introducing Euro 2 - 4 emission and fuel regulations by 2010.

Broadly, the policy gives a roadmap for achieving various vehicular emission norms over a period of time and the corresponding fuel quality upgradation requirements. While it does not recommend any particular fuel or technology for achieving the desired emission norms, it suggests, taking into account security of supplies and existing logistics perspectives, that liquid fuels should remain as main auto fuels throughout the country and that the use of CNG/LPG be encouraged in cities affected by higher pollution levels so as to enable vehicle owners to have the choice of the fuel and technology combination. The report also recommends measures for improving the present mechanism of checking pollution from in-use vehicles.

As pointed out by the Mashelkar report, sustainable improvements in air quality can be achieved only by a holistic approach, of which fuel quality will be just one component. Another important component of this holistic approach should be emissions control technology. While petrol engine emissions control technology (*e.g.* catalytic converters) is mature and readily available, significant technical progress is still required for effective diesel emissions control. Diesel emissions control is more important and relevant for India, which consumes a lot more diesel in comparison to other hydrocarbon fuels

Fuel Switching Including the Use of Waste Materials

In India, a number of industries use fuel for steam generation and/or process heat, with the choice of fuel being determined by cost, fuel availability and environmental regulations.

Waste materials (tyres, plastics, used/waste oil, solvent, sewage sludge, even meat and bones) are being used by a number of industries.

It is estimated that global CO_2 emission can be reduced by 12 per cent through increased use of waste fuels.

The steel industry has developed technology to use wastes such as plastics as alternate fuels.

Pretreated plastic wastes could be recycled in coke ovens and blast furnaces.

Incineration of wastes (tyres, municipal and hazardous wastes) in cement kilns is one of the most efficient methods of disposal of these materials. Up to 70 per cent waste materials can be used.

Cement companies in India are using non-fossil fuels such as agricultural wastes, sewage sludge, domestic refuse, used tyres, wide range of waste solvents and other organic liquids employing improved burners and burning systems.

The Clean Development Mechanism (CDM) of the UN Framework Convention on Climate Change (UNFCCC) has funded over 300 projects that incinerate garbage and produce methane for energy from landfills.

One of the less expensive and well-established technologies to produce refuse derived fuel (RDF) from MSW is mechanical biological pre-treatment (MBT). An MBT plant separates out metals and inert materials, screens out organic fractions (for stabilisation using composting processes, either with or without a digestion phase), and separates out high-calorific fractions for RDF.

RDF can also result from a 'dry stabilization process' in which residual waste (after separating out metals and inert materials) is dried through a composting process leaving the residual mass with a higher calorific value.

Where the cost of incineration is high and where source separation is well developed, the MBT route is still more favorable. This partly reflects the fact that moderate sized composting equipment (20,000 tonnes) tends to cost half the cost of incineration. Indeed, it may become a cheaper option even where RDF is combusted in dedicated incinerators since the mass reduction achieved and the higher calorific value of the remaining material may 'pay for' the separation and treatment process.

There are additional reasons for a municipality to chose MBT as a more flexible solution to mass-burn incineration. Not only can the biological treatment aspect of the process be made modular (to allow switching away from treatment of mixed waste to composting of source separated waste) but also the use of RDF in co-incineration plants removes the need to invest in capital intense, dedicated incineration (or thermal treatment) facilities.

There is a need to study the increase of heavy metals in cement and other by-products from co-incineration facilities to investigate possible environmental consequences this may cause.

The gasifier plants were funded by the UN's Global Environment Facility, and supported by the Ministry of Environment and Forests of the Government of India, the Government of Karnataka, and UNDP.

In the gasifiers, wood or coconut shells are reduced to small pieces and burned in a reactor that converts them to combustible gases, a mixture of carbon monoxide and hydrogen. This so-called 'producer gas'(syn gas) runs the engines, which produce power

It is now necessary to substitute wood or coconut shells with waste derived fuel.

The first plant under the project was inaugurated in the village of Kabbigere on 24 January and has provided around 10,000 kilowatt-hours of electricity to four villages since then.

A second plant was commissioned under the project and has the capacity of delivering 250 kilowatts of electricity, with excess power to be sold to the Bangalore Electric Supply Company, according to UNDP.

A third plant, producing 250 kilowatt-hours, is being commissioned in Seebirayanapalya and another in Chinnenahalli.

Syngas

Syngas, a mixture of carbon monoxide and hydrogen, is produced by partial combustion of biomass, that is, combustion with an amount of oxygen that is not sufficient to convert the biomass completely to carbon dioxide and water. Before partial combustion the biomass is dried, and sometimes pyrolysed.

The resulting gas mixture, syngas, is itself a fuel. Using the syngas is more efficient than direct combustion of the original biofuel; more of the energy contained in the fuel is extracted. In addition, emission of unburnt black carbon is eliminated.

Syngas may be burned directly in internal combustion engines or turbines. The wood gas generator is a wood-fueled gasification reactor mounted on an internal combustion engine. Syngas can be used to produce methanol and hydrogen, or converted via the Fischer-Tropsch process to produce a synthetic diesel substitute, or a mixture of alcohols that can be blended into gasoline. Gasification normally relies on temperatures >700°C. Lower temperature gasification is desirable when co-producing pyrochar but results in a Syngas polluted with tar.

Fischer-Tropsch Process

The Fischer-Tropsch process (or Fischer-Tropsch Synthesis) is a catalyzed chemical reaction in which synthetic gas, a mixture of carbon monoxide and hydrogen, is converted into liquid hydrocarbon of various forms. The most common catalysts are based on iron and cobalt, although nickel and ruthenium have also been used.

The principal purpose of this process is to produce a synthetic petroleum substitute, typically from coal, natural gas or biomass, for use as synthetic lubrication oil or as synthetic fuel.

Chapter 30

Development of Low-Cost Fish Feed using Byproduct and Nutritious Food Industry Wastes

Santosh Zargar, G.V.Mulmule and T.K.Ghosh[*]

Environmental Biotechnology Division,
CSIR-National Environmental Engineering Research Institute,
Nehru Marg, Nagpur

ABSTRACT

Feed is one of the major inputs in aquaculture, and the success of fish farming depends primarily on the provision of adequate quantity of nutritionally balanced feed in a form, which is acceptable to fish. The traditional feed, presently used in composite fish culture of carps, comprises of rice bran and groundnut/soybean oilcake in equal proportions. These are being used only as supplementary feeds and the availability of natural protein in the form of plankton is essential. Moreover, traditional feed lacks desired levels of vitamins/minerals and other items. Fish Feed products in India are mostly based on ingredients derived from crop residues. On the other hand, India has rich resources of slaughterhouse/ poultry/silkworm by-products which are also the better source of proteins and minerals. In the present study, four prepared diets (feeds I, II, III, IV) were formulated using locally available ingredients and nutritious wastes while taking into account certain practical consideration (acceptability of mixture to the animal, ability of the mixture to be palleted, cost, etc.) and an experiment was designed to determine the acceptability of different formulated diets by select fish (Catla catla). The feed V was used as control feed, procured from CIFE, Mumbai. Growth in terms of net weight gain in grams (50.0± 5.0 per cent) in fish feed V and least in feed IV (13.3± 1.0per cent) during the study period ($P< 0.05$). The values of NWG, PER, and FCE were found in this order: Feed V > Feed II > Feed III > Feed I > Feed

IV. The growth of the carp was hardly influenced by the variation recorded in select physico-chemical parameters of water during the experiment conducted. It may be added the experimental results, under natural condition (natural fish ponds), indicated 100 per cent increased growth of fish fed by NEERI's feed (feed II) over those fed by control feed (conventional feed). The cost of the feed has been evaluated about two-third of the similar kind of feed, available in the market. The study could provide the technology to local people for their employment generation, pollution free environment caused due to scattering nutritious waste materials, utilized in the fish feed and solution to minimize the incorporation of grains in fish feed by replacing these with nutritious food industry waste materials leading to scope in solving the problem of grain paucity in India.

Keywords: *Formulation, Byproducts, Nutritious wastes, Fish feed, Catla catla, NWG, PER, FCE.*

Introduction

Fish, like any other animal, require nutrients to remain healthy, and active for which nutrients play a vital role. Nutrients can be provided by various types of food items, which are determined both by the feeding behavior or preference of the animal and the availability of items. Growth and/or net nutrition deposition is the most accurate and important tools in studying fish feed efficiency and nutrient requirements (Das *et al.*, 1991). Feed is one of the major inputs in aquaculture, and the success of fish farming depends primarily on the provision of adequate quantity of nutritionally balanced feed in a form, which is acceptable to fish. The traditional feed, presently used in composite fish culture of carps, comprises of rice bran and groundnut/ soybean oilcake in equal proportions. These are being used only as supplementary feeds and the availability of natural protein in the form of plankton is essential. The shortcoming of such feeds in terms of quality becomes obvious when the plankton availability in the pond is insufficient. Moreover, traditional feed lacks desired levels of protein, carbohydrate, vitamins and minerals.

Since the livestock, especially poultry, pig and fish compete with human for high quality protein; the demand for grains and cereals is high. India faces nearly 50 per cent deficiency in concentrate feed which is expected to show further increase by 2010 AD (www.tifac.org.in/offer/tlbo/rep/TMS162.htm). This deficiency in food production can be removed only when there are alternate feed resources for reducing the composition of crop residues in the animal feed and in substituting with the unconventional ingredients like slaughter house waste materials. The use of animal waste or its content in fish feed should be enhanced. This will not only reduce the consumption of grains but will also increase the nutritional value of feedstuff. The animal tissues contain numerous proteolytic enzymes, which help in increasing the digestibility of the feed.

Fish Feed Products in India are mostly based on ingredients derived from crop residues. On the other hand, India has rich resources of slaughterhouse/poultry/ silkworm by-products which are also the better source of proteins and minerals. However, most of these are going waste. Utilization of these so called waste materials, in the production of fish feeds, will not only solve the problem of nutritional needs of fish but also helps in reducing the environmental pollution which is caused by the unused animal tissues lying scattered all over the rural as well as the urban areas, leading to public nuisance and vector breeding.

The present study aimed at formulating fish feeds, comprising of byproduct and nutritious food industry waste-based materials and also to improve the quality of conventional feed used by pisciculturists. The study could provide a technology to local people for their employment generation, provide a pollution free environment by preventing scattering of nutritious waste materials and channeling them to the fish feed manufacture and a solution to minimize the incorporation of grains in fish feed by replacing these with nutritious food industry waste materials leading to scope in solving the problem of grain paucity in India.

Material and Methods

Fish Feed Formulation and Preparation

Feed formulation is the process of designing a mixture that will meet the nutritional requirement of the animal while taking into account certain practical consideration (acceptability of mixture to the animal, ability of the mixture to be pelleted, cost, etc.). The four prepared diets (feeds I, II, III, IV) were formulated using locally available ingredients and nutritious wastes as shown in Table 30.1. 50 g of corn starch was boiled in oven gas stove at 100°C with 250 ml water and grinded ingredients were mixed with it. Grinding increases the nutrient availability of feed, and by reducing particle size, augments activity of proteolytic enzyme released during tissue grinding. Prior to use soybean seeds, the seeds were soaked (overnight: 6 hours), blanched (in boiling water: 20 minutes), de-husked and sun-dried. Subsequent treatment destroyed Trypsin inhibitors of soybean meal, thus increasing its nutritional value. Nutritious wastes, *viz.* poultry and silkworm, were sun-dried before processing. Dough was prepared and extruded through 2 mm diameter die of a hand operated Extruder. The extruded moist noodles were sun-dried for a day, palletized sun-dried noodles manually and graded of fish feed by using sieves, and then stored in airtight

Table 30.1*:* Per cent Composition of the Ingredients Used for Preparing Experimental Feeds.

Ingredients	*Feed I (per cent)*	*Feed II (per cent)*	*Feed III (per cent)*	*Feed IV (per cent)*
Soybean	35.0	45.0	35.0	–
Rice bran	30.0	15.0	30.0	50.0
DOC Soybean	–	–	–	50.0
Maize Powder	10.0	10.0	10.0	–
Bakery waste	14.0	14.0	14.0	–
Poultry waste	–	5.0	–	–
Silk-worm waste	–	5.0	5.0	–
Vitamin mineral mixture	0.1	0.1	0.1	–
Starch (binder)	5.0	5.0	5.0	–

jars/polythene bags at room temperature. It may be mentioned that for the preparation of the pelleted fish feed, the equipments *viz.*, Pulverizer, Extruder and Packaging machine were used. Proximate analysis of major ingredient was undertaken in the laboratory following the method of Association of Official Analytical Chemists (Anonymous, 1980).

It may be mentioned that a feed was procured from CIFE, Mumbai containing soybean 44 per cent, fish meal 10 per cent, corn meal 4.5 per cent, coconut oil 9.5 per cent, wheat flour 4.5 per cent, cod liver oil 2.0 per cent, sunflower seed 2.0 per cent, CMC binder 2.0 per cent and vitamin C 0.1 per cent. The feed was used as control feed (Feed V).The crude fats were 12.2, 12.8, 13.7, 9.7 and 9.0 per cent respectively in the feeds.

Design of Experiment

An experiment was designed to determine the acceptability of different formulated diets by select fish (*Catla catla*). The fishes (average weight 20± 2.0 g) were procured from Mansar hatchery farms, Nagpur. Experiment was conducted in triplicate in plastic cisterns of 500 L capacity with continuous aeration. Each cistern was stocked with fifteen fishes. Prior to the commencement of experiment, the pools were thoroughly cleaned, filled with water and stocked with pre weighed acclimatized fish fingerlings. Special care was taken to avoid escape of the fish during the experiment. The feeding experiment was continued for 30 days. Intermittently the pools were cleaned at every 5 days by siphoning of water. The feeding frequency plays an important role in the growth of the fish. The fish were fed daily once at the rate of 2 per cent of its body weight throughout the experimental period. Any uneaten food or fecal matter from each pool was removed daily by siphoning method. About one third water from each cistern was changed every morning to keep the water quality as good as possible.

Dietary Performance Evaluation

Important water quality parameters like temperature, pH, hardness, alkalinity and DO were analysed (APHA, 1998) during the experiment on alternate days. For evaluating the dietary performances, various nutritional indices, *viz.* Net Weight Gain (NWG), Specific Growth Rate (SGR), Feed Conversion Ratio (FCR), Protein Efficiency Ratio (PER) and Food Conversion Efficiency (FCE) were used (Rajan *et al.*, 1996).

Results and Discussion

Water Quality

The water quality during the study period remained in following ranges: pH 7.8 ± 0.2 to 8.0 ± 0.2; alkalinity 256.1 ± 9.5 to 276.2 ± 3.4 mgl^{-1}; dissolved oxygen 5.3 ± 0.5 to 6.0 ± 0.4 mgl^{-1}; and total hardness 120 ±5.0 to 136 ± 6.0 mgl^{-1}. Since fish are poikilotherm, water temperature plays an important role in energy partitioning, protein assimilation and growth (Swamy *et al.*, 1994). Water temperature was varied from 23.8 ± 4.0 to 26.7 ± 0.5 °C. All selected water quality parameters were in the ranges recommended for aquaculture practices wherein the values of pH: 6.7-9.5; alkalinity: 50-300 mgl^{-1}; dissolved oxygen: 5-10 mgl^{-1} and total hardness: 30-180 mgl^{-1} (CIFE, Mumbai, 2003). The growth of the carp was hardly influenced by the variation recorded in select physico-chemical parameters of water during the experiment conducted.

Dietary Performance

Growth in terms of net weight gain in grams (50.0± 5.0per cent) in fish feed V and least in feed IV (13.3± 1.0per cent) during the study period ($P< 0.05$) (Table 30.2). SGR and FCE were also greatest in case of fish feed V (0.61, 15.0 and 83.33 respectively (Table 30.2). SGR can be taken as an index of growth evaluation of diets; however, higher FCE signifies better utilization of the consumed matter. FCR and PER were least in fish feed V (1.20) and feed IV (0.85) respectively. In the study FCR were found to be varied from 1.20 (feed V, 30.0 per cent protein) to 4.5 (feed IV, 26.2 per cent protein) in different diets. However, Choudhury *et al.* (2002) reported FCR varied from 1.29 to 1.95 in different formulated diets and at different feeding frequencies. Different authors have reported values if FCR on different fishes, such as, 1.12 to 1.35 for *Cyprinus carpio* with formulated diet (38.74 per cent protein level) at five times feeding (Capper *et al.*, 1982), 3.68 to 3.82 for *Labeo rohita* (23.9 per cent protein level) in cages (Ahmed *et al.*, 1983), 3.08 in *L. rohita* (Rangacharyulu *et al.*, 1991) and 1.79 to 2.16 per cent for *L. rohita* (Ahsan, 1995), for diet containing 33.92 per cent protein, the values were closer to the findings of the present study.

Table 30.2: Select Feed Utilization Parameters for the Study .

Diet	*NWG (per cent)*	*SGR*	*FCR*	*PER*	*FCE*
Feed I	26.7	0.33	2.25	1.90	44.44
Feed II	46.7	0.56	1.28	2.8	77.77
Feed III	33.3	0.44	1.80	2.04	55.55
Feed IV	13.3	0.19	4.5	0.85	22.22
Feed V	50.0	0.61	1.20	2.77	83.33

A superior growth performance was noted in fish fed with feed I when compared to fish feed IV in spite of the lower protein content (23.1 per cent) in the former. This observation draws attention to two essential inferences; one the crude protein per cent in feed IV might be having higher content of indigestible protein and, two the lower crude lipid percentage in feed IV might have played a crucial role in feed utilization. The values of NWG, PER, and FCE were found in this order: Feed V > Feed II > Feed III > Feed I > Feed IV. The minimum growth was found in fish fed with feed IV, which contained rice bran (50 per cent) and DOC soybean (50 per cent). There was no mortality of fish during the period of experiment conducted. The results indicated that fish feed prepared with poultry (feed II)/silkworm (feed III) waste were digested efficiently and more comparable to commercial fish feed (V) than the other feed (I and IV).

In the study, four types of byproducts/wastes, namely, rice bran, bakery, poultry and silkworm were used. Composition details of the byproducts/wastes are mentioned in Table 30.3. Rice bran is produced in great quantities as a by-product of the rice milling industry. Rice bran is nutritious, as it contains a higher content of protein, having better amino acid profile. However, the byproducts like poultry/silkworm/bakery are the better source of proteins and minerals, and are available in comparatively less cost in the market. Maize is a relatively poor cereal since it has limiting amounts of two essential amino acid, lysine and tryptophan (Hanifa *et al.*, 1987).

Table 30.3: Composition detail of Byproducts/Wastes.

Sl.No.	Parameters	Rice Bran	Silkworm Waste	Poultry Waste	Bakery Waste	Slaughter House Waste
1.	Protein (per cent)	15.3	54.2	59.9	10.7	64.5
2.	Fat (per cent)	15.92	30.3	17.1	12.7	27.6
3.	Crude fibre (per cent)	8.44	3.9	2.1	0.4	2
4.	Ash (per cent)	6.7	5.2	15.5	3.8	4.9
5.	Carbohydrate (per cent)	53.64	6.4	5.4	72.4	1.0

Among the protein sources of plant origin, soybean (*Glycine max*) is reported to have greater value as fish feed ingredient because of its easy availability, better chemical composition, amino acid balance and relatively low cost (Smith and Akiyama 1987). Soybean was found deficient in two essential amino acids- lysine and methionine (Viola, 1982). Raw soybean is also known to have certain anti-nutritional factor responsible for growth depression (Smith and Akiyama, 1987; Wilson and Poe, 1985). Studies on Essential Amino Acid (EAA) requirement of Indian major carps are already in progress using synthetic diets. Earlier, Singh (1984) observed that maximum total EAA level required in the synthetic diet for rohu fry was 27.4 per cent having 1.58 per cent methionine along with cystine and 3.83 per cent lysine. These are very similar to that used by Nose *et al.* (1974) for common carp, that is, total maximum EAA 28.95 per cent having methionine and cystine 1.76 per cent and 3.57 per cent lysine. This indicated that EAA requirement of rohu might be very similar to common carp.

The growth rate of rohu fingerlings was found to be significantly increased on addition of water-soluble and fat-soluble vitamins (Singh *et al.*, 1986). Four fat-soluble and eleven water-soluble vitamins and vitamin-like compounds are essential to fish. Minerals play a great role within the animal body. Minerals are important nutrients as these are required for normal bone, tissue, blood plasma and hemoglobin formations and also for many enzymatic reactions. Calcium (Ca) and Phosphorus (P) are required for the formation of skeletal tissues. Fishes absorb a good quantity of calcium from water and rest from diet but most of phosphorus should be provided in the diet. For common carp, the minimum requirement of calcium in diet is about 0.028 per cent and that of phosphorus is 0.6–0.7 per cent. Trace elements, required in traces, are growth stimulants. Elements like manganese, copper, iron, cobalt, iodine and zinc are required in minute quantities in balanced ration, mainly for improving protein assimilation and survival rates (Das and Kaviraj, 1994)). Fish obtain minerals from their diet and also from their environment.

Feeding Trial by State Fishery Department

In order to assess the quality of the developed feed (II) in natural condition and also to have an opinion from State Fishery Department on the feed, an experiment was conducted in natural ponds for 52 days by District Fisheries Department, Bhandara (DFDB). For the purpose, 100 kg of feed was provided to DFDB. The feeding trial experiment was conducted on two varieties of matured fish (*Labeo rohita* and *Catla catla:* 10 nos. each) in two natural ponds (area: each of 51m x 26m x 1m) at fish hatchery farm located in Iteadoh, District Gondia for 2 different types of feed (NEERI's feed and conventional feed).Both the feeds were dispensed into the respective ponds at the rate of three per cent of total fish body weight per day during the experiment. The results indicated 100 per cent increase in growth of fish fed by NEERI's feed over those fed by control feed (conventional feed). Further, release of eggs after artificial breeding quantified eleven per cent more in fish, grown on feed provided by NEERI, Nagpur. It may be mentioned that hatching percentage was 65 per cent in both of NEERI's and control feed.

Evaluation of Cost and Quality

It may be mentioned that production cost of the developed feed (Rs. 12/- per kg) was about two-third or less of the similar kind of feed, available in the market [Amrut fish feed, Aurangabad: Rs. 18/-(per kg); Fish feed from CIFE: Rs. 20/- (per kg); Tokyu, Japan: Rs. 500/- (per kg), etc.] during the commencement of experiment. The cost of the feed procured from CIFE was higher, since none of the ingredients (soybean 44 per cent, fish meal 10 per cent, corn meal 4.5 per cent, coconut oil 9.5 per cent, wheat flour 4.5 per cent, cod liver oil 2.0 per cent, sunflower seed 2.0 per cent, CMC binder 2.0 per cent and vitamin C 0.1 per cent) was waste-based. It may be added that a byproduct is a secondary and incidental product from a manufacturing process and is not the primary product. A byproduct or waste can be procured at a reduced price or free of cost, and be used as raw material of other product. The cost of feed based on prawns/shrimps is still higher. The cost of the developed feed was reduced by substituting animal wastes and byproducts instead of costly ingredients, normally being used in prevailing industry. The invented pelletized fish feed has essential nutrients for proper growth of fish.

Conclusion(s)

Present study dealt with different formulated fish feeds using select byproducts and nutritious food industry waste based materials. The cost of the developed feed appears to be about two-third of the similar kind of feed, available in the market. The study could provide the technology to local people for their employment generation, and pollution due to scattered nutritious waste materials, can be reduced substantially.

References

1. Ahmed, G.U., Haque, A.K.M., Islam, M.A. and Haque, M.M. (1983). Intensive culture of *Labeo rohita* (Hamilton) in floating ponds with special references to different stocking densities. Bangla. J. Fish., 6: (11-17).

2. Ahsan, M.N. (1995). Growth of rohu *Labeo rohita* (Ham.) fingerlings originating from hatcheries and those from wild sources fed on two formulated diets., M.Sc. thesis. Bangladesh Agricultural University. Mymensingh, pp: (59).

3. American Public Health Association (APHA) (1998). Standard Methods of examination of water and wastewater, Washington DC.

4. Capper, B.S., Wood, J.F. and Jackson, A.J. (1982). The feeding value for carp of two types of mustard oil cakes from Nepal. Aquaculture, 29: (373-377)

5. Choudhury, B. B. P., Das, D. R., Ibrahim, M. and Chakraborty, S. C. (2002).Relationship between feeding frequency and growth of one Indian major carp Labeo rohita (ham.) finerlings fed on different formulated diets.Pakistan journal of biological sciences 5(10):1120-1122

6. Course manual (2003). Biochemical Technology in Fisheries, Central Institute of Fisheries Education (CIFE), Mumbai

7. Das, B.K. and Kaviraj, A. (1994).Influence of Potassium Permanganate, Cobalt Chloride, and Dietary Supplement of Vitamin B Complex on the Histopathological Changes in Gill Epithelium of Common Carp Exposed to Cadmium *The Progressive Fish-Culturist* 1994; 56: 265-268

8. Das, K.M., Mohanty, S.N. and Sarker, S. (1991). Optimum dietary protein to energy ratio for *Labeo rohita* fingerlings. Fish Nutrition- Research: In Asia. Proceedings of the fourth Asian Fish Nutrition Workshop. Silva, S.S. de (ed.) No. 5: (69-73)

9. Hanifa, M.A., Murugesan, A.G. and Fleming, A.T. (1987). Influence of plant animal food on food utilization of the freshwater carp *Labeo rohita* (Ham.) Curr, Sci., 56: (846-848)

10. Nose, T.S., Arai, D.L., and Hashimoto, Y. (1974). A note on amino acids essential for growth of young carp. Bull.Japn.Soc.Sci.Fish. 40: 903–908.

11. Anonymous (1980). Official Methods of Analysis of Association of Official Analytical Chemists, 13th Ed., W.Hertiz, Washington, DC.

12. Rajan, M.R., Thilijath, B. and Thilagavathy, D. (1996). Feed utilization and growth of *Cyprinus carpio* var communis in relation to different protein levels. Environ. Ecol. 14: (60-66)

13. Rangachoryulu, P.V., Sarkar, S., Mohanty, S.N. and Mukhopadhyay, P.K. (1991). Growth and protein utilization in rohu (*Labeo rohita*) under different feeding levels. National Symposium on New Horizons in Freshwater Aquaculture: (95-97)

14. Singh, B.N. (1984). Nutrition and feed formulation for carps. Souvenir, Fourth Advisory Committee Meeting of NACA, FAO/UNDP Project 3–6 Dec. 1984, Bhubaneswar, Orissa, India, pp. 53–67.

15. Singh, B.N., Singh, V.R.P., Kumar, K and Swamy, D.N. (1986). Obserations on the feed formulations and fortification of conventional fish feed for rohu and mrigal fingerlings. Int. J. Acad. Ichthyol 7: 31-34.

16. Smith, R.R. and Akiyama, M. (1987). Soybean and wheat flour by-products in trout feeds. American Soybean Association 541, Orcherd Rd, 11-03 Liat Towers, republic of Singapore 0923, (8).

17. Swamy Vijaykumar, H.V. and Devraz, K.V. (1994). Growth response to fry of the fish *Catla catla* fed on three formulated feeds. Environ. and Ecol. 12 (3): (519-523).

18. Viola, S. (1982). Partial and complete replacement of fish meal by soybean meal in feeds for intensive culture of carp. Aquaculture 26: (223-226).

19. Wilson, R.P. and Poe, W.E. (1985). Effects of feeding soybean meal with varying trypsin inhibitor activities on growth of fingerling channel catfish. Aquaculture 46: (19-25).

Chapter 31

Performance Validation of an Innovative Instrument: Total Organic Carbon (TOC) Analyzer

S.K. Lokhande[1], A.P. Vanerkar[2], J.B. Helonde[1], D.M. Dharmadhikari[1] and N.P. Thacker[1]

[1]CSIR-National Environmental Engineering Research Institute, Nehru Marg, Nagpur

[2]National Institute of Miners' Health (NIMH), JNARDDC Campus, Nagpur

ABSTRACT

Total organic carbon analysis is becoming very important technique for many aspects of environmental monitoring. Total organic carbon (TOC) analyzer is highly automated and sophisticated microprocessor based system. For quick and accurate measurement of pollution load in water and wastewater this technique is most popularly used worldwide. Hence it becomes very crucial to generate valid and accurate data to determine quality of water. As with all measurements made by Total Organic Carbon (TOC) analyzer, there are many variables that affect the total response of the system. The two main variables to be verified for all technologies are oxidation and detection technique. TOC measurements must be compliance with USEPA regulations. The validation study in terms of accuracy, reproducibility, oxidation efficiency, correlation coefficient of calibration curve becomes essential for any TOC analyzer. This chapter envisages the methodology developed in our laboratory for performance validation of TOC analyzer to ensure quality data production.

Keywords: *Validation, Accuracy, Reproducibility, Oxidation efficiency.*

Introduction

Ground and surface water is the prime need for human beings. Disposal of waste generated from industries and municipal sewerage system contaminates the natural water bodies and needs prior treatment before discharging which is depending on the nature of pollutants. The load of carboneous material, determines the pollutional strength of water and wastewater. Chemical Oxygen Demand (COD) and Biochemical Oxygen Demand (BOD) are the conventional methods, which are routinely used to measure the pollutional load of wastewater from industries and the wastes of domestic origin (1). However, BOD and COD tests are cumbersome, time consuming and is affected by the other interfering elements such as Hydrogen, Nitrogen, Phosphorus, and Sulphur. TOC is becoming increasingly important as screening technique for many aspects of environmental monitoring (2, 3). The principal of working of TOC analyzer is that the organic matter is oxidized to carbon dioxide, which is then estimated by various analytical techniques. It has to be considered that TOC depends on the types of considered water, but it also affected by several parameters such as temperature, salinity, pH, microbial activity, surrounding vegetation; as a consequence, TOC value is highly variable (4). However all instruments must go through a process shown in Figure 31.1.

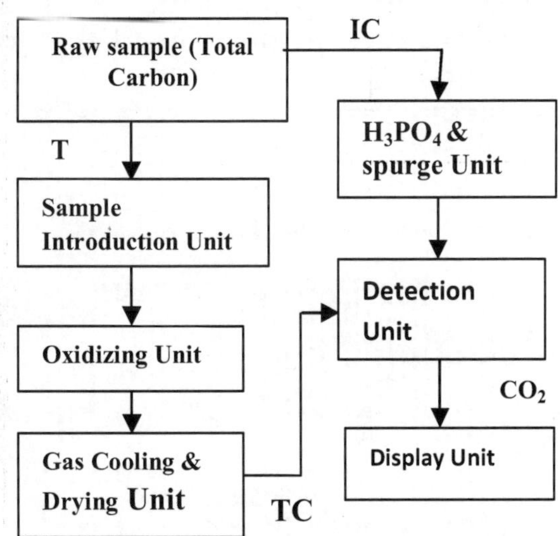

Figure 31.1: Representation of Basic Units of TOC Analyzer.

Modern TOC analyser is used to measure Total Carbon (TC), TOC, Inorganic Carbon (IC), Purgeable Organic Carbon (POC), and Non-purgeable Organic Carbon (NPOC). Among the detectors used in TOC analyzers, the most common detector is the Non Dispersive Infra Red (NDIR) detector. Generally, the commercially available instrument can operate in the near IR (800 to 2500 nm) or in the mid IR (2500 to 15384 nm) (5). Sensitive to the absorption frequency of CO_2, the NDIR generates a non-linear signal that is proportional to the instantaneous concentration of CO_2 in the carrier gas. That signal is then liberalized and integrated over the sample analysis time. The resulting area is then compared to stored calibration data and a sample concentration is calculated.

The TOC analyzer in the central facility is used for the performance validation study. It can be effectively used to detect total organic carbon in liquid samples (6). Depending on sample pre-treatment and required parameter TC, TOC, NPOC, and IC can be measured. It is equipped with liquid auto sampler. Automatic calibration lines build up and regular checks are possible. The software fully controls the operation of the instrument.

The instrument analyses the sample under controlled conditions, using separate inert and combustion zones in the furnace to promote complete combustion of organically bounded carbon-to-carbon dioxide. With a boat injection the sample is introduced into the furnace. During the movement to the furnace, the sample is flushed with argon (carrier gas). The volatile component will go directly to the furnace where the sample oxidizes with oxygen. The gas flow is led through a copper oxide scrubber and transferred to the IC scrubber. The gas is dried in a perma pure dryer, and passes via particle filter. After this process, the gas flows into the NDIR detector, where the CO_2 concentration is measured in ppm. TOC concentration will be calculated on the basis of area of the signal.

The working of the instrument can be summarised by the flow diagram shown in Figure 31.2.

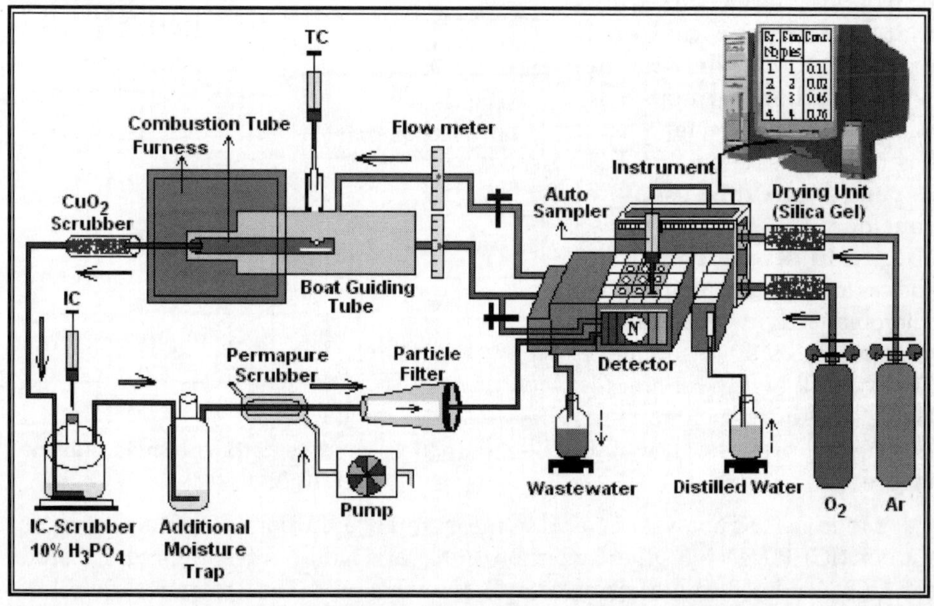

Figure 31.2: Schematic Diagram of TOC Analyzer.

In present study an attempt has been made to perform the validation study of TOC analyser to determine the performance of the instrument in terms of accuracy, repeatability and oxidation efficiency at variable temperatures, since before reporting validation of data becomes very crucial for many aspects of environmental monitoring

Materials and Methods

Stock solutions are prepared using analytical grade Potassium Hydrogen Phthalate (KPH) and Sodium Carbonate (Na_2CO_3) chemicals and working standards were accurately prepared by stock solutions and two different calibration lines are drawn for TC and IC. Calibration curves with correlation coefficient obtained are shown in Figures 31.3 and 31.4 respectively. Standard of 100 ppm TC and IC is used throughout the experiment as a reference known concentration to check performance and validation of the instrument. Triplicate readings were obtained for known concentration to determine accuracy and repeatability of the instrument.

To determine the oxidation efficiency/per cent recovery of TOC analyzer, combustion temperatures are varied from 200 ^{0}C to 1000 ^{0}C in increasing steps. For each given temperature settings first analysis is performed with the empty boat then known concentration is injected and analysed. Thus the interference of the previously injected samples to the next sample due to incomplete combustion at lower temperatures is prevented. This experiment is repeated with the presence of the catalyst and without catalyst shown in Figures 31.5 and 31.6 respectively.

Results and Discussion

Results obtained for assessment study of TOC analyzer relative to the parameters *i.e.* accuracy and repeatability are shown in Table 31.1.

Table 31.1: Representation of Accuracy, and Repeatability.

Standard Concentration 100 ppm	*Measured Value (ppm)*	*Mean*	*SD*	*RSD*
TC	93	97.66	7.234	7.40
	94			
	106			
IC	110	112.33	2.51	2.23
	115			
	112			

It can be observed that instrument yields good results for the KHP standard. The overall average relative accuracy over a period of seven days for total 11 measurements for KHP and Na_2CO_3 were found 108.5 per cent ± 14.17 per cent with a range of 93 per cent–138.8 per cent and 118.9 per cent ± 7.6 per cent with a range of 110 per cent–135 per cent respectively. The repeatability obtained also passes the norms which is 10 per cent Relative Standard Deviation (rsd) for a triplicate. The study conducted to determine the oxidation efficiency at different temperatures with the presence of catalyst and without catalyst. In both the study no peaks were found up to 200 ^{0}C but at 400 ^{0}C and onwards instrument produces significant corresponding peaks. Peak height and peak area were increasing progressively with the increase in temperature, which represents that percentage recovery is increasing with temperature. In the study it is also found that above 600 ^{0}C baseline of the instruments shifts slightly up side that is due to addition of oxidation of impurities in the carrier and oxidant gases which was unable to oxidised below 600 ^{0}C. The results show that oxidation efficiency is in the range of 84 per cent to 90 per cent for KHP standard at 600 ^{0}C while it reaches to more than 100 per cent above 600^{0}C. The peaks obtained to study oxidation study at different temperatures with and without catalyst are shown in Figures 31.5 and 31.6.

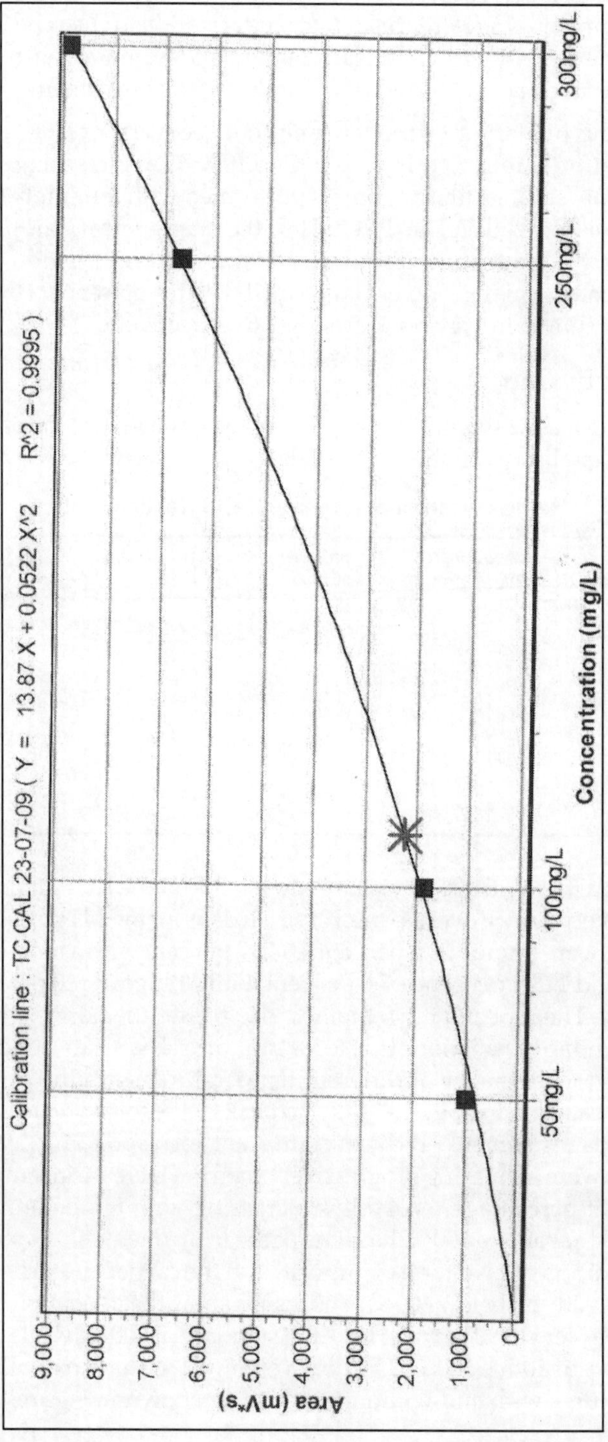

Figure 31.3: Calibration Curve for TC.

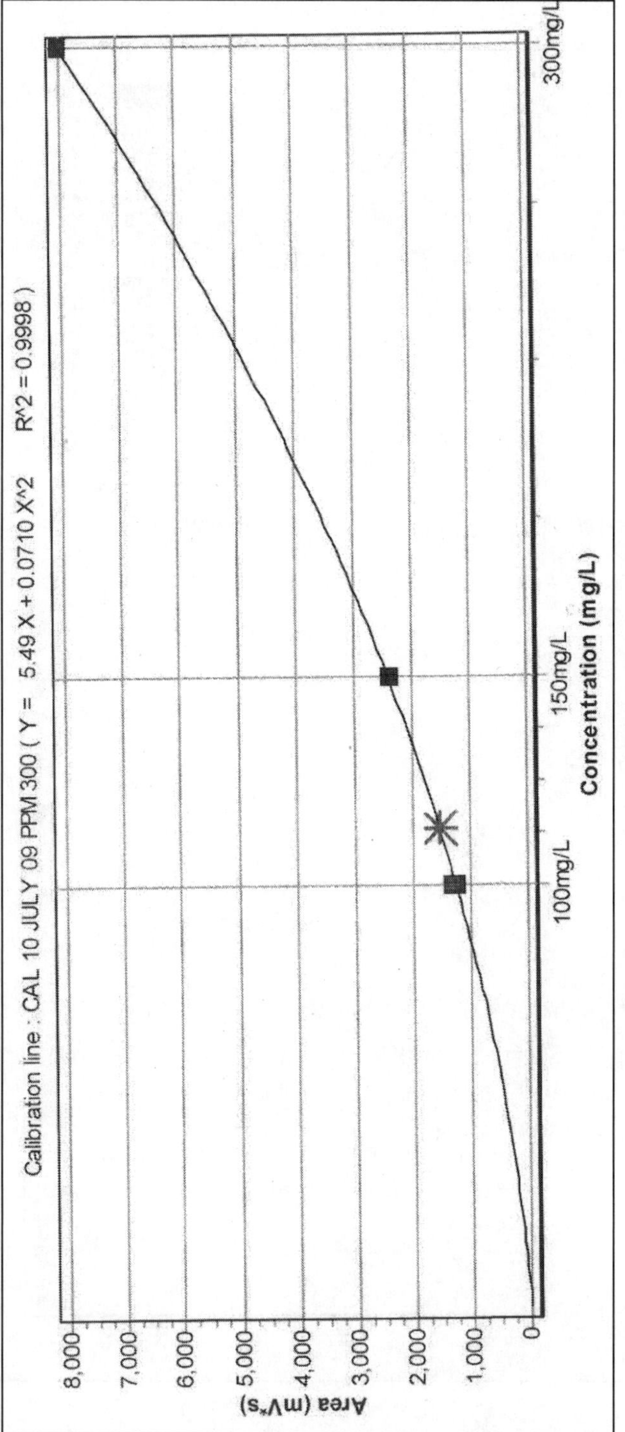

Figure 31.4: Calibration Curve for IC.

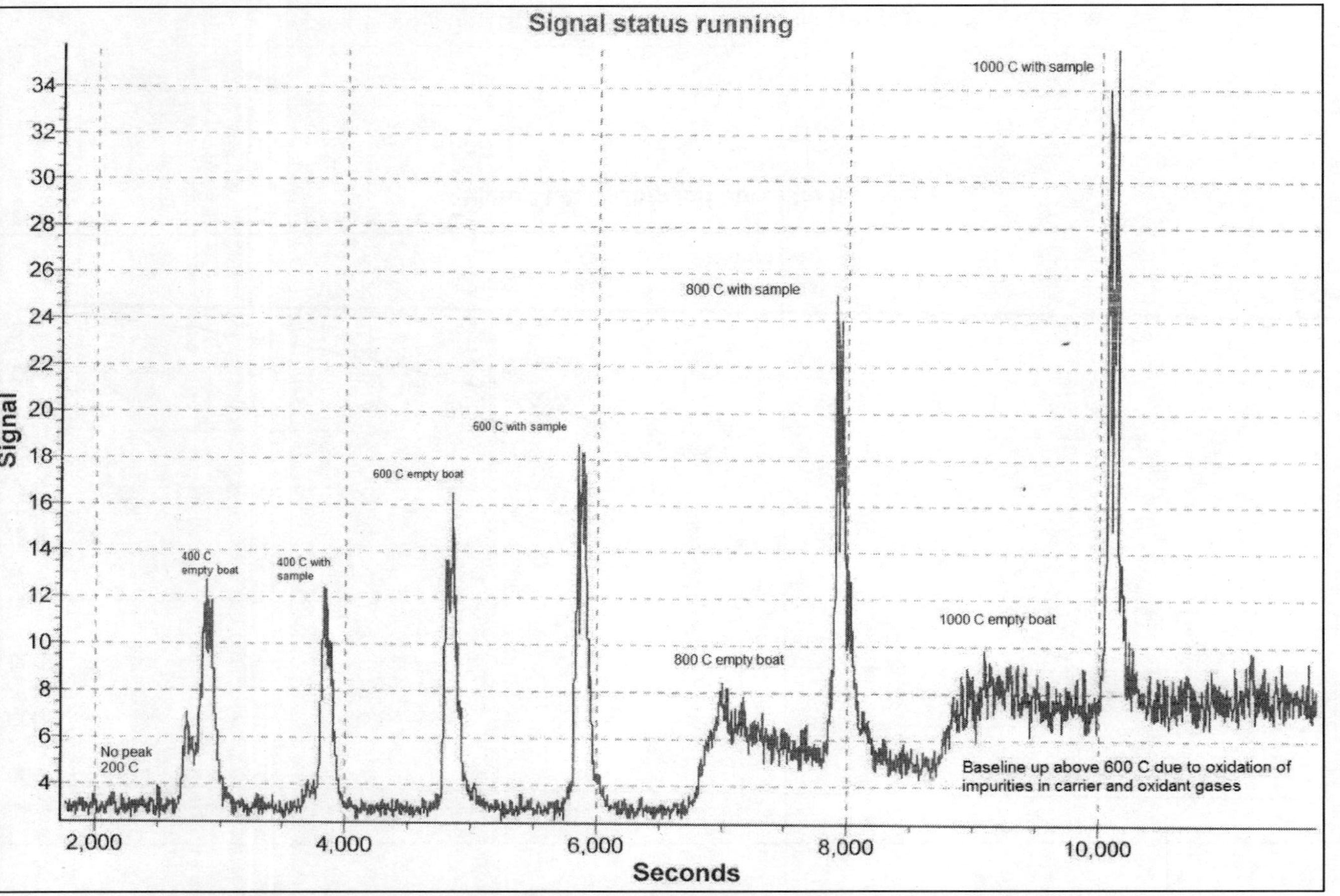

Figure 31.5: Peaks Obtained with Catalyst for 100 ppm KHP at Different Temperatures.

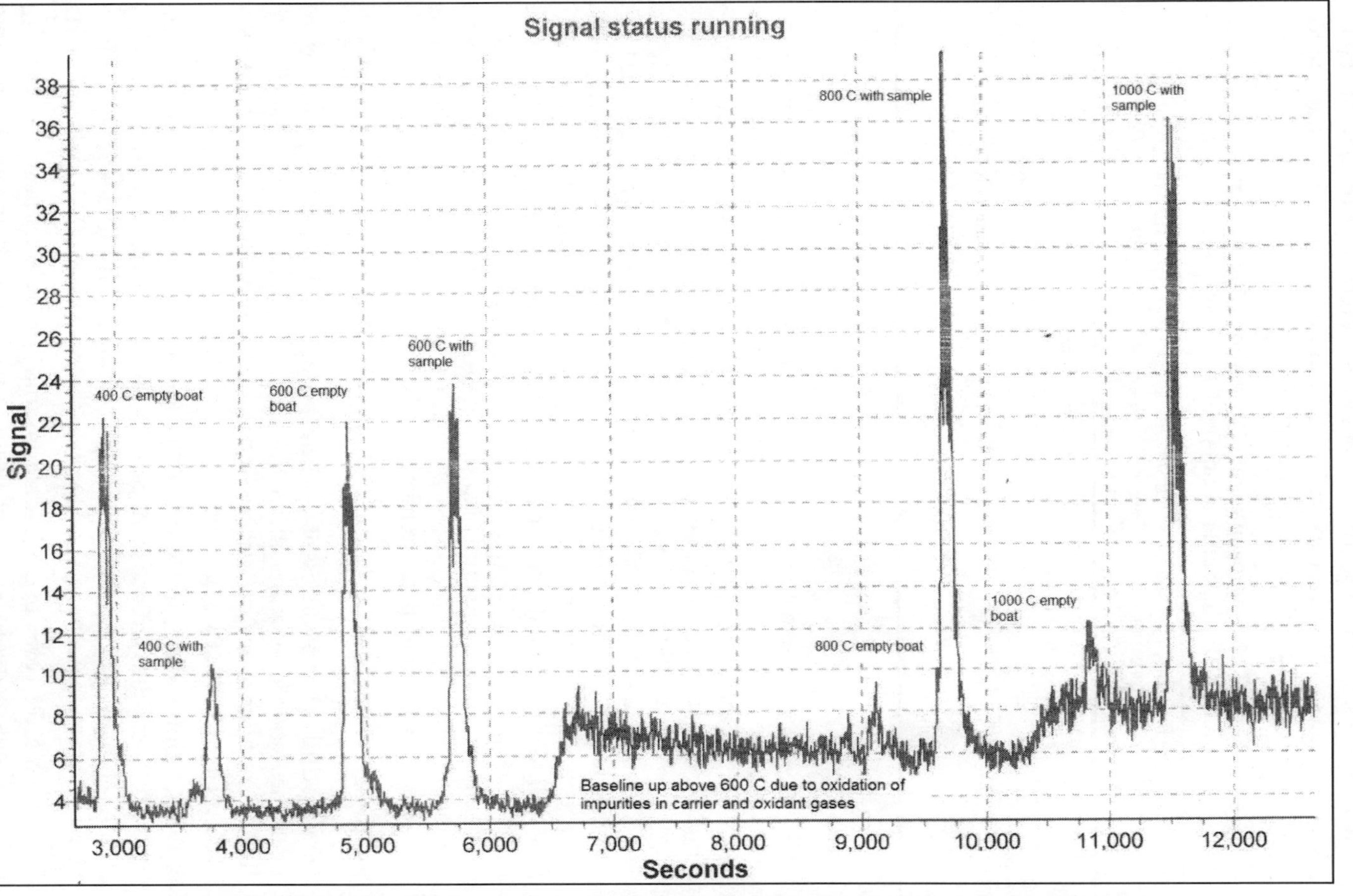

Figure 31.6: Peaks Obtained without Catalyst for 100 ppm KHP at Different Temperatures.

Study also reveals that for KHP standard solutions results obtained using catalyst percentage recovery is better up to 600 °C when compared to the results obtained in the absence of catalyst. However above 600 °C both experiments show more or less similar oxidation efficiency as summarized in Table 31.2.

Table 31.2: Representation of Oxidation Efficiency of KHP Standards at different Temperature.

Sl.No.	Combustion Temperature in °C	Oxidation Efficiency Obtained in Per cent with Catalyst	Oxidation Efficiency Obtained in Per cent without Catalyst
1.	200	4.54	2.33
2.	400	55.87	49.51
3.	600	84.75	82.82
4.	800	103.44	102.83
5.	1000	105.32	103.31

Conclusion(s)

TOC experimental and theoretical values are obtained and the comparison between these two values shows difference not higher of 20 per cent and compliance with the USEPA certification range for relative accuracy. The linearity/regression value of the calibration curves obtained for TC and IC are $R^2 = 0.9995$ and 0.9998 respectively shows reliable results can be obtained with the instrument. Oxidation efficiency is poor and slightly improves in the presence of catalyst up to 600 °C. However above 600 °C both experiments show more or less similar oxidation efficiency. This may be reason for combustion technique commercially available TOC analyzers uses heat (680 °C) or higher. It is considered that any organic matter may be converted to carbon dioxide; contamination is a constant threat to the accuracy of TOC measurement. Obtaining high purity reagents, gases and solvent water is therefore a paramount importance in making accurate measurement. The quality of the blank is crucial to ensuring accurate measurement.

References

1. Dharmadhikari D M. Present position, progress and future prospects of instrumentation for quantification of pollution load in industrial effluents. Indo Italian Conference on Green and Clean Environment, March 20-21, MAEER's MIT College of Engineering, Pune.
2. Visco G, Campanella L, and Nobili V. The International Norms for the Measure of the TOC and the Carbons in Waters Chronological and Critical Analysis. XI Italian - Hungarian Symposium on Spectrochemistry, New Challenges in Human Health Protection Anthropic and Remote Areas. October 20-23, palazzo Zorzi.
3. Urbansky E T. Total organic carbon analysis as a tools for measuring carbonaceous matter in natural waters, Journal of Environmental Monitoring, February 2001, 3(1), 102–112.

4. Canadian water quality guidelines, "parameter specific background information", Canadian Council of Research and Environmental Ministers, 1996.
5. Banwell C N and McCash E M. Fundamentals of molecular spectroscopy, IV edition, McGraw-Hill Publishing Company, England, 1994.
6. User manual TOC 1200 Eng. V4.0 okt2001.

Chapter 32

Novel Environmentally Friendly Silicone Based Waterborne Polyurethane Coatings for Corrosion Protection of Aluminum Alloys

Akanksha Sharma

Corrosion Science and Engineering,
Department of Metallurgical Engineering and Materials Science,
Indian Institute of Technology Bombay, Mumbai

ABSTRACT

Paint and coating industry has been continually undergoing changes in technology due to changing consumer's expectations regarding improvement in corrosion protection, appearance and most importantly environmental protection. The diverse effects of organic solvent and metal chromates on human health have forced the paint industries to move towards new "compliant" coating technology involving high solids coatings, waterborne coatings, radiation curable coatings and powder coatings. Lion's share of research and development is going on for improving the performances of environmentally friendly waterborne coatings that utilize water as the main volatile liquid component. Hybrid inorganic-organic sol-gel coating technology involves a synergistic combination of inorganic metal alkoxides like silicones, which are environmental friendly, and organic polymers (acrylics, alkyds, polyesters, epoxies, polyurethanes, and vinyl), thus combining the properties typical of both the constituents. Thus, some inferior properties of waterborne polyurethanes such as low mechanical strength and thermal stability can be improved by this technology. In the present chapter, formulated coating materials and their precursors were characterized for their structural analysis, thermal behavior and light sensitivity using Fourier Transform

Infra-Red Spectroscopy (FTIR), Thermogravimetric Analysis (TGA) and UV-visible spectroscopy, respectively. Surface morphology of coated substrates was characterized by Scanning Electron Microscopy. Corrosion resistance properties of the coatings were evaluated using Tafel polarization studies. Pencil hardness, impact test and bend test was done to determine the adhesion, hardness and flexibility of the coatings. Results showed that the coatings on the substrates were smooth, continuous, crack free and had excellent adherence to aluminum substrate through chemical bond. Also the coatings showed good corrosion resistance against the uncoated bare aluminum substrates. Thus, an attempt was made to develop an environmentally friendly single step coating that combines the functions of conversion coating, primer and top coat, with improved corrosion resistance and thermal and mechanical properties, to broaden the scope of polyurethanes in aerospace, automotive and marine applications.

Keywords: *Waterborne coatings, Corrosion, Polyurethane, Silicone, Aluminum.*

Introduction

Corrosion[1] is a destructive attack on materials which leads to loss of their engineering properties as well as aesthetic value. Aluminum and their alloys undergo pitting and stress corrosion cracking under continued exposure to corrosive agents such as chloride ions, dissolved oxygen, moisture, SO_2 and NO_X [2]. Restriction on migration of these corrosive agents towards aluminum substrates by coating can be a means of corrosion protection. The paint and coatings industry is a mature industry that has been undergoing a continual change in technology. Development of raw materials for coatings is dominated by three major driving forces, which are improvement in corrosion protection properties, improvement in appearance and most importantly, protection of the environment. Conventional coatings for aluminum and their alloys are chromate based coatings having [3] excellent cost/efficiency ratio. Chromate conversion coatings provide very good adhesion and corrosion protection to aluminum, however, they are also carcinogenic and endocrine disrupters and are regarded as very hazardous soil and groundwater pollutants [4]. In addition, the organic solvents present in the coatings are toxic to human beings and their minor inhalation may cause nausea, vomiting, headache, dizziness etc. Over the years, the use of conventional organic solvent based paints has come under increasing environmental pressure. More recently, the toxicity of these solvents to humans has also become an area of concern. In addition to voluntary reductions in waste, the introduction of various forms of environmental regulation has forced the paint and coatings industry to move towards zero waste generation.

The solutions that address the diverse effects of organic solvent and metal chromates on human health pertain to the development of new "compliant" coating technology involving high solids coatings, waterborne coatings, radiation curable coatings and powder coatings. Waterborne coatings [4] utilize water as the major carrier liquid. Lion's share of research and development is being done for improving the performance of waterborne coatings and now this system is found to perform at a level equal to or even superior to solvent borne coating systems. Almost all types of resins are now available in waterborne formulations, which include acrylics, alkyds, polyesters, epoxies, polyurethanes, and vinyl — meaning there are a wide array of waterborne solutions available to meet specific needs. These coating alternatives can reduce emissions of VOCs and, in so doing, reduce the generation of hazardous wastes and decrease worker exposure to toxic air emission.

The resins commonly used in conventional solvent borne coatings are usually modified by increasing their hydrophilic nature prior to being incorporated into a waterborne coating formulation. The advantages exhibited by waterborne formulations include low viscosities, very low VOC content, reduced flammability, less odor, easy to clean up, long term durability, easy application using conventional equipment and protection in most applications. Waterborne coatings also provide significant advantages over high solids coatings in thin film applications.

Polyurethanes are the newest and potentially one of the largest polymer families used in coating industry and have found use in the aerospace and wood industries. Polyurethane coatings are known for their excellent outdoor stability and fast curing. Today, there is a range of high performance waterborne polyurethane coatings, and development is still in progress. However, their thermal and mechanical properties have limited their applications, so they need some modifications.

Sol-gel coatings on aluminum have shown excellent adhesion and exceptional barrier properties against chemical and mechanical stresses. Sol-gel coating formulation involves a synergistic combination of metal alkoxides like silicones (R′-Si-OR) and organic polymers (acrylics, alkyds, polyesters, epoxies, polyurethanes, and vinyl) by hydrolysis and condensation of molecular precursors in aqueous or non-aqueous medium. Silicones [5] are conducive to environment and are widely used as adhesion promoters, surfactants and hydrophobic agents in various coating formulations. Various attempts have been made to develop silicone based environmentally acquiescent conversion coatings [6]. Recently, silanes as alkoxysilanes have shown wide concern for improving corrosion resistance, adhesion and mechanical properties as well as resistance to heat, moisture and ultra-violet degradations of conventional organic resins used for coating formulations [7-13]. A number of literatures [14-15] are available on the modifications of conventional organic resins such as acrylics, alkyds, polyesters, epoxies, polyurethanes, and vinyl with silicones. Silicone modified conventional resins offer the prospects of combining the mechanical toughness and flexibility of the organic resin components with the hardness, UV and thermal stability of the silicone components [16-17].

Polyurethanes modified with silicones can be used to formulate coatings having high hardness with excellent flexibility, thermal stability, better adhesion and corrosion resistance. In recent years, great efforts have been made to develop a true single step coating combining the functions of conversion coating, primer and topcoat.

Chen *et al.* [23] prepared polyester/silica hybrid resins and polyester/silica/polyurethane hybrid via *in situ* sol-gel process and simple blending method using different silica sols. It was found that both the methods formed agglomerates of silica-rich phases and primary the silica-rich phase in the hybrid films, however, the *in situ* sol-gel process resulted in enhanced interaction between silica phase and the polymer as well as among the silica phases themselves resulting in much bigger agglomerates and compact structure. Furthermore, the extent of interaction among polyester, polyurethane and silica increases with the number of ethoxy groups at silica. Yang *et al.* [24] synthesized waterborne polyurethane and added varying amounts of colloidal silica from 0 to 50 wt per cent to prepare waterborne polyurethane–silica hybrids. Results showed much better thermal stability and mechanical properties of hybrid films than pure waterborne polyurethane resin indicating effective interaction between silica particles and urethane polymer chains. Prabu *et al.* [25] studied the thermal and morphological properties of silicone-polyurethane-epoxy resin system and found the formation of intercrosslinked networks. It was reported that silicone modified polyurethane-epoxy matrice showed improved thermal stability than an unmodified epoxy (diglycidyl ether of bisphenol-A) followed by polyurethane modified epoxy resins. Jeon *et al.* [26] prepared water-dispersible polyurethane based hybrid composites through sol–gel process. Here also it was found that with increasing silica content, thermal, mechanical properties and water resistance of the composite improved due to chemical network formations between PU and silica.

In recent years great efforts have been made to develop a true single step coating combining the functions of conversion coating, primer and top coat. The proposed work deals with the development of silicone incorporated waterborne polyurethane coating with improved corrosion resistance and thermal and mechanical properties to broaden the scope of polyurethane in aerospace, automotive and marine applications.

Materials and Reagents

3-glycidoxypropyltrimethoxysilane (GPTMS) and methyltrimethoxysilane (MTMS) were the silicones received from Dow corning. HMMM and paratoluenesulphonic acid (p-TSA) catalyst were purchased from Kings Industries and were used in as received form. Waterborne polyurethane, UD-322, (PU) was also received from Kings Industry, USA. The aluminum alloy AA6011 substrate used for coating was degreased in toluene and methanol and deoxidized in 1.0 M NaOH solution for 1-2 min before use.

Experimental

Coating Formulation and Film Deposition

Primary sol was prepared by addition of silicones (MTMS and GPTMS) in molar ratios of 1:2 to the aqueous solution of 0.05M acetic acid [27], followed by stirring at a rate of 100 rpm for 50h at 30°C. The silicone-polyurethane coating (SiPU) solutions were formulated by addition of conventional waterborne polyurethane resin in various concentrations (10, 20, 30, 40, 50, 60, 70, 80 and 90 wt per cent) and 25 wt per cent of HMMM, as reported in our previous study [16] to the primary sol. The coatings were designated as SiPU XX1YY1 where; Si stands for silicone in the coating, PU stands for polyurethane resin in the coating, XX1 stands for wt per cent of silicone in coatings and YY1 stands for wt per cent of polyurethane resin in coatings. For example, SiPU 4060 stands for coating composed of 40 wt per cent silicone (MTMS and GPTMS in

1:2 molar ratios) and 60 wt per cent of polyurethane resin. 1 wt per cent p-TSA was also added for fast curing and catalyzing the sol-gel reaction. Cleaned aluminum alloy substrates were dipped in coating solution for 1 min and placed in furnace to cure at 180°C for 45 min. The average dry coating thickness was found in the range of 5 ± 0.4μm using Elektrophysik coating thickness gauge (Model ExactoFN, Type 1800202).

Characterization of Structural Features of the Coatings

Coatings were characterized for their structural and morphological features using Fourier Transform Infra-Red spectroscopy (FTIR) and scanning electron microscopy (SEM) respectively. FT-IR spectra of coated aluminum substrates were measured using specular reflectance accessory at 60° angle. All spectra were recorded between 4000 and 400 cm^{-1} wave numbers using Magna-IR TM spectrometer 550 spectrometer (Nicolet) equipped with a DTGS detector and averaging of 32 scans. SEM (Hitachi S3400N) were used to see the surface morphology of coated substrates.

Thermal and UV Stability

Thermal stability of coatings in the temperature range of 20°C-700°C was studied using thermogravimetric analysis (DQCQ10) at a heating rate of 10°C/min under nitrogen atmosphere. UV sensitivity of PU and coating solution was examined using Jasco V-570 UV/VIS/NIR spectrophotometer.

Corrosion Resistance

The corrosion resistance of bare and coated substrates was evaluated by means of potentiodynamic polarization studies. For this purpose potentiodynamic polarization tests were conducted by polarizing the working electrode in anodic and cathodic direction with respect to the free corrosion potential at a scanning rate of 5mV/s. The exposed surface area, for a volume of 200ml of aqueous extract, was 1 cm^2. The electrochemical system was completed with a platinum counter electrode and a saturated calomel reference electrode. Corrosion current (Icorr), corrosion potential (Ecorr) and polarization resistance (Rp) values were calculated.

Pencil Hardness and Adhesion to Substrate

Pencil hardness test (ASTM D-3363), impact resistance test (BS 3900) and bent test on coated substrates were carried out to determine the adherence of the coating to the substrate, flexibility and their load bearing capability under tension or compression.

Results and Discussion

Structural Characteristics

The polymer network formation of silicone modified polyurethane system was ascertained from FT-IR spectra (Figure 32.1). The reaction between hydroxyl groups on silanol and aluminum ($Al(OH)_x$) substrate is confirmed by the disappearance of peak at 2850 cm^{-1} of–$Si–OCH_2CH_3$ and formation of Al–O–Si at 936 cm^{-1} [38]. All spectra of coated aluminum show an absorption band at 1036 cm^{-1} due to the Si–O–

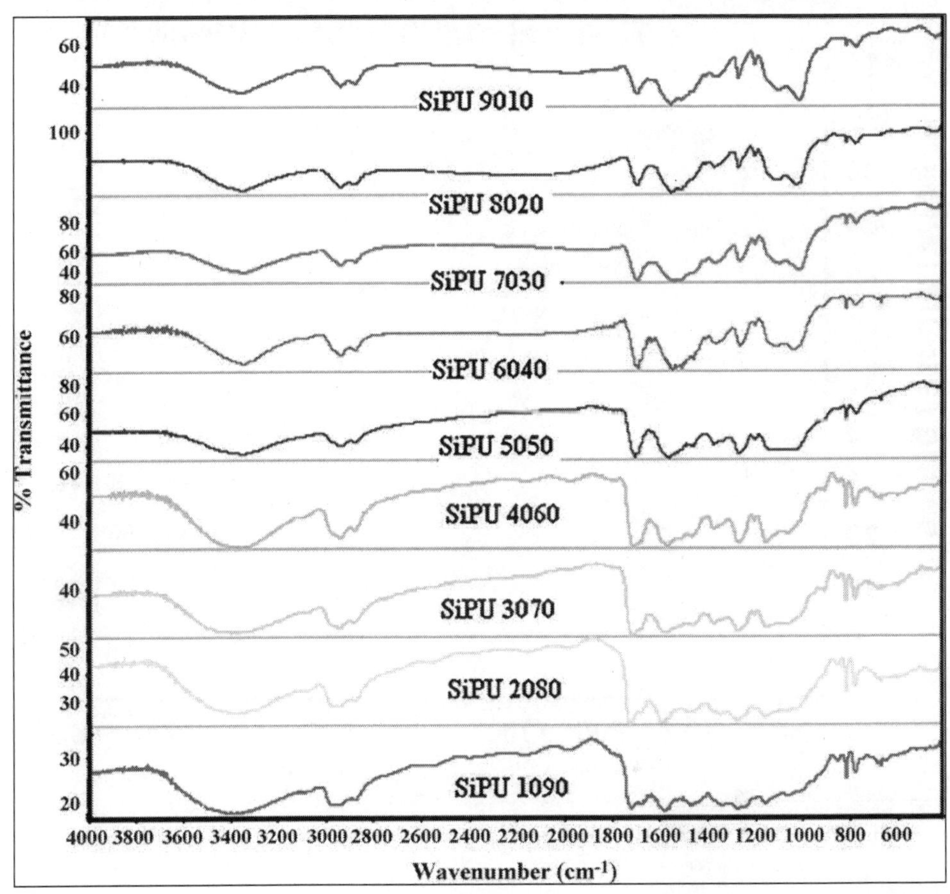

Figure 32.1: FT-IR Spectrum of Silicone Incorporated Waterborne Polyurethane.

Si linkages [34] and the intensity of this peak increases with increasing silicone concentration in the coating. Thus, the resins in the coatings have chemical (covalent bond) and physical (hydrogen bond) interaction amongst themselves and the coating is also covalently bonded to aluminum substrate through Al-O-Si bond.

Thermal Behavior of the Coatings

Thermogravimetric analysis were proposed to determine the thermal stability of coating materials. Figure 32.2 compares the TGA curves of PU and SiPU coatings in the temperature range of 30-700°C under nitrogen atmosphere. The degradation of PU started at 288°C which is likely due to presence of thermally weak urethane linkages in the coating [39]. It was found that the modification of PU by silicones improved the thermal stability and increased the degradation temperature. The presence of siloxane network (-Si-O-Si-) in the coatings delays the degradation process since high thermal energy is required to attain the same per cent weight loss than that required for PU. The delay in degradation caused by silicone moieties may also be attributed to its ablative behavior imparted by partial ionic nature, high bond energy

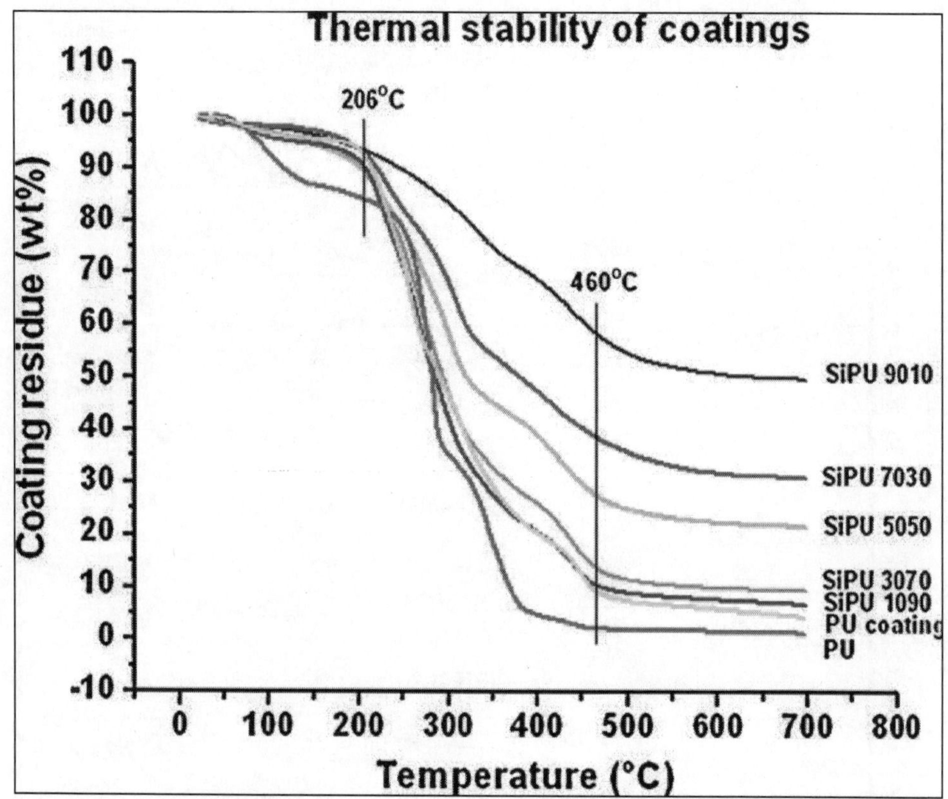

Figure 32.2: TGA Curves of PU and SiPU Coatings Incorporated Sol-Gel Coating.

and thermal stability of–Si–O–Si–linkages [40]. All the weight losses took place in three steps; the first step with gradual weight loss from room temperature to around 200°C is due to the evaporation of physically absorbed solvents and residual monomers, the second weight loss in the range of 200-450°C is ascribed to the decomposition of polymers and the third minor weight loss at 450-700°C may be caused by the partial decomposition of silicone entrapped polymeric segments. The weight loss in the interval of 200–300°C is attributed to the decomposition of aliphatic n-propyl segments decomposition of the glycidoxypropyl units [41]. The percentage weight loss between 210 to 500°C agrees well with the relative amounts organic resins present in the starting compositions. Thus, the thermal studies revealed that coatings were thermally stable up to 200°C.

Surface Morphology of the Coated Substrates

FT-IR studies of coated substrates suggested the possibility of covalent bonding between GPTMS, MTMS, HMMM and PU resins. Presence of chemical linkage decides the phase homogeneity of coating [34].

Figure 32.3 shows the SEM micrograph of SiPU 9010, SiPU 5050 and SiPU 1090 coated aluminum alloy which reveals that the coating is uniform and homogeneous.

Figure 32.3: SEM Micrograph.

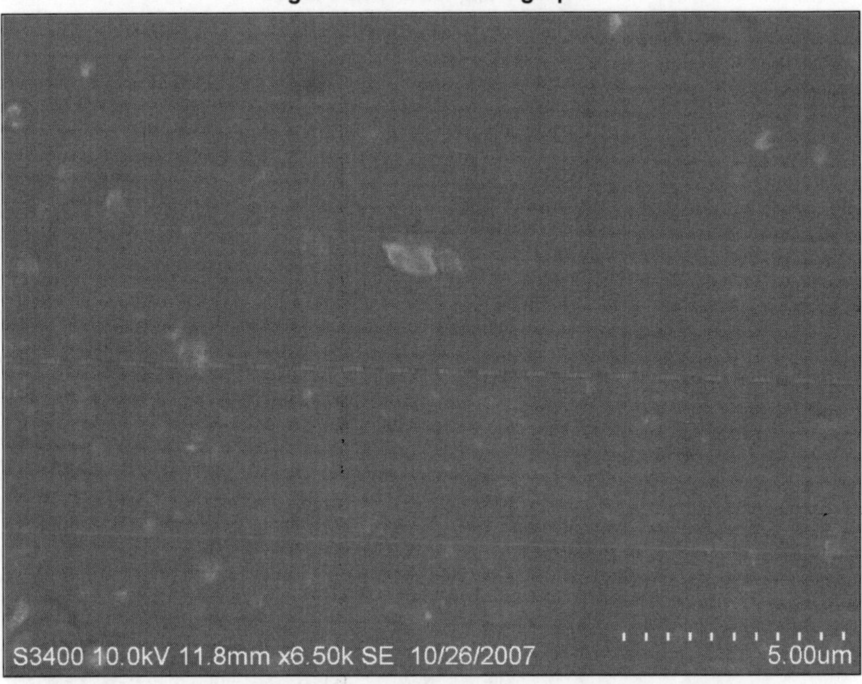

(a) SiPU 9010

(b) SiPU 5050

Contd...

Figure 32.3–*Contd...*

(c) SiPU 1090

However, it could be seen that some particles at the surface of the coating still existed. Easton *et al.*, also observed some white polymeric particles in PU coating. These particles are probably the unreacted resin or resin connected through hydrogen bond formation [43].

Corrosion resistance

Figure 32.4 present potentiodynamic polarization measurements used to assess barrier and corrosion protection of silicone coating, PU coating and SiPU coatings. The corrosion current values of SiPU coated substrates (1.21-2.71$\times 10^{-6}$A) were observed in between the corrosion current value for silicone coating (9.07 $\times 10^{-7}$A) and PU coating (3.1 $\times$ 10^{-6} A). The corrosion currents of the coatings were about one order of magnitude lower than that of bare aluminum substrate (9.2$\times$ 10^{-6} A). The lower corrosion current value of SiPU coated substrates than PU coating can be attributed to the improvement in corrosion resistance of polyurethane on silicone modification. This shows that incorporation of silicones improves the corrosion resistance properties of PU. Increased corrosion resistance of coated substrates is possibly attributable to the production of denser microstructure caused by Si-O-Si network in the coating. Furthermore, aluminum-oxygen-silicon (Al-O-Si) interface formed by surface hydroxylation reaction between silanol (-Si-OH) and aluminum substrate ($Al(OH)_x$) acts as a geometric blocking layers against chloride ions and water [44].

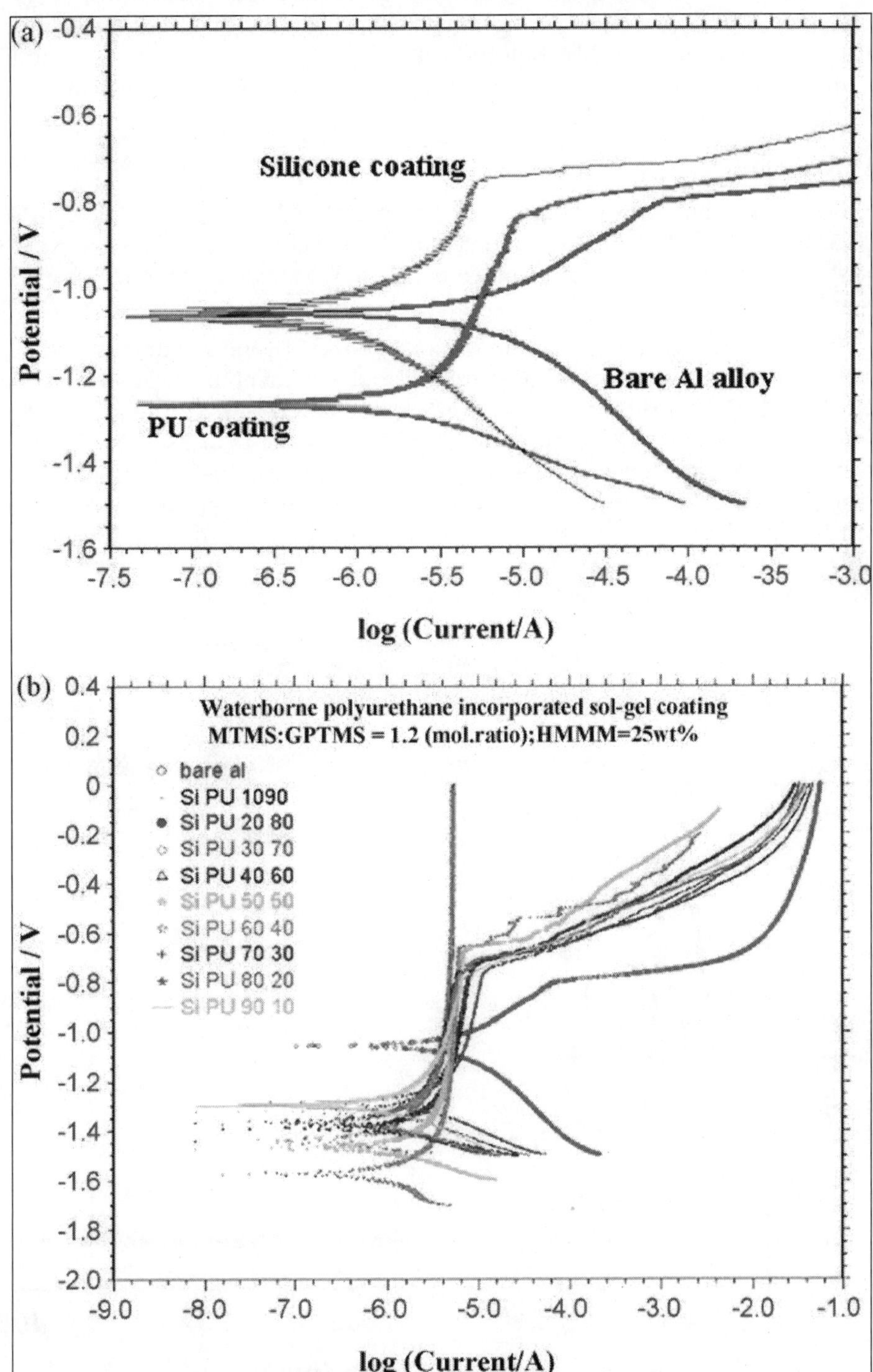

Figure 32.4: Potentiodynamic Polarization Curves of Bare and Coated Aluminum Alloys in 3.5 per cent NaCl Solution.

It was also observed that the corrosion potential of PU and SiPU coated substrates shifted towards cathodic direction indicating effective suppression of cathodic reaction. Thus, silicone modification improves the corrosion resistance of polyurethane.

UV Resistance

Figure 32.5 shows the UV-visible spectra of the SiPU 5050 coating solution, PU and PU with HMMM (PU + HMMM) evaluated using UV-visible solution and PU + HMMM show strong spectrophotometer. PU shows a broad absorption band at around 204 nm.

The SiPU 5050 coating absorption bands around 218 and 213 nm respectively which were probably due to carbon-nitrogen double bond of the melamine units of HMMM and HN-C=O moieties of PU. It is believed that HMMM forms a number of different radicals due to homolytic cleavage of valency bonds and hydrogen abstraction during photolysis The labile methoxy end groups (–O–CH_3) in the HMMM molecule are considered to be responsible for photodegradation [46]. These photoactive moieties are initiation centers for photo degradation, if any, of resins in coatings. There was no absorption peak up to 277 nm. Radiations emitted by the sun having wavelength below 280nm is almost completely absorbed by atmosphere and

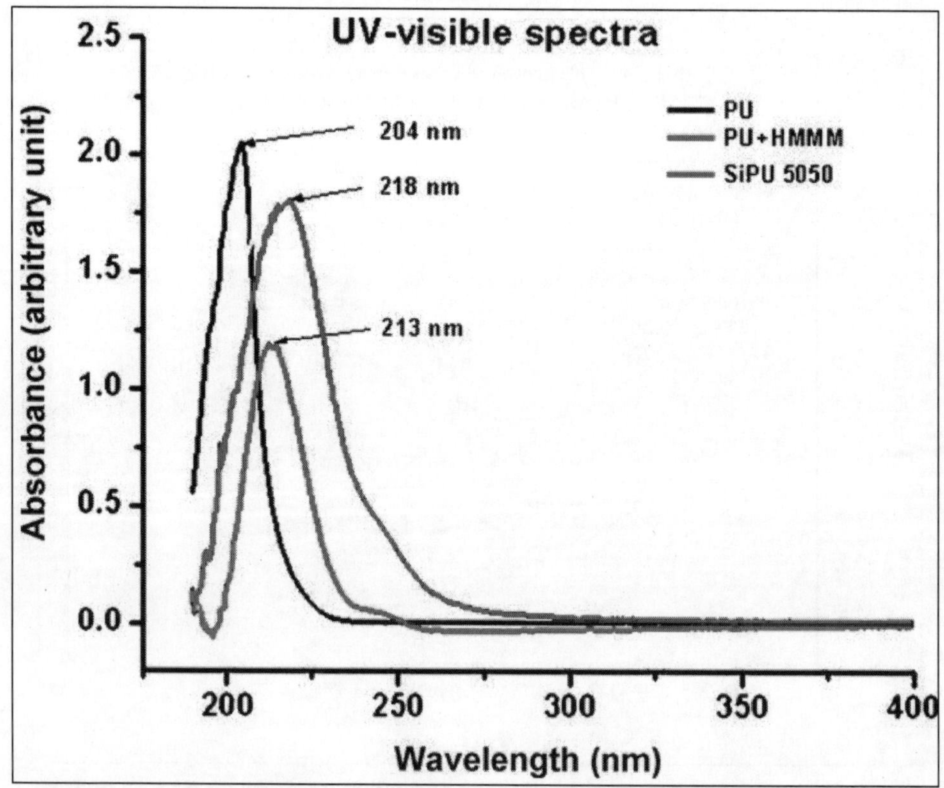

Figure 32.5: UV-Vis Spectra of Pure PU, PU with HMMM and SiPU 5050 Coating.

thus the coating solution does not have photodegradable moieties which may result in degradation of coating in outdoor exposure.

Mechanical Properties of the Coatings

Pencil Hardness

Figure 32.6 shows the variation of coating hardness (pencil hardness) with increasing silicone concentration in the coating. Pencil hardness test (ASTM D-3363-92a) for SiPU coatings varied from 3H to above 6H whereas PU coating showed hardness of only 2H. Addition of silicone to PU makes coating harder which is likely due highly crosslinked siloxane (Si-O-Si) network within the coating.

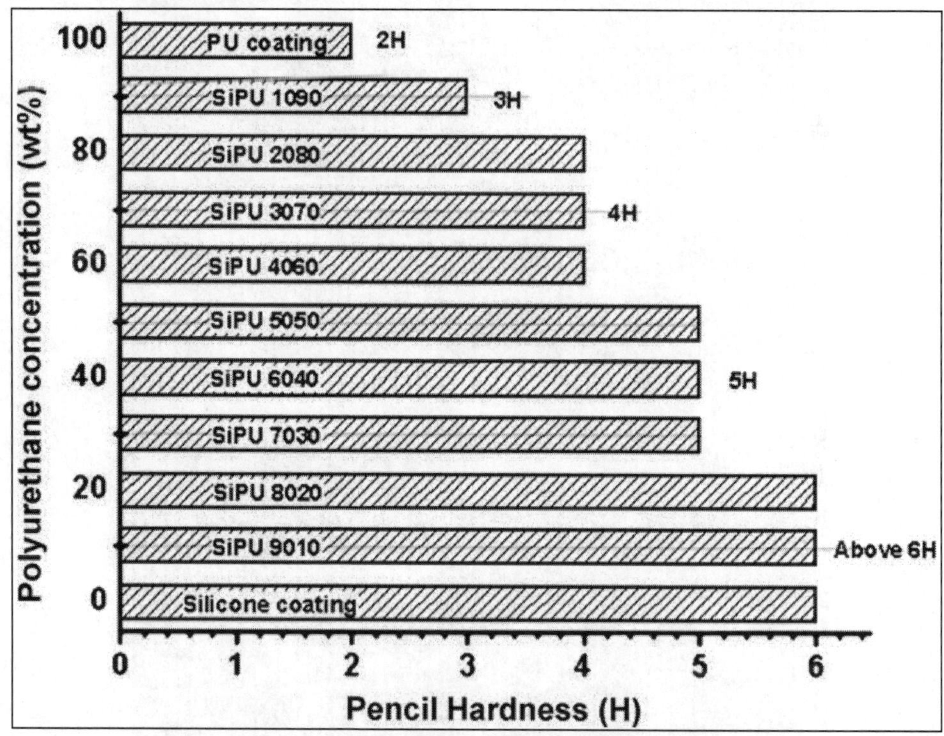

Figure 32.6: Pencil Hardness of Coatings on Aluminum.

Adhesion and Flexibility

Bend test (ASTM D-3359-02) as shown in Figure 32.7 were carried out on SiPU 5050 coated sample which showed no crack or loss of adhesion of the coating in the region of bent. This may be explained from the siloxane moiety functioning as an internal plasticizer due to its flexible–Si–O–Si–linkages.

Impact test (BS 3900) as shown in Figure 32.8 also revealed that there was no disbonding or cracking of the coating from the substrate which shows that the coatings are resilient and can withstand rapid deformation and stresses during fabrication.

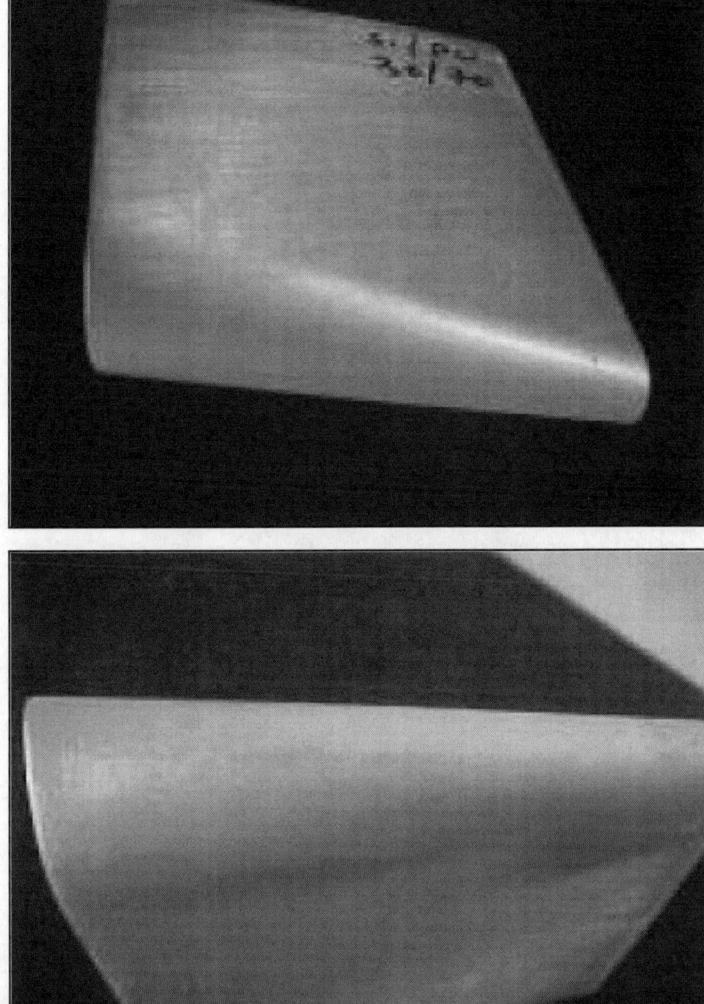

Figure 32.7: Mandrel Bent Test.

This shows that the coating does not lose its adhesion either in tension or in compression. Thus, these coatings are very flexible and hence can withstand the stresses of fabrication.

Conclusion(s)

Waterborne silicone-polyurethane coatings (SiPU coatings) were developed and their performance in terms of corrosion resistance, thermal stability and hardness were evaluated. Silicone modification increased the corrosion resistance of polyurethane. Significant decrease in corrosion current was investigated for PU

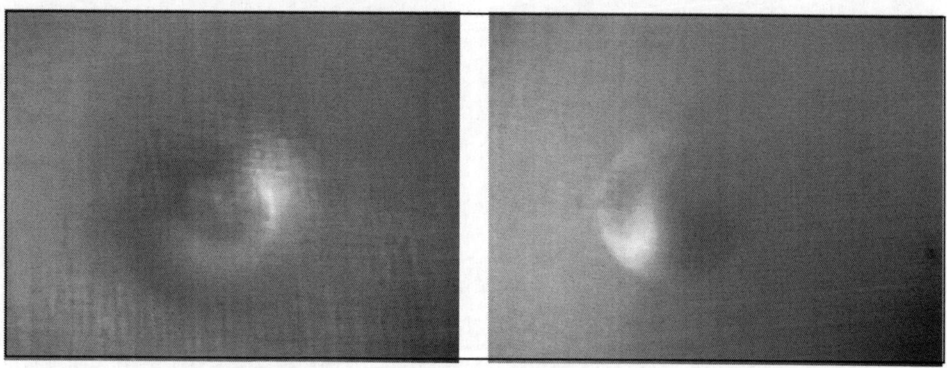

Figure 32.8: Impact Test.

coating (3.1×10^{-6}A), silicone coating (7×10^{-7}A) and SiPU coatings ($1.21 - 2.71 \times 10^{-6}$A) compared to the bare aluminum (9.2×10^{-6}A). Improvement in corrosion resistance was observed due to the formation of aluminum-oxygen-silicon (Al-O-Si) interface which acts as geometric blocking layer against the migration of chloride ions and water towards the substrate.

The pencil hardness of PU coating was found 2H while it was enhanced from 3H to above 6H with increasing silicone concentration in SiPU coatings. Furthermore, the presence of siloxane (Si-O-Si) moieties in SiPU coatings made them flexible and resilient enough to sustain the mechanical stresses during service. Coatings were found thermally stable up to around 200°C and the degradation of SiPU coatings was delayed by silicone modification. Thus, silicone modification boosts the corrosion resistance, thermal stability and mechanical properties of waterborne PU which makes them suitable for marine, automotive and aerospace applications.

References

1. M. G. Fontana, N. D. Greene, Corrosion Engineering, 3rd Edition, 1986, McGraw Hill, New York.
2. C. Vargel, Corrosion of Aluminum, Elsevier, 2004.
3. T.L. Metroke, R. Parkhill, E.T. Knobbe, Proceedings of the Material Research Society Symposium, 576 (1999) 293.
4. T.L. Metroke, O. Kachurina, E.T. Knobbe, Prog. Org. Coat., 44 (2002) 295.
5. J. M. Zeigler, F. W. Gordon Fearon, Silicon-based polymer science, a comprehensive resource, Washington, DC:ACS, 1990.
6. X. F. Yang, D. E. Tallman, V. J. Gelling, G. P. Bierwagin, L. S. Kasten, J. Berg, Surf. Coat. Technol., 140/1 (2001) 44.
7. C. Yan, O. I. Jude, Chem. Mater., 11 (1999) 1218.
8. T. Ogoshi, H. Itoh, K. M. Kim, Y. Chujo, Macromolecules, 35 (2002) 334.
9. X.Y. Shang, Z. K. Zhu, J. Yin, X.D. Ma, Chem. Mater., 14 (2002) 71.
10. S. S. Pathak, A. S. Khanna, Jan 27-30, PACE 2008, Los Angeles, 2008.

11. S. S. Pathak, A. S. Khanna, Prog. Org. Coat., 10.1016/j.progcoat.2008.02.008.
12. Y. Y. Yu, C.Y. Chen, W. C. Chen, Polymer, 44 (2003) 593.
13. S. X. Zhou, L. M. Wu, J. Sun, W. D. Shen, Prog. Org. Coat., 45 (2002) 33.
14. S. S. Pathak, A. S. Khanna, T. J. M. Sinha, Corros. Rev., 24/5 (2006) 281.
15. U. Wienhold, U. Westerwelle, Eur. Coat. J., 7 (2006) 41.
16. S. S. Pathak, A. S. Khanna, T. J. M. Sinha, Prog. Org. Coat., 60 (2007) 211.
17. T. L. Metroke, R. L. Parkhill, E. T. Knobbe, Prog. Org. Coat., 41 (2001) 233.
18. E.P. Giannelis, Adv. Mater., 8 (1996) 29.
19. A. Sellinger, P. M. Weiss, A. Nguyen, Y. F. Lu, R. A. Assink, W. L. Gong, C. J. Brinker, Nature, 256 (1998) 394.
20. P. Innocenzi, G. Brusatin, Chem. Mater., 13 (2001) 3126.
21. R. Lomax, "Polyurethane dispersion: Application and potential", Eur. Coat. J., 10 (1991) 630.
22. W. D. Davies, "Some property aspects of aqueous polyurethane dispersion", in Additive for waterbased coatings, D. R. Karsa, Ed. Royl society of chemistry, Cambridge, England, 1992.
23. C. H. Hare, Protective coatings, Fundamentals of Chemistry and Composition, SSPC 94-17, 1994, Pittsburgh.
24. D. K. Weiss, Paint and Coatings: A mature industry in transition, Prog. Polym. Sci., 22 (1997) 203.
25. A. D. Wilson, Waterborne coatings, surface coatings-3, Elsevier science publisher Ltd. 1990, England. Science Pub; 2001.
26. J. D. Wright, N. A. J. M. Sommerdijk, Sol-gel Materials: Chemistry and application, Amsterdam:Gordon and Breach
26. J. D. Wright, N. A. J. M. Sommerdijk, Sol-gel Materials: Chemistry and application, Amsterdam:Gordon and Breach.
26. J. D. Wright, N. A. J. M. Sommerdijk, Sol-gel Materials: Chemistry and application, Amsterdam:Gordon and Breach.
27. D. K. Chattopadhyay, K.V.S.N. Raju, Prog. Polym. Sci., 32 (2007) 352.
28. H. T. Jeon, M. K. Jang, B. K. Kim, K. H. Kim, Colloids and Surfaces A: Physicochem. Eng. Aspects, 302 (2007) 559.
29. A.N. Khramov, V.N. Balbyshev, N.N. Voevodin, M.S. Donley, Prog. Org. 397 Coat. 47 (2003) 207–213.
30. X. Hong, Prog. Org. Coat., Vol. 26, 121-129, 1995.
31. D. A. Jones, Principles and prevention of Corrosion, 2nd Ed., Prentice-Hall, New Jersey, 1996.
32. C. J. Brinker, George W. Scherer, Sol-Gel Science: The Physics and Chemistry of Sol-Gel Processing, Boston San Diego, New York.

33. X. Hong, Q. Chen, M. Chen, J. Chin, H. Liu, Prog. Org. Coat., 45 (2002) 165.
34. C. Li, G. L. Wilkes, Chem. Mater., 13 (2001) 3663.
35. C. S. Wu, Y. L. Liu, J. Polym. Sci. Part A: Polym. Chem., 42/8 (2004) 1868.
36. S. Frings, H. A. Meinem, C. F. V. Nostrum, R. V. D. Linde, Prog. Org. Coat., 33 (1998)126.
37. A. Rekondo, M.J. Fernandez-Berridi, L. Irusta, Eur. Polym. J., 42 (2006) 2069.
38. R. Zandi-zand, A. Ershad-langroudi, A. Rahimi, Prog. Org. Coat., 53 (2005) 289.
39. K. P. O. Mahesh, M. Alagar, S. Ananda Kumar, Polym. Adv. Technol., 14 (2003) 137.
40. Y. H. Han, A. Taylor, M. D. Mantle, K. M. Knowles, Journal of Non- Crystalline Solids, 353 (2007) 313.
41. K.H. Wu, C.M. Chao, T.F. Yeh, T.C. Chang, Surf. Coat. Technol., 201 (2007) 5782.
42. Y. Wei, D. Jin, D. J. Brennan, D. N. Rivera, Q. Zhuang, N. J. DiNardo, K. Qiu, Chem. Mater., 10 (1998) 769.
43. T. Easton, S. Poultney, J. Coat. Technol. Res., 4/2 (2007) 187.
44. S. I. Seok, J. H. Kim, K. H. Choi, Y. Y. Hwang, Surf. Coat. Technol., 200/11 (2005) 3468.
45. J. H. Hwang, B. I. Lee, V. Klep, I. Luzinov, Materials ResearchBulletin bdoi:10.1016/j.materresbull.2007.10.045.
46. N. J. W. Gamagea, D. J. T. Hilla, C. A. Lukeyb, P. J. Pomerya, Polym. Degrad. Stab., 81 (2003) 309.

Chapter 33

Recent Water Quality Scenario of Damodar River in Coal Mining and Associated Industrial Areas

R.K. Tiwary, D.B. Singh, S. Lata Dora, Binu Kumari and P.K. Arya

CSIR-CIMFR, Dhanbad

ABSTRACT

India is being endowed with a wide network of rivers that can meet a various demand of water requirement of the country. Damodar is an important freshwater resource in Jharkhand and West Bengal. With rapid increase in the coal demand the number of mines and coal based industries along with the population has been increased in the region during the last decade, which has lead to an increase in water consumption and deterioration of available water resources due to discharge of untreated sewage and industrial effluents. Therefore an attempt has been made to assess the industrial effluent and water quality of the Damodar river in the industrial belt.

The study revealed the organic pollution load is predominant. All most all the sampling sites are contaminated to a large extent by coliform groups of bacteria that make them unfit for human consumption unless disinfected. The grossly polluted sampling sites are located at down stream of tributaries carrying Domestic effluents of the townships situated in the region. Industrial effluents emanated from coal washaries, Thermal power and steel plants also play an important role in making the river polluted and unfit for designated use.

Introduction

Damodar river basin is a repository of various mineral resources and therefore comprises of many industrial activities. Growth in Industrial sector and population of Damodar river basin assume great importance in the context of water pollution since the wastewater discharges from these two sources are mainly responsible for the deterioration of water quality of the Damodar River. The inevitable increase in the industries and population in the recent years has aggravated the pollution load on the Damodar River, the life line of the region.

Previously various studies have been carried out regarding the water quality of Damodar River in terms of heavy metal pollution and impact of mining (R.K.Tiwary, 2005, 2004, 1997, 1995, 1994). But very few study on biological characteristic of the Damodar River with reference to MPN has been done. Level of MPN is the most important parameter required to assess the water quality of the river to see its different usages.

Bacterial contamination in water is indicated by the presence of coliform bacteria that find their way into river mostly through untreated sewage and cause water borne diseases (Ramya Swayamprakasham, 2009) Coliforms, including *Escherichia coli* are members of the family Enterobacteriaceae and it make upto 10 per cent of the intestinal normal flora of human and others animals.

The coliform group includes *E.coli*, *Enterobacter aerogens*, and *Klebsielle pneumoniae*. They are facultatively anaerobic, gram negative, rod shaped, non spore forming bacteria and they ferments lactose and produce gas within 48hrs at 35°C. Fecal streptococci and *Clostridium* are normal flora of the large intestine of human and other animals and are also present in feces. Thus the presence of any of these bacterial species in water is evidence of fecal pollution of human or animal origin. The presence of these organisms in water indicates the presence of intestinal pathogens. Their presence also means that organic matter of animal origin is present in the river water bodies which harms the river ecosystem lowers the dissolved oxygen (DO) levels due to decomposition of fecal matter and increase BOD level. This chapter illustrate the present water quality of Damodar river in reference to MPN, DO, BOD and other physico-chemical characteristic of the river vis-à-vis industrial effluent characteristic discharged into the river in the stretch between Tenughat to Durgapur Barrage. The results so generated were compared with water quality criteria for various uses of freshwater developed by (CPCB, 2005). Such informations are necessary for determining the cost effective and long lasting solution to the water related problems. The goal is to provide the current water quality status and trends, so as to facilitate the identification of emerging issues and future priorties.

Study Area

The Damodar River Basin (DRB) is a sub-basin and part of the Ganges River System spreading over an area of about 23.371 sq. km in the states of Jharkhand and West Bengal. The geographical boundary of the basin lies between 22^0 15′ to 24^0 30′ N latitude and 84^0 30′ to 88^0 15′ E longitude. The region is richly endowed with mineral resources of coal, mica and bauxite. Industrial activities comprising of about

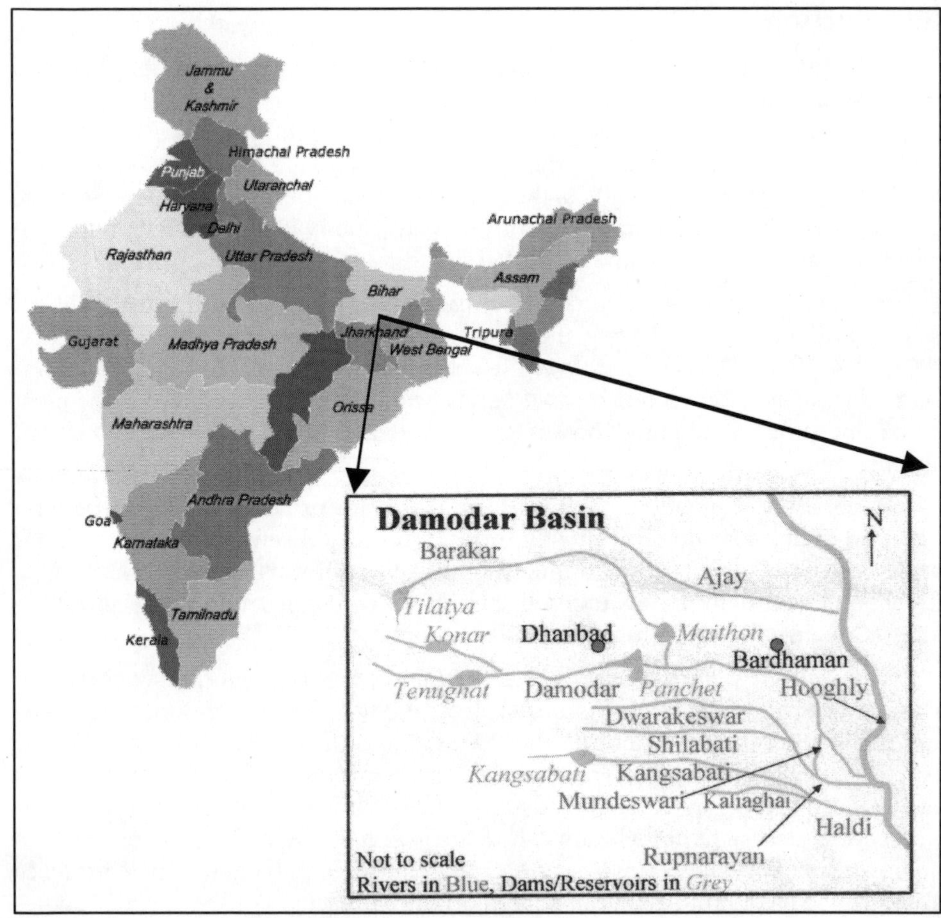

Figure 33.1: Drainage Pattern of Damodar River.

310 coal mines, 180 non coal mines, 80 urban centers and 80 other industrial centers have grown up in the region. The Damodar is Principal River of the Damodar river basin. The total stretch of the river is approx 541 km. For the study the stretch between Tenughat Dam to Durgapur Barrage has been taken covering 235 km falling in Jharkhand and West Bengal Drainage pattern of the Damodar river basin is shown in the Figure 33.1. The main tributaries of the Damodar are Konar, Bokaro, Jamunia, Katri, Barakar and small streamlets like Noonia and Tamla joins the river. The entire stretch between Tenughat Dam to Durgapur barrage has divided into two parts, stretch I and II. Stretch I covers from Tenughat Dam to Panchet Dam of Jharkhand state and Stretch II from D/S Panchet to Durgapur Barrage falling in West Bengal. Eighteen sampling stations in the main stream of Damodar were selected for the study during the two sampling period *i.e.* winter and summer. The sampling stations with its codes are shown in Figures 33.2 and 33.3 respectively.

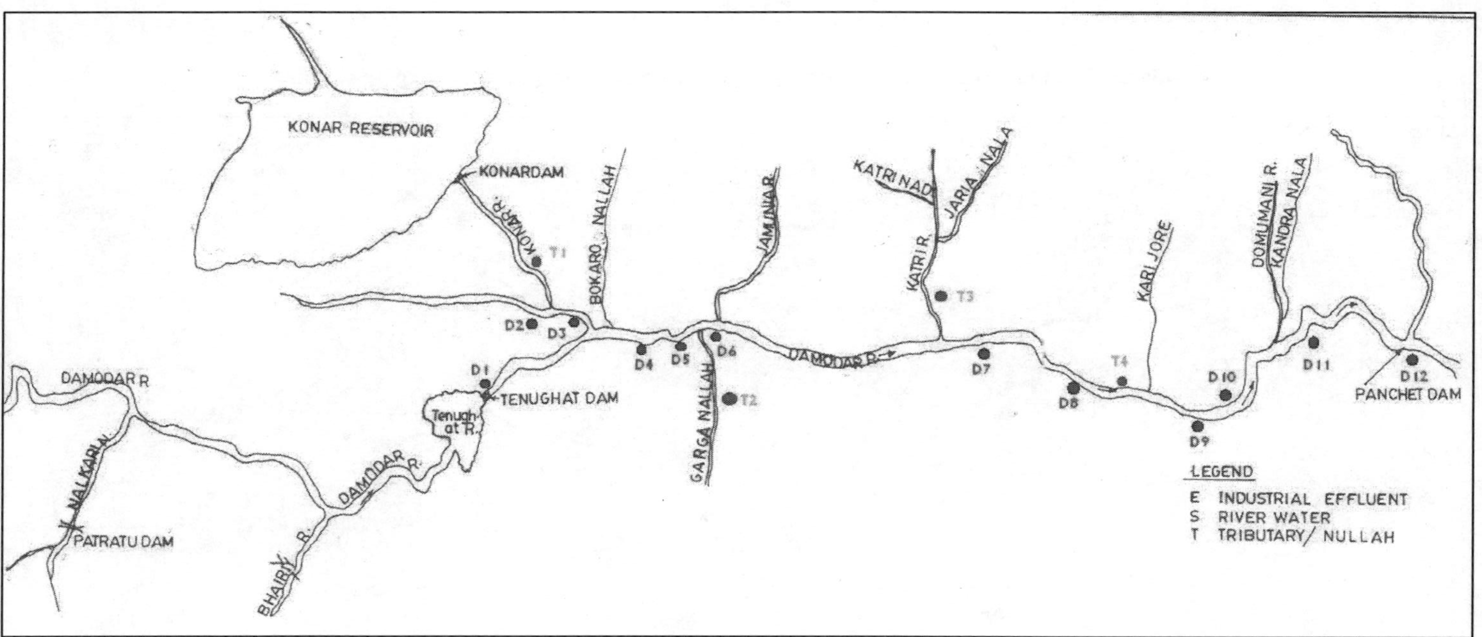

Figure 33.2: Sampling Location in the Stretch between Tenughat Dam to Panchet Dam.

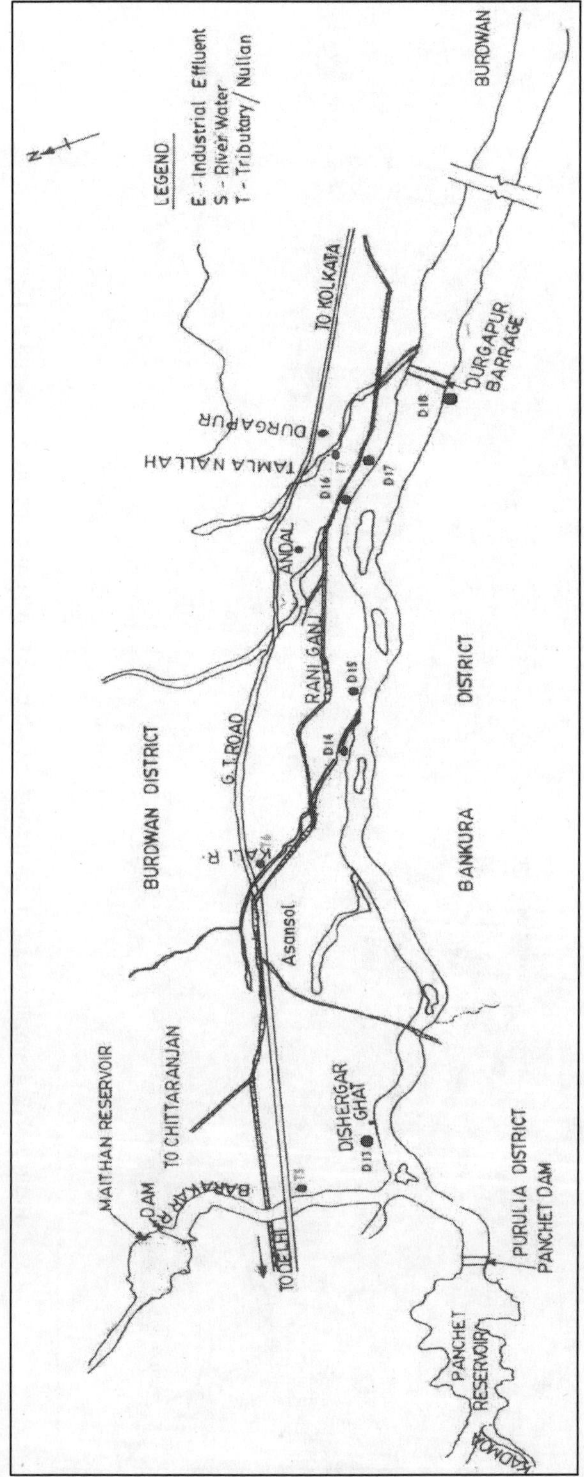

Figure 33.3: Sampling Location between D/S Panchet Dam to Durgapur Barrage.

Materials and Methods

Physico-chemical characterization of water samples were carried out by following Standard methods of examination of water and wastewater (APHA, 1998). MPN technique is followed for determining bacterial contamination. Samples were collected from the river stretch at 18 sampling locations covering winter (Feb-March, 2009) and summer (May-June, 2009).

Results and Discussion

Characteristic of Industrial Effluents

Damodar River basin comprises of several coal based industries and discharge effluents to Damodar River directly or indirectly. Damodar River receives mainly wastewater including Industrial effluent, domestic wastewater and mine wastewater discharge. The major industries present between Tenughat and Durgapur include coal washaries, thermal power plants, steel plants. Coal washeries such as Kathara, Moonidih, Sudamdih, Patherdih, etc mainly discharge high amount of TSS which reduces the usage of streams. Oven Plants are mainly responsible to cause high BOD, COD, and TDS in the river water.

Industrial effluents including steel plants, thermal power plants, coal washaries in between the stretch from Tenughat Dam and Durgapur barrage are collected and analyzed for winter and summer and results are presented in the Tables 33.1–33.3.

Table 33.1: Physico-chemical Characteristics of Industrial Effluents. Coal Washeries

Parameters	*Sudamdih Coal Washery*	*Patherdih Washery*	*Bhojudih Coal Washery*	*Moonidih Coal Washery*	*Kathara Coal Washery*
pH	5.9	6.2	6.6	6.0	5.9
TSS	800.0	950.0	890.0	650.0	1050.0
TDS	560	460	510	630	1500
BOD	3.1	3.0	2.5	2.9	3.2
Mn	0.398	0.784	0.514	3.121	1.106
Fe	4.45	8.96	6.61	10.35	9.23
Cu	0.147	0.172	0.103	0.112	0.073
Zn	1.02	0.98	0.43	0.64	2.00
Cl	76.0	64.0	76.0	108	48

All values are expressed in mg/l except pH.

Coal Washery

TSS and TDS values were found very high and vary between 650.0 mg/l to1050 mg/l and 460 mg/l to 1500mg/l respectively. Chloride ranged between 48 to 108mg/l during 2009 found within the permissible limits of 1000mg/l of the Effluent discharge IS: 2490. Whereas heavy metals such as Fe and Mn were found in high concentration with maximum of 10.35 and 3.121 mg/l respectively which is above the prescribed limits of 2 and 3 mg/l respectively.

Thermal Power Plant

Characteristic of the effluent from the thermal power plant shows high temperature, TSS, TDS, Study during 2009 showed TSS ranging between 52.0 mg/l to 500 mg/l with maximum in the effluent from Durgapur Thermal plant. TDS was observed maximum in effluent from Chinakuri thermal power plant, with the value of 3038.8 mg/l,. chloride ranged between 28 to 480 mg/l which was observed much above the prescribed limits in Thermal Power Plant effluent. Heavy metals concentrations were below prescribed limits except for Fe and Mn during 2009.

Steel Plants

Maximum TSS and TDS in the effluent from Bokaro Steel plant were found above the threshold value during 2009. The value of TSS and TDS were 225.0 mg/l and 1500mg/l respectively. Rests of the parameter were below the permissible threshold value. Chloride ranged between 108 to 320 mg/l. with the maximum at Alloy steel plants. Among heavy metals Fe exceeds the prescribed limits with maximum value at Bokaro steel plant (11.01mg/l).

Biological Characteristic of Damodar River

Physicochemical and biological parameters of water quality of Damodar River have been carried out during winter and Summer season in 2009 and results are presented in Tables 33.1 and 33.2 respectively. The water quality assessment has been carried out in respect of DO, BOD and MPN.

Table 33.2: Physico-chemical Characteristics of Industrial Effluents. Thermal Power Plant

Parameters	*Bokaro Thermal*	*Chandra-pura Thermal*	*Tenughat Thermal*	*Chinakuri Power Plant*	*Dishergarh Thermal*	*Durgapur Thermal*
pH	6.8	5.8	6.0	7.6	9.2	8.4
TSS	320.0	290.0	300.0	108.4	52.0	500.0
TDS	120	390	160	3038	458	113.6
BOD	3.1	3.1	3.8	4.1	0.2	1.2
Mn	0.398	4.19	3.15	0.15	1.51	1.08
Fe	4.60	8.55	10.54	7.57	10.63	10.11
Cu	0.04	0.04	0.05	0.120	0.104	0.160
Zn	0.27	0.60	0.86	0.067	0.091	0.426
Cl	28	36	72	860.9	480.5	150

All values are expressed in mg/l except pH.

The level of Do is observed more than 4mg/l at all the sampling sites during both the study period. The values for DO during winter was found in the range of 4.3 to 8.3

mg/l, while in summer it ranges from between 4.2 to 8.3 mg/l. The sampling location at MADA pump house and U/s Noonia nalla showed higher values for DO during summer which may be due to high growth of phytoplankton at the sampling site. Dissolve oxygen is a vital parameter for survival for aquatic animal and plants. DO in water bodies are aided by autotrophic aquatic plants as a result of photosynthesis. Presence of algal community in water bodies (high nutrients availability) increase the DO level in daylight by photosynthetic activity and also depleted oxygen concentration at night. Temperature and flow are important factor for dissolved oxygen concentration in water. Cold and fast flowing water holds more oxygen than warm water. But high microbial contamination decreases the DO level. The optimum DO value for good water quality is 4-6mg/l DO value which reflects the healthy aquatic life of the river.

Table 33.3: Physico-chemical Characteristics of Industrial Effluents. Steel Plants

Parameters	*Bokaro Steel Plant*	*Alloy Steel Plant*	*Durgapur Steel Plant*
pH	5.9	8.1	8.7
TSS	225.0	171.2	144.0
TDS	1500	152.8	194.4
BOD	3.5	4.4	1.2
Mn	2.71	0.65	1.27
Fe	11.01	8.98	10.11
Cu	0.08	0.10	0.17
Zn	0.28	0.30	0.27
Cl	108	320	190

Values are in mg/l except pH.

The Biochemical oxygen demand, one of the most important indicator of the pollution, was observed at D/s Garga nalla during both the sampling period (11.2 and 10.9 mg/l) respectively.

The values for the BOD was observed high during summer season than the sampling done during pre-monsoon which was palpable due to high biological activities at elevated temperatures, that consumes lots of oxygen during the oxidation of organic matter that increase the demand for oxygen in the water. Lowest values for the BOD was observed at Tenughat due to least organic matter pollution.

The number of observed BOD values less than 3 mg/l were 38.88 16.66 per cent during pre-monsoon and summer respectively in 2009 the total number of observations having more than 3mg/l is 83.33 per cent during summer. The above value was high in contrast to 61.11 per cent which was observed during pre-monsoon. It was observed from the results that there is gradual increase in the concentration of organic pollution load in the river.

Total numbers of observation having total coliform number 50 or less than 50 MPN/100 ml is 16.11 per cent, between 50-500 MPN/100 ml is 38.88 per cent and in between 500-5000 MPN/100 ml is 33.33 per cent during pre-monsoon. Similarly the number of observations having total coliform number 50 or less than 50 MPN/100ml is 16.66 per cent, between 50-500 MPN/100ml 22.22 per cent and in between 500-5000 MPN/100 ml was 61.11 per cent, during summer season. On comparing the two observed values obtained during the two sampling seasons showed that there is gradual increase in the total coliform numbers making the river water unsuitable even for bathing purpose, as it may cause skin related diseases or allergies. Highest coliform contamination is observed during the study period was at D6 *i.e.* Damodar River D/s to Garga nalla that joins the Damodar river carrying heavy domestic sewage load. Values observed at Tenughat showed least MPN values and BOD values during both the study period showing clean water.

The biological water quality with respect to distance from the Tenughat and flow at different sampling location are shown in Tables 33.4 and 33.5 and graphical representation is shown in Figures 33.4 and 33.5.

The above observation when compared with the water quality criteria that has been developed by CPCB (R.M.Bhardwaj, 2005) Table 33.6.

It was found that the water at Tenughat falls under class "A" for having MPN less than 50, DO above 6 and BOD less than 2 mg/l. The sampling site U/s Konar, Garga nalla, DTPS and Domgar ghat falls under the class B, whereas rest all the sampling site falls under the class "C". for having MPN values 5000 or less.

The flow of water in the river is another important parameter that regulates the pollution level of the river, and flow is maintained by the water release pattern by the dams. The water release pattern by various dams constructed on the river Damodar is given in Table 33.7. It is observed that during summer period release of water from Dams are reduced by 30 to 40 per cent and thus environmental flow and self

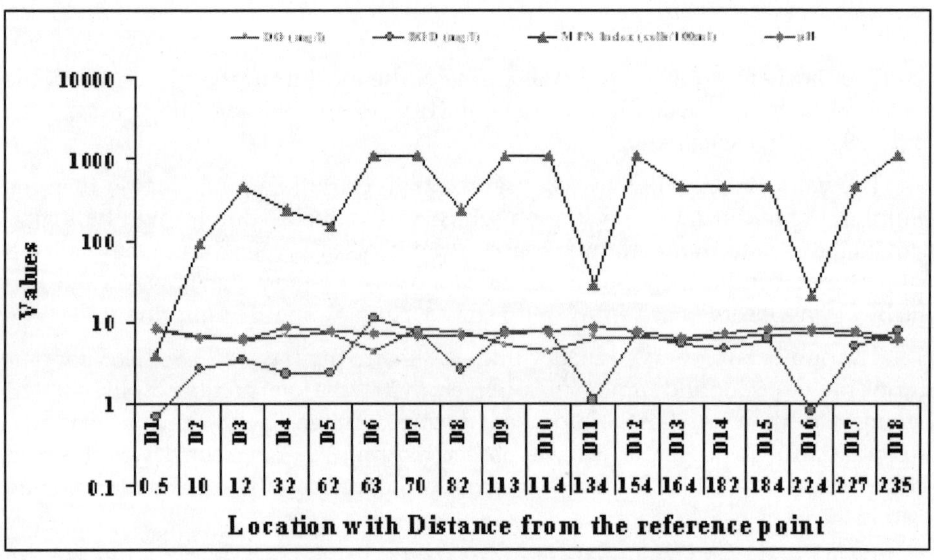

Figure 33.4: Study on MPN with respect to BOD and DO.

Table 33.4: Biological Characteristics of Damodar River (Feb-Mar, 2009).

Location	*Distance (km) Reference Point*	*Sampling Code*	*DO (mg/l)*	*BCD (mg/l)*	*pH*	*MPN Index (cells/100ml)*	*Flow (cumecs)*
Damodar River D/S Tenughat Dam	0.5	D1	8.3	0.7	8.44	4	6.86
Damodar River U/S Konar River Confluence	10	D2	6.8	2.7	6.4	93	–
Damodar River D/S Konar River Confluence	12	D3	6.2	3.4	6.2	460	7.23
Damodar River at Phusro	32	D4	6.7	2.3	8.72	240	33.47
Damodar River U/S of Garga nalla Confluence	62	D5	7.2	2.4	7.73	150	7.58
Damodar River D/S of Garga nalla Confluence	63	D6	4.3	11.2	7.42	1100 (3 times dilution)	23.11
Damodar River moonidhi, Kapat ghat	70	D7	8.3	7.5	6.9	1100	–
Damodar River Jamadoba, MADA pump house.	82	D8	7.6	2.6	6.7	240	34.905
Damodar River U/S Mohalbani Ghat	113	D9	5.2	7.6	7.1	1100	55.88
Damodar River D/S Mohalbani Ghat	114	D10	4.3	7.8	7.5	1100	–
Damodar River at Domgarh Ghat	134	D11	6.6	1.1	8.5	28	17.33
Panchet Reservoir	154	D12	6.6	7.2	7.89	1100	19.51
Damodar River D/S confluence of Barakar and Damodar	164	D13	6.2	5.2	6.4	460	39.63
Damodar River U/S Noonia nalla	182	D14	6.4	4.6	7.1	460	–
Damodar River D/S Confluence of Noonia nalla	184	D15	6.7	6.2	8.2	460	134.35
Damodar River U/S DTPS	224	D16	7.1	0.8	8	21	–
Damodar River D/S DTPS	227	D17	6.9	4.9	7.8	460	329.27
Damodar River Durgapur Barrage	235	D18	5.7	7.6	6.2	1100	–

Table 33.5: Biological Characteristics of Damodar River (May-June, 2009).

Location	*Distance (km) Reference Point*	*Sampling Code*	*DO (mg/l)*	*BOD (mg/l)*	*MPN Index (cells/100ml)*
Damodar River D/S Tenughat Dam	0.5	D1	6.7	1.0	43
Damodar River U/S Konar River Confluence	10	D2	5.4	2.1	150
Damodar River D/S Konar River Confluence	12	D3	5.4	3.6	1100
Damodar River at Phusro	32	D4	6.3	3.7	1100
Damodar River U/S of Garga nalla Confluence	62	D5	5.2	4.1	460
Damodar River D/S of Garga nalla Confluence	63	D6	4.2	10.9	
Damodar River moonidhi, Kapat ghat	70	D7	7.6	6.9	1100 (1 times dilution)
Damodar River Jamadoba, MADA pump house.	82	D8	9.4	4.30	460
Damodar River U/S Mohalbani Ghat	113	D9	5.0	7.30	1100
Damodar River D/S Mohalbani Ghat	114	D10	5.3	7.7	1100
Damodar River at Domgarh Ghat	134	D11	6.5	4.2	240
Panchet Reservoir	154	D12	6.6	7.7	1100
Damodar River D/S confluence of Barakar and Damodar	164	D13	7.1	6.4	23 (1 times dilution)
Damodar River U/S Noonia nalla	182	D14	8.3	6.5	7 (1 times dilution)
Damodar River D/S Confluence of Noonia nalla	184	D15	5.2	5.0	150 (1 times dilution)
Damodar River U/S DTPS	224	D16	6.9	3.0	15
Damodar River D/S DTPS	227	D17	6.4	5.9	43
Damodar River Durgapur Barrage	235	D18	5.5	7.9	1100

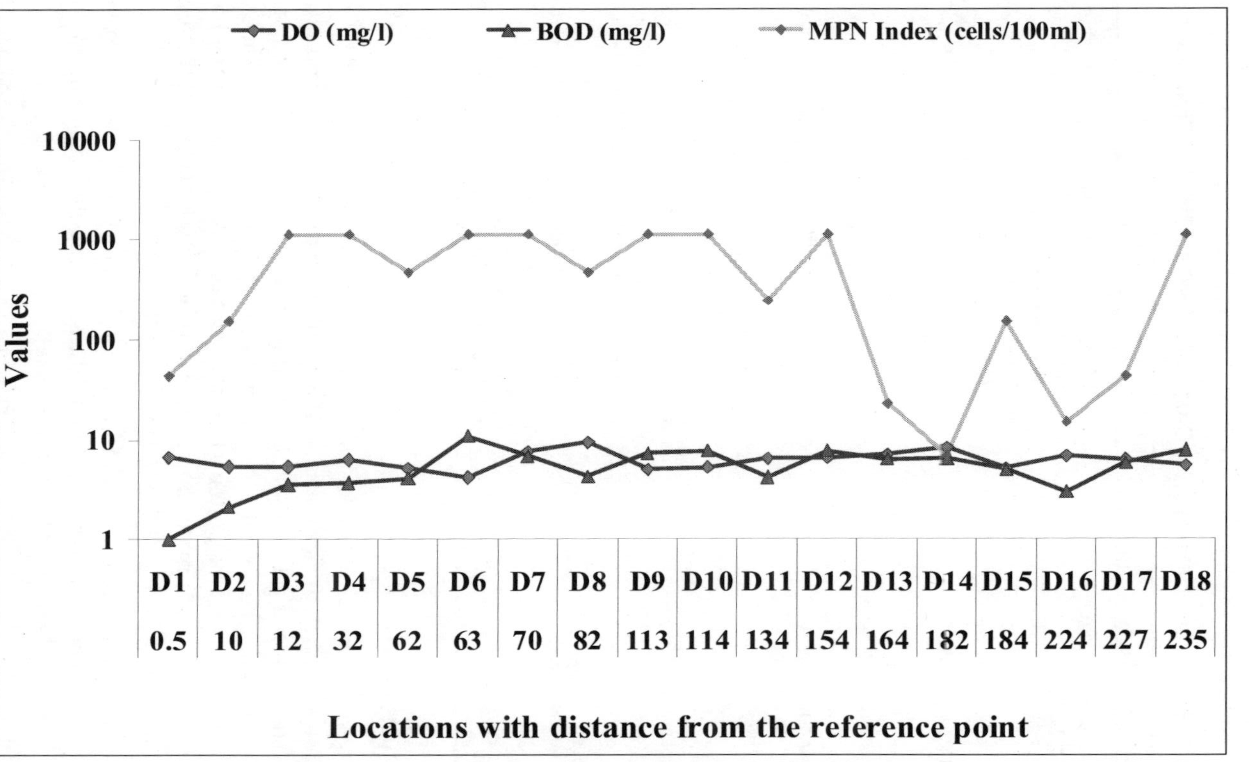

Figure 33.5: Study on MPN with respect to BOD and DO.

purification capacity of the river gets affected and increase the pollution load of the river

The maximum and minimum percentage values for the two sampling seasons are shown in Figure 33.6.

Table 33.6: Primary Water Quality Criteria for Various Uses of Freshwater.

Designated Best Use	*Class*	*Criteria*
Drinking water source without conventional treatment but after disinfections	A	☆ Total coliform organism MPN/100 ml shall be 50 or less. ☆ pH between 6.5 and 8.5 ☆ DO 6 mg/l or more ☆ BOD 2mg/l or less
Outdoor bathing (organized)	B	☆ Total coliform organism MPN/100 ml shall be 500 or less. ☆ pH between 6.5 and 8.5 ☆ DO 4 mg/l or more ☆ BOD 3mg/l or less
Drinking water source with conventional treatment followed by disinfection	C	☆ Total coliform organism MPN/100 ml shall be 5000 or less. ☆ pH between 6 and 9 ☆ DO 6 mg/l or more ☆ BOD 2mg/l or less
Propagation of wild life, fisheries	D	☆ pH between 6.5 and 8.5 ☆ DO 4 mg/l or more ☆ Free ammonia (as N) 1.2 mg/l or less
Irrigation, industrial cooling, controlled waste disposal	E	☆ pH between 6.0 and 8.5 ☆ Electrical conductivity less than 2250 micro mhos/cm ☆ Sodium adsorption ratio less than 26 ☆ Boron less than 2 mg/l

Conclusion(s)

The physico-chemical and biological study was carried out in the part of Damodar river that flows through the industrial belt to examine type and level of pollution in the stretch. The study indicates that organic pollution continues to be the predominant in the river Damodar. The organic pollution measured in terms of BOD and Total coliform count shows that maximum sampling locations are contaminated with high fecal coliform bacteria. The main sources are the tributaries that carries the domestic sewage of the towns situated at the bank of the river.

Lowest dissolved oxygen was found at the D/s of the Garga nalla carrying domestic sewage of Bokaro township. Maximum sampling station of Damodar River showed DO value to be greater than 6mg/l.

The BOD value varies from 0.7-11.2mg/l. Damodar river Down stream to Garga nalla (D6) showed maximum BOD value during winter and summer, the values were 11.2 mg/l and 10.7 mg/l. Lowest BOD was found at D1 (Tenughat).

Table 33.7: Monthly Average Discharge (Ham) (1 ham = 10,000m^3) of 2009 of Tilaiya, Tenughat, Maithan, Panchet and Konar Dam.

Year 2009	*Tenughat*		*Konar*		*Panchet*		*Tilaiya*		*Maithon*	
	M. Av. Inflow	*M. Av Outflow*	*M. Av. Inflow*	*M.Av. Outflow*	*M. Av. Inflow*	*M. Av Outflow*	*M. Av. Inflow*	*M. Av Outflow*	*M Av. Inflow*	*M. Av Outflow*
January	10.74	44.64	33.54	79	66.80	125.45	1.64	88.61	123.09	231.45
February	40.5	2	27.53	79	66.89	220.78	0.17	3.82	191.67	398.82
March	8.61	220.03	32.35	102.35	200.61	376.22	0.22	2.45	179.61	243.12
April	0.00	39.33	39.47	86.00	10.63	229.87	0.00	1.57	57.80	272.10
May	17.23	38.90	52.32	106.52	68.97	65.48	65.48	9.48	73.58	135.65

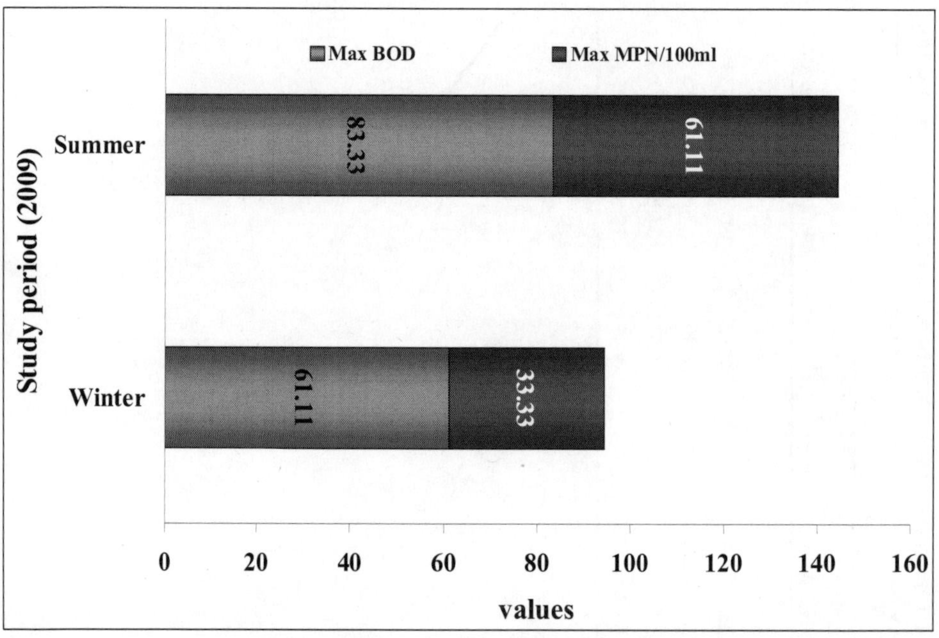

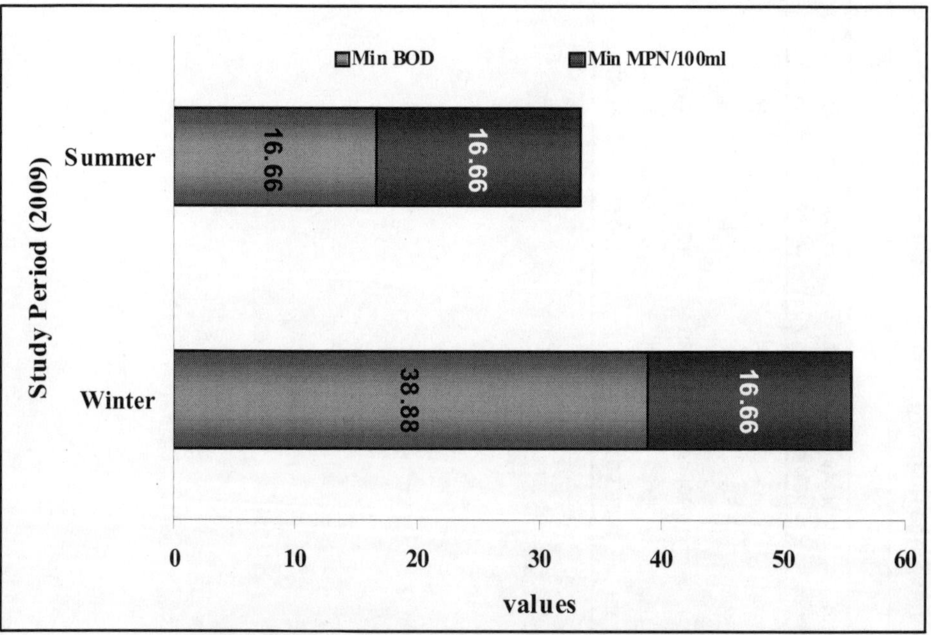

Figure 33.6: showing Max and Min BOD and MPN Values in Percentage during Two Study Period.

Highest number of coliform is found at the sampling site D6 which is the D/S of Garga nalla and lowest coliforn was observed at D1 (Tenughat). Maximum sampling station is highly contaminated with coliform bacteria except few stations.

Effluents from the industries showed higher level of TSS, TDS, Fe and Mn, which are being discharged into the river and increases the level of these pollutants. It also causes the high sedimentation of the river bed which reduces the flow rate and self purification capacity of the river. The main source for TSS and TDS are coal washaries and thermal power plants.

References

1. APHA-AWWA-PCF(1998).Standard methods for examination of water and wastewater, 20th Edition, New York.
2. CPCB, New Delhi, India. Primary water quality criteria for designated best use classes
3. R.K.Tiwary and Abhishek, Impact of Coal Washeries on Water Quality of Damodar River in Jharia Coalfield, to Indian J. Environment Protection, 25(6) 518-522 June, 2005.
4. R.K Tiwary, G.P Rajak, Abhishek and M.R Mondal. Water Quality Assessment of Ganga River in Bihar region, India. J. Environmental Science and Engineering, Vol.47, No-4, pp. 326-335, (Oct 2005).
5. R.K.Tiwary, M. Ahmad, M.K. Chakravorty and Asha Gupta, Impact of Mining and Other Industries on the major Rivers with Special Reference to River Damodar, Indian J. Env. Protection, 23(11):1208-1216 (2003)
6. R.K.Tiwary and B.B.Dhar, Environmental Pollution from Coal Mining Activities in Damodar River Basin, India, Mine Water and Environment, Wollongong, NSW 2500, Australia (1994) Vol. 13, June-December issue, pp.1-10.
7. R.K.Tiwary and B.B.Dhar, Effect of Coal Mining and Coal Based Industrial Activities on Water Quality of the River Damodar with Special Reference to Heavy Metals. Int. J. of Surface Mining, Reclamation and Environment, 8 (1994): 111- 115.
8. R.K.Tiwary, Environmental Impact of the oil spillage into Damodar river, India, Journal of the Institution of Public Health Engineers, India, Published on the occasion of National seminar on Environmental Engineering with special emphasis on Mining Environment, NSEEME ISM, Dhanbad, 19-20 Mar, 2004.
9. R.K.Tiwary, A. Jamal and B.B. Dhar, Prediction of Water quality Drainage in Coal Mines–Agrochemical approach in Mine Environment (Eds) Bharat B. Dhar and D.N.Thakur, Oxford and IBH Publishing Co. Pvt. Ltd., New Delhi, pp. 155-162. 1997.
10. R.K.Tiwary, Environmental Impact of coal mining on water regime and its management, Int. J. Water, Air and Soil Pollution, 132, 185-199, 2001.
11. R. Swayamprakasham. "Coliform will count" Science and Environment online. Down to Earth. Jan 31, 2009.
12. Prescoott, Herley and Klein's Microbiology by Christopher J. Woolurton, Joanne Willey, Linda M. Sherwood, 2007. Published by Mc Graw-Hill Science Engineering.

Chapter 34

Metal Contaminants in Soil and Sediments of an Indian Coal Mining Site

B.P. Baruah, Puja Khare and P.G. Rao

Coal Chemistry Division,
CSIR-North-East Institute of Science and Technology, Jorhat

ABSTRACT

Acid mine drainage (AMD) from coal mines presents a significant environmental and financial liability to the mining industry. It is caused by the aqueous oxidation of sulfides in mine wastes and coals resulting in acidic discharge and mobilization of metals to the nearby environment. In acid mine-drainage contaminated rivers, sediments consist of a complex mixture of various geochemical phases containing potentially toxic metals.

The sub-bituminous coals of northeastern (NE) region of India, which are of Tertiary origin, has high sulfur, and prone to acid mine drainage formation. Liquid run offs from the overburden and stockpiles of coal have contaminated land and water bodies in and around the mining areas, threatening plant and animal lives and having a devastating effect on the people's life and rural economy. The soil nearby these coal-mining areas has become highly acidic and is unsuitable for cultivation. The dissolved trace and heavy metals (THEs), some of which are also toxic, contaminate the land and water bodies. The agricultural productivity, which is the main source of rural economy, is therefore, dwindling day by day.

In the present investigation, the coal, soil and sediment samples from the active mine sites having high sulphur contents from Assam, India have been characterized and their physico-chemical characteristics and trace and heavy elements present has been reported. Soil and sediments play a key role in the transport and sorption of trace metals in an aquatic environment. The acidic nature of soil and sediments degraded by AMD is evidenced from pH range 3.1% 6.5. Acidity of soil is related to the leaching of sulphates from AMD water. A strong negative correlation between pH and sulphur (r=-0.61) indicates the influence of acidic mine effluents on soil acidity. Ca and Mg are the major neutralizing species in the soil. Toxic and heavy metal concentration of the soil, sediment and water samples were found quite high. Concentrations of Cr, Mn and Zn in soil and sediment samples collected for study were found higher than other elements. All soil samples have high enrichment factor (EF) values for As, Bi and Cd and slightly enriched with Pb, Co and Ge, the situation is quite similar to many other sites that are affected by mine tailing, leaching and AMD production. The sediments are also found to have high EF values for As, Cd, Hg with slightly increased Pb. Soil pollution index (SPI) shows that the soil has been dangerously polluted with the Cr, Cu, Cd, Pb. It may be due to accumulation of metals from the aqueous leachates in the coal mines.

Introduction

Coal mining activities have a significant negative impact on the environment. The sulphide minerals, mainly pyrites and marcasites present in coal and mine rejects are largely responsible for acid mine drainage (AMD) formation due to aqueous weathering. A number of reactions between the acidic discharge and many other mineral phases present give rise to enhanced dissolution and mobilization of sulphur and hazardous toxic elements during coal mining. Soil and sediments play a key role in the transport and sorption of these toxic metals in an aquatic environment and plants. Where, they become biovailable (Forstner and Wittmann, 1981). Hence, characterization of these elements in soil and sediments is necessary to assess the level of contamination.

The present investigation reports trace and heavy metal characterization of soil and sediments affected by AMD waters. Enrichment factor and Soil Pollution Indices has been calculated to assess the metal contamination.

Experimental

Soil samples from the mining sites were collected from a depth of 2-3 and 6-8 inches below the surface of the soil depending upon the depth of soil horizon (Herr and Gray, 1997).

Sediment samples were collected by submerging sample containers to the stream bed and scooping sediment directly into containers. Three sedimentary stones of equivalent sizes were treated as one replicate sample (Herr and Gray, 1997). A minimum of three replicate samples were taken from each location. The samples are air dried, sampled and sieved through 72BS (-0.211mm) and analyzed for pH and metals.

The pH and sulphur analyses of the soil and sediment samples were done by Microprocessor based Water and Soil Analysis Kit model 1160E and Sulphur (S-144DR, Leco, USA) analyzers, respectively.

The trace and heavy elements present in soil and sediment samples were determined by Energy Dispersive X-Ray Fluorescence (polarized) [ED-XRF (P)] spectrometer, Spectro, Xepos with Pd window X-ray tube. For the analysis, pellets were prepared by mixing 4 g of powder of soil/sediment with 0.9 g of Hoechst wax in a homogenizer (Swing Mill MM 301) and then pressed in a manual hydraulic press with 10 tonnes of pressure to a 32 mm pellet.

Results and Discussion

Soil and sediments affected by acid mine drainage have acidic pH range from 3.1- 6.5. Acidity of soil is related to the leaching of sulphates from AMD water. A strong negative correlation between pH and sulphur (r=-0.61) indicates the influence of acidic mine effluents on soil acidity (Figure 34.1). Ca and Mg were the major neutralizing species in the soil. Al, Si and Fe were found to be the dominating elements in the soil and sediments.

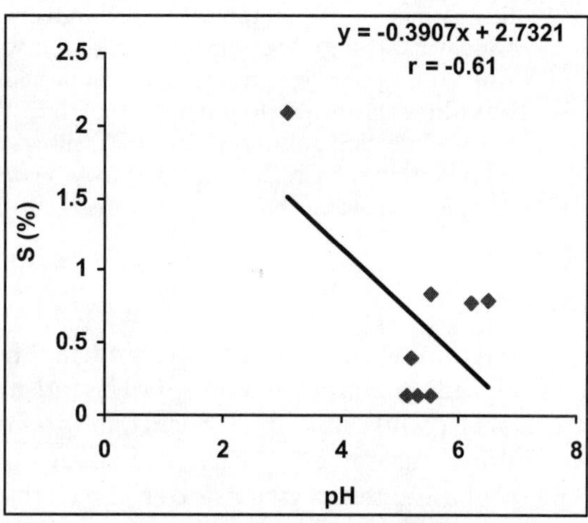

Figure 34.1: Regression Plot of pH with Sulphur Content of Soil Samples.

Si and Fe are the dominating elements in the soil and sediments of AMD affected area of Makum coalfields. Order of abundances of trace and heavy elements is MnH ≈ Cr > Ni > Zn > Co > Cu > Pb > Sb > As > Cd.

Enrichment Factor (EF)

The average concentration of an element in the earth's crust is termed as the Clarke value. Enrichment factors (EF)are obtained by dividing the metal contents in soil and sediment by crustal abundance (Klein *et al.,* 1975).Enrichment factors calculated for soil and sediments for Makum coalfields are given in the Table 34.1. Elements whichhave EF values less than 2 are deficient to minimal enrichment. Zn, Cu, V, Mn and Ni in some samples show deficiencyto minimal enrichment. Soil is moderately enriched with Pb and Co, their EF values ranges between 2 to 5. AMDaffected soil is observed as moderate (EF 2 to 5) to significantly enriched (EF 5 to 20) with As concentration.Similarly soil is significantly enriched to very highly enriched (EF > 20) with Sb. Enrichment factor of Cd is in very high to extremely high range. It is interesting to note that there is similar trend among the EF of Coal, Reject/OB and soil/sediments. In all the cases, Zn, Cu, V, Mn and Ni shows low EF values and Sb and Cd shows high EF values. As there is no other source of these elements, these metals leached out from coal, reject/OB is accumulated in the soil (Figures 34.2 and 34.3.)

Soil Pollution Index (SPI)

Soil Pollution Index (SPI) was developed by relating the existing pollution levels of nine trace elements to their critical soil concentration level; the critical levels being

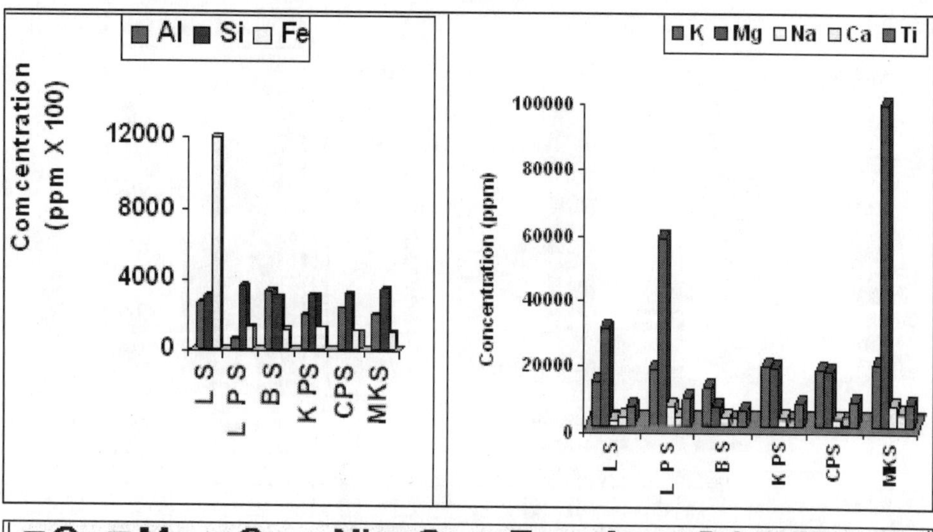

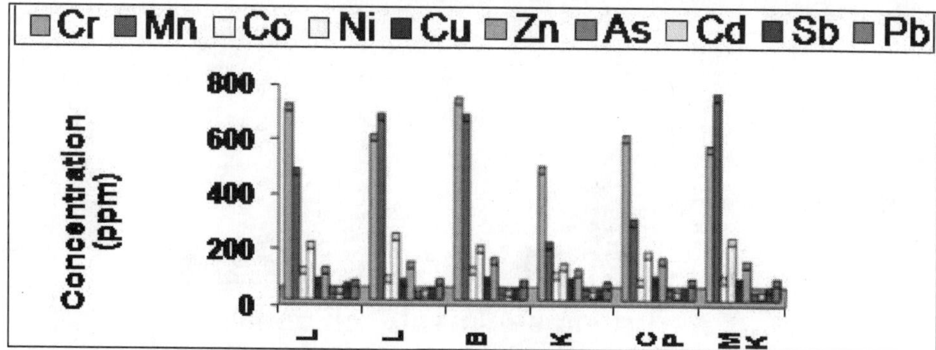

LS: Ledosoil, LPS: Ledo Pathar Soil, BS: Baragolai Soil, KPS: Kalpara Pathar Soil, CPS: Chipenala Pathar Soil, MKS: Moulung Khamti Pathar Soil.

Figure 34.2: Elemental Concentrations in Soil of AMD Affected Areas of Makum Coalfield.

Table 34.1: Enrichment Factors of Elements Present in Soil and Sediments of Makum Coalfield.

	Sb	Pb	Mn	Ni	Cu	V	As	Cd	Co	Zn
Ledo Soil	33.9	2.8	0.5	2.3	0.5	0.03	9.0	110.7	3.9	0.5
Ledo pathar soil	15.3	3.1	0.7	2.7	0.5	0.03	2.8	58.7	2.6	0.6
Baragolai soil PS	17.6	2.9	0.7	2.1	0.6	0.03	8.6	77.3	4.0	0.6
Kalpara pathar soil	12.1	2.6	0.2	1.3	0.5	0.03	6.6	33.3	3.2	0.4
Chipenala soil	18.3	3.2	0.3	1.9	0.6	0.03	7.0	33.3	2.2	0.6
Moulung Khamti pathar soil	23.2	3.3	0.8	2.5	0.5	0.03	0.7	33.3	2.7	0.6
Tipong sediment	6	2.5	0.2	2.7	0.6	0.03	3.8	33.3	2.1	0.6
Chipenala sediment	8.5	2.1	0.2	1.0	0.4	0.03	4.5	33.3	6.5	0.4

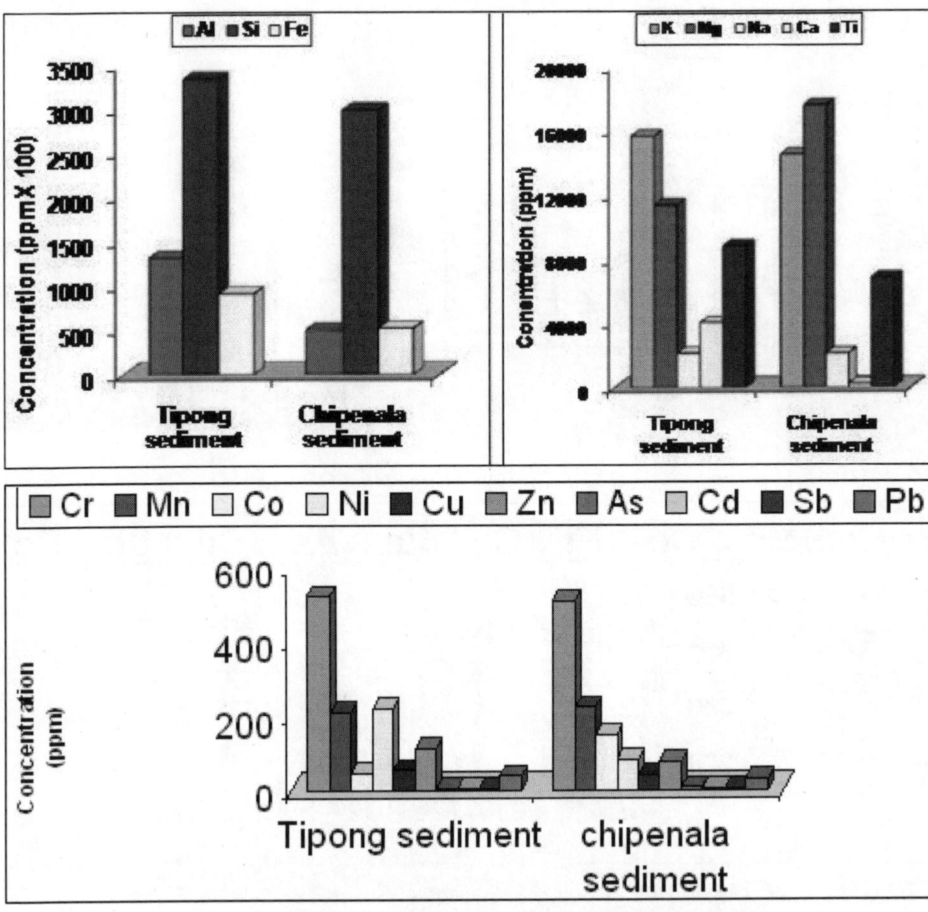

Figure 34.3: Elemental Concentrations in Sediments of AMD Affected Areas of Makum Coalfield.

Table 34.2: Soil Pollution Index (SPI) in the Mining Areas of Makum Coalfield.

	Cr	Cu	Mn	Ni	Cd	Pb
Ledo Soil	913.3	83.2	30.0	181.2	553.3	44.5
Ledo pathar Soil	769.3	81.7	43.4	214.4	293.3	49.3
Baragolai soil PS	946.7	92.0	43.3	170.4	386.7	45.6
Kalpara pathar soil	613.3	86.7	12.3	106.5	166.7	40.9
Chipenala pathar soil	766.7	101.3	17.9	151.1	166.7	50.8
Moulung Khamti pathar soil	714.7	86.2	48.4	200.0	166.7	52.7

<10: Unpolluted; 10-20: Slightly polluted; 20-30: Moderately Polluted;
30-40: Severely Polluted; >40: Dangerously Polluted.

assumed as reference base line for each pollutant (element) and then converting the concentration of elements into the percentage of the standard (Singh and Srivastava, 1997; Kabata-Pendias and Pendias, 1984). The pollution index was used to evaluate the degree of metal contamination. SPI (Table 34.2) shows that the soil has been dangerously polluted with the Cr, Cu, Cd, Pb, which may be due to accumulation of metals from the aqueous leachates in the coal mines.

The decreasing tendency of SPI of these five metals from the point sources also attributed their incorporation from the mine sites which are retained by the different soil components during the course of time.

Conclusion(s)

The soil and sediments is affected by acid mine drainage. Acidity of soil is related to the leaching of sulphate from AMD. Al, Si and Fe were found to be the dominating elements in the soil and sediments. Order of abundance of trace and heavy elements was MnH"Cr>Ni>Zn>Co>Cu>Pb>Sb>As>Cd. Sulphates, aluminium oxides and clay minerals might have played a significant role in trace element sorption resulting in subsequent enrichment of trace elements in the existing environment. Soil was moderately enriched with Pb and Co and extremely enriched with Cd. Soil Pollution Index shows that the soil has been highly contaminated with the Cr, Cu, Cd, Pb due to accumulation of metals from acidic leaching in coalmines.

References

1. Förstner, U.; and Wittmann, G. (Eds.), *Metal Pollution in the Aquatic Environment*, Springer, Berlin, 1981.

2. Herr, C.; and Gray, N. F., *Environ. Geol.*, 1997, 29, 37-45.

3. Kabata-Pendias, A.; and Pendias, H., In: *Trace Elements in Soils and Plants*, CRC Press, Boca Raton, FL, 1984.

4. Klein, D. H.; Andren A. W.; Carter, J. A.; Emery J. F.; Feldman, C.; Fulkerson, W.; Lyon, W. S.; Oge, J. C.; Talmi, Y.; Van Hook, R. L.; and Batton, N., Environ. Science and Tech., 1975, 9 (10), 973-976.

5. Singh, G.; and Srivastava, B.K., *Energy Environment Monitor*, 1997, 13(1), 33-39.

Chapter 35

Greenhouse Gases, Climate Change, Ecological and Carbon Footprints (EF and CF) and Green Building

A.P. Mishra, M. Tembhare, J.S. Pandey, R. Kumar and S.R. Wate

CSIR-National Environmental Engineering Research Institute (NEERI), Nagpur

ABSTRACT

Buildings have a significant contribution to make in terms of environmental impact as they are strongly dependent on highly energy-intensive materials like cement and steel etc. "Ecological Foot-printing (EF)" and "Carbon Foot-printing (CF)" can be very useful tools for ranking and designing green buildings. The present paper illustrates some model development exercises based on realistic and relevant parameters from the point of view of green buildings. Moreover, these parameters are easy to measure and monitor. Role of green buildings in taking recourse to appropriate adaptation and mitigation measures is also important in terms of climate change research. How these activities should be integrated and furthered is discussed at the end. In short, the article recommends the kind of activities which need regular pursuance, refinement, modification and application in regard to evolving site-specific, region-specific and ecosystem-specific environmental management plans aimed at combating the climate regulated environmental crisis which is unfolding before us every day with a newer dimension.

Introduction

Apart from being resource centers, cities are also the centers for resource degradation and residential, commercial and industrial buildings are the main resource-consumers as well as waste-producers. Water used in industrial, commercial and residential units is usually returned to rivers, lakes or the sea in a form which is of sufficiently low quality. Part of the solid waste collected from industrial, commercial and residential units is dumped on the land-fill-sites. Much of it finds its way into various water-bodies, adding to their already existing pollution. Even air pollution emanating from the cities finally becomes a water pollution problem like the problem of acid rain (Pandey *et al.*, 2001-2) which has a far wider ramifications leading to ecological impacts through the entire food chain.

Improper waste disposal or management, often times, results in creating many environmental as well as human health related problems. It may result in contaminating either groundwater (through soil-percolation) or surface water (through run-off from waste sites to streams, lakes and wetlands). In addition, surface water can get contaminated indirectly also through groundwater discharges.

In synergistic combination with changing climatic factors, these problems are likely to aggravate. For instance, in the year 2000, seven people died and about 2000 took ill because of improper disposal of animal wastes in Walkerton, Ontario. Many important drinking water sources world over have suffered badly because of improper waste disposal. In the town of Smithville, Ontario, Poly chlorinated Bi-phenyl (PCB) waste leaked into groundwater from a waste transfer facility. In the densely populated areas of Sydney, hazardous wastes from tar ponds have contaminated groundwater and surface water (http://www.nwri.ca/threatsfull/Ch12-I-e.html; http://www.ec.gc.ca/INRE-NWRI/default.asp?lang=En and n=235D11EB-1 and offset=13 and toc=show and printerversion=true).

Almost all landfills eventually release leachate and would therefore always have an impact on water quality of the surrounding ecosystems. The ever increasing release of bacteria and viruses of pathogenic nature, contamination by endocrine disrupting substances and animal pharmaceuticals in rural water may be a significant cause of concern in future. Therefore, the most important tasks for any environmental manager would be (a) predicting the spatial and temporal extents of the contaminant releases and their impact on ecosystem and human-health (Pandey and Joseph, 2001); (b) developing methods to minimize and mitigate these adverse impacts and (c) to evolve, design and formulate an effective regulatory framework that will ensure effective solid waste management.

Carbon Foot-printing (CF) (en.wikipedia.org/wiki/Carbon_footprint) and Ecological Foot-printing (EF) (Rees and Wackernagel, 1996) are some of the recent tools, which not only help in understanding and quantifying environmental impacts due to various activities like solid waste disposal, wastewater treatment and air pollution control etc., but also help in evolving appropriate cost-effective environmental management plans. Awareness has now significantly increased in respect of greenhouse gases (GHGs), global warming, climate change and carbon footprints (CFs). Institutions like US Environmental Protection Agency (USEPA, 2005) and Water Utility Climate Alliance are already working vigorously in this direction. CO_2 (carbon dioxide), CH_4 (methane), N_2O (nitrous oxide) and fluorinated gases such as hydro fluorocarbons, per fluorocarbons and sulfur hexafluoride are the main GHGs, which contribute significantly to total CF. These GHGs have widely different global warming potentials according to the Intergovernmental Panel on Climate Change (IPCC, 2006).

Buildings have significant environmental impacts as they utilize and consume over their entire life cycle sizable amounts of natural resources like forest and mining products iron, steel, cement, lime-stones and water etc. Thus it adds to the further depletion of natural resources, which already are under serious threat due to excessive industrialization and commercialization. While designing a green building, one has to, therefore, keep in mind that this kind of resource depletion and the consequent environmental impacts are minimized. Electricity consumption and solid waste generation by buildings can be two very important parameters on the basis of which green buildings can be designed (Pandey *et al.*, 2009a).

Role of green buildings in taking recourse to appropriate adaptation and mitigation measures is very important in terms of climate change research. How these activities should be integrated and furthered is also discussed at the end. In short, the article recommends the kind of activities which need regular pursuance, refinement, modification and application in regard to evolving site-specific, region-specific and ecosystem-specific environmental management plans aimed at combating the climate regulated environmental crisis which is unfolding before us every day with a newer dimension. The paper, while analyzing the environmental impact due to solid waste generation (SWG) and electricity consumption (EC), studies the inter-dependence of CF (Carbon Footprint), EF (Ecological Footprint), SWG and EC.

Ecological Foot-printing (EF) (Figure 35.1)

Together with the environmental and ecological risks, the other significant danger that faces us today is galloping population growth combined with food and material requirements, which are growing enormously day by day and, in turn, are causing substantial problems regarding the management of air, land and water (Pandey *et al.*, 2001a, b).

Ecological foot-printing (EF) is a methodology (Rees and Wackernagel, 1996), which estimates the ecological costs accruing from all the inputs that support a city, the raw materials required in the process, the intermediate goods and final products. In short, EF represents the land area necessary to sustain current levels of resource consumption and waste discharge by that population.

The sustainability (Rees and Wackernagel, 1999; Pandey, 2009) of our future society is directly proportional to the rationality it adopts in the use of natural resources. The most important need of the hour today is to evolve a sound understanding of ecology and economy, and to integrate them in such a manner as to ensure sufficient feed-back controls, which finally help ecosystems in enhancing their self-rectifying capacity. Ecological Foot-printing (EF) is one such example of integration between ecology and economics. The basic premises on which EF is based can be summarized as follows (Rees and Wackernagel, 1996):

- ✰ Every individual, process, activity, and region has an impact on the earth. And these impacts (due to resource use and waste-generation) can be quantified in terms of biologically productive area.
- ✰ Consumption patterns as well as waste-generation-potentials can be estimated in terms of the land-area needs for different human ecosystems.
- ✰ Estimates for biological productivity of these systems can also be worked out as ecosystems' supportive (resource-utilization) and assimilative (pollution-sink) capacities.

As far as ecological wealth of any region is concerned (Pandey *et al.*, 2004), it depends on two important factors: one accrues because of the ecological-goods the ecosystems in the region contain and the other one is due to the ecological-services they provide (Brismar, 2002). In particular, wetlands (Nichols, 1983; Gehrels, 1990; Mitsch and Gosselink, 2000; Costanza *et al.*, 1989; Pandey *et al.*, 1997) happen to be such important ecosystems, as provide a suitable link between terrestrial and aquatic ecosystems, and thus have a significant and sensitive ecological role especially in terms of bio-geo-chemical cycling of carbon, nitrogen, phosphorus and sulfur etc. More about wetlands and their specific role in climate change, will be discussed in subsequent sections.

Carbon Foot-print (CF) (Figure 35.2)

CF is defined as an estimate of the amount of greenhouse gases, expressed in CO_2-equivalent units (CO_{2e}) that are released into or removed from the environment as a result of any activity during a specified time-period (en.wikipedia.org/wiki/Carbon_footprint). CO_2 (carbon dioxide), CH_4 (methane), N_2O (nitrous oxide) and fluorinated gases such as hydro fluorocarbons, per fluorocarbons and sulfur hexafluoride are the main GHGs, which contribute significantly to total CF. These GHGs have widely different global warming potentials according to the Intergovernmental Panel on Climate Change (IPCC, 2006).

Recently, awareness has tremendously increased about greenhouse gases (GHGs), global warming, climate change and carbon footprints (CFs). Thus, it has become very important to improve the methods of estimating carbon footprints due to various industrial, commercial and residential activities. Institutions like US Environmental Protection Agency (USEPA) and Water Utility Climate Alliance are already working vigorously in this direction. In fact, CF is fast becoming an alternative tool for assessing environmental impacts (Pandey *et al.*, 2009a, b).

Activities that lead to GHG-emissions are said to be carbon-positive, while those that remove GHGs from the environment are known as carbon-negative or carbon-sinks. When GHG-emissions equal GHG-assimilation or absorption, the activities are known as carbon-neutral.

GHG-emissions can be of the following kinds:

- Direct,
- Indirect,
- Operational, or
- Embodied emissions.

Combustion of fossil fuels, vehicular emission and methane emission from wetlands come under the category of direct emissions. Indirect emissions include electricity use/purchase, transportation of people, goods, material (chemical) and waste. Operational and embodied emissions are also used alternatively in place of direct and indirect emissions respectively. It is also possible sometimes that operational and embodied emission overlap leading to double counting ultimately resulting in a higher CF than the activity may actually entail.

Interconnections between EF (Ecological Footprint), CF (Carbon Footprint), EC (Electricity Consumption) and SWG (Solid Waste Generation) (Figures 35.3-35.5)

Our next attempt was to see if the concepts of CF and EF could be used for ranking or environmental auditing of green buildings. Of all the parameters which we studied, we found that the maximum contribution to the EF and CF (Figures 35.1–35.3) comes from electricity generation and solid waste generation (Pandey *et al.*, 2009a-b).

Information on electricity consumption for various countries was collected for various countries (http://en-wikipedia.org/wiki/List_of_countries_by_electricity_consumption) for which values of EF were reported (www.footprintnetwork.org). These values were subsequently correlated (Figure 35.4). Similar exercise was done for solid waste generation (http://www.nationmaster.com/graph/env_mun_was_gen-environment-municipal-waste-generation) (Figure 35.5). Correlation with electricity consumption was of the type EF (per capita) = $0.602e^{0.0002\,(\text{Electricity consumed per capita})}$ ($R^2 = 0.8145$), while the same w.r.t. solid waste generation was EF (ha/ca) = $1.2126e^{0.0031(\text{solid waste generated per capita})}$ with $R^2 = 0.9369$. This exercise was done in order to explore the possibility of using these correlations for doing a quick assessment of the environmental impact due to any activity or an event which entails electricity consumption and solid waste generation.

Concluding Remarks

The article presents a glimpse (snap-shot) of the kind of researches which are needed in connection with green buildings and climate change. Side by side, it also presents some of those case studies which have been carried out and presented in recent scientific forums. These researches need to be extended and pursued further in terms of the dynamics of Ecological Footprints (Pandey *et al.*, 2001a); analysis of Environmental Risks by way of developing models which deal with the issues like Temporal Risk Gradients (TRG) (Pandey *et al.*, 2001b) and Ecological Economics of Natural Resources (Pandey *et al.*, 2004). There is also a need for quantifying region-

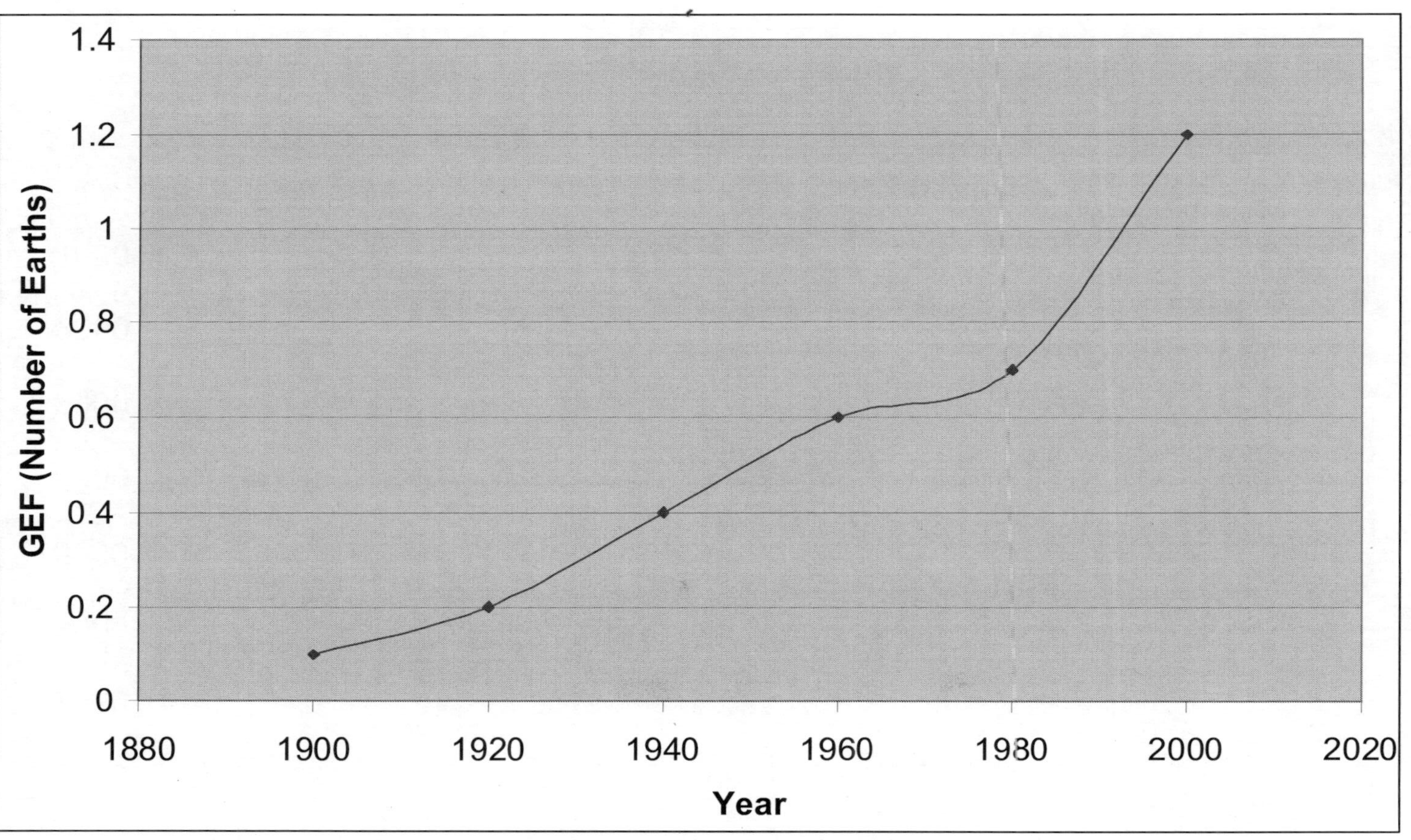

Figure 35.1: Global Ecological Footprint.

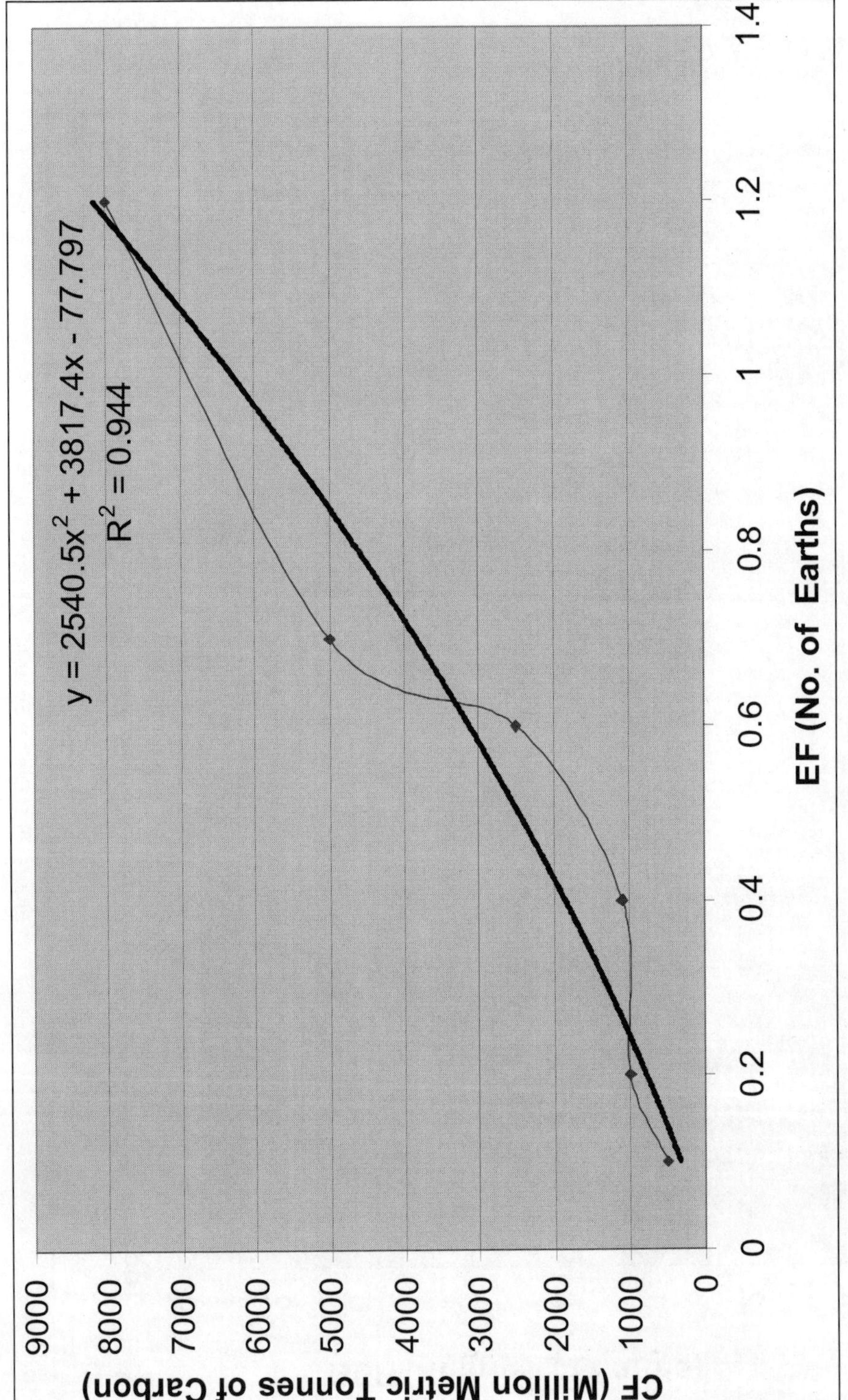

Figure 35.2: EF vs. CF.

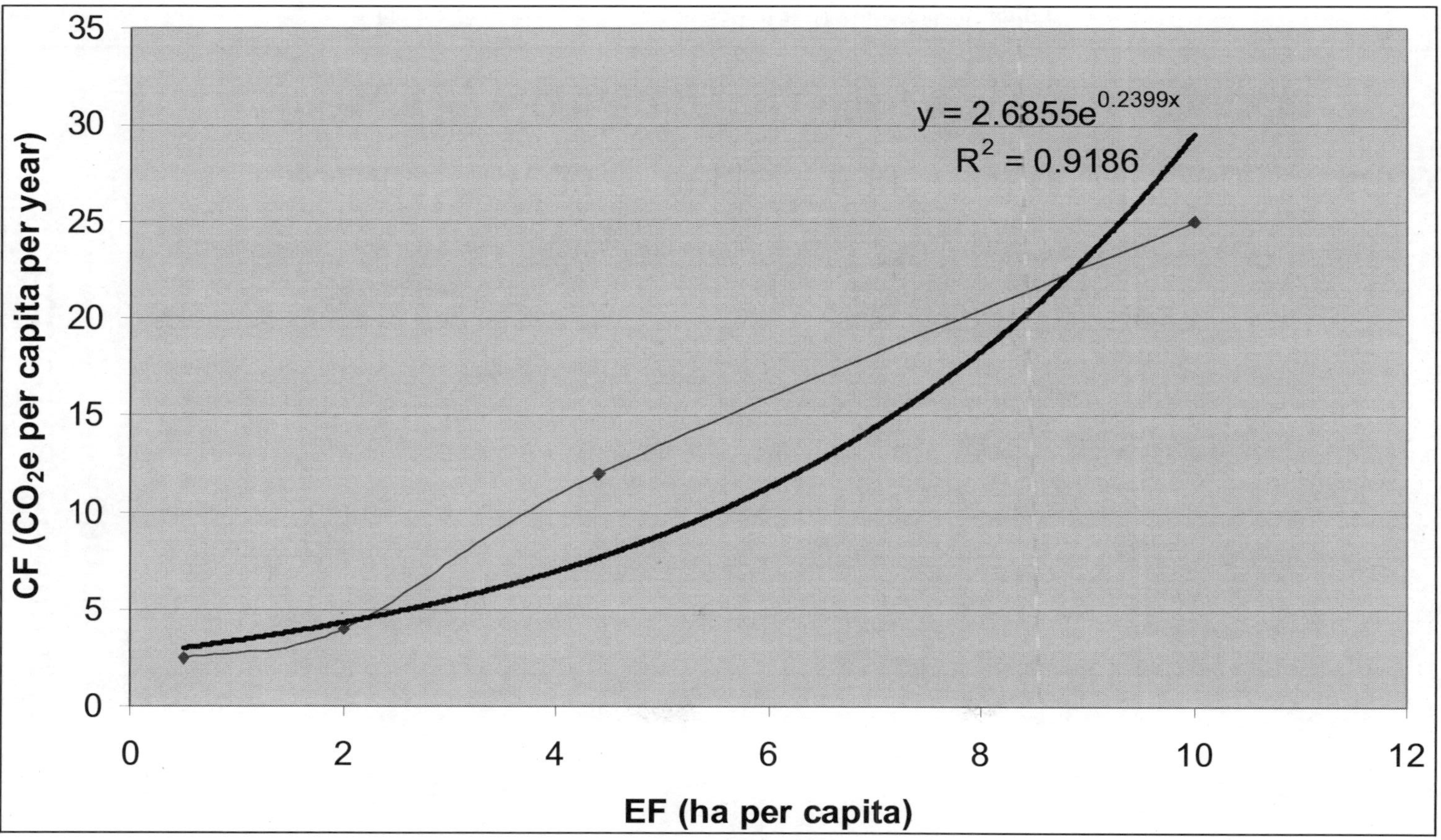

Figure 35.3: Per capita EF vs. Per capita CF.

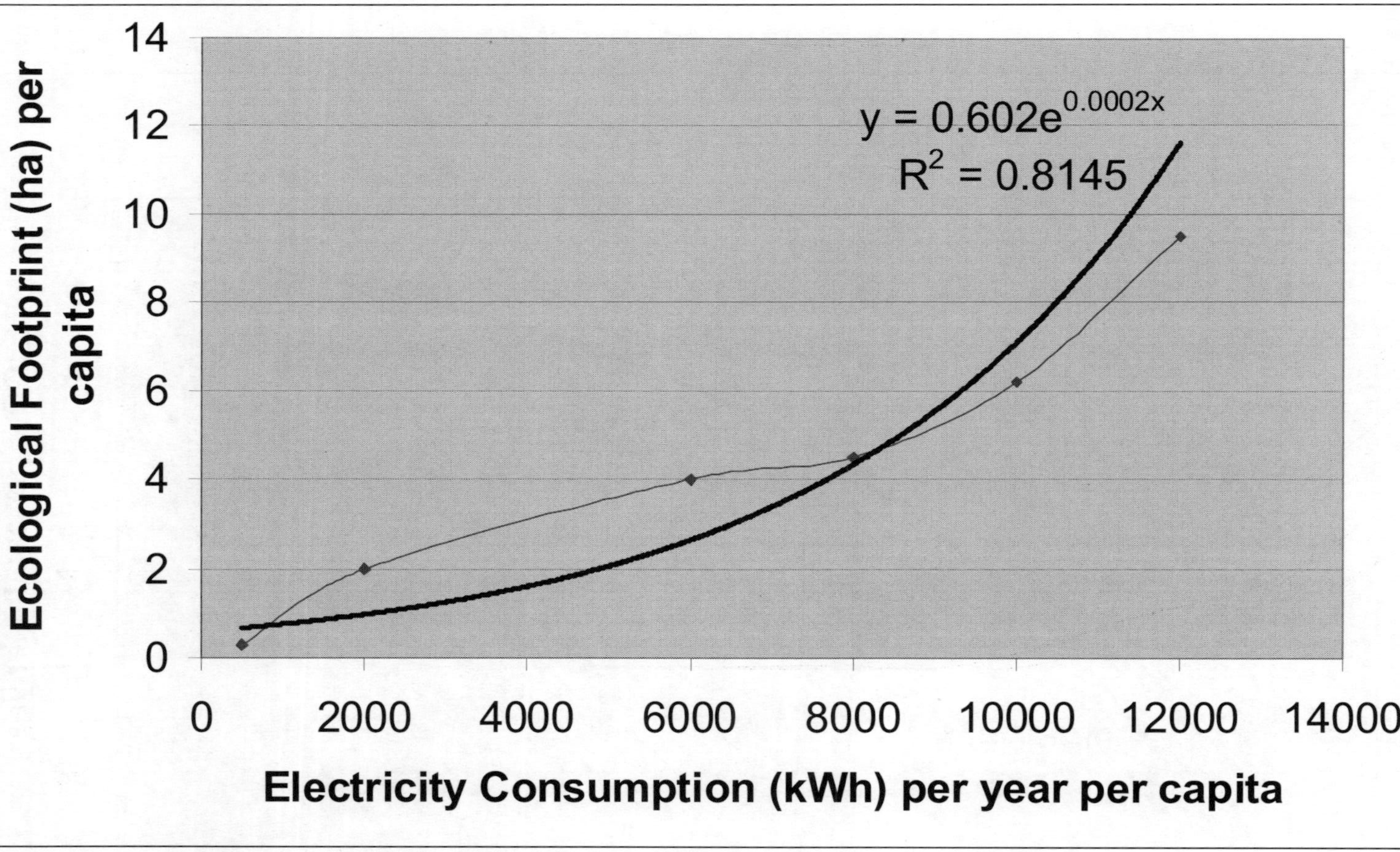

Figure 35.4: Electricity Consumption and Ecological Footprints.

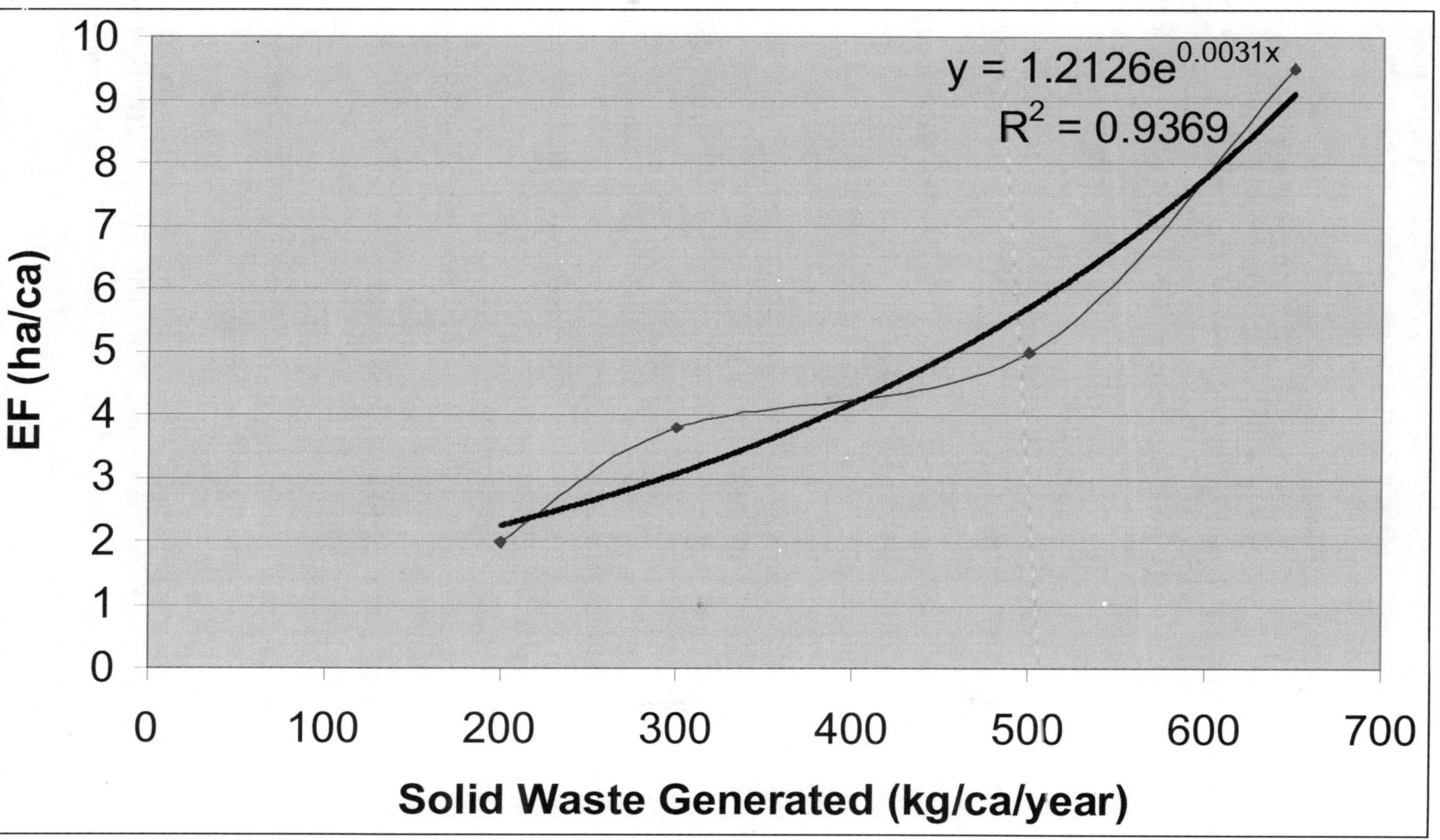

Figure 35.5: Solid Waste Generated and Ecological Footprints.

specific-emission factors for different GHGs (Pandey *et al.*, 2007). On the basis of these emission factors, region-specific ecosystem health and human health risk assessment (Pandey *et al.*, 1991-94, 2005) can be carried out.

While dealing with the issues of impact of acid rain and ecological risk assessment (Pandey *et al.*, 2002; Pandey *et al.*, 2001c), there is also a need for investigating the role of hydrological cycle in terms of environmental water demand (part of the environmental flow) (Shields and Good, 2002; Pandey *et al.*, 2006), greenhouse mitigation potential of wetlands (Pandey *et al.*, 1997), development of PFT's (Plant Function Types) (Pandey and Khanna, 1995) and methane emissions from hydroelectric reservoirs.

References

1. Brismar, A. 2002. *River Systems as Providers of Goods and Services: A Basis for Comparing Desired and Undesired Effects of Large dam Projects.* Environmental Management 29 (5): 598-609.
2. Costanza, R., Farber, S.C. and Maxwell, J. 1989. *Valuation and Management of Wetland Ecosystems.* Ecological Economics 1: 335-361.
3. Gehrels, J. and Mulamootil, G. 1990. *Hydrologic Processes in a Southern Ontario Wetland.* Hydrobilogia 208: 221-234.
4. IPCC. 2006. IPCC In: Guidelines for National Greenhouse Gas Inventories. Available at http://www.ipcc-nggip.iges.or.jp/public/2006gl/vol. 5.html.
5. Mitsch, W.J. and Gosselink, J.G. 2000. *Wetlands.* John Wiley and Sons, Inc. New York.
6. Nichols, D.S. 1983. *Capacity of Natural Wetlands to Remove Nutrients from Wastewater.* Journal of Water Pollution Control Federation 55: 495-505.
7. Pandey, J.S., Moghe, S. and Khanna, P. 1991. *Greenhouse Gases, Environmental Stress and Ecological Analysis.* IGBP Report No. 18: 2, 1991, p. 103-108.
8. Pandey, J.S. and Khanna, P. 1992a. *Speed-Dependent Modeling of Ecosystem Exposures from Vehicles in the Near-Road Environment.* Journal of Environmental Systems 21 (2): 185-192.
9. Pandey, J.S., Pimparkar, S. and Khanna, P. 1992b. *Micro-Environmental Zones and Occupancy Factors in Jharia Coal-Field: PAH-Health Exposure Assessment.* Journal of Environmental Systems 21 (4): 349-356.
10. Pandey, J.S., Pimparkar, S. and Khanna, P. 1993. *Health Exposure Assessment and Policy Analysis.* International Journal of Environmental Health Research 3: 161-170.
11. Pandey, J.S., Mude, S. and Khanna, P. 1994. *Comparing Indoor Air Pollution Health Risks in India and United States.* Journal of Environmental Systems 23 (2): 179-194.
12. Pandey, J.S. and Khanna, P. 1995. *Development of Plant Function Types for Studying Impact of Greenhouse Gases on Terrestrial Ecosystems.* Journal of Environmental Systems 23 (1): 67-82.

13. Pandey, J.S., Deb, S.C. and Khanna, P. 1997. *Issues Related to Greenhouse Effect, Productivity Modelling and Nutrient Cycling: A Case Study of Indian Wetlands.* Environmental Management 21(2): 219-224.

14. Pandey, J.S. and Joseph, V. 2001. *A Scavenging-Dependent Air-Basin Ecological Risk Assessment (SABERA) - Model Applied to Acid Rain Impact around Delhi City, India.* Journal of Environmental Systems 28(3): 193-202.

15. Pandey, J.S., Khan, S., Joseph, V. and Singh, R.N. 2001a. *Development of a Dynamic and Predictive Model for Ecological Footprinting (EF).* Journal of Environmental Systems 28 (4): 279-291.

16. Pandey, J.S., Khan, S. and Khanna, P. 2001b. *Modeling and Quantification of Temporal Risk Gradients (TRG) for Traffic Zones of Delhi City in India.* Journal of Environmental Systems Vol. 28(1): 55-69.

17. Pandey, J.S., Khan, S., Joseph, V. and Kumar, R. 2002. *Aerosol Scavenging: Model Application and Sensitivity Analysis in the Indian Context.* Environmental Monitoring and Assessment 74: 105-116.

18. Pandey, J.S., Joseph, V. and Kaul, S.N. 2004. *A Zone-wise Ecological-Economic Analysis of Indian Wetlands.*Environmental Monitoring and Assessment 98: 261-273.

19. Pandey, J.S., Kumar, R. and Devotta, S. 2005. *Health Risks of* NO_2*, SPM and* SO_2 *in Delhi (India).* Atmospheric Environment 39: 6868-6874.

20. Pandey J S. and Devotta, S. 2006. *Assessment of Environmental Water Demands (EWD) of Forests for Two Distinct Indian Ecosystems.* Environmental Management 37 (1), 141-152.

21. Pandey, J.S., Wate, S.R. and Devotta, S. 2007. *Development of emission factors for GHGs and associated uncertainties.* PROCEEDINGS: 2nd International Workshop on Uncertainty in Greenhouse Gas Inventories. International Institute for Applied Systems Analysis, A-2361 Laxenburg, Austria, 27-28 September, 2007.

22. Pandey, J.S. 2009. *Inter-disciplinarity of Issues Connected with Climate Change, Food Security and Energy Alternatives.* The International Journal of Climate Change Impacts and Responses, Volume1 (In Press)

23. Pandey, J.S., Wate, S.R. and Chakrabarti, T. 2009a. Invited Lecture: *Carbon and Ecological Foot-printing in the Residential, Commercial and Industrial Context with particular Emphasis on Designing Green Buildings.* Proceedings: Seminar on Green Building 23rd January, 2009, Organized by Chief Engineer Bareilly Zone Under the Aegis of Chief Engineer Central Command, Military Engineer Services, Bareilly, U.P. India.

24. Pandey, J.S. 2009b. Invited Lecture: Role of *Carbon and Ecological Foot-printing (CF and EF) in the Context of Innovative Technologies and Sustainable Development.* Proceedings: Seminar on Environmental Issues of Jharkhand June 5, 2009, Organized by Jharkhand State Pollution Control Board, Ranchi, India.

25. Rees, W.E. and Wackernagel, M. 1996. *Urban Ecological Footprints: Why Cities Cannot be Sustainable–and Why They are a Key to Sustainability.* Environmental Impact Assessment Review 16: 223-248.

26. Rees, W.E. and Wackernagel, M. 1999. *Monetary Analysis: Turning a Blind Eye on Sustainability.* Ecological Economics 29: 47-52.

27. Shields, J., and Good, R. 2002. Environmental Water in a Regulated River System: The Murrumbidgee River Planning Approach to the Determination of Environmental Needs. Water Science and Technology 45: 241–249.

28. USEPA. 2005. Global Anthropogenic Non- CO_2 greenhouse Gas Emission: 1990-2020 Available at http://www.epa.gov/methane/pdfs/global_emissions.pdf

Chapter 36

Influence of Open Caste Mining the Change of Climate of West Central Arid Region of India

P. Kumar

MIT College of Engineering, Pune

ABSTRACT

Yearly expansion of mining region in the arid zone of west central India, as observed by satellite imageries, has been studied in relation to the incessant decrease in the rainfall and the dust storms, over the region. With the increase in mining activity in the region, the total yearly rainfall in this region shows significant decrease. But for the winter months, all other seasons indicate the same trend. Twelve years (1990-2002) data of dust storm shows significant drop in number. This decrease shows good correlation (r = 0.8) with the rise in mining area over these years.

The cause may be attributed to extra dust loading of the atmosphere, increasing the depletion of incoming SWR. Yearly shrink in the arid region, change in land salinity and increase in the forest cover, which changes the surface absorption parameters could possibly be the other causes.

Keywords: *Climate change, Open caste mining, West Central India, Arid zone.*

Introduction

West central India is comprised of extremes in its stretch from Aravali ranges in the west to Thar Desert in the east. Aravali ranges, which stretch from Delhi to

Ahemedabad, possess average height of 737m (2418ft) with maximum height of 1720m (5650ft) near mount Abu. These ranges have not only barricaded the advance of Thar desert to further east but they also provide the watershed zone resulting into Luni river which traverses across 320 km, supplying water up to the head of Rann of Kutchch.

Region is extremely rich[5] in minerals, oil, natural gas and rare arid zone vegetation resources. It is also the leading producer of calcite, lead concentrate, ball clay, fireclay, ochre, phosphorite, silver and steatite. But it is best known for its production of marble, sandstone, marble and other stones. It produces 10 per cent of the world's and 70 per cent of India's output of sandstone. Ajmer, Bhilwara, Bikaner, Dungarpur, Jaipur, Pali, Rajsamand, and Udaipur are its main mining districts. Rapid advancement in technology in yesteryears has seen accelerated growth of opencaste mining in the region. Also human interference through arid-horticulture and floriculture in the natural settings of this region has lead to phenomenal morphological changes.

Massive aforestation campaign at governmental and NGO's level by the plantation of *Vilayati Babool* (Prophylaxis Juli Flora) and increase in the irrigated land under the Indira Gandhi Canal through Bikaner - Jaisalmer - Barmer region and its sub-canals spreading up to Jodhpur, has converted large chunk of arid and semi-arid region of west central India into fertile land. Effect is also reflected in the eco-system of the region.

Due to the shortening of the diurnal range of the temperature and moderation of the extreme temperatures in the region, vis-à-vis changes in the regional climate; it has now turned in to suitable breeding environment for the mosquitoes and mosquito–borne diseases[1,2,3,4.6], which were unheard in the region. Mean decadal frequency of more than 40^0 C temperature in the month of May and June over Jodhpur, during 1981-90 and 1990-2000 has reduced from 25 days and 19 days to 19 days and 14days, respectively.

Present study aims to highlight this phenomenal change in climatic features vis-à-vis the growth of mining in western Rajasthan.

Landscape

Figure 36.1 shows the west central region of India. Figure 36.2 shows the enlarged satellite picture of Rajasthan. Due to the centralized location of Jodhpur it has been taken as the representative region for the present study.

Sparse vegetation cover of Rajasthan is primarily located in north or western regions. Mining is spread all around due to mineral rich region.

Mining in Rajasthan

All the mines of Rajasthan are surface mines only. Poor technology still under operation causes sufficient quantity of dust spewed up in the atmosphere after every blasting. Post opencaste-mine blast cloud is shown in Figure 36.3.

Figure 36.1: West Central Region of India.

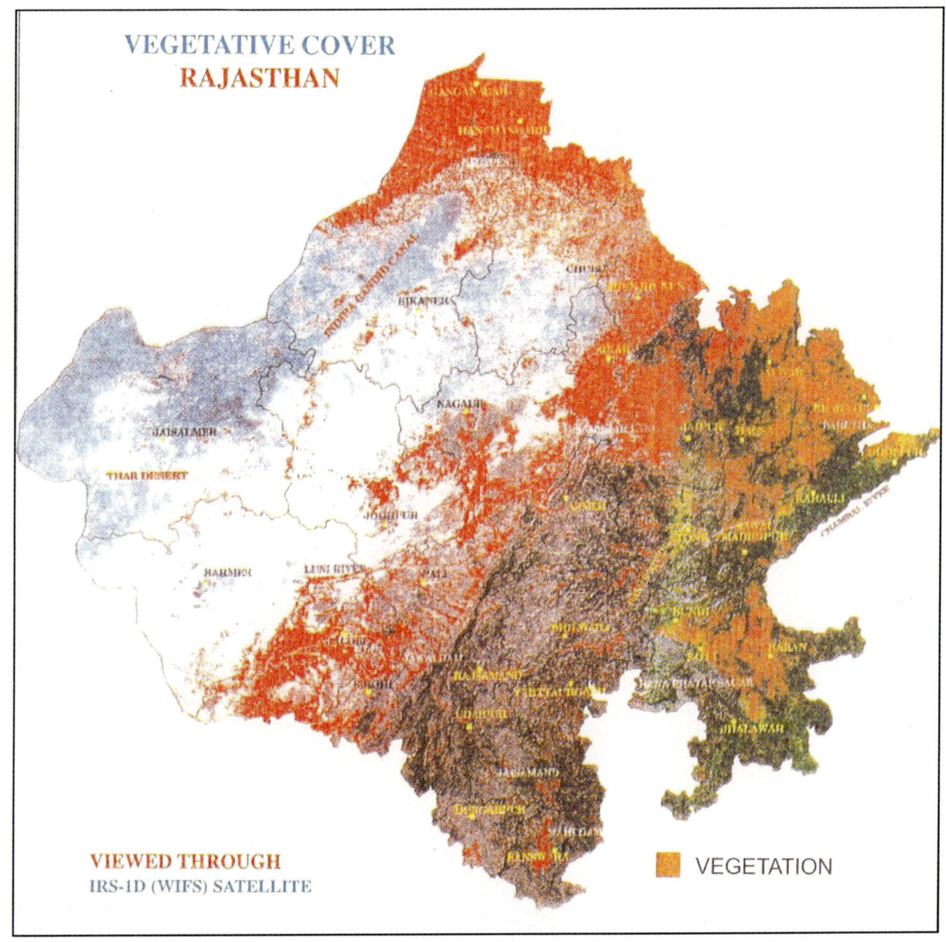

Figure 36.2: Vegetation Cover of Rajasthan.

Figure 36.4 shows the extensive mining taking place in these regions. Aravali range acts as a natural buffer checking the Thar Desert from expanding towards northern and eastern agricultural plains. The increasing gaps in the Aravali range due to unchecked mining activities add to the worries of ecologists. Loose dust out of the mining region goes up in atmosphere obstructing the incoming short wave solar radiation.

Satellite Observation

Figure 36.5 shows IRS LISS-2 satellite imageries of the mine areas of Jodhpur. Observe the rapid growth of the mining region from 6.8 sq.km in 1988 to 12 sq. km in 1999. Area was 33.6 sq. km in 2003, as per the latest National Remote Sensing Agency publication.

Figure 36.3: Cloud of ust ta ing arge uantit of oil u into t os here.

Growth of Mining and its Relation with the Climate

Figure 36.6 shows the line diagram of yearly growth of mine area in Jodhpur. Note the steep rise in the mining region between 1999 and 2003.

Drop in Climatic Temperature

Normally May is the month of highest temperature in Jodhpur (Figure 36.7).

It has been observed that there has been rapid drop in the decadal average of hot days having maximum temperature more than 40^0 C in the month of May and June. During 1981-90 the average for the months of May and June was 25 and 19 days whereas during 1991-2000 this average came down to 19 and 14 only (Figure 36.8).

Decrease in Dust Storms

Yearly numbers of dust storms are shown in Figure 36.9. Note marked decline in the trend during the last decade.

Figure 36.10 shows the graph between growth of mine area and number of dust storms over Jodhpur. Correlation coefficient(r) is–0.83.

Heaps of waste dumps

Figure 36. : tensive ining in Rajasthan.

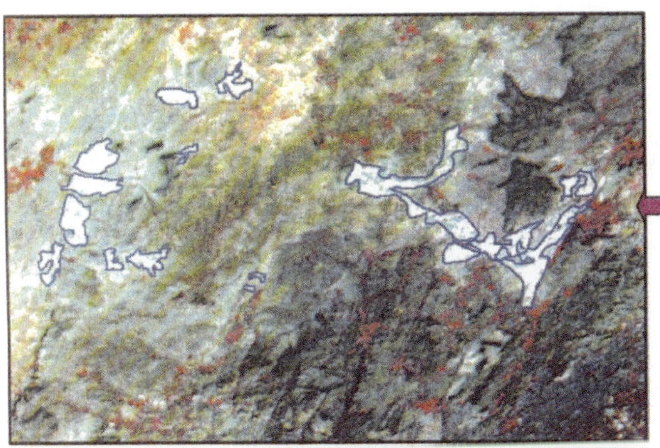

Figure 36. : IR atellite i ture.

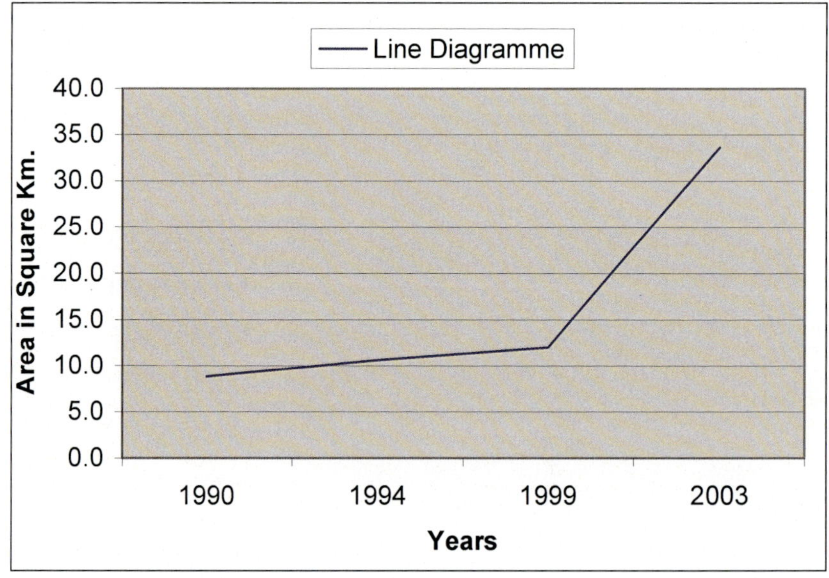

Figure 36.6: ro th of ining rea in odh ur.

Discussion

Mine rich Rajasthan has the highest number of mine leases in the country–1,324 leases for major minerals, 10,851 for minor minerals and 19,251 quarry licenses for mining stones. As per the data provided in the report, Rajasthan holds reserves for 44 major and 22 minor minerals and is the only producer of garnet, jasper, selenite, wollastonite and zinc concentrates. High negative correlation ($r \approx -0.83$) between growth of mining area in Jodhpur and gradual decline in the dust storms could be due to large quantity of dust being spewed up during blasting processes. Table 36.1 below shows the resident time of the various life of SPM's in atmosphere.

a le 36.1: ife i e of .

Size Range (μm)	*Effective Radius (μm)*	*Life Time (Hours)*	*Source Strength (MT/Yr)*	*Mass Loading Percentage*
0.1–0.18	0.15	231	2	0.5
0.18-0.3	0.25	229	18	4
0.3-0.6	0.4	225	52	12
0.61-1	0.8	219	150	31
1-1.8	1.5	179	250	32
1.8-3	2.5	126	250	13
3-6	4	67	250	5
6-10	8	28	250	2

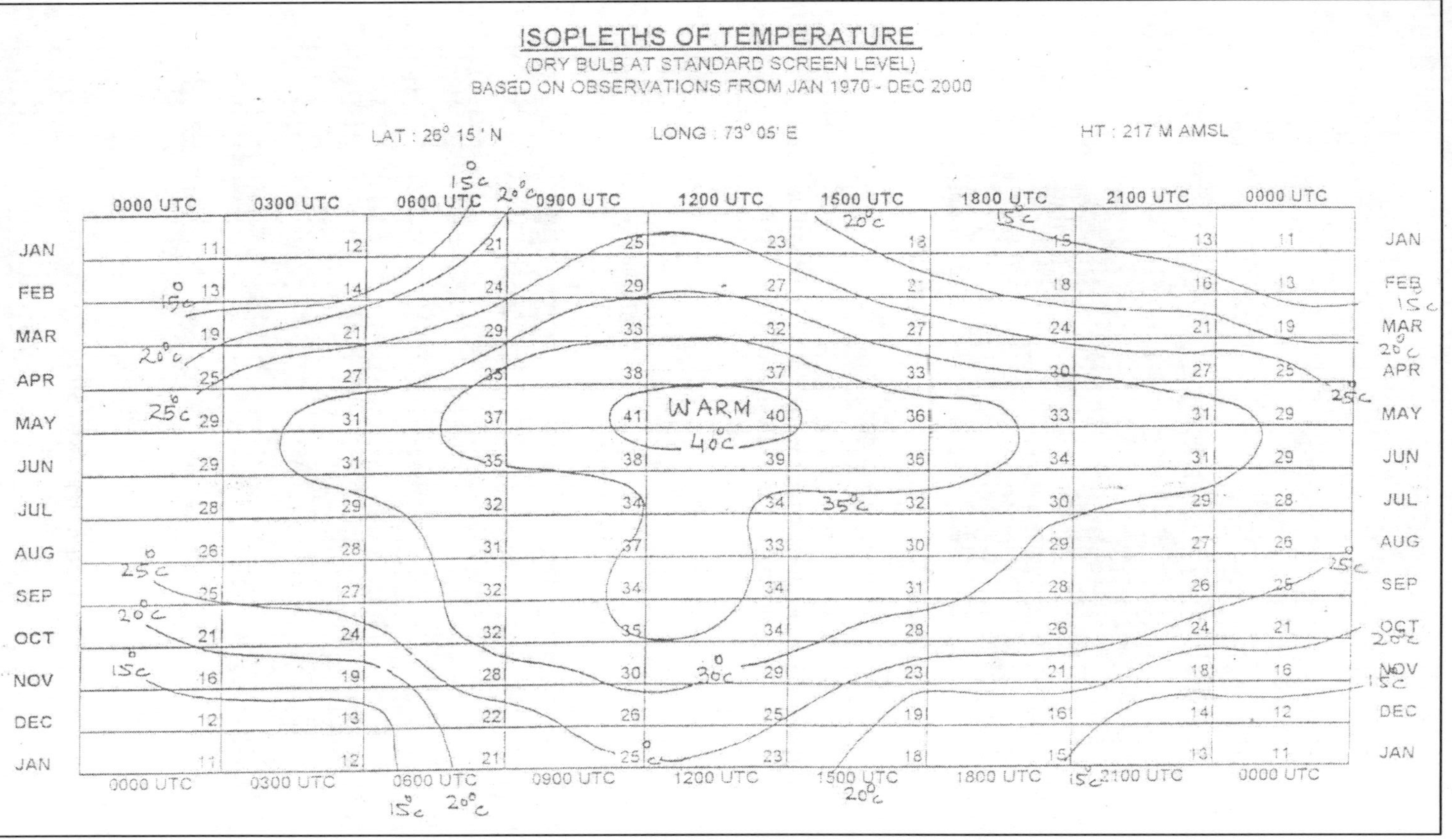
ISOPLETHS OF TEMPERATURE
(DRY BULB AT STANDARD SCREEN LEVEL)
BASED ON OBSERVATIONS FROM JAN 1970 - DEC 2000
LAT : 26° 15 ' N
LONG : 73° 05' E
HT : 217 M AMSL
0000 UTC
0300 UTC
0600 UTC
0900 UTC
1200 UTC
1500 UTC
1800 UTC
2100 UTC
0000 UTC
JAN
FEB
MAR
APR
MAY
JUN
JUL
AUG
SEP
OCT
NOV
DEC
JAN
WARM

Figure 36.7

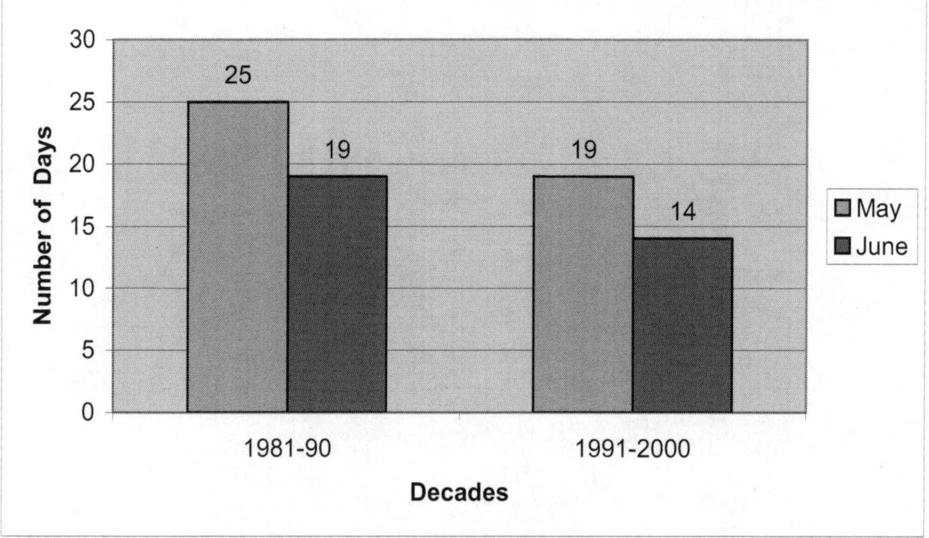

Figure 36.8: Decadal Average Numer of Days of >40 degree in May and June Over Jodhpur.

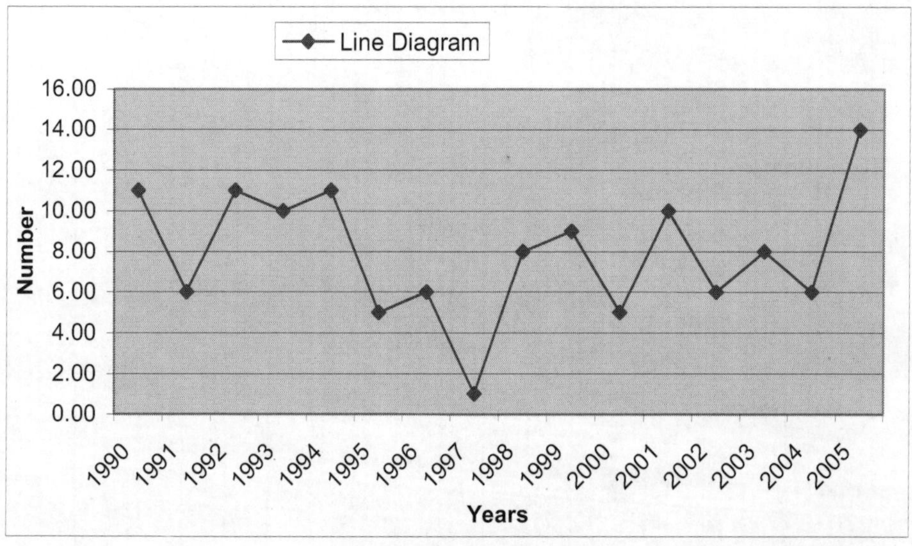

Figure 36.9: Dust Storms.

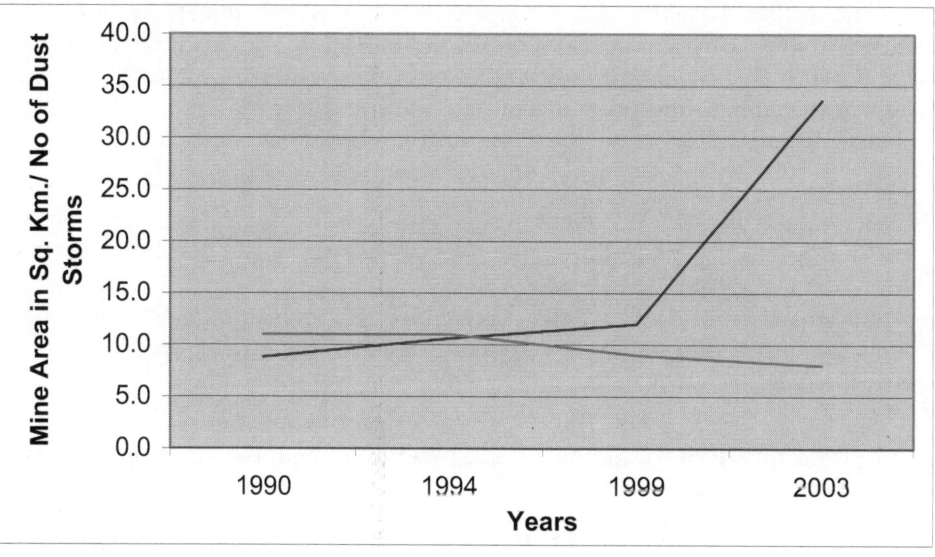

Figure 36.10: Dust Storm and Mine Area.

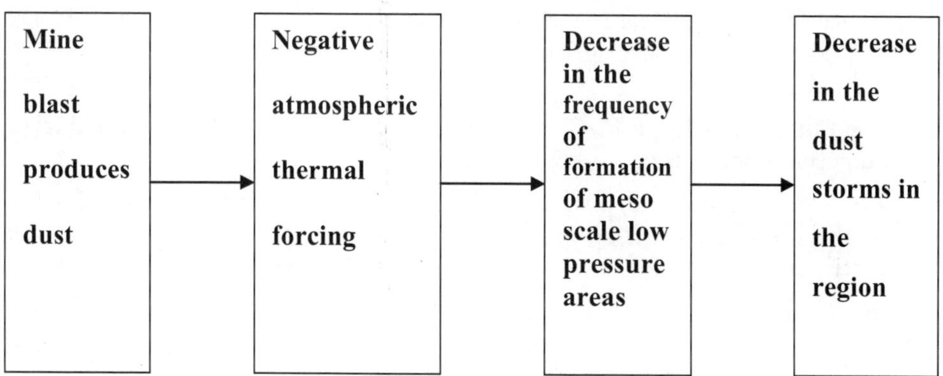

Figure 36.11: Schematic of Mining and decrease of Dust storms in Rajasthan.

Table 36.2 shows the range and the average thermal forcing of different types of SPM's. It may be noted that mineral dust forcing ranges from–0.6 to + 0.4.

Table 36.2

Aerosols	*Range of Thermal Forcing Wm-2*	*Average*
Sulphate	–0.2 to–o.8	–0.4
Biomass Burning	–0.03 to–0.3	–0.1
Fossil Fuel Organic Carbon	+0.1 to +0.4	+0.2
Mineral Dust	–0.6 to + 0.4	

Poor quality of blasting technology used by the opencaste miners seems to cause sufficient mineral dust loading in the atmosphere affecting the forcing factor towards negative end. This could have resulted in reducing the number of meso-scale low-pressure systems during premonsoon period. The lesser are the meso-scale low pressure systems the lesser are the dust storms in the region.

Schematic of mining and decrease of dust storms in Rajasthan was shown in Figure 36.11 explains the process.

References

1. Bansal, S.K. Singh, K.V., Relative susceptibility of some common mosquito vector larvae to synthetic compounds in north-western Rajasthan. J Environ Biol 28 (2007) 829.
2. Bansal, S.K. and Singh, Karam V., Laboratory evaluation for comparative insecticidal activity of some synthetic pyrethroids against vector mosquitoes in arid region. J Environ Biol 27 (2006) 251.
3. Bansal SK, Singh KV., Efficacy of different organophosphate and synthetic pyrethroid insecticides to the larvae of malaria vector *Anopheles stephensi*, Liston. *J Environ Biol.* 2004;25:485- 8
4. Bansal SK, Singh KV., Insecticide susceptibility status of *Aedes aegypti* to DDT and dieldrin in desert and non-desert parts of Rajasthan. *J Environ Biol.* 2003;24:113- 6
5. Chaudhari, J.S. Dhir, R.P., Clay mineralogy of Arid zone soil developed from quaternary deposits of west Rajasthan.
6. Singh, K.V. and Bansal, S.K., Larvicidal potential of microbial agents *Bacillus thuringiensis* var. *israelensis* and *Bacillus sphaericus* against different mosquito vectors prevalent in arid areas. Ann Arid Zone 46 (2007) 217.

Chapter 37

Influence of Inorganic Carbon on Fluorescence Measurement of Dissolved Organic Matter in Water Bodies

Nivedita P. Kulkarni, D.M. Dharmadhikari and S.R. Wate

CSIR, National Environmental Engineering Research Institute, Nagpur

ABSTRACT

Molecular spectroscopy has wildly been accepted for measurement of dissolved organic carbon (DOM) in water bodies. Present research work tried to find out the effect of inorganic carbon (IC) concentration on total fluorescence measurement of water bodies. Study has been carried out on four different types of water bodes *viz.* domestic sewage, lake water, dairy wastewater and herbal pharmaceutical wastewater. It shows that concentration of IC along with its intimacy with organic carbon alters the total fluorescence intensity of the sample. IC plays a major role in measurement of DOM using fluorescence spectroscopy.

Keywords: *Inorganic carbon, Organic matter, Fluorescence spectroscopy.*

Introduction

With rapid industrialization and population increase, the pollution of water resources is on the increase, which is caused by the discharge of industrial wastewater and sewage in to natural water bodies. Under the ecological point of view, organic and inorganic carbon measurements give quantitative information about the carbon cycle and productivity of natural bodies of water (1). Dissolve organic matter (DOM) concentration and composition both directly and indirectly influence the biology (microbial and plankton ecology, *e.g.* Williamson *et al.*, 2001), chemistry (trace metal speciation and transport, e. g. Berault *et al.*, 1996; polycyclic aromatic hydrocarbon toxicity *e.g.* Diamond et.al, 2000) and physics (optical properties, *e.g.* Bricaud., 1981) of aquatic environment.

Quality of the water has long been defined in terms of concentration of organic carbon present. Hence, it is of quite important parameter to organizations concerned with water supply and effluent treatment plants. In pollution abatement work, assessment of carbon load in natural water bodies is necessary to control the level of pollution. Standard methods for the determination of levels of organic matter in water are chemical oxygen demand (COD), biochemical oxygen demand (BOD) and total organic carbon (TOC) analysis. These methods are slow and tedious. The use of these testes severely limits the number of samples which can be analyzed. (6)

Present study tried to find the influence of inorganic carbon salt matrix which is highly associated with organic carbon on the fluorescence study of DOM. Present research work includes the study of effect of inorganic carbon on fluorescence measurement of four types of water-bodies *viz.* sewage water, lake water and tertiary treated wastewater from dairy and ayurvedic herbal pharmaceutical wastewater. DOM due to its complex polymeric nature is responsible for the fluorescence, but inorganic carbon (IC) which is embedded with it must have some role on total fluorescence. Hence, in the measurement study of DOM with fluorescence spectroscopy the effect of IC on total fluorescence must taken into consideration.

Experimental

Source of Sample

Four different group of water samples were collected for the fluorescence analysis from various sources in and around Nagpur (India). The first sample was a wastewater sample from a settled domestic sewage. Second one was from the natural source of Ambasary Lake. Third and fourth samples were tertiary treated wastewater effluents from dairy industry and baidhyanath ayuervedic medicine industry as an herbal pharmaceutical wastewater respectively.

Gross sampling was done weekly for a period of six months. Samples were collected using high density polyethylene bottles and were stored below 4°C in the dark prior to analysis. All samples were allowed to reach at room temperature before analysis. Analysis was done within 24 hr.

Instrumental

Fluorescence measurements were made on the Hitachi; F-4500 fluorescence spectrophotometer. The intense 150 watt xenon source provides maximum light energy over the entire 200-900 nm wavelength range of the spectrophotometer. The cuvette used for the spectral analysis was of high grade quartz composition with a 10 mm path length. The spectra (excitation and emission) were corrected for instrument biases. A rhodamine B quantum counter is used to check the spectral accuracy on excitation side and emission correction spectrum derived using ground quartz diffuser (as recommended by manufacturer). Comparison of the integrated Raman spectra of Milli-Q water over excitation wavelength reveled no change in lamp intensity during the measurement period

Analysis

Samples were acidified to pH 2 with orthophosphoric acid and spurge for 8-10 min with ultra pure air to remove inorganic carbon. Proper care of pH change has been taken by maintaining the original pH value by using alkali. All the samples were studied in normal condition of pH and temperature. As the pH range of all the water-bodies was in the range of 5-8 and fluorescence intensity remain unaffected in this range (7).

Results and discussion

The aim of the study is to find out the relation of fluorescence intensity with total dissolved carbon of the water bodies where organic carbon is embedded with inorganic carbon matrix. Study of fluorescence measurement for all the samples was carried out after selection of excitation and emission wavelength at optimized instrumental parameters which are kept constant throughout the experiment. Optimized instrumental parameters for all the samples are given in the Table 37.1.

Table 37.1: Optimized Instrumental Parameters of Sample.

Source of Sample	*Excitation Wavelength (Ex.)*	*Emission Wavelength (Em.)*	*Ex./Em. Slit Width (nm)*	*PMT Voltage (V)*
Domestic Sewage	290	350	5.0/10.0	700
Lake water	304	405	2.5/5.0	700
Dairy Wastewater	280	338	5.0/10.0	950
Herbal Pharma Wastewater	373	480	2.5/5.0	700

Response–Auto, Scan Speed (nm/min) - 240, were kept constant for all samples.

Fluorescence measurements before and after removing IC from the samples were carried out. Results are presented in Figures 37.1–37.4. It is seen from the figures; in all cases fluorescence intensity were decrease after removing IC. It shows that fluorescence intensity of DOM has some influence on concentration of IC. Detail discussion on fluorescence study and influence of IC on total fluorescence measurement have been discussed below.

Sewage Sample

For sewage sample major fluorescence signals was obtained at about 350nm when excited at 280nm.From Figure 37.1 it was observed that after removing IC from the sample fluorescence intensity was decreases. But, at wavelength of 430 nm fluorescence intensity was increases. Peak at 430nm is responsible for non-biodegradable organic compounds (8). From the study it was seen that IC has great

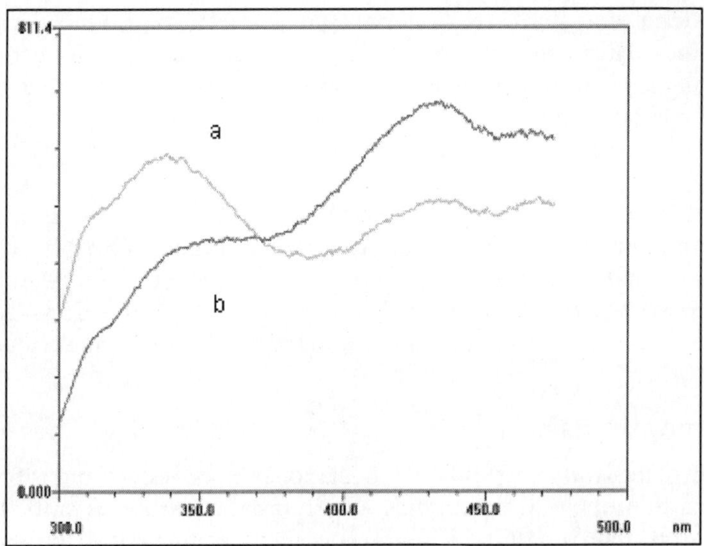

Figure 37.1: Influence of IC Removal on Sewage.
a: Before removing IC; b: After removing IC.

influence on fluorescence intensity of sewage. For biodegradable compounds concentration of IC also contributes in the total fluorescence measurement. But, for non biodegradable compounds presence of IC quenches the total fluorescence.

The presence of dissolved oxygen is known to quench some fluorescesing species. However, previous unpublished research has shown that sewage fluorescence is not quenched by the presence of dissolved oxygen and will not be affected by aeration (Reynolds, D.M; Ahmad, S.R. 1997).

Lake Water

A normalized fluorescence spectrum of lake water was obtained at 405nm. For lake water we are getting weak fluorescence, because aquatic DOM in lakes and rivers is often quite dilute and present with an inorganic salt matrix (Westerhoff Paul, *et al.*, 2001).

Influence of IC study on lake water also shows decrease in fluorescence intensity of the sample after removing IC (Figure 37.2). Lake water is normally associated with inorganic salt matrix and responsible for weak fluorescence signals as per previous study but it was observed that after removing IC fluorescence intensity shows decrease in total fluorescence. From the study it was seen that IC also plays a major role in fluorescence measurement.

No interference will be obtained from chlorophyll or any other common natural pigment, chlorophyll fluoresces strongly in the far red part of the spectrum with maximum at 665 nm. (Yentsch and Menzel, 1963). Measurement of DOM from natural water bodies are of interest as interaction of humic substances with some charged pesticides, herbicides, xenobiotics and some metal ions in order to predict rate and

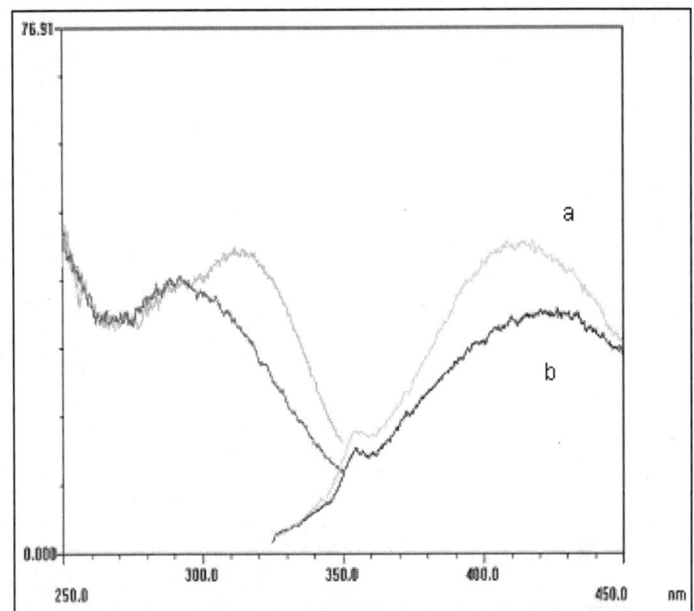

Figure 37.2: Influence of IC Removal on Lake Water.
a: Before removing IC; b: After removing IC.

transport of these chemicals (Frimmel, F 2002).In remote sensing it is important due to their effect on light propagation in natural waters.

Dairy Wastewater

Dairy wastewater comprises of fluorophores of protein fractions of amino acid which are responsible for fluorescence at 338 nm with excitation at 280nm. A liquid chromatographic study with fluorescence detection showed that the two fluorescent acids tyrosine and tryptophan are present in the soluble fraction of milk (13). Tyrosine and tryptophan are responsible for the intrinsic fluorescence of the wastewater from the dairy industry. Standard tyrosine and tryptophan are highly fluorescent compounds, which in turn responsible for intense fluorescence obtained from dairy wastewater.

After removing IC from the dairy sample highly decrease in fluorescence intensity was observed (Figure 37.3). Dairy wastewater as mainly comprises of protein fraction which is intern responsible for the fluorescence of the sample must associated with inorganic salt matrix. Study shows a concentration of IC should taken into consideration while studying the DOM of dairy sample using fluorescence spectrophotometer.

Herbal Pharmaceutical Wastewater

The brownish yellow color of wastewater is a hindrance for analysis by fluorescence spectrophotometer. This is due to the different additives used to various

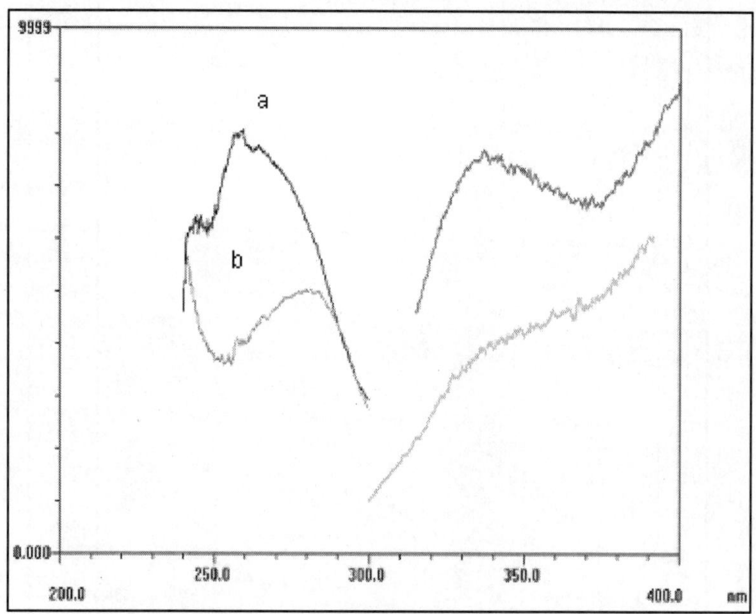

Figure 37.3: Influence of IC Removal on Dairy Wastewater. a: Before removing IC; b: After removing IC.

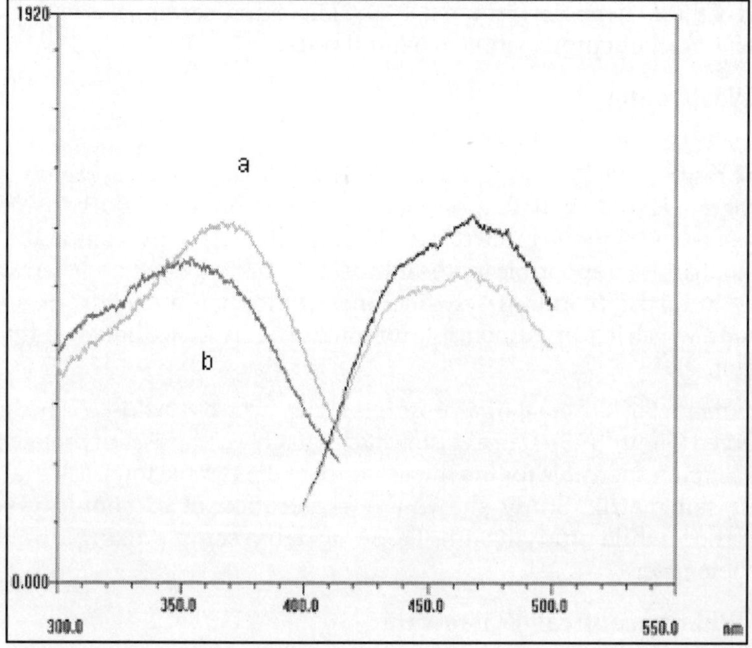

Figure 37.4: Influence of IC Removal on Pharmaceutical Wastewater. a: Before removing IC; b: After removing IC.

drug preparations The raw material used can be listed as the part of medicinal plants like herbs, skin, stumps, fruits, flowers, root, seeds, gum, oils etc. the synthetic chemicals added are alcohol, sugar, gelatin, lactose, solvents and special type of minerals and clays. High density and colored samples of herbal pharmaceutical wastewater were diluted by using 0.01M KCl solution in order to nullify inner filter effect due to absorbance at lower wavelength region and also helps to maintain the pH of the samples. Ionic strength in the range of 0-1M KCl and humic acid concentration in the range of 5-100 mg/L had little effect on fluorescence spectral characteristics. (Jarafshan J. M., *et al.*, 1996).Dilution factor was considered at the time of fluorescence intensity measurements.

Fluorescence study of the herbal pharmaceutical wastewater shows weak fluorescence. Excitation wavelength was in the range of 300 - 380 nm and emission wavelength was in the range of 350–480nm. These figures agree closely with the emission maxima found for the wide range of colored samples in the United States reported by Ghassemi and Christman (1968). Most of the time the Ex and Em was around 373nm and 480 nm respectively. But, it was noted that composition of DOM of wastewater accompanies with the nature of the products i. e. liquids, solids and semisolids and raw materials used. Study of influence of IC on total fluorescence of herbal pharmaceutical wastewater also shows decrease in fluorescence intensity. Though sample consists of high concentration of IC, removal of IC from the sample does not shows much decrease in total fluorescence. From this we can say that IC was not intimately associated with organic carbon.

References

1. Fadini, P.S.; Jardim, W.F.; Guimaraes, J. R. *J. Braz. Chem.Soc.*2004, 15 (1), 131-135.
2. Williamson, C.E.; Neale, P.J.; Grad, G.;De Lange, H.J.Hargreaves, B.R. *Ecol.Appl.* 2001,11,1843–1857.
3. Berault, R.F.; Colman, J.A.; Aiken, G.R.; McKnight, D.M. *Environ.Sci.Technol.* 1996, 30, 3477-3486.
4. Diamond, S.A.;Mount, D.R., Burkhard, L.P.Ankley, G.T.;Makynen, E.A.Leonard, E.N. *Environ. Toxicol. Chem.* 2000, 19,1386–1396.
5. Bricaud, A., Morel, A., prieur, L *Limnol.oceanogr*. 1981, 26, 43 -53.
6. Smart, P.L; Finlayson, B.L; and Rylands, W.D and Ball, C.M. *Water Research.* 1976, 10, 805-811.
7. Dharmadhikari, D.M.; Barhate, N.M. *J. Indian Association for Environmental Management.*2003, 30, 198–204.
8. Reynolds D.M *J. of Chemical Technology and Biotechnology.* 2002, 77 (8), 965 -972.
9. Reynolds, D.M; Ahmad, S.R. *Water Research.* 1997, 31(8), 2012- 2018.
10. Westerhoff, P; Chen, W; Esparza, M. *J. of Environmental Quality*.2001, 30, 2037-2046.
11. Yentch, C.S; Menzel, D. W. *Deep Sea Res.* 1963, 10,221-231.

12. Frimmel, F.H; Abbt-Braun, G.; Heumann, K. G.; In *Refractory Organic Substances in the Environment;* Hock, B.; Ludemann H.- D.; Spiteller, M: Federal Republic of Germany, 2002;Chapter 2, pp 200- 214.

13. Pulgarin, J. A. Murilo; Molina, Aurelia Alanon; Pardo, M.Teresa Alanon. *Analytica Chimica Acta*. 2005, 536,153-158.

14. Jarafshan, J. M.; Hemmingsen, S. L; Autry, J. L.; McGown, L. B. *Environmental Science and Technology*. 1996, 30 (10), 3061–3065.

15. Ghassemi. M; Christman R.F *Limnol. Oceanogr*. 1968, 13, 583-597.

Chapter 38

Size Distribution of Heavy Metals Associated with Particulate Matter in Commercial-cum-Residential Area of Nagpur City

P.P. Pipalatkar, D.G. Gajghate, V. V. Khaparde, V.M. Shinde and C.V. Chalpati Rao

CSIR-National Environmental Engineering Research Institute (NEERI), Nehru Marg, Nagpur

ABSTRACT

A study of the atmospheric particulate size distribution of PM10 and associated heavy metal concentrations has been carried out in residential cum commercial area of Mahal at Nagpur city. 24 hourly sampling of size fraction of particulate matter was performed during winter season using eight-stage cascade impactor with a pre-separator and back-up filter. Samples were analyzed in nine-particle size fractions with reference to toxic metals using inductively coupled plasma- optical emission spectroscopy (ICP-OES).The average concentration of PM 10 and fine particulate matter (effective cut of aerodynamic diameter ≤ 2.2um) was found to be 300 μg/m^3 and 136.7 μg/m^3 exceeding the limit of CPCB and USEPA respectively. Maximum 15 per cent of percentage mass concentration in size range of 9.0-10.0μm and minimum 6.8 per cent in size range of 2.1-3.3 μm were observed.

Among all size fractions, maximum contribution of Sr, Ni As, Pb, Zn, Pb, Fe, Si and Cu was observed in size range of 0.4-0.7 μm, 0.4-0.7 μm, 0.7-1.1 μm, 1.1-2.1 μm, 2.1-3.3 μm, 3.3-4.7 μm, 4.7-5.8 μm, 5.8-9.0 μm and 9.0-10 μm respectively.

Metals (Sr, Ni and Zn) have large proportions below 0.7 pm diameter and could therefore pass directly into the alveoli. Other metals could conceivably penetrate at least as far as the bronchi and some as far as the bronchiole. Emission Inventory of PM 10 for point and area sources were also carried out and result indicates that thermal power plant and construction activity was the major contributor towards PM 10 emission. This study shall be helpful to identify problems of different size fraction of PM 10 along with associated toxic metals and their contribution sources for formulation of appropriate control measures and mitigation strategies for minimizing the levels of particulate matter.

Keywords: *Size fractions, Cascade impactor, Trace metals, Emission inventory.*

Introduction

In ambient air, there are numerous particles whose aerodynamic diameters range from less than 0.01 µm to larger than 100 µm. The amount of hazardous compounds in the various particles varies considerably and is dependent on their aerodynamic diameters. Aerosol size is vital in understanding its impact on human health and dynamics of the aerosol associated processes (Morawska *et al.,* 1999).Recent epidemiological studies suggest that inhalation of fine particulates of less than 2.5 µm aerodynamic diameter (PM2.5) can cause respiratory diseases [Trijonis, 1983;Reichhardt T, 1995; Pope CA *et al.*, 1995 (a, b)]. Fine and ultra-fine particles can be deposited deep in the lungs and their residence time will be very long, maybe up to several months [WHO, 1996]. Particles smaller than 10 µm in diameter get respired inside our body (Mc Cornac 1971). Particles with aerodynamic diameter <5 µm are filtered in the nose for the most part and the particles, less than 1–2 µm in aerodynamic diameter, are mostly deposited in the alveolar region of the lung where the adsorption efficiency for trace elements is 60–80 per cent (Infante R, 1991)and can affect lung physiology. Hence size distribution of the particulate matter is therefore the most important property with respect to its penetration and effects in the respiratory system (Lodge, 1981).

Size distribution of atmospheric particles with trace metals are found more important since this not only influences the toxicity of a metal when inhaled but also controls the extent to which metals may be dispersed via atmospheric transport and hence is a prerequisite for the determination of rates of deposition of metals to the Earth's surface. The distribution of trace metals within atmospheric particles in all over world has been studied at the North Sea (Otten *et al.*, 1994; Ottley and Harrison, 1993; Xhoffer *et al.*, 1991), Europe (Gysels and Van Grieken, 1999; Horvath *et al.*, 1996; Migon *et al.,* 1997), United Kingdom (C. McDonald 1979, Allen et, al, 2001) Poland (Krzemi ska-Flowers M., *et al.*, 2005), Pittsburg, USA (Juan C. Cabada, 2005).Several investigations have also been made on the measurements of size distribution of aerosols and associated elemental concentrations in urban areas (Zoller *et al.*, 1974; Spengler and Thurston, 1983; Orsini *et al.*, 1986; Anderson *et al.*, 1988; Infante and Acosta, 1991). In Indian context studies on size distribution of PM10 are reported by Khemani *et al.* (1982) for Deccan plateau in India. Later Mishra (1988) determined the size distribution of atmospheric aerosols and its chemical composition in Mumbai city. Similar study was also conducted at Agra using an eight stage anderson cascade impactor (Kulshrestha *et al.*, 1998). Some more important studies were carried out by Gokhale and Patil, 2004 in Mumbai; Yadav and Rajamani (2006) in the region of N-NW, India and Srivastava and Jain (2007) in the city of Delhi for size distribution of toxic metal on particulate matter. However the size distribution of PM 10 and associated metal in the city of Nagpur have not been yet reported. Therefore, the present study is undertaken to assess the distribution of metals in various particle size fraction of particulate matter in commercial cum residential area of Nagpur city. Emission inventory is also undertaken for estimation of PM 10 emission loads from area and point sources for identified sources and their contribution.

Material and Methods

Distribution of size fraction of PM10 was carried at sampling location in Mahal which is residential cum commercial area of Nagpur city. The sampling location is shown in Figure 38.1. Site location represents densely populated and congested area with narrow roads having high traffic density of two/four wheeler vehicles, commercial activities which includes gold smelting shops, cloth markets, grocery shops and many more shops which results in increasing traffic and crowd. Diesel generator (DG) sets are also used during power shortage in market areas.

Figure 38.1: Location of Sampling Site in Nagpur City.

Nagpur is the largest city in central India having a population of 2.5 million and with the area of approximately 220 km² (latitude 21°06' N, longitude 79°03' E). It has continental type of climate with three distinct seasons: summer, monsoon and winter. The monsoon season is from June to September, winter months are from October to February and summer months are from March to June. During monsoon period, the temperature ranges between 26°C and 39°C with average annual rainfall of 1,205 mm (47.44 in) during June to September. The winter months are cool and temperature is generally goes down below 10°C. The summers are characterized by high daytime temperature ranging between 35°C- 46°C. The annual windrose shows distributed winds from NW-N and W with calm wind conditions around 30 per cent. Wind directions in winter season dominated from N-NE-E with calm wind conditions around 41 per cent.

Sampling was performed during winter season, in the month of October-November, 2008 using eight-stage cascade impactor (Model No. 10-710, Anderson samplers Inc., Atlanta, Georgia) with a pre-separator and back-up filter. The sampler is designed to operate at a flow rate of 28 L min"1 and having effective cut diameters at the stages 10, 9, 5.8, 4.7, 3.3, 2.1, 1.1, 0.7, and 0.4 μm with back-up filter collects aerosols of diameter 0.4 μm and smaller. Cascade impactor was operated at a nominal designed flow rate of 28.3 L/min for each sampling. The pump for the cascade impactor was calibrated by using a dry gas meter. The initial and final flow rate was checked by this dry gas meter. Flow rate was found to be within ± 1 per cent.Particles were collected on glass microfibre filters (Whatman, GF/A). Blank filters were also weighed for QA/QC protocol. All filter paper was conditioned before and after weighing using dissector. Sample filters were weighed to a high accuracy on a Mettler AE 163 Balance. To determine the toxic metals of the size segregated particulate matter collected on each stage, the filters were digested with concentrated nitric acid (12N). Each stage filter paper was digested using 10 ml of concentrate HNO_3 in Teflon vessel. For digestion purpose microwave digestion lab station (ETHOS 900 milestone) is used. Total programme for digestion is 20 minute, after digestion cool it and then filtered through Whatman 41 filter paper into acid cleaned beaker. In a similar fashion blanks were prepared taking an unexposed filter. The acid solutions were analyzed for all the fractions of the particulate matter for element by inductively coupled plasma- optical emission spectroscopy (ICP-OES) (Make- Perkin Elmer, USA)

Results and Discussion

The airborne particulate of each size fractions 0.0-0.4 up to 10.0 um were collected. The average PM10 concentration and average fine particulate matter (effective cut of aerodynamic diameter ≤ 2.2um) concentration was found to be 300µg/m^3 and 136.7µg/m^3 respectively The higher concentration of PM may be contributed due to the conjunction in traffic, highly commercial activities and old and high rising building prevents dispersion of particulate matter apart from thermal plant emissions downwind to the sampling location. Figure 38.2 shows per cent mass distribution of each size fraction of particulate matter. The result indicates that highest percentage mass concentration 15 per cent in size range of 9.0-10.0µm and minimum per cent mass concentration of 6.8 per cent in size range 2.2-3.3 µm was observed. In size fraction of 0.7-1.1 µm per cent mass concentration was 12.5 per cent and in fine size range of 0-0.4 µm and 0.4-0.7 µm, per cent mass concentration 10.9 per cent and 10.4 per cent were observed. The results of the analysis are expressed as a percentage of the total measured over all stages was given in Figure 38.3. Among all size fractions, maximum contribution of Sr, Ni As, Pb, Zn, Pb Fe Si and Cu was observed in size range of 0.4-0.7 µm, 0.4-0.7 µm, 0.7-1.1 µm, 1.1-2.1 µm, 2.1-3.3 µm, 3.3-4.7 µm, 4.7-5.8 µm, 5.8-9.0 µm and 9.0-10 µm respectively.

It can be seen that the more toxic metals have large proportions below 0.7 pm diameter and could therefore pass directly into the alveoli. These are Sr (47 per cent), Ni (43 per cent), and Zn (30 per cent). The other metal fractions which could enter the alveoli are As, Mg, Ba, Cu Fe and Mo. Other metals could conceivably penetrate at least as far as the bronchi and some as far as the bronchiole. The metals like Ba, Ni,

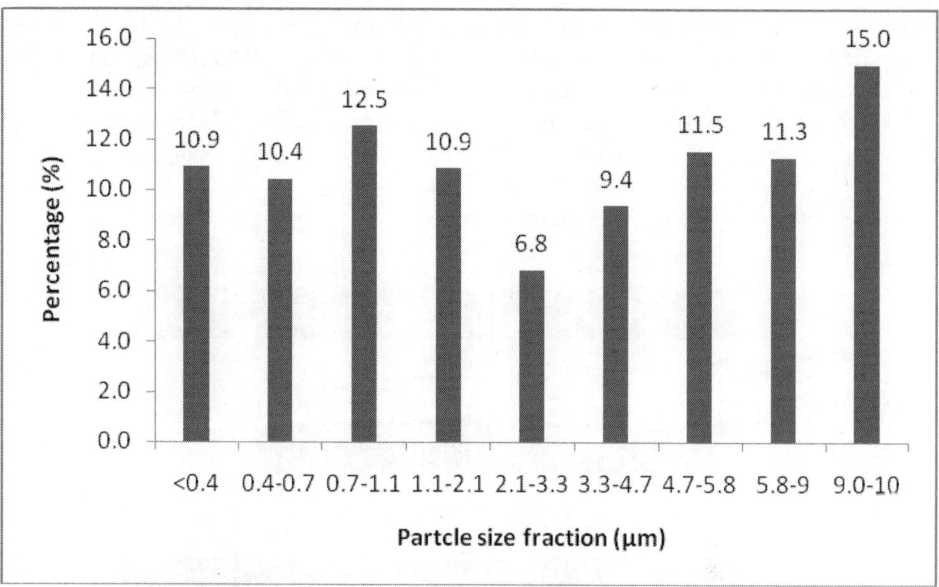

Figure 38.2: Particle Size Distribution as Percentage of PM 10.

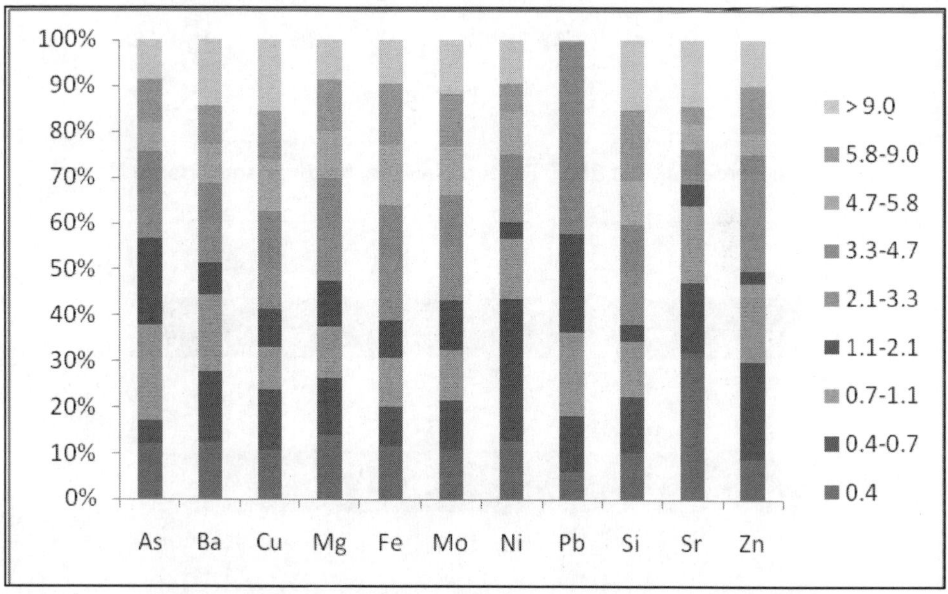

Figure 38.3: Per cent Distribution of Individual Metals According to Size Fraction.

and Zn are associated with fine fraction while Si shows major portion in the coarse fraction. However, the metals Mo, Fe, and Mg and Cu have no much variation in distributions among all size fractions.

According to the American Conference of Governmental Industrial Hygienists (ACGIH), [ACGIH, 1995] and [ISO, 1995], particles over 4 μm are classified as coarse particles, and those between 1–4 μm are medium and below 1 μm are designated as fine particles. In the present study medium cut-off points are at 4.7 and 1.1 μm, which closely match the ISO classification and hence we categorized a coarse range of 10–4.7 μm, medium range of 4.7–1.1 μm and a fine range below 1.1 μm. According to these classification, distribution of PM 10 is shown in Figure 38.4 and the result

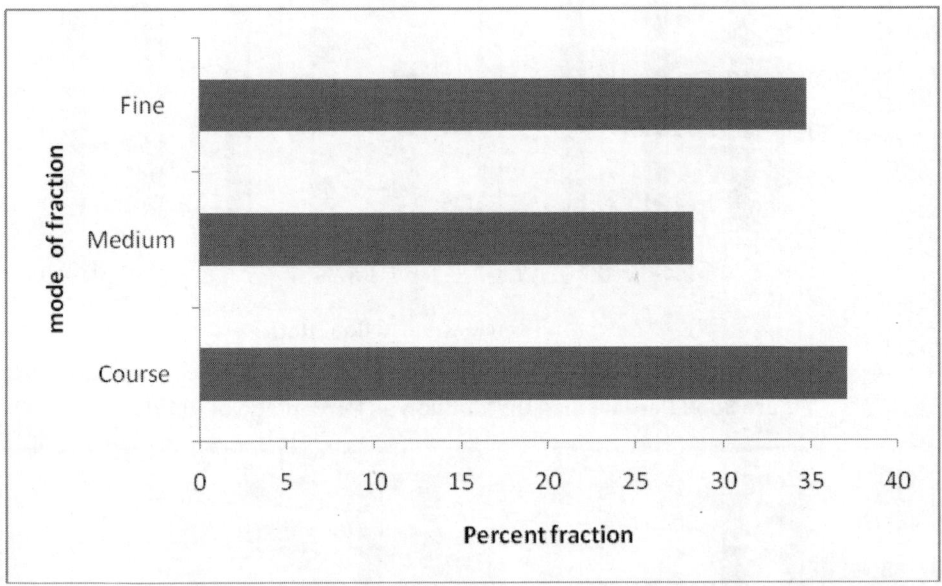

Figure 38.4: Percentage of PM10 Fraction in Fine, Medium and Coarse Modes.

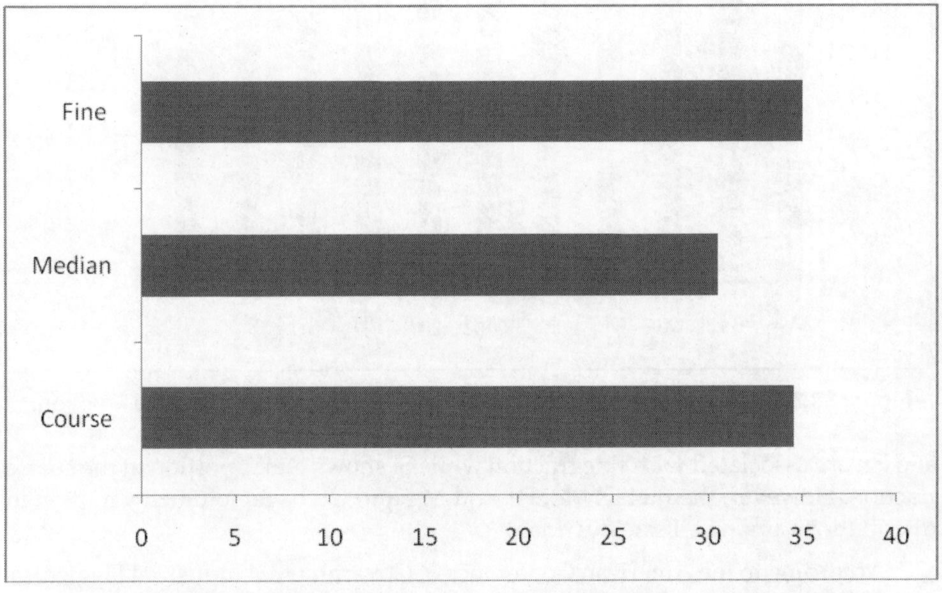

Figure 38.5: Percentage of Total Metals in Fine, Medium and Coarse Modes.

indicate that percentage mass concentrations in the coarse region is around 37 per cent and in fine mode, it is about 34 per cent and 28 per cent of mass concentration is of with medium mode. Similarly distribution of total metal classified as fine, medium and coarse mode and represents in Figure 38.5. The results show that total metal contributes 34.5 per cent in coarse mode, 34.9 per cent in fine mode and 30.4 per cent in median mode.

Emission Inventory of PM 10 for Point and Area source was carried out for Nagpur city. Secondary data relating type and quantity of fuel used etc. was obtained from MPCB, DIC and Mineral Department; Collectorate Office, Civil Lines, Nagpur. The major identified air polluting units are two large-scale power plants concentrated in Nagpur city -Khaperkheda and Koradi. Emission load of PM 10 was calculated using AP-42 factors (USEPA, 1996). The total estimated emission load for point source was found to be 1396777 MT/Year and for area source 1109.4 MT/Year. The category wise emission load of PM 10 for point source was also calculated as shown in Figure 38.6. The results shows that maximum PM 10 emission is contributed from thermal power plant (88 per cent) followed by metallurgy and other industry (both 5 per cent) then textile and chemical industries (1 per cent). In area source, based on the activity, PM 10 emission load was calculated represented in Figure 38.7. In this area source, the major contribution of PM 10 emission from construction activity (65 per cent) followed by domestic activity (29 per cent), hotels and restaurant (5 per cent) and crematoria (2 per cent) was observed. In addition to above, emission of PM10 from line sources also contribute to total emission load which presently is not included in this study.

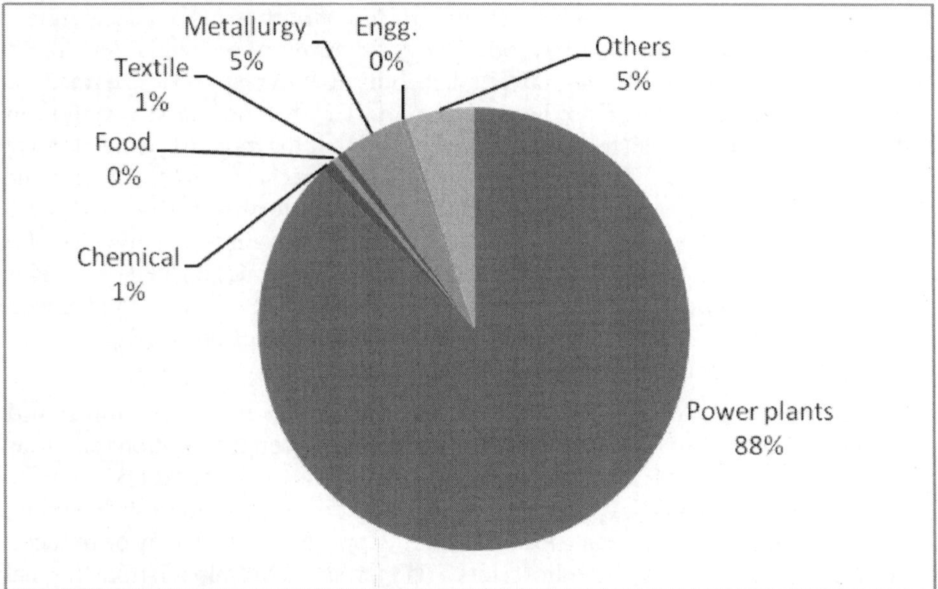

Figure 38.6: Contribution of Category-wise PM 10 Emissions for Point Source.

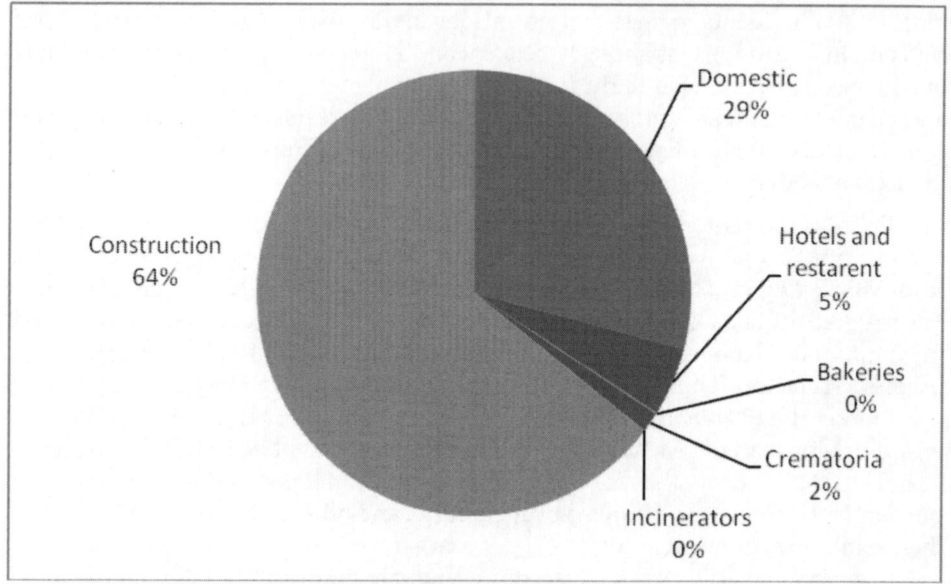

Figure 38.7: Contribution of Sector-wise PM 10 Emissions for Area Sources.

Conclusion

This study reveals that the total 24 hourly average of PM 10 concentration levels are above CPCB limit (100 µg/m^3) and levels of average concentration of fine particulate matter (effective cut of aerodynamic diameter ≤ 2.2um) was found to be 136.7µg/m^3 violating the USEPA limit(65 µg/m^3). Major portion of PM concentration was found in coarse mode and fine mode. The concentrations of metals (As, Zn, Ba, Ni, Sr and Si) show invariably significant distribution with respect to size fractions of PM10. However, Metals (Mo, Fe, and Mg and Cu) has no much variation in distributions among all size fractions. It can be seen that the more toxic metals have large proportions below 0.7 μm diameter and could therefore pass directly into the alveoli. As trace metals are more harmful for human health, further study on identification of sources and their contribution is in process. Emission inventory for point and area source suggest that, the major contributor of emission load of PM 10 is from thermal power plant followed by urban infrastructural development and related construction activities are contributing to particulate pollution in Nagpur city. However, line source also contribute PM 10 emission.

The study indicates a need for appropriate urban traffic planning and management to prevent build up of fine PM concentration on the locations and thus safeguard public health from ill-effects of fine particulate matter and toxic metals. Area sources such as DG sets, use of biomass and kerosene for domestic cooking, refuse burning; bakeries etc create local PM quality problems which may be reduced by supplying cleaner fuel to domestic sectors. High levels of PM emission from thermal power plant can be minimize using appropriate control equipment and use of alternative cleaner fuel like natural gas (NG) are suggested.

References

1. ACGIH, (1995).TLVs for Chemical Substances and Physical Agents and Biological Exposure Indices. American Conference of Governmental Industrial Hygienists, ACGIH Publications, Kemper-Woods Center, Cincinnati, pp. 42–46.
2. Allen, AG, Nemitzb, E, Shia, JP Harrisona, RM JC, (2001). Greenwood Size distributions of trace metals in atmospheric aerosols in the United Kingdom. Atmospheric Environment, 35, 4581–4591.
3. Anderson, JR, Agget, FJ, Buseck, P., Germani, MS, Schattuck, T,(1988). Chemistry of individual aerosol particles from Chandler, Arizona, an arid urban environment. Environmental Science and Technology, 22, 811–818.
4. Cabada, JC, Rees, S, Takahama, S., Khlystov, A., Pandis, SN, Cliff, I, Davidson, CI., Robinson, AL, (2004). Mass size distributions and size resolved chemical composition of fine particulate matter at the Pittsburgh supersite. Atmospheric Environment, 38, 3127–3141.
5. Gokhale SB, Patil RS (2004). Size distribution of aerosols (PM10) and lead (Pb) near traffic intersections in Mumbai (India). Environmental Monitoring and Assessment, 95, 311-324.
6. Gysels, K, Van Grieken, R, (1999). Field evaluation of a wind tunnel–impactor system for sampling ambient aerosols. Journal of Aerosol Science, 30, 639–650.
7. Horvath, H, Kasahara, M, Pesava, P, (1996). The size distribution and composition of the atmospheric aerosol at a rural and nearby urban location. Journal of Aerosol Science, 27, 417–435.
8. Infante, R, Acosta, IL (1991). Size Distribution of Trace Metals in Ponce, Puerto Rico Air Particulate Matter. Atmospheric Environment, 25B, 121–131.
9. ISO, (1995).Air Quality-Particle Size Fraction Definitions for Health Related Samplings. International Organization for Standardization, ISO 7708, ISO Publications, 1st edn.
10. Khemani, LT, Momin, GA, Naik, MS, Vijay, KR, Murthy, BHV, (1982). Chemical composition and size distribution of atmospheric aerosols over the Deccan Plateau, India. Tellus, 34, 151–158.
11. Krzemi ska-Flower, M Bem, H., Górecka, H, (2006).Trace Metals Concentration in Size-Fractioned Urban Air Particulate Matter in £ódŸ, Poland. I. Seasonal and Site Fluctuations. Polish Journal of Environmental Studies, 15 (5), 759-767.
12. Kulshrestha, UC, Kumar, N, Saxena, A, Kumari, KM, Srivastava, SS, (1995). Identification of the nature and source of atmospheric aerosols near the Taj Mahal, India, Environmental Monitoring and Assessment, 34, 1–11.
13. Lodge, JP, Waggoner, AP, (1981). Non-health Effects of Airborne Particulate Matter.Atmospheric Environment, 15,431–482.
14. Mc Cornac, BH (1971). Introduction to the scientific study of air pollution, Reide Dordreet. Holland.

15. Mcdonald C. and Duncan HJ, (1979). Particle size distribution of metals in the atmosphere of Glasgow. Atmospheric Environment, 13, 977-980.

16. Migon, C, Journel, B, Nicolas, E, (1997). Measurement of trace metal wet, dry and total atmospheric fluxes over the Ligurian Sea. Atmospheric Environment, 31, 889–896.

17. Mishra, UC, (1988). Studies in aerosol size distribution and composition. Journal of Aerosol Science, 19, 1165–1169.

18. Morawska, L., Thomas, S., Jamriska, M., and Johnson G. (1999). The modality of particle size distribution of environmental aerosol. Atmospheric Environment, 33, 4401–4411.

19. Orsini, CQ, Tabacniks, MH, Artaxo, O, Andrade, MF, Kerr, AS, (1986). Characteristics of fine and coarse particles of natural and urban aerosols of Brazil. Atmospheric Environment, 20, 2259–2270.

20. Otten, P, Injuk, J, Van Grieken, R, (1994). Elemental concentrations in atmospheric particulate matter sampled on the North Sea and the English Channel. Science of the Total Environment, 155, 131–149.

21. Ottley, CJ, Harrison, RM, (1993). Atmospheric dry deposition flux of metallic species to the North Sea. Atmospheric Environment, 27A, 685–695.

22. Pope, CA, Bates, DV, Raizenne, ME 1995(b). Health effects of particulate air pollution time for Reassessment. Environmental Health Perspective, 103,472-480.

23. Pope, CA, Thun, MJ, Namboodiri, MM, Dockery, DW, Evans, JS, Speizer, FE, Heath CW.1995 (a): Particulate air pollution as a predictor of mortality in a prospective study of US adults. American Journal of Respiratory and Critical Care Medicine, 151, 669-674.

24. Reichhardt, T, (1995). Weighing the health risks of airborne particulate. Environmental Science and Technology, 29, 360A-364A.

25. Spengler, JD, Thurston, GD, (1983). Mass and elemental composition of fine and coarse particles in 6 United-State cities. Journal of Air Pollution Control Assessment, 33, 1162–1171.

26. Srivastava, A, Jain, VK (2007). Size distribution and source identification of total suspended particulate matter and associated heavy metals in the urban atmosphere of Delhi. Chemosphere, 68, 579–589.

27. Trijonis, J. (1983).Development and application of methods for estimating inhalable and fine particle concentrations from routine hi-vol data. Atmospheric Environment, 17, 999–1008.

28. WHO, (1996). Air Quality Guidelines for Europe: Particulate Matter. Final edited draft, 19 September, p.84.

29. Xhoffer, C, Bernard, P, Van Grieken, R, Van der Auwera, L, (1991).Chemical characterization and source apportionment of individual aerosol particles over

the North Sea and the English Channel using multivariate techniques. Environmental Science and Technology, 25, 1470–1478.

30. Yadav S, Rajamani V (2006). Air quality and trace metal chemistry of different size fractions of aerosols in N-NW India - Implications for source diversity. Atmospheric Environment, 40, 698-712.

31. Zoller, WH, Glatney, ES, Duce, RA, (1974). Atmospheric concentration and sources of trace metals in the South Pole. Science, 183, 198–200.

Chapter 39

Use of Fly Ash in Road and Embankment Construction

Virendra Singh

Civil Engineering,
Indian Institute of Technology (BHU), Varanasi

ABSTRACT

Fly ash and bottom ash are the solid residues from coal combustion process for power generation. Continuously growing problem of fly ash disposal has been compelling to find out easily available and low cost technique to curb this problem. Different methods are being applied to use the fly ash as engineering material. One of them is, use of fly ash with an alluvial soil in different proportions for embankment construction. The current rate of generation of coal ash from 80 thermal power stations in India is about 120 million tons per year. In spite of best efforts the percentage utilization of fly ash in this country is around 13 per cent. Efforts are being made to maximize the utilization of fly ash in dykes, highways, embankments and retaining walls, which require large quantity of select fill materials. Fly ash fulfills the technical properties required for fill in these uses. However, experience has been that inadequate awareness among the users at various levels has resulted in a limited use of fly ash materials fill. The need for evaluation of properties of fly ash and quality control during construction are important aspects.

The fly ash in its finest form was collected from Anpara Thermal Power Project, Anpara (U.P.) In the Anpara Thermal Power Project total generation of fly ash is about 2.5 million tons per year and utilization of fly ash out of these is about 0.2 million ton per year. The chemical composition and physical properties of the fly ash were studied. From the chemical studies it was observed that the silica and iron oxide contain is slightly more. It is depending upon type of coal and source of coal used. Physical properties of fly ash were determined by performing tests, i.e. particle size analysis, Standard Proctor Compaction test, Atterberg limit test, permeability test, direct shear test and unconfined compression strength test. From these tests it was found that Anpara fly ash can be safely used in local embankment construction for fly ash disposal.

The maximum dry density and optimum moisture content of the fly ash are 13.58 kN/m^3 and 18.6 per cent respectively. The permeability of the fly ash worked out to be 5.32×10^{-4} cm/sec. The value of cohesion and angle of shearing resistance in dry condition are 0.05 kg/cm^2 and 30°, and in the wet condition 0.075 kg/cm^2 and 23° respectively. The CBR value of soil and fly ash (soaked) was found to be 3.63 per cent and 6.13 per cent respectively. When adding fly ash in alluvial soil the CBR value is increased due to pozzolonic effect. By adding 15 per cent fly ash in alluvial soil the CBR value was found to be 8.61 per cent. When the soil is stabilized with 15 per cent of lime and fly ash admixture having ratio of 1:3, the C.B.R. of admixture is 21.87 per cent. Since fly ash soil admixture is cheaper than lime fly ash-soil admixture, therefore it is recommended that fly ash soil admixture can be used safely in sub-base of road construction.

Factors, which Effect the Utilization of Fly Ash

There are some factors due to which, utilization of fly ash is not increasing at required rate.

- ☆ Lack of awareness of fly ash properties.
- ☆ High transportation cost.
- ☆ Non-availability of dry fly ash.
- ☆ Lack of proper co-ordination.
- ☆ Top soil is easily available.
- ☆ Wide variation of fly ash in quality.

Different Way of Utilization of Fly Ash

Efforts are now being made in India in major way, to increase the utilization o0f fly ash in different ways. Some of which are given below:

1. Bulk utilization as fill material in the construction of dam, embankment, retaining structure, and as sub-base material.
2. In concrete, fly ash used as a partial replacement of cement, which renders the concrete many advantageous properties like, improved workability, resistance to chemical attack, impermeability.
3. Good quality of building bricks can be produced with lime and fly ash using sodium silicates or cement as a binder.
4. For the stabilization of sub-base and sub grade in pavement construction. Pozzolonic concrete pavement distributes load over a large area by slab action and have more or less like rigid pavement.
5. Used for soil stabilization of railway embankment, in U.S.A. and India. The track on expansive clays, need repetitive attention and extra ballast which lost due to preparation into the soil and maintenance cost, by using fly ash as sub ballast this problem can be overcome.

6. Used for reclamation of low lying waste lands and refuse dumps, falling of mines, treatment of polluted waters and unsuitable soils for agriculture.
7. To improve foundation soil.
8. Use as canal lining.
9. Use as in modification of soil pH, in agricultural application.
10. As a very good industrial absorbent for successful removal of phonetic compound trinitrotoluene (T.N.T.), alkyl benzene etc.

The boom in the construction industry, in the developed countries and soil conservation efforts have contributed a great deal in making the fly ash industry a success in these countries. In present study certain characteristics of fly ash from Anpara Thermal Power project fly ash is to be determined and its use in embankment construction is to be recommended.

Use of Fly Ash in Geotechnical Application

In order to utilize fly ash in bulk quantities, ways and means are being explored all over the world to use it for the construction of dykes, embankments lagoon and roads. Fly ash satisfies major design requirements of strength and compressibility except for its susceptibility to erosion and possible liquefaction under extreme conditions (Guru Vittal and Murthy, 1998). However, when it is used as sub-base in flexible pavements, it is completely confined and also the thickness of such layer is relatively small whereby the above problems can be estimated. Further provision of a reinforcing layer facilitates drainage, thus reliving the excess pore water pressure (Balksdale *et al.*, 1989, Rao, 1996, Murthy, 1999).

Commonly the lagoons, dykes are formed by constructive earthen bounds which consume large quantity of top fertile soil. In the recent years efforts are mooted to use the pond as to rise the height of existing earthen bunds by U/S, D/S and C/L methods of tailing dam construction (Gandhi and Mathew, 1996, Ghosh, 1999). Such attempts enable bulk utilization of fly ash at the pond itself thus increasing the life of lagoon for the same area.

As thermal power plants are spatially distributed all over the country, utilization of fly ash from these plants for the dykes, embankments, roads constructions not only help the consume bulk quantities of fly ash solving its disposal problem to a certain extent but also to satisfy the construction requirements.

Table 39.1: Physical Properties of Anpara Fly Ash.

Colour	Grey
Physical State	Powder with traces of unburnt carbon
Sp. Gravity	2.02
Particle Size	
Clay per cent (<0.002mm)	4.5 per cent
Silt per cent (0.075 - 0.002mm)	63.5 per cent
Sand per cent (0.075 - 0.002mm)	32 per cent

Table 39.2: Chemical Properties of Anpara Fly Ash.

Constituents	*Percentage (by weight)*
SiO_2	68.0
Al_2O_3	24
Fe_2O_3+Fe_3O_4	2.18
TiO_2	2.64
CaO	1.49
MgO	0.06
SO_4	–
Loss of Ignition	1.63

Soil

The soil in the present study is clay of low plasticity (CL).

Table 39.3: Index Properties of Soil.

Liquid Limit (per cent)	27
Plastic limit (per cent)	17
Plasticity Index	10.0
Grain size distribution	
Clay per cent (<0.002mm)	13
Silt per cent (0.075 - 0.002mm)	75
Sand per cent (>0.075mm)	12
Classification	CL

Lime

Calcium Oxide (CaO) was obtained by market. This material has been referred as lime.

Experimental Results (Figures 39.1-39.8 and Tables 39.1-39.4)

Compaction Characteristics

The maximum dry density (MDD) and optimum moisture content (OMC) were determined by light compaction (Standard Proctor) tests. The MDD and OMC are respectively, 18.51 kN/m^3 and 14.8 per cent for the Varanasi soil, and 13.58 kN/m^3 and 18.6 per cent for the Anpara fly ash.

The MDD and OMC of all the fly ash-soil mixtures are summarized in Table 39.4.

Permeability

By falling head method the coefficient of permeability of fly ash was found to be 5.32×10^{-4} cm/sec.

Table 39.4: MDD and OMC of all the Fly Ash-Soil Mixtures.

Mix Designation	*MDD (kN/m³)*	*OMC (per cent)*
Varanasi soil (VS)	18.51	13.8
Anpara Fly ash (F)	13.58	18.6
95 per cent soil + 5 per cent fly ash (FVS_1)	18.29	14.1
90 per cent soil + 10 per cent fly ash (FVS_2)	18.01	14.4
85 per cent soil + 15 per cent fly ash (FVS_3)	17.70	14.8
80 per cent soil + 20 per cent fly ash (FVS_4)	17.51	15.2

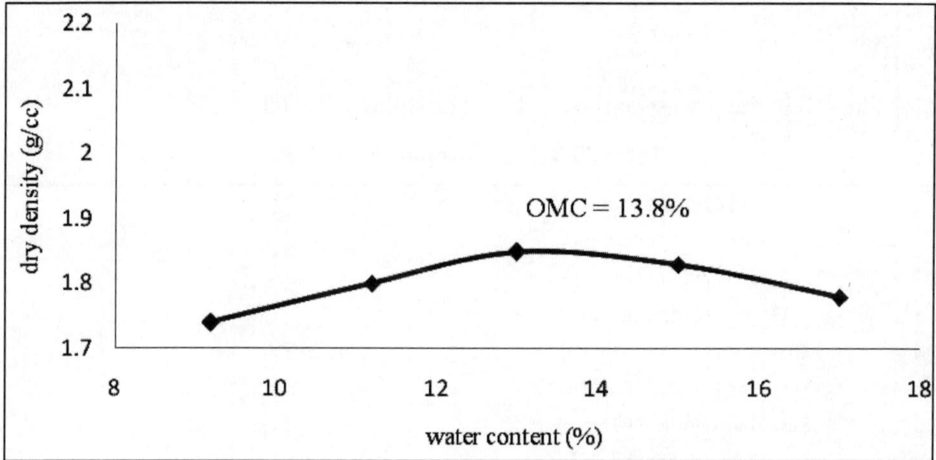

Figure 39.1: Compaction Curve of Soil.

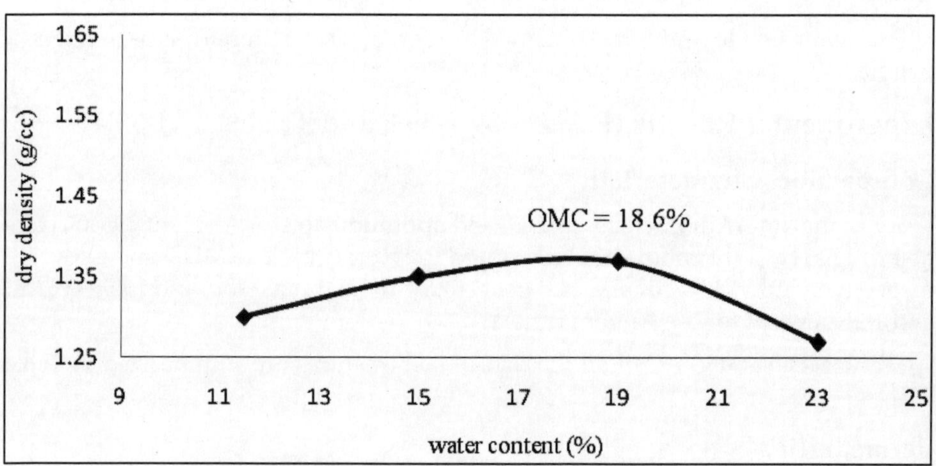

Figure 39.2: Compaction Curve of Anpara Fly Ash.

Direct Shear Test

The shear parameters of Anpara fly ash are given below:

In Dry Condition

Cohesion (C) = 0.05 kg/cm^2

Angle of shearing resistance (φ) =30°

In Wet Condition

Cohesion (C) = 0.075 kg/cm^2

Angle of shearing resistance (φ) =23°

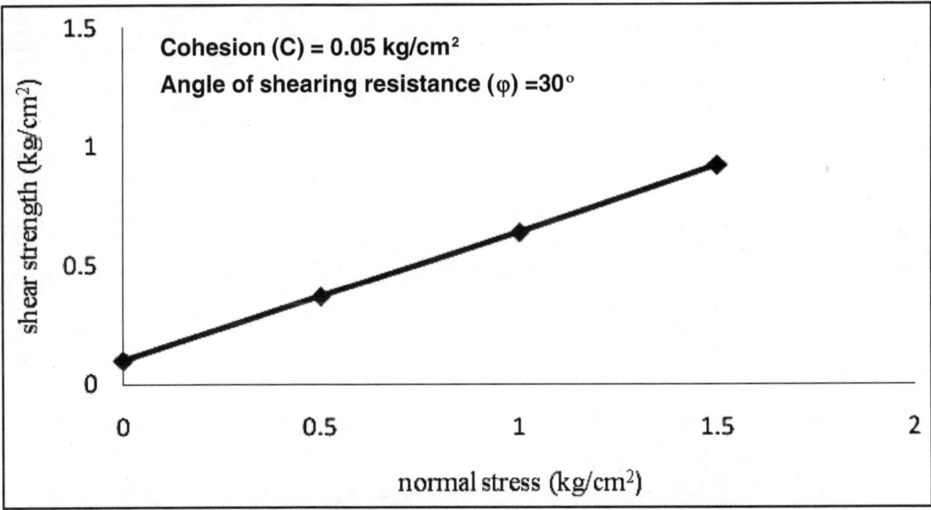

Figure 39.3: Direct Shear Test on Anpara Fly Ash in Dry Condition.

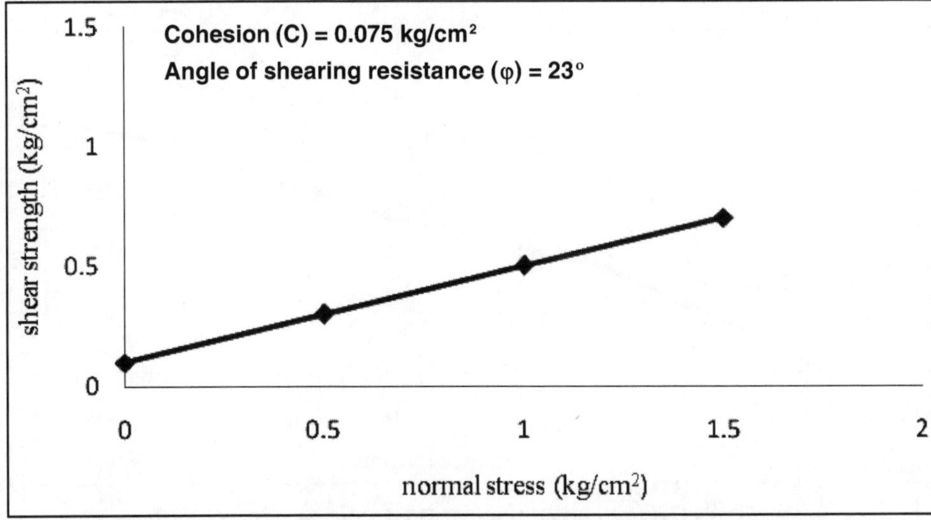

Figure 39.4: Direct Shear Test on Anpara Fly Ash in Wet Condition.

Fly ash in wet condition inhibits lower strength than in dry condition, because of decrease in angle of shearing resistance.

CBR Test Results

Table 39.5: CBR Test Results of all the Fly Ash-Soil Mixtures

Mix Designation	*C.B.R. Values (per cent)*
Varanasi soil (Soaked)	3.41
Anpara fly ash (Soaked)	6.13
95 per cent soil + 5 per cent fly ash (Soaked)	7.75
90 per cent soil + 10 per cent fly ash (Soaked)	8.09
85 per cent soil + 15 per cent fly ash (Soaked)	8.61

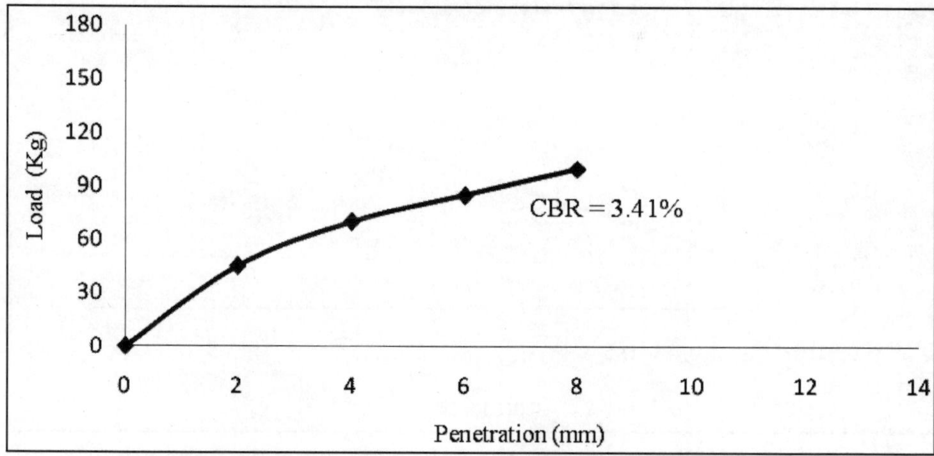

Figure 39.6: CBR Value of Varanasi Soil (Soaked).

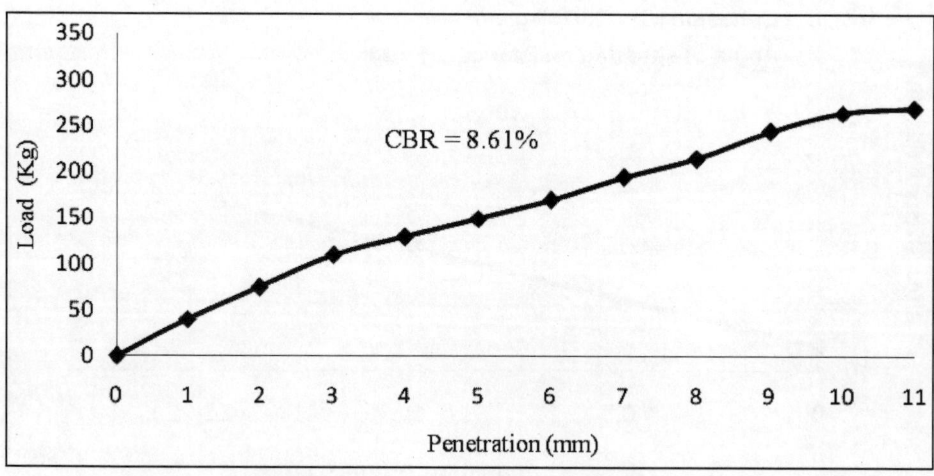

Figure 39.7: CBR Value of 85 per cent soil + 15 per cent Fly Ash (Soaked).

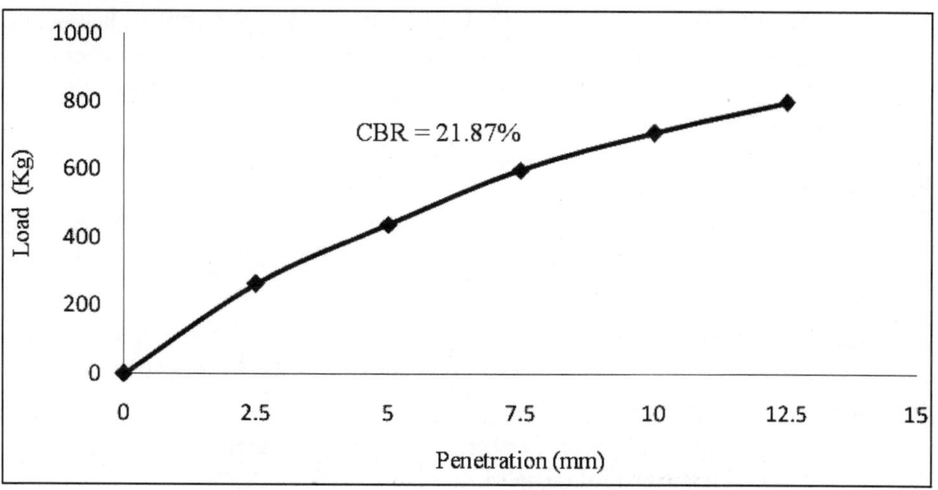

Figure 39.8: CBR Value of Soil Stabilized with 15 per cent of Lime and Fly Ash (Ratio of lime to fly ash 1:3).

Conclusion(s)

Following conclusions can be drawn from the test results obtained in present investigation.

1. Maximum dry density of fly ash soil admixture is decreasing linearly with increasing fly ash content and optimum moisture content increases as the fly ash content increases.
2. For the soil used in present investigation the best amount of fly ash is 15 per cent by weight and this admixture give the maximum value of CBR. The CBR and unconfined compressive strength was found to be 8.61 per cent and 93.03 kN/m^2 respectively.
3. When the lime fly ash is used with soil 15 per cent by weight, the best ratio of lime to fly ash for maximum CBR value is 1:3.for the admixture maximum CBR value is 21.87 per cent.
4. With 15 per cent fly ash adding by weight, the CBR value increases nearly 2.5 times. The CBR value of soaked soil is only 3.41 per cent and with fly ash it is 8.61 per cent. The Varanasi alluvial soil having low CBR values is not suitable for use in sub-base of the road. The required value of CBR for sub-base is 7-20 per cent. Therefore 15 per cent fly ash can be safely used with soil in sub-base of road.
5. If we use the fly ash as a fill material, these efforts would certainly helps in mitigation of geo-environmental hazards and at the same time top fertile soil can be used for the crop production.
6. For embankment construction physical characteristics of fly ash is determine, shear strength parameters for saturated ash is c'=0.075 kg/cm^2 and ö š"= 23°, coefficient of permeability is 5.32×10^{-4}cm/sec. and MDD is

13.58 kN/m^2 and OMC is 18.6 per cent. Therefore, physical and chemical properties indicate the Anpara fly ash can be used safely in construction of local embankment for the disposal of fly ash and it can also be safely used in the body of dyke.

References

1. Akhilesh Kumar (2004). "Study of Fly Ash and Fly Ash Soil Admixture for Sub-Base of Road and Embankment Constructions", M. Tech. Dissertation, IT BHU, Varanasi.
2. Gandhi, S.R. and Mathew, G.V. (1996). "Granular filter for Ash Dykes", Proceedings of Indian Geotechnical Conference, Chennai, pp. 532-535
3. Havukainen, J. (1983). "The Utilization of Compacted Coal Ash in Earth Works", Proceedings of the European Conferences on Soil Mechanics and Foundation Engineering, Helsinki, Finland, Vol.-II, pp. 773-775.
4. Kaniraj S.R. and Havanagi V.G. (1999). "Geotechnical Characteristics of Fly Ash Soil Mixtures", Geotechnical Engineering Journal, Vol. 30, No. 2, August 1999, pp. 129-146.
5. Martin J.P., Collins R.A., Browning J.S. and Biehl F.J. (1990). "Properties and Use of Fly Ash for Embankment", Journal of Energy Engineering, ASCE, Vol. 116, No. 2, pp. 71-86.
6. Virendra Singh, Narendra Kumar and Devendra Mohan (1996). "Use of Fly Ash in Soil Stabilization for Roads", Proceedings of Indian Geotechnical Conference, 96, Madras, pp. 411-414.

Chapter 40

Fly Ash: Utilization and Characteristics

D.B. Singh, R.K. Tiwary, P.K. Arya and Amalendu Sinha

CSIR-Central Institute of Mining and Fuel Research,
Barwa Road, Dhanbad

ABSTRACT

Fly ash is generated from coal-fired power stations and most of the ash is disposed off in the vicinity of the plant as a waste material covering several hectares of valuable land. With the increase in commissioning of several super thermal power plants with large capacity and with the increasing use of low grade coal of high ash content, the amount of ash generated from them is becoming very large. The deposit of a large amount of fly ash is a serious problem disrupts ecological cycles and set off environmental hazards. The quality and quantity of fly-ash depend on several variables which may cause characteristics of the ash to vary, invariably from one plant to another and some times even at the same power plant. These power plants all over the world face serious problems of handling and disposal of the ash produced. For this, an in-depth understanding of the physical and chemical properties, and engineering and leaching behavior of fly ash are required before handling and disposal. This paper reports a brief review of the fly ash utilization and little work which has been carried out to see the leaching behaviour and physical and chemical properties of flyash generated by some thermal power plants.

***Keywords**: Fly ash, Characteristics, Utilization, Leaching.*

Introduction

Coal-based thermal power plants using pulverized coal or lignite as fuel generate enormous quantities of fly-ash as a by-product of combustion. Fly ash is a very fine powder and by virtue of their weightlessness, tends to travel in the air and causes air, water and soil pollution, disrupt ecological cycles and set off environmental hazards, if not properly disposed off. Not only this, it settles on leaves and crops in agriculture fields around the power plant, it lowers the yield.

About 70 per cent of total power generation in India, is produced by thermal power plants (TPPs). Majority of plants (84 per cent) are run on coal [1] of which the reserve is 70 billion and rest on gas (13 per cent) and oil (3 per cent).About 260 million tones (MT) of coal (65 per cent annual coal produced in India) is being used by TPPs [2]. Presently, the annual coal ash production is nearly 118 million tones and it is estimated that this will reach 175 million tones by 2012 and may exceed 225 million tones by 2017 [3].

Fly ash is generally captured from the chimneys of coal-fired power plants, and is one of two types of ash that jointly are known as coal ash; the other, bottom ash, is removed from the bottom of coal furnaces. The quality and quantity of fly-ash depend on several variables such as coal source, degree of pulverization, design of boiler unit, loading and firing conditions, handling and storage methods etc. These factors may also cause characteristics of the ash to vary, invariably from one plant to another, and some times even at the same power plant. Depending upon these factors, the components of fly ash vary considerably, but all fly ash includes substantial amounts of silicon dioxide (SiO_2) and calcium oxide (CaO). The toxic constituents include beryllium, boron, arsenic, chromium, chromium VI, cobalt, cadmium, lead, manganese, mercury, molybdenum, thallium, selenium, strontium, and vanadium, along with dioxins and PAH compounds [4,5]. The range values for different phsyico-chemical properties including the mineral contents, major secondary nutrient and micronutrients, and trace and heavy metals along with radionuclides of Indian fly ash are given in Table 40.1 [6].

These power plants all over the world face serious problems of handling and disposal of the ash produced. The high ash content (35–55 per cent) of the coal in India makes this problem more complex. With the increase in commissioning of several super thermal power plants with large capacity and with the increasing use of low grade coal of high ash content, the amount of ash generated from them is becoming very large. Most of the ash generated from the power plants is disposed off in the vicinity of the plant as a waste material covering several hectares of valuable land. This poses serious ecological problems. All the three environmental risks, such as, air, surface water and groundwater pollution are caused by this. It is also affecting the plant growth and their survival. Looking into the nuisances poses by fly-ash, it has been well appreciated by everybody and various Government agencies, the need for the safe disposal and effective utilisation of fly-ash.

Directly related to these concerns is the additional environmental goal of aesthetically enhancing ash disposal facilities. Mine void filling is a potential area which can provide scope for environmentally safe and large volume utilization of fly ash. For this, effective transportation of the fly ash and creation of fill area with acceptable strength characteristics are the two main issues.

Fly Ash Disposal

Dry or Wet disposal procedure is mostly used for the fly-ash disposal. In dry disposal method, the fly-ash is transported by truck, chute or conveyor from the

Table 40.1: General Characteristic of Indian Fly Ash.

Physical Characteristics	*Values*	*Chemical Characteristics*	*Values*
pH	6.0-8.0	SiO_2	38-63
Specific gravity	1.66-2.55	Al_2O3	27-44
Bulk density (g/cc)	0.85-1.2	TiO_2	0.4-1.8
Grain size distribution	Sandy silt to Silt loam	Fe_2O_3	3.3-6.4
Porosity (per cent)	45-55	MnO	0.1-0.5
WHC(per cent)	45-60	MgO	0.01-0.5
EC (dS/m)	0.15-0.45	CaO	0.2-8.0
CEC (mol(p^+)/kg	—	K_2O	0.04-0.9
		Na_2O	0.07-0.43
		L.O.I	0.2-3.4

Parameter	*Total and available Micronutrients*		*Parameter*	*Total and Available Trace and Heavy Metals (ppm)*	
	Total (ppm)	*Av. (ppm)*		*Total (ppm)*	*Av. (ppm)*
Cu	40-80	0.5-1.6	Se	0.6-2.6	1.1-0.4
Zn	50-150	0.4-1.8	Cr	50-50	0.3-0.6
Mn	500-750	0.9-1.5	Pb	10-70	BDL
Fe	3.3-6.4	10-15	Co	10-50	0.05-0.15
B	17-38	0.5-0.8	Ni	50-0	0.15-0.25
Mo	2.2-67	0.1-0.6	Cd	5-10	0.03-0.07
			As	1.0-4.0	BDL
			Hg	BDL	BDL

Radioactivity Level (Bq/kg)	
226Ra	30-110
228Ac	30-110
40K	180-500

power plant to the site of disposal where these are disposed off by constructing a dry embankment (dyke). Air pollution is caused by direct emissions of toxic gases from the power plants as well as wind-blown ash dust from ash mound/pond. The air-borne dust can fall in surface water system or soil and may contaminate the water/soil system. In wet disposal system the fly-ash is transported as slurry through pipe and disposed off in impoundment called "ash pond". Most of the power plants use wet disposal system. The long storage of ash in ponds under wet condition and humid climate can cause leaching of toxic metals from ash and contaminate the underlying soil and ultimately the groundwater system. The environmental aspects of ash disposal should be taken care of in order to minimize air and water pollution. Most of the above environmental problems can be minimized by incorporating engineering measures in the design of ash ponds and continuous monitoring of surface and groundwater water systems. The coal ash can be utilized in bulk only in geotechnical engineering applications such as construction of embankments, as a backfill material, as a sub-base material, construction materials etc. Hence attempts are being made to utilize the ash rather than dump it in the vicinity of the power plants in the ash ponds. As stated above that the components of fly ash vary considerably depending upon the source and makeup of the coal being burned. For this, an in-depth understanding of the physical and chemical properties, and engineering and leaching behavior are required. This chapter reports a brief review of the fly ash utilization and little work which has been carried out to see the leaching behaviour and physical and chemical properties of flyash generated by four thermal power plants

Utilization of Fly Ash

High disposal cost and large area of land requirement for disposal of fly ash demands its utilization to the maximum extent in different uses.

Global Scenario

During 2005 (Table 40.2), Italy, Denmark and Netherlands, which have annual fly ash generation of 2 MT, show 100 per cent utilization, whereas USA, Germany, which produce more fly ash (10-75 MT/y), use around 50-85 per cent fly ash only. Major applications are in cement, concrete, mine fill and bricks etc. Fly ash generation in India (112 MT/y) and China (100 MT/y) is utilized only up to 38 and 45 per cent respectively. In Poland and Germany, bulk quantity of fly ash is used for mine filling, cement and concrete. Fly ash (80 per cent) in developed countries is used for brick manufacture, road construction, landfill, dam construction, agriculture etc. [7].

Indian Scenario

At present in the under-developed country like India, the focus is on demonstration of coal ash utilization related technologies which include fly-ash characterisation, agriculture related studies and application, hydraulic structures, handling and transportation, ash ponds and dams, roads and embankments, reclamation of ash ponds for human settlement, underground mine and opencast voids fills, utilized in concrete, brick making, soil-stabilization treatment etc. Utilization at 3 per cent from of the total 40 MT production of fly ash in 1994 has increased to 38 per cent (42 MT) of 112 MT produced during 2004-05. Earlier utilization was limited predominantly for ash bund construction, landfill and partly for production of bricks and cement. Presently, cement manufacturing and road construction are major consumers of fly ash [7].

Ministry of Environment and Forest Govt. of India has brought out a Gazette Notification for the use of fly ash. No single utilization has the potential to provide solution for safe disposal and gainful utilization of fly ash. A number of applications have been demonstrated and are in use by different agencies to increase the percentage utilization and environmentally safe practices of fly ash disposal. The acceptance of fly ash products and its applications are increasing day by day. Now efforts are

Table 40.2: Present Status of Fly Ash Utilization in India.

Name of Industry	*Utilization Sectors*	*Implementation Place*
N.T.P.C	a) Cement manufacturing b) Brick manufacturing c) Back-filling of mines d) Waste land development and Agriculture e) Structural fill f) Embankment of fly over bridge	J.P. Rewa, Maihar Cement Century cement, Hyderabad. NTPC plant at Singrauli, Korba, Ramagundam, Vindhyachal, Badarpur, Balco and TPP, Talcher.
Neyveli Lignite Corpn.	a) Building bricks b) Cement manufacturing c) Building materials	Neyveli Neyveli Neyveli
PunjabStateElec. Board	a) Fly ash Dyke b) Agriculture	Roper Roper, Bhatinda
Ahmedabad Elec. Corpn.	a) Bricks b) Slabs and tiles c) Lime Pozzolana Cement d) Land reclamation	Ahmedabad
Karnatak a Power Corpn.	a) Bricks and Asbestos cement Sheets b) Road	M/s. ACC at Wadi By CBRI at Raichur
Tata Electric Company	a) Fly ash aggregate	Trombay
Damodar Valley Corpn.	a) Fly ash bricks b) Abandoned opencast mine filling	Durgapur By CTPS at Damodar Mines of BCCL

being made to utilize more and more of fly ash in different sectors. The major areas of utilization are in the concrete and cement industries, structural and land fills, embankments, filling of mines and quarries, road construction building materials, agriculture and high value materials. Many industries are utilizing fly ash some of which are listed in Table 40.2 [6]. The effective utilization of fly ash is in construction material industry. It has been mainly used in brick making and cement industries. The utilization of fly ash concrete is also gaining importance. In other sectors like road construction, agriculture, wasteland development and stowing of underground and open cast mines, efforts are being made to utilize fly ash at the maximum possible amounts. Government is now giving focus and importance to ash utilization and special groups are being created. Formulation/upgradation of BIS standards is being expedited and would facilitate acceptance of fly ash and its products.

Constraints in Utilization of Coal Ash

In spite of all efforts from the Government, the utilization of fly ash in India is very less compared to other advanced countries. The process of fly ash generation, collection and disposal has restricted the scope of utilization of coal ash in our country. There is a long list of constraints which make fly ash unsuitable for its consumption. Certain steps are needed to be taken up to overcome the constraints and efforts should be made to utilize fly ash at the maximum possible amount. Some of the constraints are discussed here;

1. Fly ash generated need to be transported long distance for its utilization, this adds cost to the end product.
2. The cost of fly ash based materials are normally same as the generally used materials or are more due to ash separation, handling and transportation.
3. Thermal power stations spend crores of rupees for buying land for ash disposal but refuse to plan out the same expenditure for the useful utilization of their own product.
4. More than 80 per cent of Thermal power plant in our country have wet ash disposal system. When fly ash comes in contact with water, it looses its reactivity due to decrease of glass content, aggregation of ash particles, leaching of CaO, Na_2O etc. In wet collection system, fly ash and bottom ash are mixed together and disposed off in the slurry form, into ash disposal ponds. This makes fly ash heterogenous in nature and unsuitable for many demanding application.
5. The different physico-chemical properties of fly ash vary not only from different parts of the country but also in the same thermal power plant. Day to day variations in the properties of fly ash is one of the major constraints in its utilization.

Experimental

Fly ash samples were collected directly from the hoppers of Electrostatic Precipitators (ESPs) of the following four thermal power plants and brought to the laboratory.

1. IB Thermal Power Plant, Banharpali, Jharsuguda, Orissa -(OPGC)
2. Talcher Thermal Power Plant, NTPC, Talcher, Orissa -(TTPS)
3. Patratu Thermal Power Plant, Ranchi, Jharkhand -(PTPS)
4. Super Thermal Power Plant, NTPC, Kanhia, Angul, Orissa -(STPP)

The samples of each thermal power plant were mixed properly and after quartering and conning the samples were used for different analysis and experiment. Mineralogical, physical and chemical characterization of fly ash of all the four power plants and the leaching characteristics of IB Thermal Power Plant were conducted in the laboratory.

It is a common fear of leachability of trace metals into the undergroundwater due to disposal of fly ash. So the flask shake and column experiments were carried out to investigate the leachability behaviour of OPGC fly ash. The ash samples were evaluated for its chemical composition by conducting X-ray fluorescence (XRF) analysis. The specific gravity of the ash samples were determined using a stoppered bottle having a capacity of 50 ml as per the guidelines provided by the American Society of Testing Materials (ASTM D 854). The specific surface area of the ash samples were determined by using a Blaine's apparatus (ASTM C 204) with Portland cement as a standard reference material.

For the leachibility study of the OPGC fly ash, 50 g of the fly ash sample was shake with 200 ml of distilled water for different time periods *i.e.* 12 hour and 24 hours; 50 g of fly ash with 200 ml of distilled water was kept for 24 hours and 72 hours without shaking and then it was filtered with What man 42 filter paper.

Table 40.3: Sample description along with the sample code

Sample Code	*Description of Sample*
FAL-1	50 g of the fly ash sample in 200 ml of distilled water shake for 12 hour
FAL-2	50 g of the fly ash sample in 200 ml of distilled water shake for 24 hour
FAL-3	50 g of the fly ash sample in 200 ml of distilled water kept for 24 hours without shaking
FAL-4	50 g of the fly ash sample in 200 ml of distilled water kept for 72 hours without shaking
FAL-5	1000 ml leachate after passing through Fly ash packed column of height 34 cm. and internal diameter 9.6 cm was collected in 24 hrs.
FAL-6	1000 ml leachate after passing through Fly ash packed column of height 34 cm. and internal diameter 9.6 cm was collected in48 hrs.
FAL-7	1000 ml leachate after passing through Fly ash packed column of height 34 cm. and internal diameter 9.6 cm was collected in 72 hrs.
FAL-8	1000 ml leachate after passing through Fly ash packed column of height 34 cm. and internal diameter 9.6 cm was collected in 96 hrs.

Further, a glass column of height 1 meter and 9.6 cm internal diameter was packed with 3.0 kg of fly ash up to the height of 34 cm for the column leaching study. The distilled water was allowed to flow with the gravitational force at a floe rate of

88.25 ml/hr and the leachtes were collected at different time intervals for measuring the concentration of trace metals in the leachates. The sample description of the collected leachets along with the sample code is given in Table 40.3. The leachates of both types of studies *i.e.* shake flask and column study were analysed for pH, total suspended solids (TSS), total dissolved solids (TDS), alkalinity, total hardness, $SO_{4,}$ Chloride and trace metals. The trace metal analysis was done by ICP.

Results and Discussion

Chemical Composition

The chemical composition (by per cent weight) of the fly ash samples is presented in Table 40.4. It can be observed from the Table 40.4 that the major constituents of these samples are; silica (SiO_2), alumina (Al_2O_3) and ferric oxide (Fe_2O_3). Minor quantities of calcium oxide (CaO), magnesium oxide (MgO), sodium oxide (Na_2O), potassium oxide (K_2O), rutile (TiO_2) and other compounds are also observed to be present in lesser quantity which are not shown in the Table 40.4.

Table 40.4: Chemical Composition (Weight per cent) of Fly Ash Samples Studied.

Sample Code	Elements (Weight per cent)								
	SiO_2	Al_2O_3	Fe_2O_3	CaO	K_2O	TiO_2	MgO	Na_2O	Others
OPGC	59.12	34.69	1.38	0.79	1.62	0.87	0.68	0.07	0.78
TTPS	61.17	32.44	2.83	1.06	0.92	0.58	0.23	0.05	0.72
PTPS	72.34	22.53	0.71	0.69	1.12	1.57	0.07	0.03	0.94
STPP	61.19	31.16	2.17	1.43	1.39	0.62	1.32	0.04	0.68

Physical Properties

The physical properties of the different samples are given in Table 40.5.

Table 40.5: Physical Properties of Fly Ash Samples Studied.

Sample Code	Specific Surface Area (m^2/gram)	Specific Gravity	Moisture Content (per cent)	Wet Density (gram/cc)
OPGC	0.4521	1.99	0.232	1.78
TTPS	0.402	2.19	0.421	1.79
PTPS	0.191	2.29	0.152	1.91
STPP	0.431	2.21	0.235	1.99

Leaching Behaviour of OPGC Fly Ash

The values of pH, total suspended solids (TSS), total dissolved solids (TDS), alkalinity, total hardness, $SO_{4,}$ Chloride measured in the leachates are given in Table 40.7 and the same has been presented in Figure 40.1. There is no major change in pH with time. The values of other measured parameters are not showing any regular

Figure 40.1: Graphs Showing the Concentrations of different Elements in the Leachates.

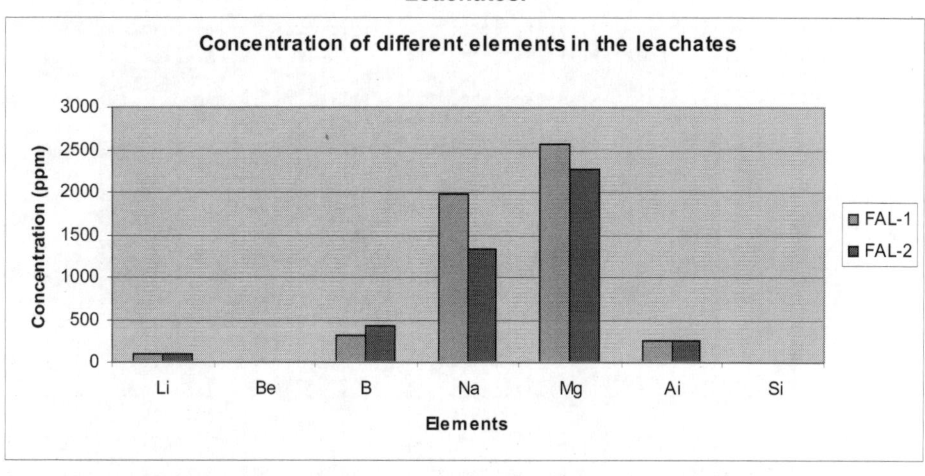

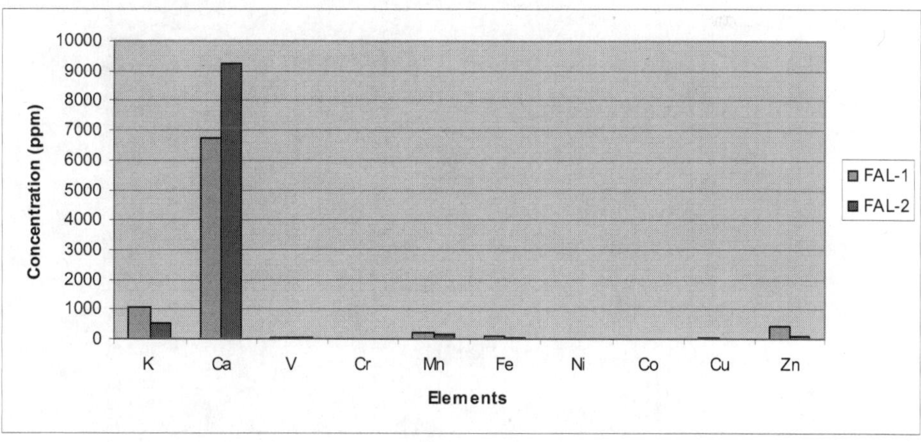

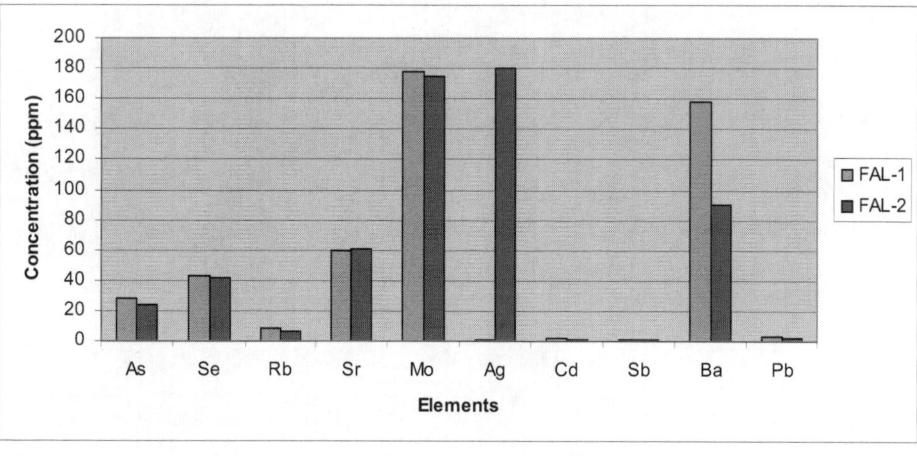

Contd...

Figure 40.1–*Contd...*

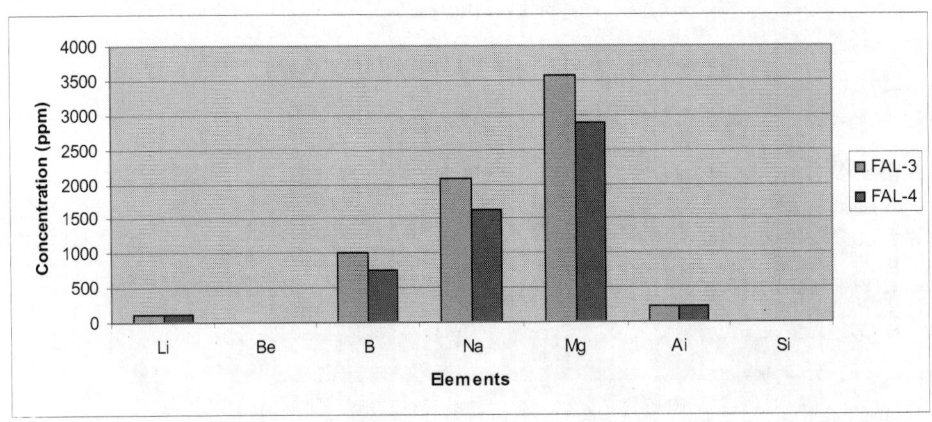

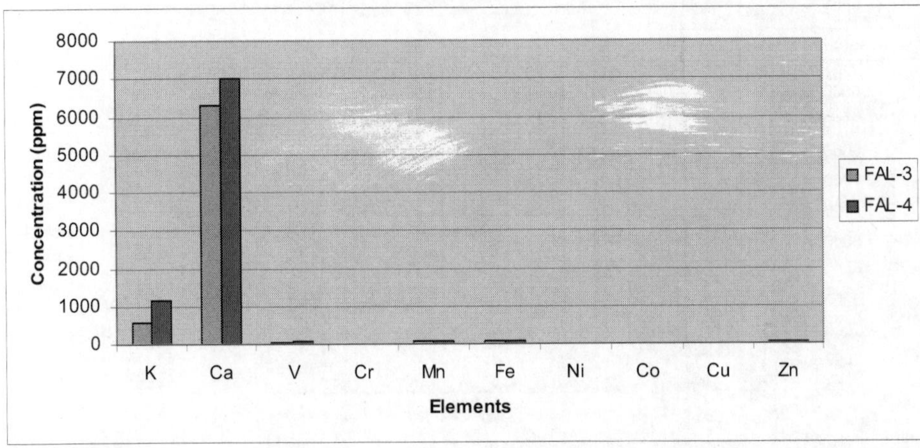

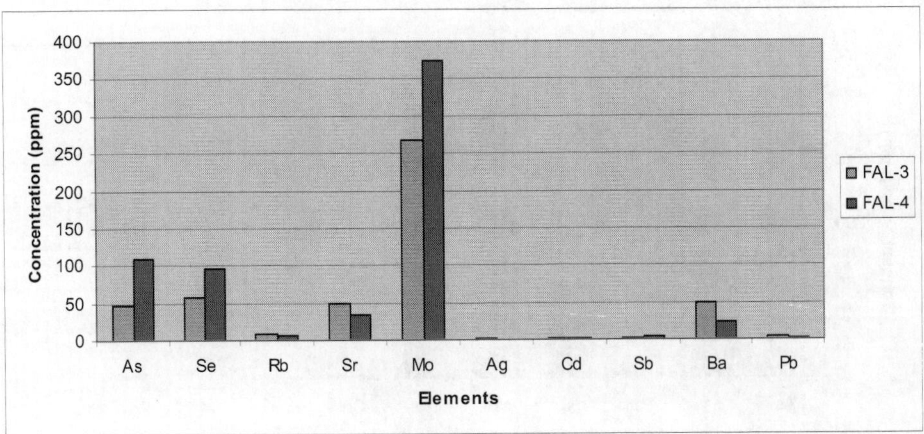

Contd...

Figure 40.1–*Contd...*

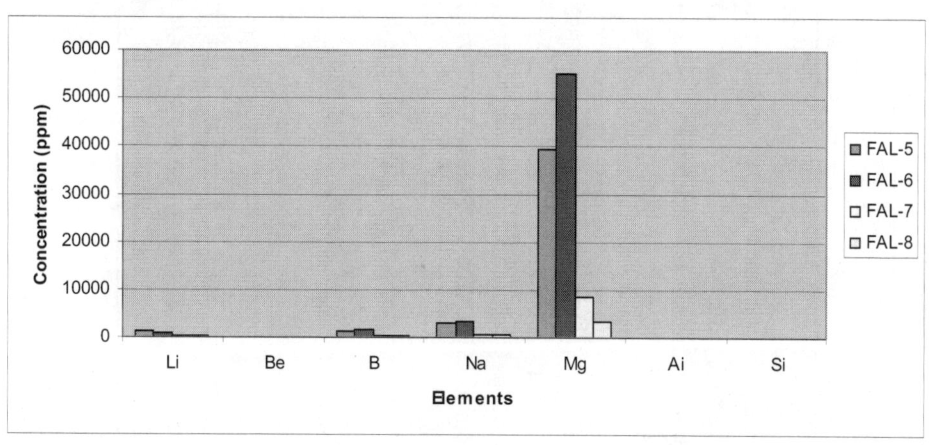

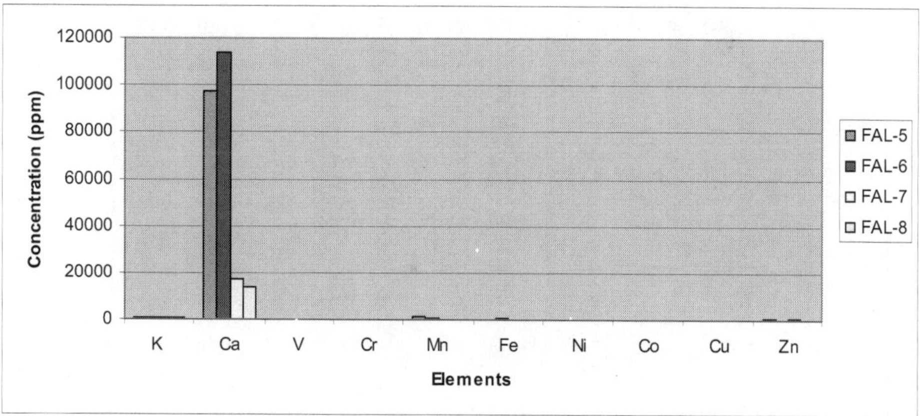

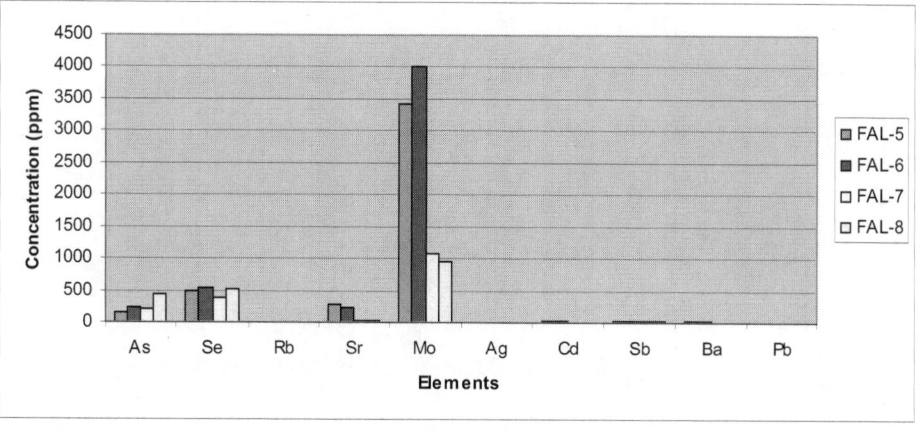

Table 40.6: Concentration of Trace Metals in the Leachates at different Time Intervals

Elements	*Concentration (ppm)*										
	FAL-1	*FAL-2*	*FAL-3*	*FAL-4*	*FAL-5*	*FAL-6*	*FAL-7*	*FAL-8*	*Average*	*Max*	*Min*
Li	101.101	100.36	108.110	120.256	1324.889	858.884	420.057	240.093	409.219	1324.889	100.360
Be	0.128	0.098	0.025	0.195	0.029	0.021	0.040	0.019	0.069	0.195	0.019
B	305.78	430.442	1004.017	748.290	1505.192	1711.409	508.109	486.244	837.44	1711.41	305.78
Na	1981.061	1325.213	2072.300	1611.729	3163.707	3500.161	827.173	626.396	1888.468	3500.161	626.396
Mg	2565.547	2282.656	3574.738	2887.299	39260.970	55352.11	8596.345	3577.722	14762.173	55353.108	2282.656
Ai	259.900	264.140	234.343	229.199	24.147	34.038	34.150	67.776	143.462	264.140	24.147
Si	4.512	3.395	3.633	5.458	10.598	8.406	10.612	8.941	6.944	10.612	3.395
K	1048.726	542.170	563.672	1136.589	569.170	712.660	562.451	634.334	721.222	1136.589	542.170
Ca	6714.750	9235.888	6318.601	7020.902	97707.88	113895.2	16910.46	13449.68	33906.666	113895.168	6318.601
V	31.898	27.962	42.370	81.613	101.858	119.890	158.862	240.400	100.607	240.400	27.962
Cr	17.345	7.651	4.006	4.387	1.703	6.036	1.726	6.271	6.141	17.345	1.703
Mn	188.287	176.567	94.435	95.718	1295.644	1005.776	153.889	56.389	383.338	1295.644	56.389
Fe	93.123	71.783	74.999	77.527	325.328	392.594	68.516	63.621	145.936	392.594	63.621
Ni	25.794	23.016	9.095	15.009	245.647	153.061	17.829	7.321	62.097	245.647	7.321
Co	3.639	3.873	0.820	1.496	42.142	23.660	2.286	0.614	9.816	42.142	0.614
Cu	43.313	10.806	7.081	11.515	1.462	2.185	1.110	1.186	9.832	43.313	1.110
Zn	411.510	82.650	34.577	55.573	345.611	191.393	370.489	80.807	196.576	411.510	34.577
As	28.583	24.424	47.355	108.602	146.354	227.656	200.599	430.216	151.724	430.216	24.424
Se	42.733	41.697	57.898	97.199	497.770	548.990	385.917	507.446	272.456	548.990	41.697

Contd...

Table 40.6–*Contd...*

Elements	*Concentration (ppm)*										
	FAL-1	*FAL-2*	*FAL-3*	*FAL-4*	*FAL-5*	*FAL-6*	*FAL-7*	*FAL-8*	*Average*	*Max*	*Min*
Rb	8.198	5.942	7.709	6.426	11.356	7.845	3.699	2.750	6.741	11.356	2.750
Sr	59.838	60.475	48.647	35.046	281.745	223.743	25.420	15.656	93.821	281.745	15.656
Mo	177.550	174.878	267.862	373.672	3420.702	4012.087	1073.985	939.344	1305.010	4012.087	174.878
Ag	1.384	179.616	2.894	0.375	0.079	0.067	0.082	0.103	23.075	179.616	0.067
Cd	1.611	1.438	1.073	1.120	15.395	14.179	3.680	3.037	5.192	15.395	1.073
Sb	1.427	1.103	2.662	4.970	24.256	25.136	15.340	17.921	11.602	25.136	1.103
Ba	158.153	90.331	49.998	23.217	19.417	14.677	5.474	3.507	45.597	158.153	3.507
Pb	3.238	2.118	3.515	2.130	0.256	0.209	0.338	0.402	1.526	3.515	0.209

Figure 40.2

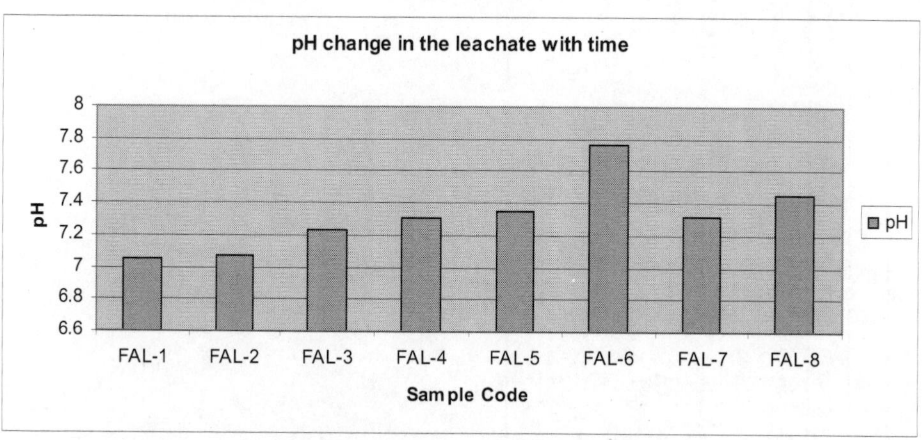

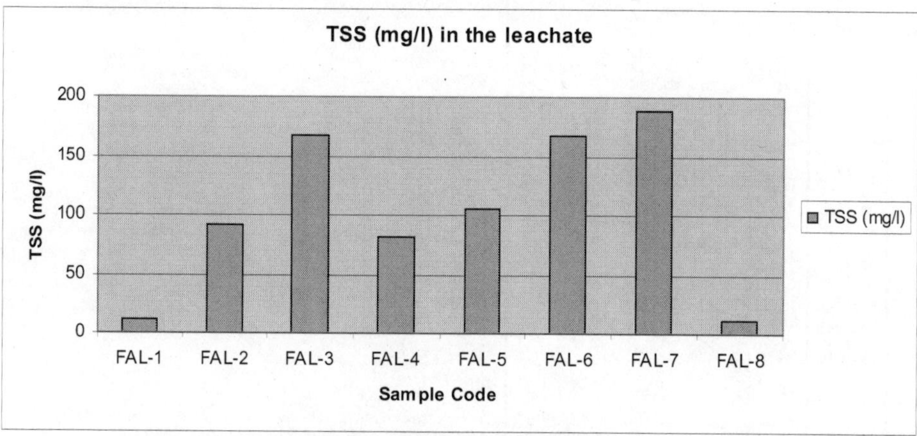

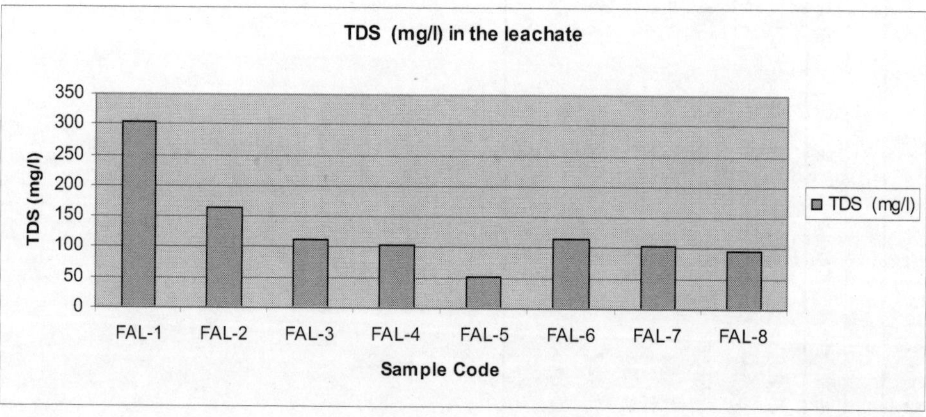

Contd...

Figure 40.2–*Contd...*

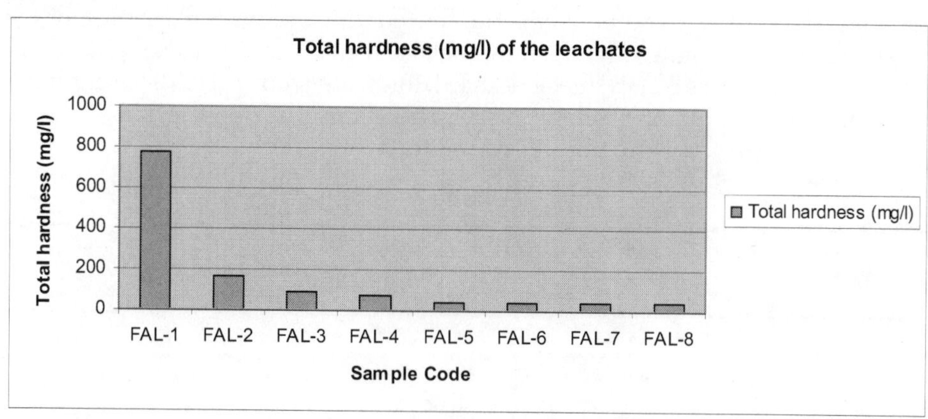

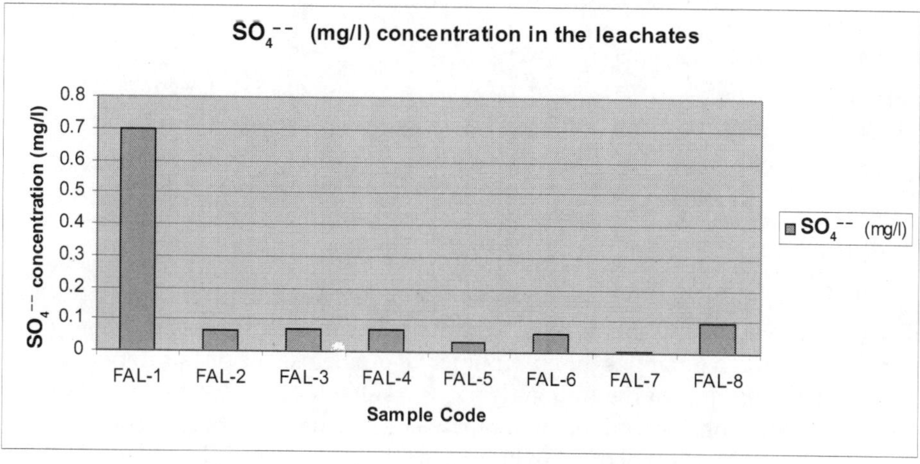

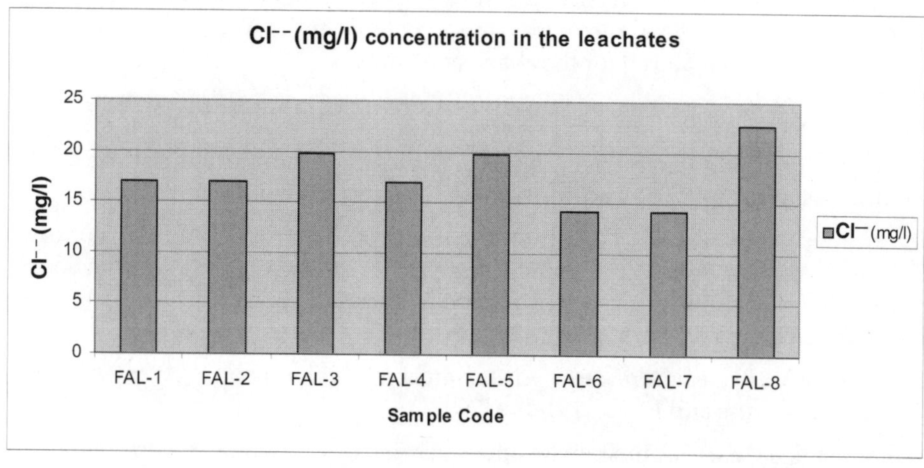

trend of variation. However, the hardness of the leachate is decreasing with increasing the contact time of fly ash and water. The concentration of the trace metals for the different leachate samples are given in Table 40.6 and reproduced graphically in Figure 40.2. This shows that the metals leached during the study period and conditions are having the concentration bellow the prescribed limit. With these results, it may be inferred that with increase of time the concentration will decrease.

Table 40.7

Sample Code	*pH*	*TSS (mg/l)*	*TDS (mg/l)*	*Alkalinity (mg/l)*	*Total Hardness (mg/l)*	*SO_4^{--} (mg/l)*	*Cl^- (mg/l)*
FAL-1	7.05	11	303	4	776	0.6975	17.02
FAL-2	7.07	91	163	8	168	0.0662	17.02
FAL-3	7.23	168	111	8	92	0.0711	19.85
FAL-4	7.31	81	103	4	76	0.0716	17.02
FAL-5	7.35	105	52	4	40	0.0329	19.85
FAL-6	7.76	167	116	4	40	0.0609	14.18
FAL-7	7.32	189	105	8	40	0.0078	14.18
FAL-8	7.46	11	96	8	44	0.0967	22.69

Conclusion

Fly ash is nuisance for the environment if not disposed off properly as it pollutes all the three major components of environment *i.e.* air, water and soil. Its more and more utilization can minimize the environmental problems. The physical and chemical characteristics of most of the Indian fly ash are suitable for cement manufacturing, brick manufacturing, back-filling of mines, waste land development and agriculture, structural fill, embankment of fly over bridge etc. pH, TSS, TDS, hardness and trace metals concentration in the fly ash leachates of IB Thermal Power Plant, Banharpali, Jharsuguda, Orissa -(OPGC) are within the acceptable limits (Table 40.7). So, this can be used as mine fill material for the abandoned mines.

References

1. Chandra A and Chandra H, Impact of Indian and imported coal on Indian thermal power plants, J Science Indian Resource, 63 (2004) 156-62.
2. Kumar V and Singh G, Fly ash–A safe resource material, 3rd Annual workshop on Flyash and Application and 8th Annual Convention of ISTE (Indian Society for Technical Education, Orissa-Section and Jagannath Insititue for Technology and Management, Paralakhemundi, Orissa), 28th January 2006.
3. Kumar, V., Fly ash: A resource for sustainable development, Proc. International Coal Congress and Expo., 191-199 (2006).
4. Managing Coal Combustion Residues in Mines, Committee on Mine Placement of Coal Combustion Wastes, National Research Council of the National Academies, 2006.

5. Human and Ecological Risk Assessment of Coal Combustion Wastes, RTI, Research Triangle Park, August 6, 2007, prepared for the US Environmental Protection Agency.

6. Prasad, B. and Singh, T.P.N., Extent of Environmental damage due to coal ash disposal and constraints in its utilization in India, Proceeding National Seminar VSE 2001, November 29-30, 2001 at CMRI, Dhanbad.

7. Dhadse, S., Kumari, P., and Bhagia, L. J., Fly ash characterisation, utilisation and government initiatives in India-A review, Journal of Scientific and Industrial Research, vol. 67, pp. 11-18, January 2008.

Chapter 41

Assessment of Trace Metal Quantity in River Mandakini at Chitrakoot (M.P. and U.P. Part)

L.N. Gupta and G.S. Gupta

Pollution Research Laboratory,
Department of Energy and Environment,
Faculty of Science and Environment,
Mahatma Gandhi Chitrakoot Gramodaya Vishwavidyalaya,
Chitrakoot

ABSTRACT

Mandakini is one of the major holy rivers of Madhya Pradesh especially of Chitrakoot area, the eastern part of Bundelkhand region -shared between the states of U.P. and M.P. It is situated at $25^0 31'$N latitude and $80^0 81'$E longitude. Water quality monitoring of river Mandakini with respect to trace metals was carried out at various stations *viz.* Sati Anisuiya (SA), Sphatikshila (SS), Vaidehi Vatika (VV), Jankikund (JK), Pramod Van (PV), Paisuni Drainage (PD), Ramghat (RG), Karwi Bridge (KB) and Suryakund (SK). The parameters investigated were Pb, Mn Cd, Cu, Zn, Co, Ni, and Fe. Analysis of water samples was made using by atomic absorption spectrophotometer. Higher concentration of lead was found in all selected stations except Sati Anisuiya (SA). The results were compared with their standards as per BIS, WHO and ICMR.

Keywords: *Mandakini river, Water quality, Trace metal, Lead.*

Introduction

Mandakini is one of the major holy rivers of Madhya Pradesh especially of Chitrakoot area, the eastern part of Bundelkhand region -shared between the states of U.P. and M.P. It is known as Paisuni in government records. It originates from Kelhaura hills about 45 km in the south-west of Chitrakoot and about 20 km in the west of Majhgawan town under Satna district. It is situated at 25^0-31′N latitude and 80^0-81′E longitude. Since the upper belt of the river is essentially seasonal therefore in non-monsoon season it is almost dried. Only the Sati Anisuiya is the first spot where Mandakini river become perennial due to recharging of the river by a number of small springs coming from Sati Anisuiya hill. Afterwards it passes through Sphatikshila, Vaidehivatika, Jankikund, Pramodvan, Paisuni drainage, Ramghat, Karwi bridge, Suryakund and finally joins the river Yamuna in Rajapur. The Mandakini river has a great historical and religious importance. It is believed that Lord Ram had lived here for 12 years during his 14 years exile. Due to this fact, numerous people come here with spiritual emotions on the occasion of Amawasya, Ram Navami, Deepawali and other auspicious days and take bath in the river Mandakini. Thus Mandakini receives much quantity of pollution due to the mass bathing.(Kannan, 1995). Besides the above, the discharge of wastewater from houses, temples, Dharamashalas, hospitals, institutes, garages, etc., and agriculture run off is also deteriorating the river water quality. Open defection by the pilgrims along the river, cattle wallowing, throwing of carcasses of half burnt dead bodies, etc. into the Mandakini, the river quality and purity is retrograding continuously. The aforesaid discharge also contributes the trace metals to the river water. It is well documented that trace metals are toxic to not only human being but also to plant, animal, soil, air, water (Lohani, *et al.*, 2007). If water has a much quantity of the trace metal as per recommended limit by WHO and other agencies, water will be highly toxic for the user. Keeping these facts in mind, the trace metal quantity in Mandakini water was measured at many sites in June, 2007.

Materials and Methods

Nine sampling sites at river Mandakini were selected for study of the presence of trace metal quantity in its water. The monitoring was performed monthly in June, 2007. Details of the sampling sites and their geographical locations are given in Table 41.1. Water samples were collected in pre-cleaned polypropylene bottles with necessary precautions (APHA-AWWA). Glasswares used in the study were of high quality borosilicate brand (Schott Duran, Germany). Chemicals used were of AR/GR grade and obtained from Qualigen/E-Merck/Hi-media.The parameters investigated were Pb, Mn Cd, Cu, Zn, Co, Ni, and Fe. Analysis of the aforesaid metallic species was made by atomic absorption spectrophotometer.

Results and Discussion

The monitoring results of trace metal quantity of river Mandakini were mentioned in Table 41.2 and depicted in Figures 41.1–41.8. The results were compared with their respective standards prescribed by various regulatory bodies *viz.* WHO, BIS, etc. (Table 41.3).

Table 41.1: Details of Sampling Stations.

Sl.No.	Sampling Station	Location	Latitude and Longitude
1	Sati Anusuiya (SA)	A upstream station,15 km in the south of Chitrakoot	25^0 4' 26.5" N and 80^0 52' 2.6" E
2	Sphatikshilla (SS)	A upstream station, 03 km in the south of Chitrakoot	25^0 8'46.5" N and 80^0 51' 25.1" E
3	Vaidehi Vatika (VV)	A midstream station, 2.0 km in the south of Chitrakoot	25^0 9' 25.9" N and 80^0 51' 46.7" E
4	Jankikund (JK)	A midstream station, 1.5 km in the south of Chitrakoot	25^0 9'31.6" N and 80^0 51' 51.4" E
5	Pramod Van (PV)	A midstream station, 1.0 km in the south of Chitrakoot	25^0 10'15.2" N and 80^0 52' 1.1" E
6	Paisuni Drainage (PD)	A downstream station, 0 km from Chitrakoot	25^0 10'32.8" N and 80^0 52' 9.1" E
7	Ramghat (RG)	A downstream station, 0.1 km in the north of Chitrakoot	25^0 10'40.8" N and 80^0 52' 15.9" E
8	Karwi Bridge (KB)	A downstream station, 10 km in the north of Chitrakoot	25^0 12'51.8" N and 80^0 54' 16.9" E
9	Suryakund (SK)	A downstream station, 15 km in the north of Chitrakoot	25^0 15'7.2" N and 80^0 53' 45.1" E

Table 41.2: Value of Trace Metal Element in the River Mandakini.

Sampling Stations	Amount of Trace Metals (mg/l)							
	Pb	Mn	Cd	Cu	Zn	Co	Ni	Fe
SA	0.0270	ND	ND	ND	0.1020	0.1420	ND	0.3460
SS	0.7253	ND	ND	ND	0.1250	0.1200	ND	0.2560
VV	0.7102	0.1176	0.0668	0.0265	0.5376	0.1600	0.0910	0.4530
JK	0.6594	0.0748	ND	0.0265	0.1075	0.1240	0.2646	0.1520
PV	1.2529	0.0802	0.0510	0.0267	0.6457	0.1020	0.1075	0.7860
PD	0.8358	0.0822	0.0380	0.0197	0.3556	0.1280	0.0480	0.0950
RG	1.2170	0.0069	0.0034	ND	0.3425	0.1260	0.0440	0.0980
KB	1.1590	0.0160	0.0042	0.0199	0.3396	0.2940	0.0500	0.1520
SK	0.6682	0.0000	0.0000	0.0000	0.1326	0.0960	0.0000	0.0820

Lead enter in water through the metallic pipes as a results of their corrosion with water. Lead compounds are very toxic to all forms of life (Manivasakam, 1996). It tends to deposit in the bones as a cumulative poison. Lead was found in the range of 1.2529 to 0.8358 mg/l (Figure 41.1). Minimum and maximum values were found at SA and PV respectively. All values were higher than their prescribed standard *i.e.* 0.05 mg/l (Table 41.3) except SA.

Table 41.3: Drinking Water Quality Standards.

Parameters	*Maximum Permissible Limit (mg/l)*		
	World Health Organisation, 1994	*Bureau of Indian Standard, 1992*	*Indian Council of Medical Research, 1975*
Lead (Pb)	0.05	–	0.10
Manganese (Mn)	0.10	0.5	0.30
Cadmium (Cd)	0.01	–	0.01
Copper (Cu)	0.05	–	1.50
Zinc (Zn)	5.00	3	15.00
Cobalt (C0)	0.20	–	–
Nickel (Ni)	0.02	–	–
Iron (Fe)	0.30	0.1	–

Manganese occurs in soil and rocks as manganese di-oxide and can be dissolved in natural water by the action of anaerobic bacteria. Manganese is undesirable domestic water supplies as it causes unpleasant taste for cooking. Its permissible limitis very low, mainly because of aesthetic and economic consideration rather than physiological hazards. It was found in not detectable amount at SA and SS stations. Maximum value (0.1176 mg/l) was recorded Vaidehivatika station (Figure 41.2). At other site it was found within permissible limit (Table 41.2).

Cadmium is highly toxic element and its minute quantity causes for adverse renal arterial changes in human kidneys. Cadmium tends to concentrate in the liver, kidneys, pancreas and thyroid of human and animals. Ones it enters the body, it is likely to remain. Cadmium was found in not detectable quantity at SA, SS and JK

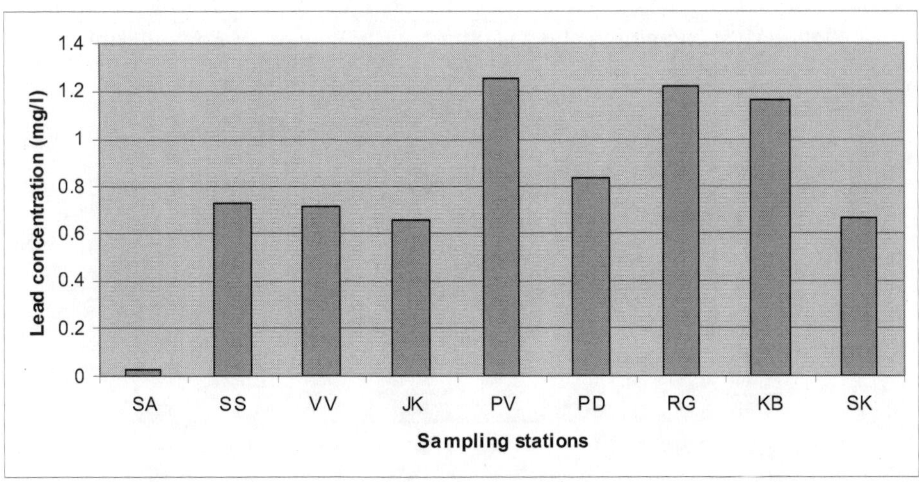

Figure 41.1: Lead Values at different Stations of River Mandakini.

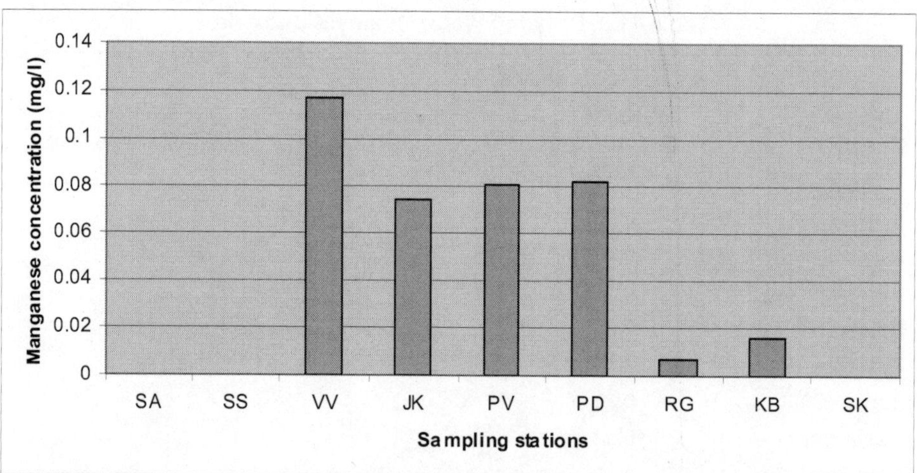

Figure 41.2: Manganese Values at different Stations of River Mandakini.

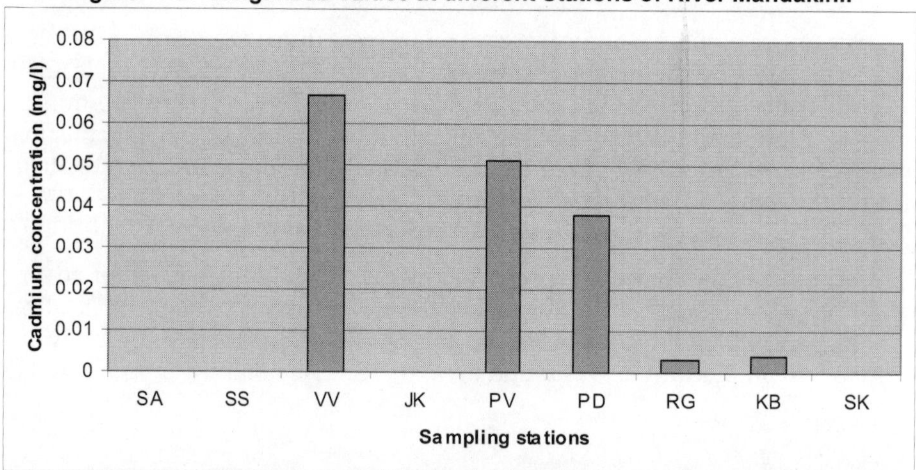

Figure 41.3: Cadmium Values at different Stations of River Mandakini.

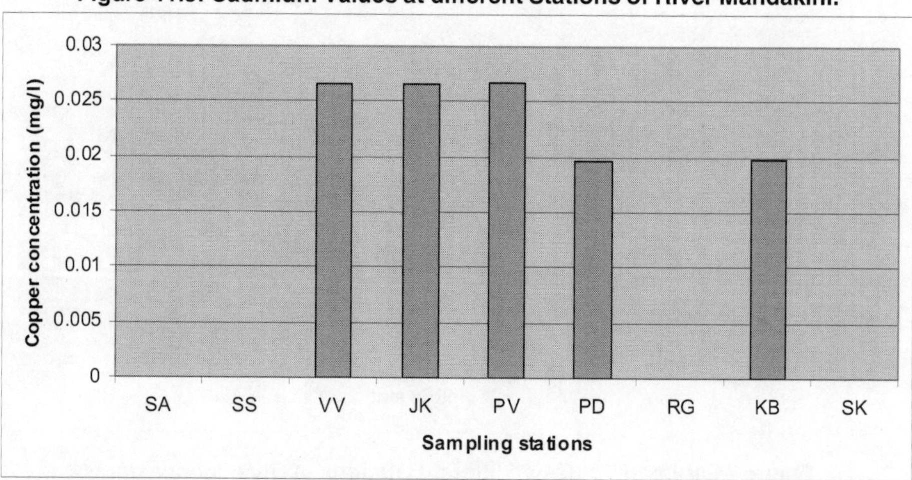

Figure 41.4: Copper Values at different Stations of River Mandakini.

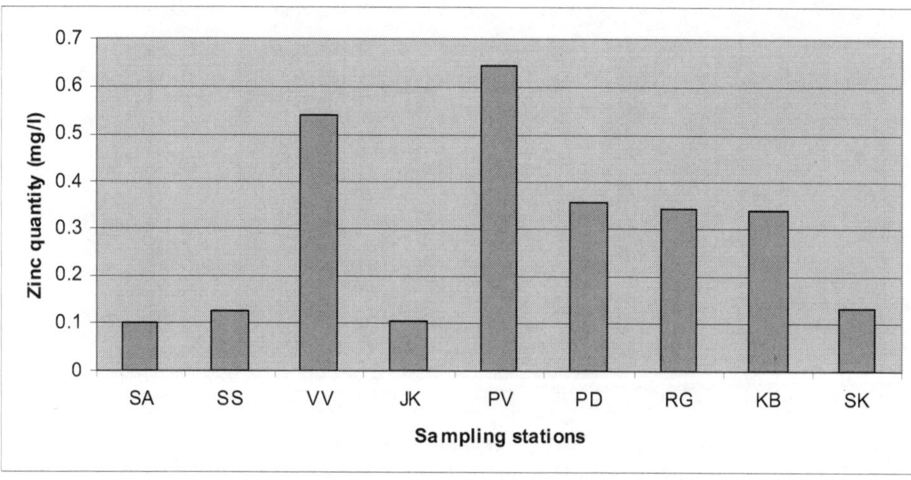

Figure 41.5: Zinc Values at different Stations of River Mandakini.

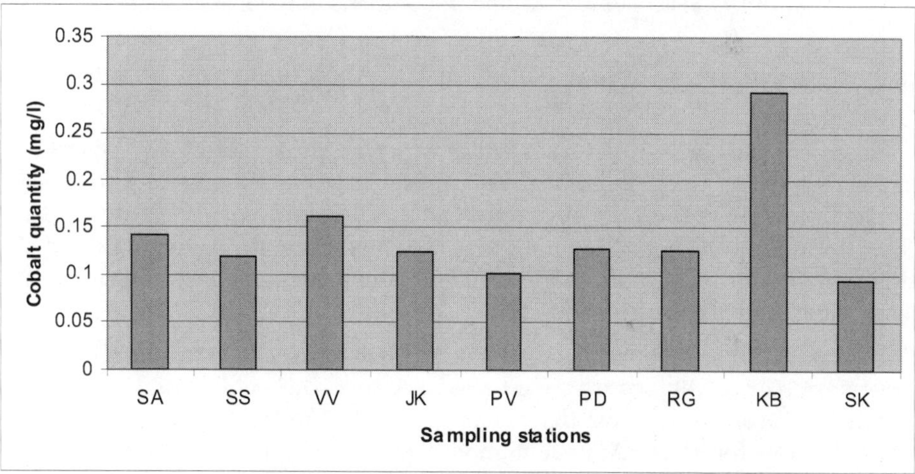

Figure 41.6: Cobalt Values at different Stations of River Mandakini.

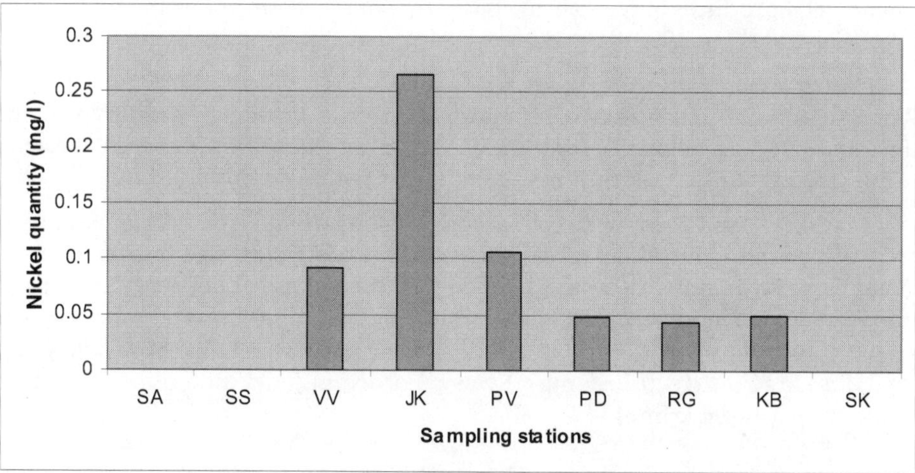

Figure 41.7: Nickel Values at different Stations of River Mandakini.

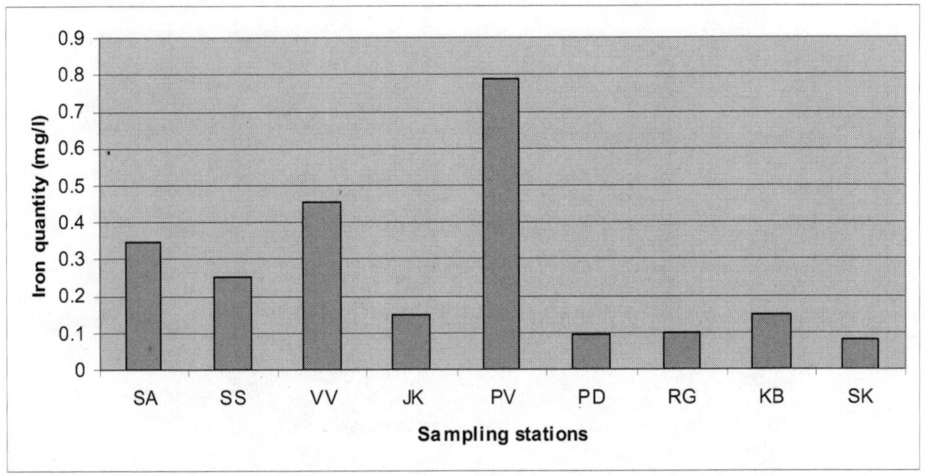

Figure 41.8: Iron Values at different Stations of River Mandakini.

stations. Maximum concentration 0.0668 mg/l was observed at VV. All values were within permissible limit except at VV and PV (Tables 41.2 and 41.3).

Copper salts occur in natural surface water in trace amount upto about 0.05 mg/l. Presence of copper renders water disagreeable to taste. Copper can be deposited in metallic form in pipes and boiler tubes with the dissolution of iron (Manivasakam, 1996). It was found in not detectable quantity at SA, SS and RG stations (Table 41.2). At Suryakund its amount was nil while at remaining stations the same was within permissible limit (Tables 41.2 and 41.3).

Zinc is commonly found in small quantities in domestic water supplies due to the corrosion of galvanized iron and brass in condensing, cooling and distributing systems. Value of zinc was found in the range of 0.1020 and 0.6457 mg/l (Table 41.2). Low value was found at SA while high value at PV (Figure 41.5). All values were within permissible limit. Although zinc has no adverse physiological effect upon human yet it produces bitter taste in water when it is present above the permissible limit (Manivasakam, 1996).

Higher concentration of cobalt in water may adversely affect to reproductive, digestive, and nervous system of human. Cobalt concentration was observed in the range of 0.096 to 0.294 mg/l (Table 41.2). Maximum extent was observed at Karwi bridge (Figure 41.6). At remaining stations it was within permissible limit.

Nickel does not occur in natural water. Metal platting water when discharged into water courses contributes nickel. Increased use of stainless steel and other nickel containing alloys in water system also contribute nickel. Nickel is not so toxic to human even above to permissible limit but it is harmful to fish and other aquatic lives. Nickel values were found in not detectable range at SA and SS (Table 41.2). Maximum amount (*i.e.* 0.2646 mg/l) was recorded at Jankikund although it was within the permissible limit.

Iron usually exists in natural water both in ferric and ferrous forms. Iron has no significant health hazards upon human but it is troublesome when present it excess *i.e.* beyond 0.3 mg/l. Its concentration in river water was found in the range of 0.082 to 0.786 mg/l (Table 41.2, Figure 41.8). It is remarkable to note that it was present beyond to its permissible value at various stations *viz.* SA, VV and PV. Higher concentration may produce the bitter taste in water.

Conclusion

Present study showed that the river Mandakini water was rich in lead, nickel and Iron. River zone from (SA) to (SS) may be designed as safe zone while zone from (VV) to (RG) as mild alarming zone.

References

1. APHA, Standard Method for Examination of Water and Wastewater, 18th ed., American Public Health Association, Washington DC (1992).
2. BIS, Indian Standard (IS:2490), Domestic Water Specification, Bureau of Indian Standards, New Delhi (1983).
3. ICMR, Manual of Standards of Quality for Drinking Water Supplies Special Report Series No.44, Indian Council of Medical Research, New Delhi (1975).
4. Kannan, G.K. Limnological Studies on River Mandakini (Paisuni) with Intention to Evolve an Approach for Conservation and Management, Ph.D. Thesis. M.G.C.G.V., Chitrakoot (1995).
5. Lohani, M.B., Rupainwar, D.C., Dhar, D.N. and Singh, A., Spectroscopic analysis of heavy metal concentration in the river Gomati at Lucknow, Indian J. Environ. Prot., 27, 162-164 (2007).
6. Manivasakam, N., Physico-chemical Examination of Water, Sewage and Industrial Effluents, 3rd ed., Pragati Prakashan, Mererut (1996).

Chapter 42

Characterization of Plastic Waste for Environmental Management and Sustainability

V.P. Sharma[1], A. Srivastava[1], V. Agarwal[1] and P.K. Mathur[2]

[1]CSIR-Indian Institute of Toxicology Research, M. G. Marg, Lucknow

[2]Chemistry Department, University of Lucknow, Lucknow

ABSTRACT

Plastics and polymeric products have a tremendous advantage over natural products as they can be tailor made for a specific use and their application ranges from packaging, housing, toys to agriculture, automobiles, biomedical and spacecrafts. Plastic finished products indiscriminately discarded in to our atmosphere after their service life or during their use or production phase. The amount of plastic waste in municipal solid waste (MSW) has grown steadily in recent years, reaching 6–12 per cent in mass and higher percentage in volume. Variations may occur with seasons and areas where MSW is collected. Several chemicals additives *viz.* plasticizers, stabilizers, colorants, heavy metals are added in plastic and polymeric products to make it resistant against degradable environmental conditions.

Xenobiotic chemicals may migrate due to leaching or biological transformation from discarded waste plastic and further percolate into surrounding soil, surface and groundwater basins. These plastic additives produce toxic effects with cumulative effect at higher concentration level under different environmental matrices. In present study, an evaluation of optimum migrational/ leaching conditions for waste plastic by using characterization results has been accomplished. Minimization of these wastes at the source has been considered as part of the solution for the proper disposal of MSW and has been encouraged by governments, communities and even by economic sectors. In addition to reducing the impact of pollution, this would represent savings of natural resources and even a source of work and income. We need to grow up with a significantly different mind set to issues of lifestyle and environment with reviewing periodically business process through the lens of ecological sustainability.

Keywords: *Polymeric products, Plastic waste, Xenobiotic, Environmental, Sustainability.*

Introduction

Plastics are high molecular weight organic polymeric materials which have molded the modern world and transformed the quality of life (Silva SA *et al.*, 2006). Plastics have many unique properties in terms of their manufacturability, production possibilities with tailor made specifications to serve the intended purpose. Plastic is convenient, lightweight, unbreakable and relatively inexpensive. The constituent polymer can be easily cast, pressed, or extruded into an enormous variety of shapes - such as films, fibres, plates, tubes, bottles, boxes etc. providing a versatile applications of plastic products in our daily life, from biomedical accessories to packaging, defence, sports, automobiles, spacecrafts, communication, scientific appliances and manufacture of household items (Lakshmi S *et al.*, 1998).

Plastic products can be easily found in human residential and occupational environments in high concentrations (Ejlertsson J et al.1996). After their service life, the plastic finished products are dispensed in the environment giving rise to large volume of inert garbage. Plastic wastes have become a subject of increasing worldwide environmental concern. The main sources of plastic waste are domestic, commercial, industrial and packaging activities. Domestic Solid Waste (DSW) and Municipal Solid Waste (MSW) composed of heterogeneous mixture of plastic and polymeric fraction, paper, cardboard, textiles, organic refuses generated by residential areas, market, commercial places etc. (De AK, 1995). Plastic and polymeric wastes are not necessarily pollutant but they have potential to become harmful to surrounding flora and fauna if not managed properly or when it occurs on a wrong place (Srivastava M, 1996). Plastic wastes clog the drains and thus hit especially urban sewage systems. Choked drains provide excellent breeding grounds for disease-causing mosquitoes besides causing flooding during the monsoons. The inert plastic wastes when reaches into rivers, streams and seas contaminate the water, sediment and marine life (Rebeiz KS *et al.*, 1995). Plastic waste can affect marine wildlife in two important ways: by entangling creatures, and by being eaten (Derraik JGB, 2002). Burning of plastic is may create skin and respiratory problems, resulting from exposure to and inhalation of toxic fumes *viz.* dioxin, furans, carbon mono-oxide, hydrocarbons etc (Fortelny I *et al.*, 2004).

With the development of the society, the waste generation is quite obvious and inevitable. The characterization of solid waste generation is an essential tool for adequate management strategy of urban solid waste in a city containing large volume of plastic waste. Inappropriate plastic waste operation, collection, storage, transportation and disposal practices may entail environmental and public health risks. Knowledge of plastic waste composition is necessary for an adequate management of urban solid waste (Alter H 1993).

Plastic Waste Characterization: Need for Environmental Management

Waste characterization includes studies related to physical composition of waste as well as identification of chemical, microbiological, or radiological constituents of waste material (Al-Salem *et al.*, 2009). Characterization of plastic waste is an important yardstick to determine reusable, recyclable and hazardous plastic fraction from large volume plastic garbage generated per day. Local government planners, rag-pickers and recyclers may estimate the amount of certain plastic waste materials in the waste stream through the available waste characterization database. This needs well planned mass dissemination programme for proper awareness about plastic waste generation and recycling in our society. Waste characterization information also helps in planning how to reduce plastic waste, set up recycling programs, conserve money and resources. Waste characterization studies have been carried out in other countries too which revealed a direct relation between plastic waste generation and income level of society (Abu Qdais *et al.*, 1997).

Plastic in Municipal Solid Waste (MSW)

The quantum of plastic waste is increasing at a rapid speed due to over consumerism, disposables and throw-away habit (Reinhart DR *et al.*, 1991). According to the report of Central pollution Control Board, India, it is estimated that approximately 10 thousand tons per day (TPD) of plastic waste is generated which is approximately 9 per cent of 1.20 lacs TPD of MSW in India. Plastic waste includes both thermoset and thermoplastics and may range from 20 per cent to 80 per cent of total post-consumer plastic waste. The country's plastic production is increasing at a rate of about 5 per cent to 20 per cent in various sectors (CPCB Report, 2006). Twenty per cent of solid municipal wastes in plastics are non-degradable and accumulate at the rate of 25 million tonnes per year. According to an estimate more than 100 million tonnes of plastic is produced every year all over the world while in India use of plastic is 3 kg per person per year. Polyolefins such as low-density polyethylene (LDPE), high-density polyethylene (HDPE) and polypropylene (PP) account for roughly 60 per cent of the total plastic wastes. Many of the thermoplastics are recyclable including Polyethylene Terephthalate (PET), Low Density Poly Ethylene (LDPE), High Density Ethylene (HDPE), Polypropylene (PP) etc. Inert portion of the total garbage mass is becoming a serious environmental concern due to composition of unclassified recycled plastic, Poly Vinyl Chloride plastic (PVC) and other plastic scrap (Christmann W *et al.*, 1989). The plastic content (20–25 per cent by weight) of medical waste is also significantly increasing and contaminating MSW (Marrack and Meyers, 1994).

The various options involved in plastic waste management process are landfilling, incineration, and recycling wastes into useful products. Recycling of plastic would not be successful unless the proper infrastructure to collect the waste is being set, the technology to economically reprocess the waste into new products is available, and the establishment of markets for the cost-effective use of recycled products is developed (Alter H 2005). The development of new construction materials using recycled plastics is important to both the construction and the plastics recycling industries. Extensive research investigated the use of resins based on recycled poly (ethylene terephthalate) (PET) plastic waste for the production of a high performance composite material, namely polyester concrete (PC). According to United States Environmental Protection Agency (USEPA) report recovery rate of MSW by weight was about 27.3 per cent, while the recovery rate of plastics was only 5.4 per cent, *i.e.*, the recovery of plastics is much lower than that of other MSW and most plastics are landfilled. Therefore plastic is becoming one of the most important components of the waste stream to occupy landfill space. To save landfill space, to preserve natural resources and to reduce expensive disposal cost, recycling of plastic waste should be increased on the basis of waste characterization data.

The sanitary landfill is an effective proposition for inert plastic waste disposal if it is properly designed and managed. Chemically plastics are consists of synthetic or naturally based inert macromolecular organic polymeric chains. Several chemical additives such as antioxidants, stabilizers and plasticizers etc. are incorporated in the polymer matrix to increase the durability, tensile strength and shelf-life of plastic product (Tullo AH *et al.*, 2000). Such landfilled plastic waste may undergo slow transformations through chemical and biological conversion (Ejlertsson J *et al.*, 1996). Since these plastic additives are not chemically bound to the constituent polymeric chain, but remain present as a freely mobile and leachable phase, they may migrate from soft plastic in the pool of organic and inorganic precursors, known as leachate (Deipser A et al.1994). Such compounds may percolate through soil and contaminate groundwater basins in the drainage area of the landfill. Leachate production and management is now recognized as one of the greatest problems associated with sanitary landfills having large quantity of plastic waste (Chu LM et al.1996). During landfill operation, leachates are produced, mainly due to the infiltration of rainwater through the refuse tips (Dass P *et al.*, 1977). Waste characterization plays an important role to determine the migrational aspects of plastic products (Foldes E 1998). On this basis a waste may categorized under hazardous waste or disposable non-hazardous waste.

Toxicity Characteristic Leaching Procedure (TCLP): Determination of Hazardous Characteristics of Wastes

This method is based on USEPA Method 1311-Toxicity Characteristics Leaching Procedure (TCLP). It is applicable to the determination of mobility of metals and semi-volatile organic compound in solid matrices and well applicable to characterize plastic waste. During the leaching procedure, primarily alkalinity of crushed and grinded plastic waste sample is determined to select appropriate extraction fluid (glacial acetic acid solutions in variable concentration). The solid sample is extracted (20:1 liquid to solid ratio) by shaking it end over end for 18 + or–2 hours at a controlled temperature. The extract also known as the leachate is then filtered and analyzed for the analyte of interest. Phthalate content and other semi-volatile organics are analyzed on High Performance Liquid Chromatography (HPLC) while inorganic metal contents can be analyzed on Atomic Absorption Spectrophotometer (AAS). The moisture content of the solid sample is determined separately and reported with the analytical results.

Conclusion

The main objective of a solid waste management system is to effectively safeguard the public health, safety, and welfare. It is necessary to carry out characterization studies for plastic waste as the generated data are useful in determination of treatment, disposal and recycling options for plastic waste. These are essential to design a sustainable and environment friendly solid waste management system. Depending on the type of plastic waste disposal guidelines for proper disposal and recycling pattern may be frame based on the experiences and success stories of developed countries. It is essential to consider the lab studies to pilot studies and thereafter converting them into cost effective high performance acceptable technologies for masses. The appropriate technologies will be the one which is easily accessible, financially viable and has the acceptance by the global community. It should meet the challenges of the market and customer satisfaction.

References

1. Abu Qdais H, Hamonda MF, Newham J. 1997. Analysis of residential solid waste at generation sites. Waste Management and Research 15: 395–406.

2. Al-Salem SM, Lettieri P, Baeyens J. 2009. Recycling and recovery routes of plastic solid waste (PSW): A review. Waste Management, 29: 2625-2643.Kushwaha R, Khadse K. 2009.

3. Alter H. 1993. The Origins Of Municipal Solid Waste: II. Policy Options For Plastics Waste Management. Waste Management and Research, 11: 319-332.

4. Alter H. 2005. The recovery of plastics from waste with reference to froth flotation. Resources, Conservation and Recycling, 43: 119-132.

5. Christmann W, Kasioke D, Kloppel KD, Partocht H, Rotard W. 1989. Combustion of polyvinylchloride—an important source for the formation of PCDD/PCDF. Chemosphere 19:387–392.

6. Chu LM, Cheung KC, Wong MH. 1994. Variations in the chemical properties of landfill leachate. Environ. Manage. 18:105-117.

7. Dass P, Tamke GR, Stoffel C. 1977. Leachate production at sanitary landfill sites. J. Environ. Eng. Div. 103(EE62): 981-988.

8. De AK. 1995. Environmental Chemistry. Third Edition, New Age International Ltd., New Delhi, p. 4-6.

9. Deipser A, Stegmann R. 1994. The origin and fate of volatile trace components in municipal solid waste landfills. Waste Manage. Res. 12:129–139.

10. Derraik JGB. 2002. The pollution of the marine environment by plastic debris: a review. Marine Pollution Bulletin, 44: 842-852.

11. Ejlertsson J, Johansson E, Karlsson A, Meyerson U, Svensson BH. 1996a. Anaerobic degradation of xenobiotics by organisms from municipal solid waste under landfilling conditions. Antonie van Leeuwenhoek 69:67–74.

12. Ejlertsson J, Meyerson U, Svensson BH. 1996b. Degradation of phthalic acid esters in municipal solid waste under methanogenic conditions. Biodegradation 7: 345–352.

13. Foldes E. 1998. Study of the effects influencing additive migration in polymers. Die Angewandte Makromolekulare Chemie, 261/262:65–76.

14. Fortelny I, Michalkova D, Kruli Z. 2004. An efficient method of material recycling of municipal plastic waste. Polymer Degradation and Stability, 85: 975-979.

15. Lakshmi S, Jayakrishnan A. 1998. Migration resistant, blood compatible plasticized polyvinyl chloride for medical and related applications. Artif Organs 22: 222–9.

16. Marrack D, Meyers D. 1994. PP/PE medical syringe recycling system: sharps away. 87th Annual Meeting and Exhibition, Air and Waste Manage. Assoc., Paper No. 94-RP-123B.01, Cincinnati, OH, June 19–24.

17. Rebeiz K S, Craft A P. 1995. Plastic waste management in construction: technological and institutional issues Resources, Conservation and Recycling, 15: 245-257

18. Silva SA, Garcýa RS, Cooper I, Franz R, Losada PP. 2006. Compilation of analytical methods and guidelines for the determination of selected model migrants from plastic packaging. Trends in Food Science and Technology 17, 535-546.

19. Srivastava M, Mathur PK. 1996. Waste Management. Proceedings of National Seminar on Waste Management 31-43.

20. Tullo AH. 2000. Plastics additives' steady evolution. Chem Eng News 78:21–31.

21. US EPA, 1996. Medical Waste Incinerator Rule. US Environmental Protection Agency Public Meeting, February 14.

22. US EPA, 1998. Characterization of Municipal Solid Waste in the United States: 1997 Update. EPA/530-SW-R-98–007. US Environmental Protection Agency.

23. USEPA 1994, Test Methods for Evaluating Solid Waste, Physical/Chemical Methods (SW-846). United States Environmental Protection Agency (USEPA), Test Method 1311-Toxicity Characteristics Leaching Procedure (TCLP).

Chapter 43

Polishing of Effluent from UASB by Surface Aerators: Monitoring by ORP, COD and BOD

***Rinku Walia*[1], *Pradeep Kumar*[1] *and Indu Mehrotra*[2]**

[1]*College of Military Engineering, Pune*

[2]*Department of Civil Engineering, I.I.T. Roorkee, Roorkee*

ABSTRACT

The performance of a good number of upflow anaerobic sludge blanket (UASB) based sewage treatment plants (STPs) has been evaluated. In India post-treatment of effluents from UASB based STPs is generally done by non–algal polishing ponds designed to have hydraulic retention time (HRT) of 1-1.6 days. However, the BOD, SS and MPN of the effluent from five STPs studied were found to be more than the limits prescribed by Indian Standards. A need has, therefore, been felt to evaluate the possibility of mechanically aerating the effluent from UASB. During aeration, ORP and DO increases where as COD and BOD decrease. Overall first order rate constant for decrease in BOD and COD varies from 3×10^{-3} to 6×10^{-3} per minute (0.18 to 0.36 per h) and 1.2×10^{-3} to 2.6×10^{-3} per minute (0.072 to 0.156 per h) respectively. The rate of DO transfer depends on ORP, DO deficit and the nature and concentration of BOD or COD. If the COD is high i.e > 150 mg/L then DO diversion towards COD assimilation is more but if COD< 150 mg/L then increase in DO is perpetually more.

Keywords: *Aeration, Anaerobic treatment, Chemical oxygen demand, Wastewater, Oxygen transfer.*

Introduction

The treatment of sewage by UASB process is an attractive and appropriate option for developing countries since it involves low initial investment and low energy for operation, easier maintenance than conventional aerobic processes, and an energy recovery through the production of methane gas (Van Haandel and Lettinga, 1994). However, the effluent quality from an anaerobic treatment unit in general, is not comparable with that of activated sludge process (ASP). It, therefore, indicates that the anaerobic process requires post treatment for the removal of residual organics, nutrients, and pathogens.

With the aim to achieve discharge standards, some researchers have explored the possibility of using aerobic post-treatment after anaerobic treatment. *e.g.*, UASB followed by ASP, rotating biological contactor (RBC), trickling filter (TF), waste stabilization pond (WSP), downflow hanging sponge cube (DHS) etc. However, these processes are difficult to implement in developing countries like India. In India all the UASB based STPs have a non algal pond with 1 to 1.6 day retention as a final polishing unit (FPU). The installation of these non-algal ponds reduced land requirement, but from treatment point of view it at best offered only removal of solids washed out of the UASB reactor. Total coliform count in the effluent from polishing ponds was in the range of from 10^6 to 10^9 MPN/100 mL is more than the maximum permissible limit of 10,000 MPN/100 mL (Foundation for Greentech Environmental Systems, 2004). Suspended solids (SS) concentration in polishing ponds ranged from 107 to137 mg/L. Also the BOD of the effluent is more than the limits prescribed by Indian Standards (Sato *et al.*, 2006).

A need has, therefore, been felt to explore technically feasible and economically viable options for the post treatment. Keeping this in view, a study has been undertaken to polish the effluent from STPs and bench scale UASB reactor with surface aerator. The aeration has been monitored by measuring ORP, DO, COD, and BOD.

Experimental Design and Analytical Methods

A 56 L bench-scale continuous fed UASB reactor was run in the laboratory of Civil department of IITR. The reactor kept in a temperature-controlled chamber, maintained at 30 ±2°C, was charged with the sludge from 38 ML/d STP at Saharanpur (S-38). During first 30 days, the reactor was fed with soluble wastewater consisting of sucrose only. Later the reactor was fed with the complex synthetic wastewater (SW) consisting of 40 per cent cellulose, and 60 per cent sucrose and peptone. Composition of the buffer and trace elements added in the feed were as per Prashant (2003). HRT was reduced stepwise from 12 hour to 6 hour. Samples were drawn for analysis after reactor attained pseudo steady-state (PSS).

Effluent (Eu) from 4 STPs and bench scale UASB reactor were aerated using a mechanical surface aerator. Details of four STPs and bench scale reactor are shown in Table 43.1. Surface aeration was carried out at the surface of 10 L effluent taken in a 12 L capacity container. The aeration device was operated through one H.P. variable speed AC/DC motor (220 volts, 4 amperes), and controlled by means of an AC/DC converter to regulate speed of motor at desired level.

Table 43.1: Sources, Sample Description and Feed Composition.

Sample	UASB	Particulate COD (CODp) per cent	Soluble COD (CODs) per cent	Total COD (CODt) mg/L
SW2	224 L/d +(Cellulose, Sucrose and peptone)	40	60	525-600
S-38	38 ML/d Saharanpur*	32	68	202-428
N-34	34 ML/d Noida*	58	42	330-490
N-27	27 ML/d Noida*	36	63	278-484
G-70	70 ML/d Ghaziabad*	34	65	384-525

+ 56 L bench scale UASB reactor at HRT of 6h

* Location: Between 28°38′ N 77°12′E and 29°58 N 77°23′E.

Experiments were conducted at temperature ranging from 15-22°C. All measurements were done in triplicate. Values given are either the average of three determinations or range of variation. Samples were aerated for 2 hours. After every 5 minutes, ORP and pH were recorded. Samples for COD, BOD, and DO were taken intermittently and analyzed in accordance with the procedures laid down in Standard Methods for the Examination of Water and Wastewater (APHA, 1998). The ORP was measured using an ORP electrode and DO was analyzed with Aqualytic OX 24 DO meter.

Results and Discussion

Effluent from four STPs namely S-38, N-34, N-27, G-70, and bench-scale UASB reactor were aerated by surface aerator. The temporal response of ORP, BOD, COD, and DO are shown through Figures 43.1(a)–(f). Perusal of data indicates an (i) increase in ORP and DO [Figures 43.1(a) and (b)] and (ii) decrease in BOD and COD [Figures 43.1(d)–(f)]. Perusal of data indicates an initial instantaneous increase in DO of SW2 and S-38. The DO of N-27, N-34 and G-70 increased slowly. Instantaneous increase in DO was not indicated in these cases. The increase in ORP can be somewhat superimposed over increase in DO. The nature of the effluent influences DO and ORP profiles. There can be two alternative patterns for increase in ORP and DO. Either both the DO and ORP increase instantaneously as the aerator starts and then tend to become asymptotic or, when the aeration is started, the oxygen concentration and potential rise steadily and DO stabilizes at a relatively low value where as the potential continues to increase slowly. The ORP and DO profiles of S-38 and SW2 follow the former trend. The degradation of the substrate (oxidation) quickly depletes the nutritive substance in the medium and thus reduces the oxygen uptake rate, causing an attendant increase in dissolved oxygen concentration and potential (Heduit and Thevenot 1989).

DO Profile

The main function of surface aerator is to enhance the transfer of oxygen into solution. The rate of oxygen mass transfer is determined by the magnitude of oxygen deficit and mass transfer coefficient (K_la, T^{-1}).

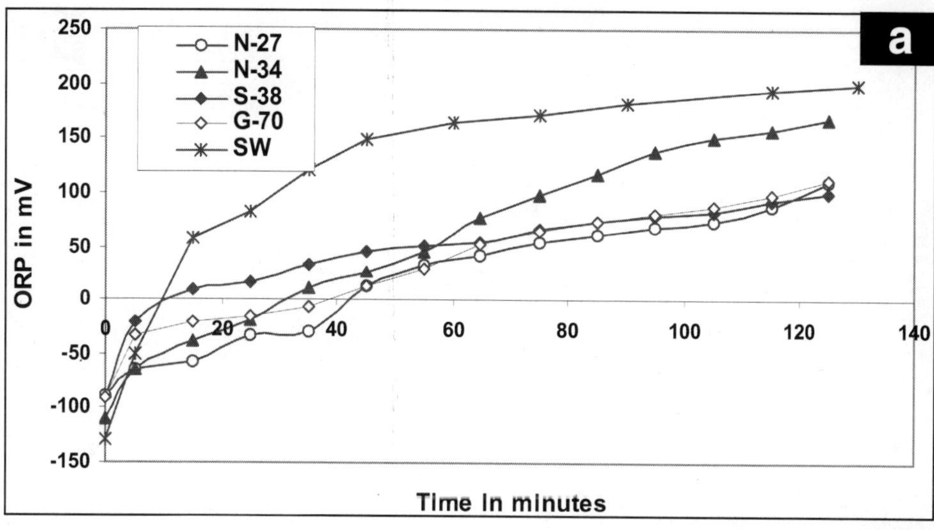

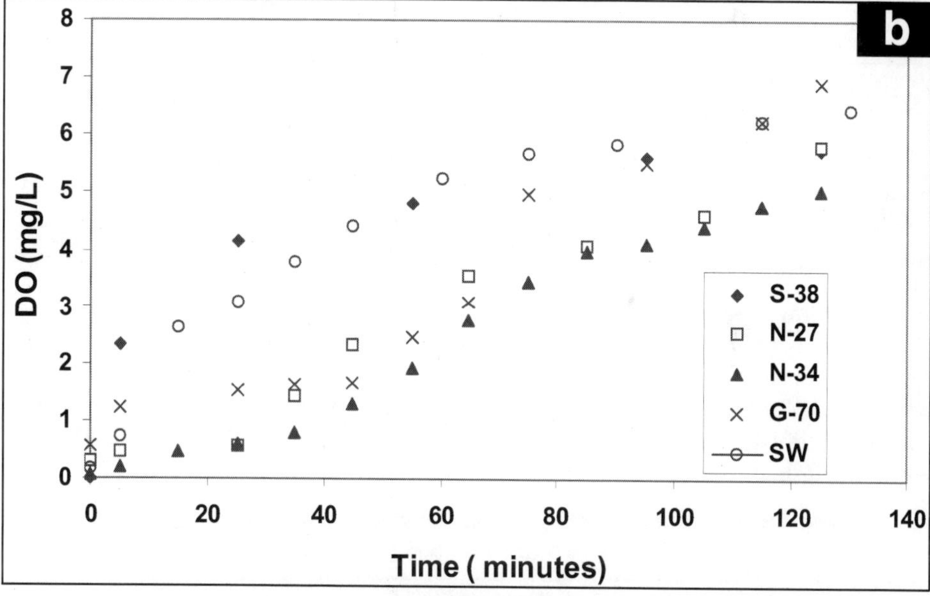

Figure 43.1: Aeration of Effluents
(a) Temporal profile of ORP; (b) Temporal profile of DO.

$$\frac{dDO}{dt} = Kla(DOs - DO) \tag{1}$$

Where DOs is the saturation concentration of dissolved oxygen in the solution and DO is the concentration of dissolved oxygen in solution. The temporal response of DO of the effluent from different STPs [Figure 43.2(b)] shows a narrow range of variation. Equation (1) has been used to model temporal observations recorded for

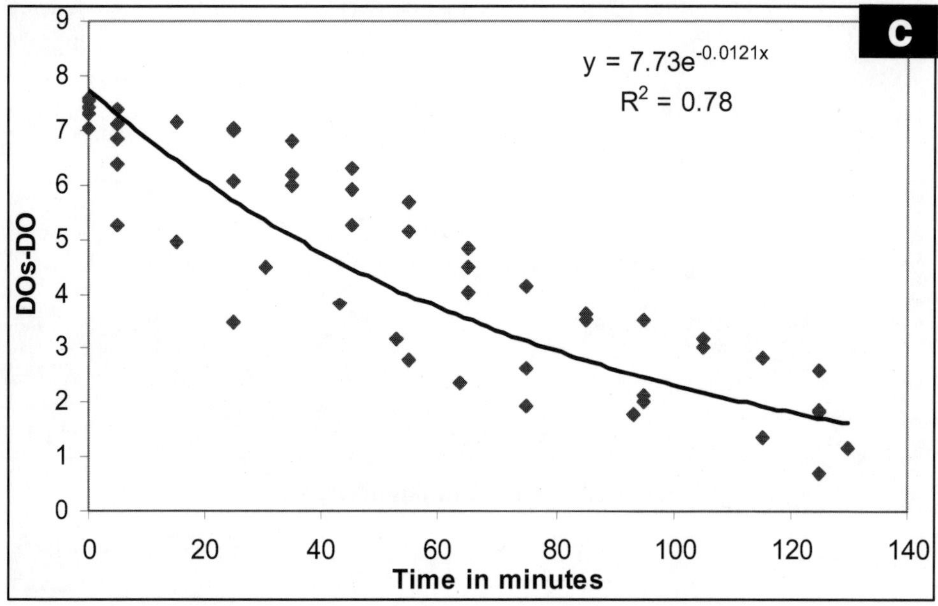

Figure 43.1(c): Temporal Variation of DO Deficit in Mean of all Data.

DO [Figure 43.2(c)]. The value of the constant (7.7 mg/L) gives deficit after 5 min of aeration. The deficit thereafter decreased at a rate constant (k_a) of 0.012/min.

BOD Profile

The temporal variation in BOD follows diphasic trend. The instantaneous decrease in first five minutes was 20-35 per cent of the total BOD removed [Figure 43.1(d)]. It is referred as immediate oxygen demand (IOD) and is probably due to negative ORP of the effluent. Thereafter, BOD decreased in a linear fashion. The correlation between BOD removal (per cent) and time are given in Table 43.2 and Figure 43.1(e).

Table 43.2(a): Empirical Correlation Between BOD Removal (per cent) and Time after 5 Minutes.

Sample Details	*Empirical Relation*	*Computed BOD Removal (per cent) in 5 minutes*	*R^2*	*Actual BOD Removal per cent in 5 minutes*
S-38	Y = 0.35x + 11.0	12.8	0.98	14
N-27	Y = 0.21x + 7.3	8.4	1	8.3
N-34	Y = 0.25x + 8.0	9.3	0.89	12.5
G-70	Y = 0.24 x + 19.0	20.2	0.99	20
SW	Y = 0.26x + 18.0	19.3	0.95	16.7

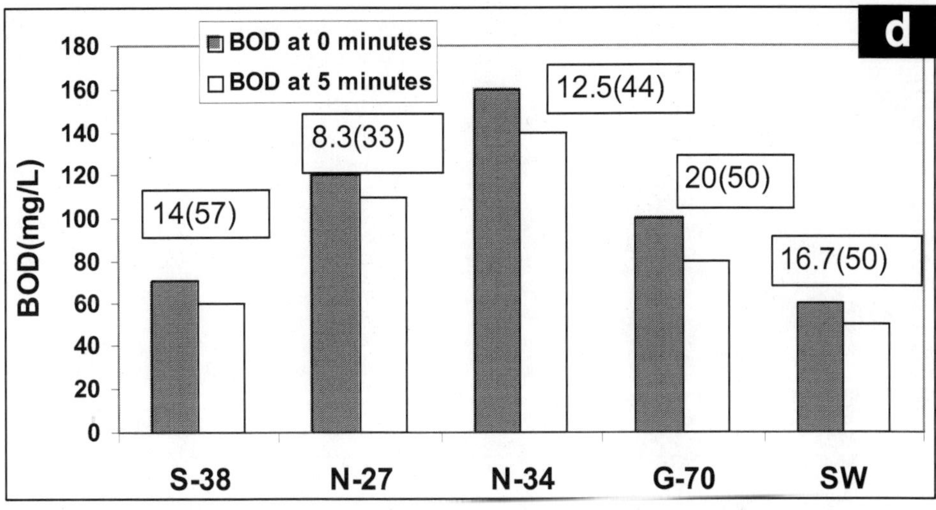

Figure 43.1(d): Temporal Variation of BOD for First Five Minutes. Values above bars indicate percentage BOD removal. Values in parenthesis indicate percentage of total BOD removal.

Table 43.2(b): Empirical Correlation Between BOD and Time.

Sample Details	*Empirical Relation*	R^2
S-38	$Y = 67.0\, e^{-0.006x}$	0.96
N-27	$Y = 116.0e^{-0.003x}$	0.97
N-34	$Y = 155.\, e^{-0.0038x}$	0.90
G-70	$Y = 90.0\, e^{-0.0046x}$	0.92
SW	$Y = 54.0\, e^{-0.0048x}$	0.92

The intercept corresponds to BOD removal without aeration. The BOD removal was computed from the correlations given in Table 43.1. The computed and observed values at t = 5 min are in fair agreement. The BOD removal (after 5 min) can also be described by exponential decrease (Table 43.1). The exponential first order rate constant for decrease in BOD varied from 3×10^{-3} to 6×10^{-3}/min (4.32 to 8.64/d).

The rate constant is high as compared to the values reported for sewage (0.2 - 0.46/d) on natural log basis. These observations suggest that BOD removal during aeration is probably non CBOD removal. Observations made here are in conformity with the observation made by Mason *et al.* (2006). They observed multiphase BOD exertion profile for the wastewater from dairy farm and found rate constant values in the range from 2.74 to 17.36 d^{-1}. These findings are contrary to the fact that rate of BOD exertion decreased with the extent of treatment. Such a statement perhaps holds good for aerobic processes. The anaerobic treatment renders a large fraction of organics and other contaminants in the reduced form in the wastewater, which are amenable at faster rate oxidation.

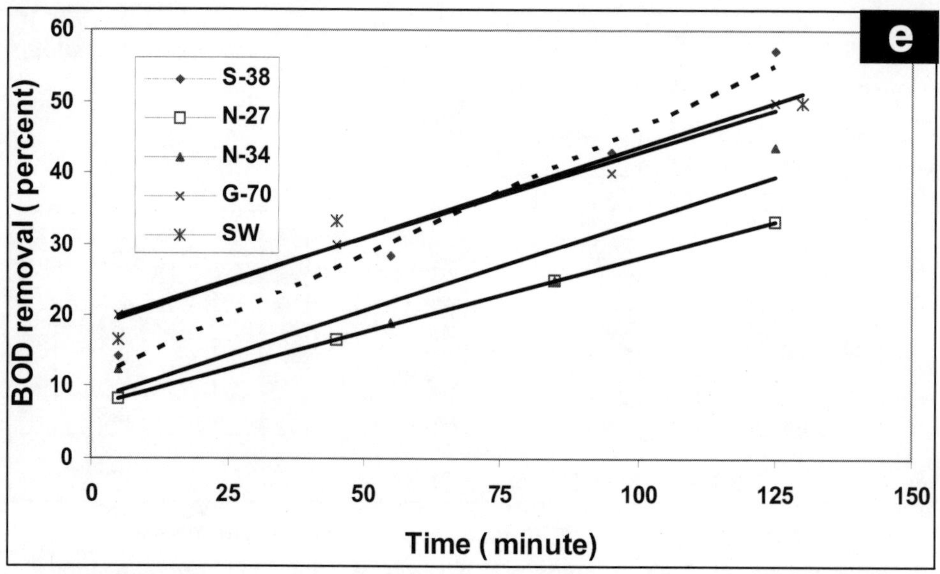

Figure 43.1e: Temporal Variation of BOD Removal per cent after 5 Minutes.

COD Profile

COD decreased exponentially with time. The values are given in Table 43.3. The rate constant for COD decrease varies from 1.2×10^{-3} to 2.6×10^{-3} per minute. The rate constant for the decrease in COD of the effluent from G-70 is higher than the others; nevertheless rate constant is independent of initial COD. Initial COD of Eu from G-70 and N-34 are same but decay rate of COD of the effluent from G-70 is twice as high as that of N-34. Decay rates of Eu from SW and S-38 are comparable. The decay rate for Eu from N-34 is minimum i.e 1.2×10^{-3}.

Table 43.3: Kinetics of COD Removal.

Sample Details	*Empirical Correlation*	k_d *(min⁻¹)*	R^2
S-38	$Y = 135.0\ e^{-0.0024x}$	2.4×10^{-3}	0.79
N-27	$Y = 230.0\ e^{-0.002x}$	2×10^{-3}	0.96
N-34	$Y = 199.0e^{-0.0012x}$	1.2×10^{-3}	0.90
G-70	$Y = 197.0\ e^{-0.0026}$	2.6×10^{-3}	0.95
SW	$Y = 101.0\ e^{-0.0021x}$	2.1×10^{-3}	0.80

From the data in Table 43.2(b) and 43.3 it appears that there is no correlation between kinetics of BOD exertion and COD removal. A relook of the values, however, does not reflect so. Data has further been scrutinized, analyzed and presented below.

BOD, COD and DO

During aeration of anaerobically treated effluent, the BOD and COD decrease at a rate much faster (~ 8/d) than the biodegradation (0.1/d-0.3/d). The biodegradation

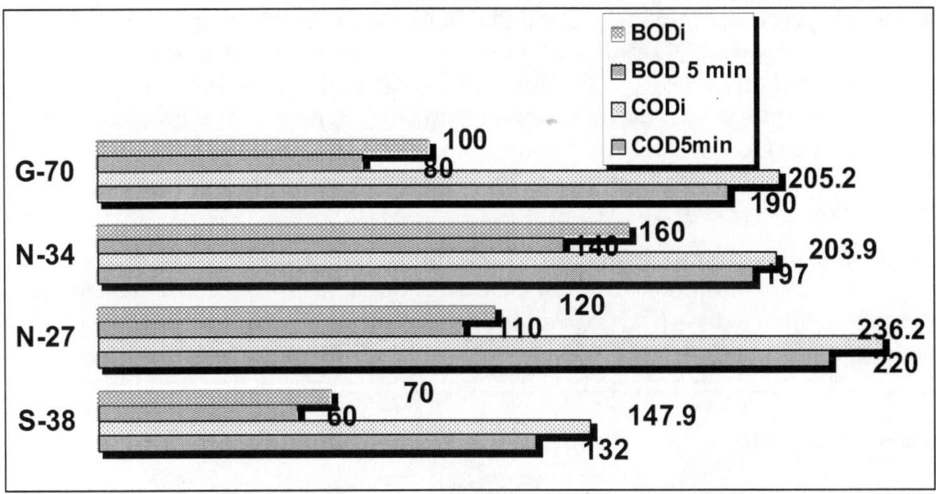

Figure 43.2a: Variation of BOD and COD in First 5 Minutes.

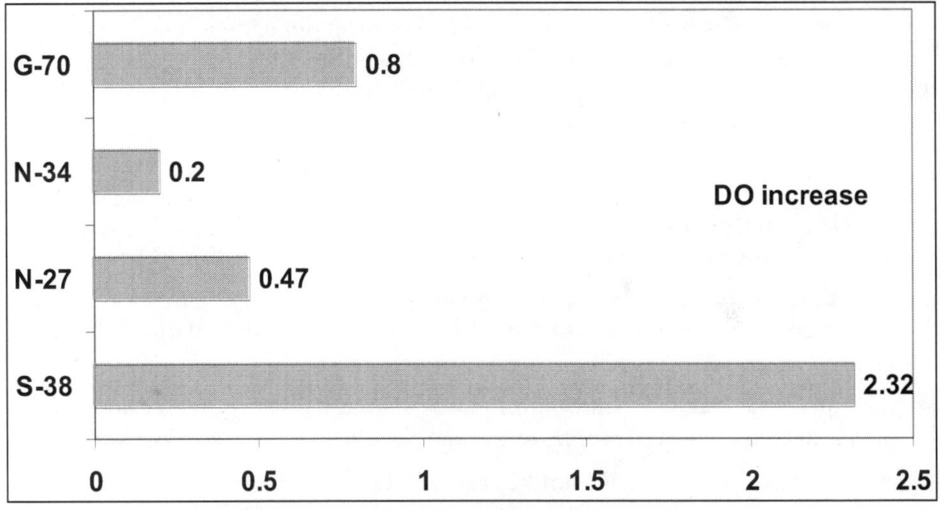

Figure 43.2b: DO Increase in First 5 Minutes.

is ruled out because aerobic bacteria would require lag time for acclimatization from anaerobic to aerobic. The decrease in oxygen demand (BOD) may be due to air stripping of volatile organics or some other physico-chemical process. In case volatile organics are removed by air stripping, then the oxygen transfer coefficient (k_a) would be comparable to that of water (3.9/h) or aerobically treated wastewater. The k_a however is 0.7/h. Also the initial instantaneous increase in DO depends on initial oxygen demand (Figure 43.2).

In order to further probe the instantaneous increase in DO and decrease in BOD and COD data from four STPs for first five minutes is presented in [Figures 43.2 (a) and (b)]. The per cent COD and BOD removal was found to range from 4 to 10 per cent and 8 to 20 per cent respectively. The initial BOD and COD did not appear to influence per cent decrease in BOD and COD. However, a distinctive difference in the rate of aeration based on initial COD has been noticed. If COD is less than 150 mg/L as in case of Eu from S-38 then the initial increase in DO (after 5 min) is more as compared to the effluents having COD in the range from 203 to 236 mg/L. The DO of the Eu from S-38 increases to 2.32mg/L where as the increase in the DO of Eu from N-34, N-27 and G-70 ranges from 0.2 to 0.8 mg/L.

In case of wastewaters the rate of DO transfer depends on DO deficit and the nature and concentration of BOD or COD. Although the initial deficit in all the four samples is equal to DOs (DOi =0) but the concentration of COD varies. If the COD is high *i.e.* > 150 mg/L then DO diversion towards COD assimilation is more but if COD< 150 mg/L then increase in DO is perpetually more.

Conclusion(s)

The effluent from four field UASB based STPs and a bench scale UASBR were characterized and aerated for two hour by surface aerators. During aeration BOD and COD decrease at a rate much faster (~ 8/d) than the generally observed BOD exertion rate for sewage (0.3/d). The rate of DO transfer depends on ORP, DO deficit and the nature and concentration of BOD or COD. If the COD is high i.e > 150 mg/L then DO diversion towards COD assimilation is more but if COD< 150 mg/L then increase in DO is perpetually more.

References

1. APHA, Standard methods for the examination of water and wastewater." 20th edition, American Public Health Association, Washington, DC, (1998).

2. Foundation for Greentech Environmental Systems, Case study on sewage treatment plants and low-cost sanitation under River action plans, New Delhi, (2004).

3. Heduit A. and Daniel R. Thevenot., Relation between Redox potential and oxygen levels in activated–sludge reactors, *Water Sci. Technol.*, 21, 947-956, (1989).

4. Mason, I. G., Mclachlan, R. I. and Gerard, D. T., A double exponential model for biochemical oxygen demand, *Biores.Technol.*, 97 (2), 273-282, (2006).

5. Prashanth, S., Treatment of sewage using UASB process, Ph.D. thesis, Indian Institute of Technology, Roorkee, India, (2003).

6. Sato, N., Okubo T., Onodera T., Ohashi, A. and Harada, H., Prospects for a self-sustainable sewage treatment system: A case study on full-scale UASB system in India's Yamuna River Basin. *Journal of Environmental Management*,1-10,(2006)

7. Van Haandel, A.C. and Lettinga, G., Anaerobic sewage treatment. A Practical guide for regions with a hot climate, Wiley, Chichester, U.K. (1994).

Chapter 44

Saccharification of Lignocellulosic Waste using Rumen Micro-organisms

S.P.M. Prince William, Kalyani Mohapatra, M. Mohanapriya, G. Prabhavati , A.N. Vaidya and T. Chakrabarti

CSIR-National Environmental Engineering Research Institute, Nehru marg, Nagpur

ABSTRACT

Saccharification of lignocellulosic biomass (LCB) is the major task in biofuel production for which bioprocessing is more beneficial. Albeit, a number of micro-organisms have already been identified as effective for lignocellulosic disruption and cellulosic ethanol production, search is still on for a suitable microbe / microbial consortia to make biofuel a dearer commodity. In the present investigation, an attempt has been made to utilize the rumen-clarified liquid (RCL) as a source of cellulolytic micro-organisms for the saccharification of organic residues in the form of vegetable waste. The results of current experiments showed that application of rumen micro flora for saccharification of LCB was encouraging, but it could not be a productive one because of the presence of non-cellulolytic microbes (scavengers) in RCL which consumed and depleted the reducing sugars produced by cellulolytic microbes leading to an overall reduction in the recovery of reducing sugars.

Introduction

Utilization of lignocellulosic biomass (LCB) for the extraction of liquid fuel (cellulosic ethanol) is the latest development in sustainable environmental technologies. Use of lignocellulosic wastes for biofuel production may improve energy availability, decrease air pollution and diminish atmospheric CO_2 accumulation. The technology involves collection of biomass, its decomposition into simple sugar followed by microbial conversion of sugar into the fuel. However, the main difficulty for the production and recovery of liquid fuel from LCB is its chemical structure *i.e.* cross linking between polysaccharides (cellulose and hemicelluloses) and lignin via ester and ether linkages that resists its degradation (Yan and Shuya, 2006; Xiao *et al.*, 2007). Presently, various pretreatment processes such as mechanical, physical, chemical and biological are being employed to make structural or compositional change in the chemical structure of lignocellulose in order to enhance digestibility, improve the rate of enzyme hydrolysis and increase the yields of desired products (Mosier *et al.*, 2005; Hendriks and Zeeman, 2009).

Compared to physical and chemical methods of processing of LCBs, biological methods are advantageous as they are efficient with low energy expenditures. In biological treatment process the whole enzyme or micro-organisms such as bacteria or fungi can be applied for LCB degradation. Commercially prepared fungal and bacterial hydrolytic and oxidative enzymes are also widely used in place of these micro-organisms. However, these purified enzymes are either costly or substrate specific in nature. Therefore, for a cost-effective production of liquid fuel from LCB, broader spectrum of microbial sources, which are relatively easy to harvest and efficient in LCW degradation, is essential. Oyeleke and Okusanmi (2008) have employed rumen micro-organisms of ruminants as a culture resource to degrade LCW and found rumen micro-organisms as effective in converting lignocellulosic biomass into fermentable sugars. Gijzen *et al.* (1988) and Barnes *et al.* (2003) have studied the utility of rumen micro-organisms for biomethanation and other bioprocesses. In the present study an attempt was made to exploit rumen microbes for the saccharification of vegetable waste. This method of LCW processing could be advantageous over other conventional (physio-chemical and biological by using purified enzymes) methods that are either energy or cost intensive in nature.

Materials and Methods

Preparation of Substrates

Vegetable waste was collected from cotton market, Nagpur and processed in laboratory. The fresh material was initially sun dried for 24 hrs followed by oven drying at 70^{0}C for 42 hrs. The dried material was powdered, sieved through a 2mm sieve and used for further experiments.

Preparation of Rumen Clarified Liquid (RCL)

Rumen waste was collected from a slaughterhouse at Mominpura, Nagpur, in an airtight glass bottle. Course particles were removed from the rumen waste through filtration aseptically in a laminar air flow chamber using muslin cloth and the clarified liquid (RCL) was used as a source of microbial inoculum.

Experimental Setup

Batch experiments were carried out using Teflon test tubes (50 ml capacity) with screw caps. Three treatments were made using 9 tubes (3X3 replication) in each treatment Nine tubes of the first treatment (T-1) were filled with 4 gm of autoclaved vegetable waste+ 30 ml of distilled water. Nine tubes of the second set (T-2) contained 4 g of vegetable waste +30 ml of RCL, and the third set contained only 30 ml of RCL. The tubes of T-1 and T-2 were connected through flexible silicon tubes for the release of gas during anaerobic process.

Analytical Techniques

Chemical Analysis

The total organic carbon content was estimated in terms of volatile fatty acid (VFA), determined by distillation method as per standard methods (1998). Reducing sugar was estimated by DNS (Dinitrosalicyclic acid) method (Miller, 1959). Concentration of cellulose was estimated by gravimetric method (Lossin, 1961).

Microbiological Analysis

The total rumen micro flora was determined as CFU (colony forming unit) on nutrient agar plate. The cellulolytic microbial biomass was determined on CMC (0.5 per cent) nutrient agar plate.

Estimation of Microbial Enzymatic Activity

The activity of cellulose was determined by filter paper assay (FPA) using the method of Ghosh (1987).

Results and Discussion

The results of the present investigation are presented in Figures 44.1–44.4. The efficacy of RCL to hydrolyze the cellulose of the substrate (vegetable waste) was assessed by taking into consideration the yield of reducing sugar after the RCL

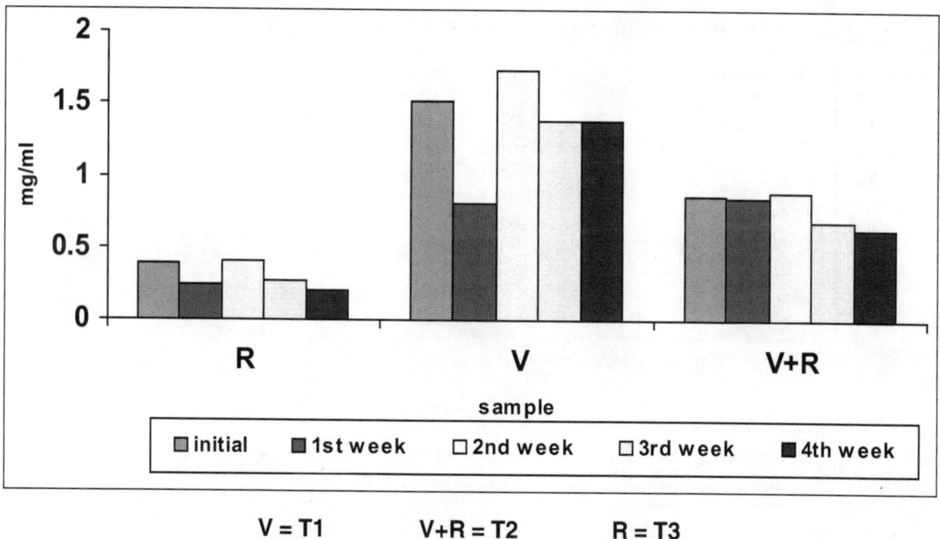

V = T1 V+R = T2 R = T3

Figure 44.1: Yield of Reducing Sugar in different Treatments.

application. Among the 3 sets of experiments, the concentration of reducing sugar was found to be higher in T-1 treatment where only raw material was present without RCL. On contrary, T-2 treatment which had both raw material and RCL yielded comparatively low concentrations of reducing sugar (Figure 44.1). These results contradicted our expectations that addition of RCL would enhance the yield of reducing sugar. However, activity of cellulase in RCL treatment was significantly higher than treatments without RCL (Figure 44.3). This indicates that RCL certainly improved the rate of cellulose degradation, however, the reducing sugar (RS) seem to be disappearing may be because of the over consumption of RS by microflora in RCL treatment (Figure 44.1).

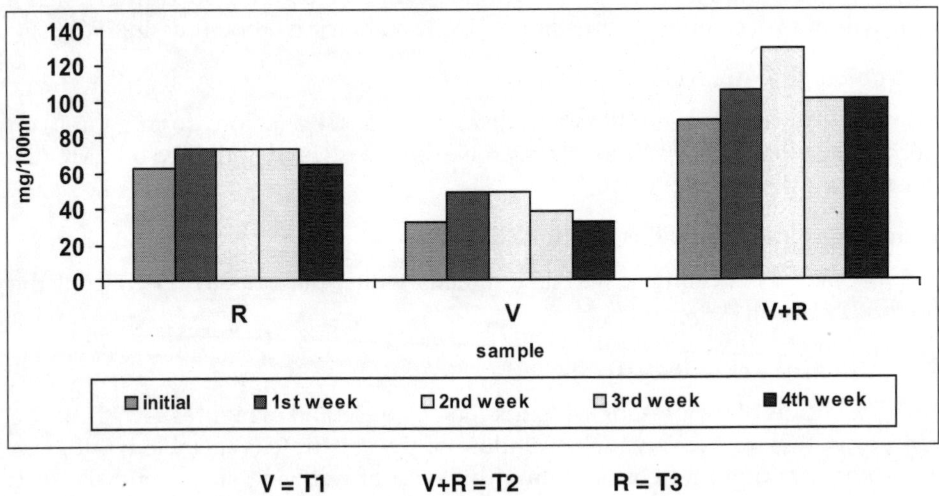

V = T1 V+R = T2 R = T3

Figure 44.2: VFA (Volatile Fatty Acids) Production in different Treatments.

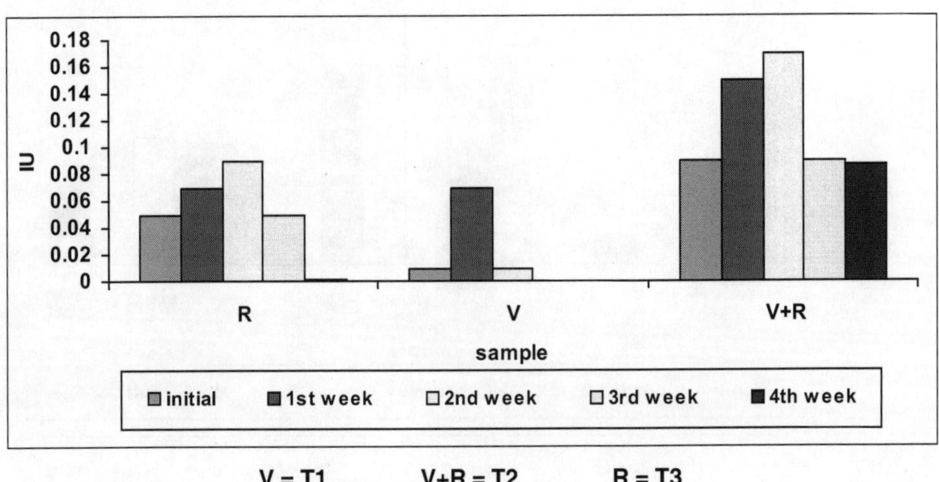

V = T1 V+R = T2 R = T3

Figure 44.3: Cellulase Activity in different Treatments.

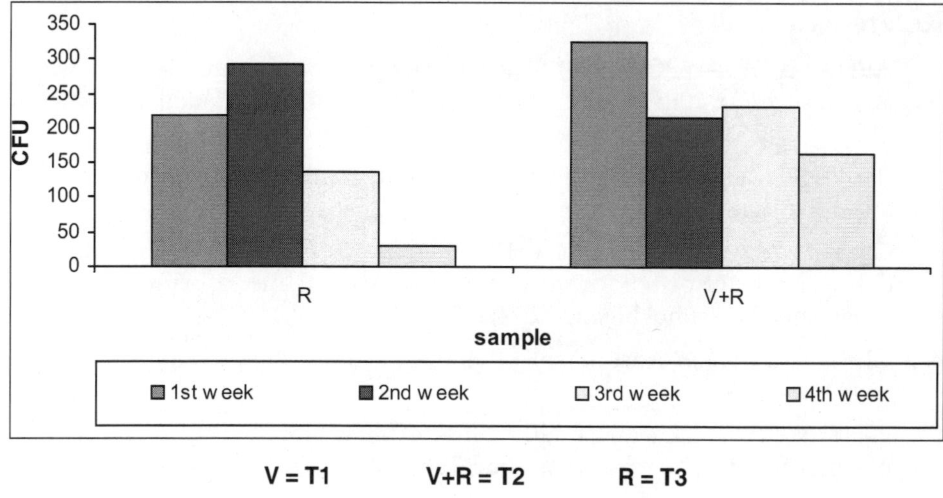

V = T1 V+R = T2 R = T3

Figure 44.4: Microbial Biomass in Treatments with and without RCL Addition.

The microbial density in case of RCL treatment was high (Figure 44.4) and it should be mentioned here that the biomass in RCL treatment is not only consisted of cellulolytic microbes, but also other substrate specific micro-organisms such as pectinolytic, proteolytic, ureolytic, sugar utilizing species, methane producing species, acid utilizing species and ammonia producing species. Hence, there is a possibility that the RS generated as a result of RCL activity could have been rapidly utilized by the non-cellulolytic microbes in RCLwhich ultimately resulted in a decline in the yield of reducing sugar. The other reason which could be cited for the reduction in RS yield is the increasing rate of VFA production in RCL treatment. Unlike other treatments (without RCL application) in case of RCL treatment the VFA production rate showed an increase (Figure 44.2). VFA consist of many fatty acid compounds and the precursor fro VFA is glucose. RCL treatment being rich with many different types of micro-organisms would have facilitated the production of VFA by converting most of the available reducing sugar into VFA.

Conclusion(s)

In the present investigation, an attempt was made to utilize the rumen-clarified liquid (RCL) for the saccharification of ligno cellulosic material (vegetable). The results clearly indicated the efficacy of rumen micro-organisms for the saccharification of vegetable waste, however, the recovery of reducing sugar was not high due to the presence to non–cellulolytic scavengers in RCL which consumed the reducing sugar produced by cellulolytic species of RCL. Further works are contemplated to selectively isolate cellulolytic microbes from RCL and to employ them for lignocellulosic disruption.

References

1. APHA-AWWA-WPCF (1998). Standard methods for the examination of water and wastewater, 20th edition, American Public Health Association, Washington D.C.
2. Barnes SP, Keller J (2003). Cellulosic waste degradation by rumen–enhanced anaerobic digestion.Water Sci Technol,48(4):155-62
3. Gijzen HJ, Zwart KB, TEunissen MJ,Vogels GD (1988). Anaerobic digestion of a cellulose fraction of domestic refuse by means of rumen micro-organisms.Biotechnol Bioeng 32:749-55
4. Ghosh T.K. (1987). Measurement of cellulase activity, Pure and Applied Chem, Vol59: 257-268
5. Hendriks ATWM, Zeeman G (2009). Pretreatments to enhance the digestibility of lignocelluloses biomass. Bioresour. Technol.100(1)10-18
6. Miller, G.L. (1959) Use of dinitrosalicyclic acid for determination of reducing sugar.Analytical chem., 31:426
7. Mosier N, Wyman C, Dale B, Elander R, Lee YY, Holtzapple M, Ladisch M (2005). Features of promising technologies for pretreatment of lignocellulosic biomass. Bioresour. Technol. 96(6): 673-686.
8. Oyeleke, S. B.* and Okusanmi, T. A. (2008).Isolation and characterization of cellulose hydrolyzing micro-organism from the rumen of ruminants. African Journal of Biotechnology Vol. 7 (10), pp. 1503-1504
9. Xiao C, Bolton R, Pan WL (2007). Lignin from rice straw kraft pulping:Effects on soil aggregation and chemical properties. Bioresour.Technol. 98(7): 1482-1488.
10. Yan L, Shuya T (2006). Ethanol fermentation from biomass resources:Current state and prospects. Appl. Microbiol. Biotechnol. 69(6): 627-642.
11. Zhen-Hu Hu, Han-Qung Yu (2005).Application of rumen micro-organisms for enhanced anaerobic fermentation of corn stover.Process Biochemistry 40:2371-2377

Chapter 45

Performance Studies of Ecorite PAC-2010 Visa-Avis Alum as Coagulant in Water Treatment

H.D. Juneja[1], Jayashri P. Kanfade[2] and Sunil P. Pande[3]

[1]*RTMNU, Nagpur*

[2]*, Priyadarshini College of Engineering, Nagpur*

[3]*CSIR-NEERI, Nagpur*

ABSTRACT

This paper describes the performance evaluation studies of ECORITE PAC–2010 of M/s. Shriram Consolidated Limited (DSCL) as a coagulant. It was observed that ECORITE PAC–2010 acts as a good coagulant over a wide range of turbidity and produces larger and more rapidly settleable flocs than the alum. It also describes the performance studies of ECORITE PAC-2010 and alum in the removal of fluorides and some heavy metals (Fe, Mn, Cd, Cr, Pb and Hg) from waters of different alkalinity. It was observed that the removal of fluoride was better with alum than with ECORITE PAC-2010. The heavy metal removal efficiency of alum and ECORITE PAC-2010 show a similar trend over a different range of alkalinity of the test water. The residual aluminium is less in ECORITE PAC–2010 treated water than that in alum treated water. The sludge volume generated by ECORITE PAC–2010 is less than that generated by alum for attaining the same quality of settled water from turbid raw water.

Keywords: *Alum, Aluminium, Coagulants, Fluorides, Heavy metals, Sludge.*

Introduction

Coagulants are those substances which are capable of removing colloidal impurities from water, and coagulation is the process by which such removal is brought about. Alum has been traditionally used as a coagulant in water treatment for over hundred years. It is commonly available, relatively inexpensive and has long been recognized as a successful compound for removal of colour and turbidity from water supplies. However, more recently attention has been directed to alternative coagulants in water treatment in efforts to reduce residual aluminium in finished water, minimize sludge production and eliminates post precipitation of aluminium residues in the distribution systems Gao, Bao Yu (2005). Alum is conventional coagulant against which performance of other coagulant is generally compared. The Poly Aluminium Chloride is an effective coagulant for removal of organic matter, fluorides and heavy metals from waters of different alkalinities.

The performance evaluation of Poly Aluminium Chloride (PAC) *vis-à-vis* alum as a coagulant in water treatment has been reported in this paper. The PAC investigated in this study was ECORITE PAC–2010 a product from M/s Shriram Consolidated Limited (DSCL). It is a coagulant which is claimed by M/s DSCL to be the latest generation Poly Aluminium Chloride based liquid coagulant/flocculant manufactured at Kota (Rajasthan) in India for the first time using state-of-the art technology imported from Italy.

The study involved extensive Jar Test experiments on different waters of natural origin as well as those prepared under laboratory conditions. The data generated in these investigations are presented in this chapter.

Materials and Methods

Preparation of Coagulant Solution

The working alum solution was freshly prepared by dissolving 10 gm of alum (ferric alum grade 2 ISI specification) in one litre of distilled water. For making 1 per cent solution of ECORITE PAC–2010 the dilution of this coagulant was done with distilled water on daily basis. The neat ECORITE–PAC 2010 was dosed by using a micro syringe.

Sampling

A large volume of raw water sample was collected at a time for the studies so that all the portions tested will be from the same source for a series of experiments. About 30 litres of water sample was collected each time to permit the requisite series of Jar Test of six portions each.

Jar Test Equipment

All the laboratory tests were carried out using Phipps and Bird Multiple Stirring Device (Jar Tester) equipped with stirring paddles and provision for controlled mixing. The floc size and its settleability were observed in illuminating device at the base of the apparatus.

Test Conditions

Measured volumes [1000 ± 10mL] of samples were flocculated using the Jar Test apparatus in 1000 ml beakers. The beakers ware placed in position on the Jar Tester.The motor of the paddled stirrer was started. After the addition of coagulant in each beaker simultaneously, rapid mixing was maintained at 100 ± 10 rpm (rotations per minute) for 30 seconds followed by slow mixing at 25 ± 5 rpm for 10 minutes. At the end of stirring period, the beakers were removed slowly from the Jar Tester platform and the contents of the beaker were allowed to settle for 20 minutes. For each series the Jar Tests were repeated and average value of turbidity recorded to eliminate subjective errors. The criteria used for the evaluation of the efficiency of the coagulants were settled water turbidity and visual appearance of flocs.

Test Waters

The test waters of different alkalinities were prepared in tap water by adding appropriate volumes of sodium bicarbonate solution. The test waters were spiked with suitable volumes sodium fluoride solution and heavy metals solutions for studying the effectiveness of fluoride and heavy metals removal by alum and ECORITE PAC-2010.

The test waters used for turbidity removal studies were natural waters or synthetic waters. The natural waters studied were river water collected from Kanhan river. In those natural waters where different turbidity ranges were studied, the highest turbid water was tested first. This highly turbid water was allowed to settle for some time and then supernatant was withdrawn. This supernatant has the same physico-chemical characteristics as the original raw water except less amount of suspended solids. Thus less turbid water of same source of natural water could be obtained. The same procedure was adopted for getting lesser turbid waters of different ranges from the same natural water of high turbidity.

The synthetic waters were prepared by preparing artificial turbid solutions. The preparation of artificial turbid solutions involved following steps:

- ✰ Air dried black cotton soil was collected from NEERI premises
- ✰ The air dried black cotton soil from previous step was again oven dried at 100°C for one day
- ✰ The soil was crushed manually using mortar and pestle and brought into powder form
- ✰ The crushed soil was passed through 100 mm sieve
- ✰ 20 gm of powder soil was added to 500 mL of distilled water, mixed thoroughly and soaked the entire soil completely. The soil was allowed to swell in a water for about 24 hours.(this allows maximum swelling of clay particles)
- ✰ The above suspension was then blended in high speed varying blends for 5 to10 minutes.(This helps in breaking the swollen clay particles to the colloidal size such that the resultant liquid when diluted with water to

form the suspension which will not allow any more settlements of particles after attaining stability)

- ☆ The blended portion was then further diluted with water to obtain suspension. This was then allowed to settle overnight
- ☆ The supernatant of the above step was then decanted without disturbing the sediments and used as a Stock Turbidity Suspension
- ☆ The standard turbidity suspension of desired turbidity was then prepared by diluting the Stock Suspension by approximate quantity of distilled water
- ☆ The quantity of soil and water can be suitably increased to prepare lager quantity of Stock Turbidity Suspension.

Analytical Methods

All the analytical estimations were done as per the procedures described in the 20th Edition of Standard Methods (1998). Nephelometric method was used for making turbidity measurements. This method is based on comparison of intensity of light scattered by the sample under defined conditions with the intensity of light scattered by a standard reference suspension under the same conditions. The instrument used for the turbidity measurements was Hach 2000A Turbidimeter. Spectrophotometric method was used for aluminium estimation using Eriochrome Cyanine R[ECR] dye.With the ECR dye, dilute aluminium solutions buffered to a pH of 6.0 produce a red to pink complex that exhibits maximum absorption at 535 nm. The optimum aluminium range was between 20 and 300 μg/L, but could be extended upward by sample dilution. The instrument used for absorbance measurement was Hach DR/ 2000 spectrophotometer. The alkalinity of the water samples was determined by titrimetric method and the fluoride analysis was done by the spectrophotometric method using SPADNS reagent and heavy metals were estimated by Jobin Yvon ICP Spectrometer and Coleman Mercury Analyser.

Turbidity Removal Studies

A series of Jar Tests were conducted using different turbid waters for evaluating the performance of ECORITE–PAC 2010 *vis-à-vis* alum. Natural water samples were collected from the Kanhan river. The turbidity of the test waters were achieved as described earlier. The optimum dose of alum and ECORITE–PAC 2010 to achieve comparable settled water turbidity from different test waters was found to vary. However, it was observed that in general ECORITE PAC-2010 produces larger and more readily settleable flocs than alum at different test water turbidities.

It was observed that for different raw water turbidities, the optimum ECORITE PAC-2010 dose in comparison to alum, to bring the settled water turbidity to 5 Nephelometric Turbidity Unit (NTU) was as under:

- ☆ On an average the dose required for ECORITE PAC-2010 was 50 per cent that of alum when it was dosed by preparing 1 per cent solution, whereas the dose was 40 per cent that of alum if applied neat.

- ☆ A typical set of comparative Jar Test results obtained for turbidity values of 50 NTU [Synthetic Water], 300 NTU [Natural Water], 500 NTU [Synthetic water] and 950 NTU [Natural Water].
- ☆ The exact dose for different waters will however, depend upon many factors such as the nature of turbidity, alkalinity, nature of organic matter and other characteristics of that particular water. Bavihashemi (2008); Wen Po Cheng (2008) and Annadurie (2004). Correct dose has to be optimised by performing Jar Test experiments on individual waters. The actual dose for the water treatment plants can be ascertained only after performing plant trials. There is always a plant factor which may vary from plant to plant and will be responsible for deviation in the Jar Test value and the actual value of coagulant to be applied in water treatment plant.

Study of Residual Aluminium

These studies were carried out using test waters having turbidities of 5 NTU, 500 NTU and 1000 NTU. Test waters were laboratory tap water and turbid waters prepared by mixing the tap water with bank sediments of Kanhan river. Jar Test experiments were performed on these waters as described earlier. It was observed that ECORITE PAC-2010 treated water always contained less residual aluminium than water treated with same amount of alum.

Fluoride Removal Studies

Studies on fluoride removal were conducted on test waters of four different alkalinities, *viz.* 250 mg/L, 400 mg/L, 600 mg/L, and 830 mg/L, $CaCO_3$.The fluoride concentration in the raw water was 5.0 mg/L. The dose of the coagulants applied to the test waters was 50 mg/L, 100 mg/L, 200 mg/L, 300 mg/L, 400 mg/L and 500 mg/L.

Heavy Metal Removal Studies

Jar Test experiments were performed on synthetic waters of different alkalinities prepared in distilled water. The test water was spiked with 1 mg of Fe and Mn and 0.55 mg of Cd, Cr, Pb and Hg. The dose of coagulant applied to the test water was 100 mg/L, 200 mg/L and 300 mg/L. Different sets of test water were subjected to Jar Test experiments as described earlier. The settled water was filtered through Whatman 42 filter paper and then analysed for heavy metals by Jobin Yvon ICP spectrometer and Coleman Mercury Analyser.

The results of heavy metals removal at test water alkalinity of 250 mg/L, 300 mg/L, 600 mg/L and 1000 mg/L as $CaCO_3$. It is observed from these Tables that there is considerable amount of reduction in the levels of metals in the test water which is subjected to Jar Test conditions without the addition of alum and ECORITE PAC-2010, indicating that under the existing pH and alkalinity conditions these heavy metals get precipitated as their corresponding insoluble complex salts. The removal of heavy metals may therefore, be attributed to the combined effect of coagulation as well as precipitation occurring at the pH and alkalinity conditions of the test water.

It is further observed that the removal of heavy metals is enhanced with the increase in the alkalinity of the test water.

Study of Sludge Volume

The sludge volume generated by alum and ECORITE PAC-2010 was compared by studying their performance on Kanhan river water and synthetic water prepared from black cotton soil. The doses of alum and ECORITE PAC-2010 were optimised by Jar Test experiments to obtain a settled water turbidity of 5 NTU. One litre of turbid water was subjected to sludge volume analysis in Imhoff cones after subjecting to half minute flocculation in the Jar Test apparatus. It was observed that with 3 hours settling period the reduction in sludge volume with ECORITE PAC-2010 as compared to that obtained with alum was 16 per cent in case of Kanhan water and 11 per cent in case of synthetic turbid water.

Conclusion(s)

From the investigations carried out in the present study it is evident that:

- ECORITE PAC- 2010 acts as a good coagulant over a wide range of turbidity
- ECORITE PAC- 2010 produces larger and more readily settleable flocks than alum at different test water turbidities
- For different raw water turbidities on an average the dose required for ECORITE PAC- 2010 is 50 per cent that of alum when it is dosed by preparing 1 per cent aqueous solution, whereas the dose is 40 per cent that of alum if dosed neat
- Slight haziness is observed in the water treated with 1 per cent solution of ECORITE PAC-2010, where as it is very clear when dosed neat
- The results obtained in the study showed that the consumption of alkalinity and drop in the pH is more in case of alum than with ECORITE PAC-2010.
- The residual aluminium is less in ECORITE PAC- 2010 treated water than that in alum treated water
- The fluoride removal is better with alum than with ECORITE PAC-2010.
- The studies with ECORITE PAC-2010 and alum show similar trends in removal of heavy metals from different test water in the alkalinity range of 250 mg/L to 1000 mg/L as $CaCO_3$.
- Heavy metal removal efficiency increases with increase in alkalinity of test waters.
- The sludge volume treated by ECORITE PAC- 2010 is less than that generated by alum for attaining the same quality of settled water from a turbid raw water.

Nomenclature

PAC: Poly Aluminium Chloride

rpm: Rotations per minute

ECR: Eriochrome Cyanine R

NTU: Nephelometric Turbidity Units

ppm: Parts per million

References

1. Annadurie, G., S.S.Sung, Due-Jong Lee, 2004. Simultaneous removal of turbidity and humic acid from high turbidity storm water, *Advances in Environmental Research*. 8, 712-725.

2. Bavihashemi, A., Mohammad Reza Alvi Moghadam, Reza Maknoon, Manouchehr Nikazar, 2008. Development of coagulation/flocculation predictive model for turbidity removal from Teheran water treatment plant, *Environmental Engineering and Management Journal*. 17 (1), 13-16.

3. Cheng, W., P, Fung Hwa Chi, Chun Chang Li, Ruey Fang Yu, 2008. A study on removal of organic substances from low water turbidity and low alkalinity water with metal Poly silicate coagulants, *Colloids and Surfaces, A Physico- chemical and Engineering Aspect*. 312, 238-244.

4. GAO Bao Yu. CHU Yong Bao., YUE Qin Yan. and WANG Bing Jian, WANG Shu-Guang, 2005. Characterization and coagulation of Poly Aluminium Chloride (PAC) coagulant with high Al_{13} content. *Journal of Environment Management*. 76, 142-147.

5. Standard Methods for the Examination of Water and Wastewater. 1998. 20th Edition, American Public Health Association/American Water Works Association and Water Environment Federation, Washington DC.

Chapter 46

Isolation and Characterization of a Phosphate Solubilizing Chitinolytic Bacterium from Soil and its Potential Application as a Bioinoculum

Divya Prakash[1], N.N. Nawani[2] and B.P. Kapadnis[1]

[1]Department of Microbiology, University of Pune, Pune
[2]Dr. D.Y. Patil Biotechnology and Bioinformatics Institute, Akurdi, Pune

ABSTRACT

Chitin is a polymer made of repeating units of N-acetylglucosamine (GlcNAc) which is efficiently broken down by chitinases into valuable by products having numerous applications in the field of medicine and agriculture. Chitinolytic bacteria were isolated from water and soil samples from various locations at Wayanad, Kerala. A potential chitin degrading actinomycete which could also solubilize inorganic phosphate was isolated. As microbes that can replace or reduce the usage of chemical pesticides and fertilizers are in great demand, this strain could provide a prospective biofertilizer with lieu to its phosphate solubilizing and chitin degrading ability. Besides, chitin is the main constituent of the exoskeleton of pests and fungi thus the strain can also be used as a biopesticide. Studies on identification of the isolate using molecular approaches and on quantitation of its phosphate solubilization ability are in progress. Using low cost carrier for agriculture, this isolate could be used as a potent bioinoculum in agriculture to improve crop yield and thus, reduce the usage of chemical pesticides.

Keywords: *Chitinolytic bacteria, Phosphate solubilization, Bioinoculum.*

Introduction

Chitin is a homopolysaccharide made of repeating units of *N*-acetylglucosamine (GlcNAc) concurrent by β-1-4 linkages. Chitin is found mostly in fungi and exoskeleton of molluscs (Cohen-Kupiec and Chet 1998; Wang and Chang 1997). Chitinases are a class of hydrolases which break down chitin to release chitooligosaccharides or GlcNAc. Based on their mode of action, chitinases are classified as endochitinases which act at random on chitin to release short chains of GlcNAc known as chitooligosaccharides and exochitinases which are further divided as chitobiosidases and *N*-acetylglucosaminidases. Chitobiosidases act on the short chains of GlcNAc to release chitobiose $(GlcNAc)_2$. *N*-acetylglucosaminidases release GlcNAc from the non-reducing end (Patil *et al.*, 2000). The action of the different chitinases is found to be synergistic and successive (Deshpande 1986; Gooday *et al.*, 1985: Shaikh and Deshpande 1993). The end products of chitin hydrolysis are found to have important applications in the field of medicine and agriculture (Patil *et al.*, 2000).

Phosphorous is the second growth limiting nutrient after Nitrogen and it is acquired by the plants in the form of phosphate anions. But since they are highly reactive with cations like Ca^{2+}, Mg^{2+}, they are easily precipitated and not available to the plants. The production of organic acids is the main mechanism by which plants utilize the organic Phosphorous from the soil and Phosphatase enzyme plays a key role in this method. Gluconic acid seems to be the most frequent agent of mineral phosphate solubilization. Also, 2-ketogluconic acid is another organic acid identified in strains with phosphate solubilizing ability. Chemical fertilizers have played an important role in the green revolution but due to their excessive usage, it has led to the reduction in soil fertility and environmental damage (Gyaneshwar *et al.*, 2002). Economic and environmental benefits can include increased income from high yields, reduced fertilizer costs and reduced emission of the greenhouse gas, N_2O as well as reduced leaching of NO_3-N to groundwater. Plant growth promoting bacteria are important in managing plant growth because of their effects on soil conditions, nutrient availability, growth and yields. The main aim of this study was to isolate chitinolytic Bacteria showing potent Phosphate solubilization which can be incorporated with low cost carriers and used as a Bioinoculum.

Materials and Methods

Collection of Samples

Water samples were collected from different locations in Wayanad, Kerala under sterile conditions. They were held at 4°C until analysis. For the collection of soil samples, the top layer of soil was removed to the desired depth with a clean spade and a thin layer of soil was removed. The area of the soil which came in contact with the spade was discarded.

Estimation of pH of the Samples

pH of the soil samples was analysed out by mixing the soil with sterile distilled water in 1:1 ratio and vigorous mixing was done. After the slurry was allowed to stand for 3-4 minutes, the pH was recorded. The pH of the water samples was also recorded using a digital pH meter (Orion Co., USA).

Preparation of Colloidal Chitin

Colloidal chitin was prepared as described by Hsu and Lockwood (1975). 10 g of practical grade chitin (SD-fine Industries, Mumbai, India) was dissolved in 400 ml of concentrated HCl and stirred for 1 h at room temperature. The chitin was precipitated by slowly adding 500 ml of water. The suspension was collected and subjected to continuous washing till the pH rose to 4.0 to 5.0. The amount of colloidal chitin was estimated by the dry weight and stored at 4°C until further use.

Isolation of Chitinolytic Bacteria

Isolation of chitinase producing Bacteria was carried out by suspending 1 g of soil sample in 10 ml sterile saline. The supernatant was serially diluted and plated on Colloidal chitin agar (CCA) having pH 5.0, 7.0 and 9.0. This was mainly done to determine the chitinase producing Bacteria at different pH range. Water samples were treated in a similar way. The composition of CCA is as follows (g/l): 4.0 Colloidal chitin, 0.2 yeast extract, 0.7 K_2HPO_4, 0.3 KH_2PO_4, 0.5 $MgSO_4$, 0.01 $FeSO_4$, 0.01 $ZnSO_4$, 0.01 $MnCl_2$, and 25.0 agar-agar. The plates were incubated at 37°C for 10 days. The isolates showing a zone of clearance around them were chosen for further studies.

Preservation of Chitinolytic Bacteria

The pure cultures were preserved on CCA slants at 4°C for actinomycete cultures and on nutrient agar for bacteria. They were also preserved at -20°C in 40 per cent glycerol with saline.

Screening of Phosphate Solubilizing Bacteria

The chitinase producing isolates were grown on Picovaskaya's (PKV) (Pikovskaya, 1948) medium to check for their ability to solubilize Tricalcium phosphate. The plates were incubated at 37°C for 2-3 days.

Genomic DNA Isolation

The bacterial cultures were cultivated in Luria-Bertani broth at 37°C in an orbital shaker (150rpm) for 3days. The genomic DNA was extracted according to the method described by Pospiech and Newmann (1950). The genomic DNA was treated with RNAse to remove the contaminating RNA. 5 µl of the genomic DNA was loaded on a 0.8 per cent agarose gel and electrophoresis was carried out at 50 V and constant current. The gel was stained with Ethidium Bromide solution (0.5 $\mu g\ ml^{-1}$) and then washed with distilled water to remove the excess Ethidium Bromide. The DNA was visualized under UV light in Gel documentation system (Alpha Innotech Co., USA). The genomic DNA having appreciable yield and good quality was employed to amplify their 16S rRNA gene to subject them for sequencing for identification at the molecular level.

Amplification of 16S rRNA Gene

The amplification of the 16S rRNA gene was carried out with specific primers as reported previously (Nawani *et al.*, 2002). The condtions of the PCR were as follows: 94°C, 5 min (Initial denaturation), 94°C, 1 min (denaturation), 65°C, 1 min (Annealing), 72°C, 1min (Extension) repeated for 30 cycles and 72°C, 10 min (Final extension)The PCR products were gel eluted and purified using Gel elute kit (Sigma-aldrich, USA). The sequencing of the 16S rRNA PCR products are being currently carried out.

Results

Isolation of Chitinolytic Bacteria

The water and soil samples were collected for isolating chitinase producing Bacteria from Wayanad, Kerala since Western Ghats especially Kerala is one of the hot biodiversity spot in India. Out of the 500 isolates enumerated, around 382 isolates showed promising chitinase activity. This indicates the presence of chitin degrading Bacteria among the different locations from which samples were collected and thus, their role in the recycling of nutrients in the environment. The large number of chitinase producers when compared to non-chitinase producers obtained from these locations is as shown in Figure 46.1 which indicates their predominance in the environment.

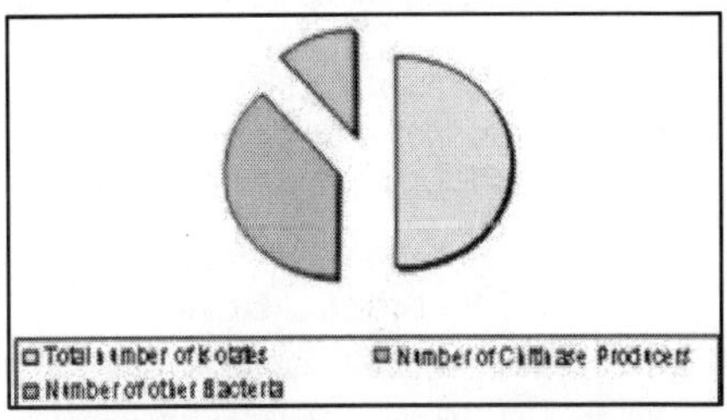

Figure 46.1: Distribution of Chitinase Producing Bacteria in the Samples Obtained from Wayanad, Kerala.

Relative Abundance of Chitinase Producing Organisms at different pH

The Bacteria which were isolated from different soil and water samples showed chitinase production in acidic, neutral and alkaline pH range. This is indicated in Figure 46.2. From this, it can be derived that chitinolytic organisms have the ability to produce chitinase at different pH range and importantly grow in various pH conditions because the soil conditions vary from one region to another. Mostly, the chitinase producing bacteria isolated grew well at neutral pH.

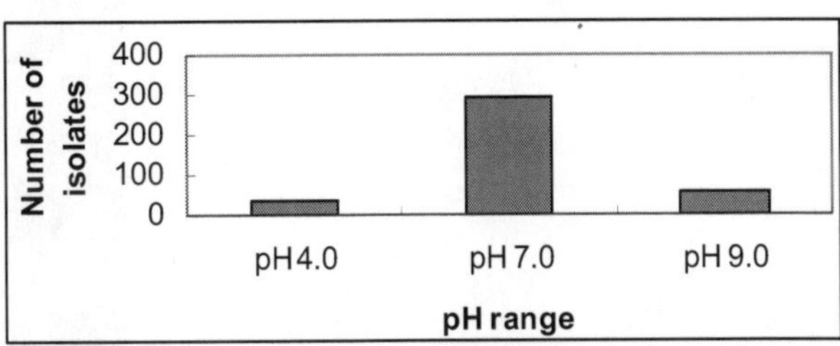

Figure 46.2: Relative Abundance of Chitinase Producers at different pH Values.

Screening of Chitinolytic Bacteria for Phosphate Solubilization

The chitinolytic Bacteria were grown on Picovskaya's medium. Out of the many isolates tested, one isolate showed predominant zone on the plate indicating Phosphate solubilization. This is shown in Figure 46.3.

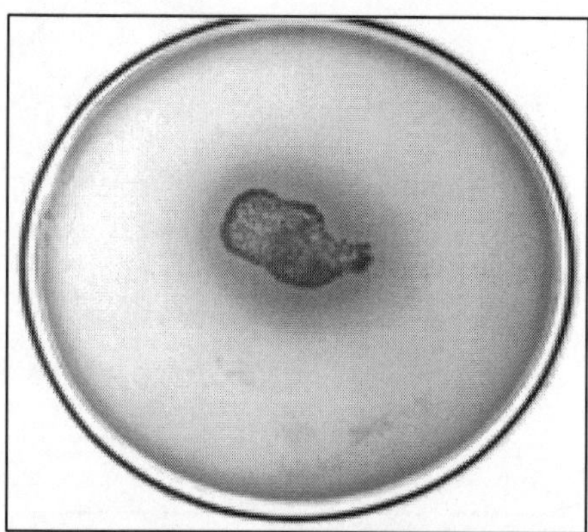

Figure 46.3: Zone of Phosphate Solubilization by One of the Isolates at 72 h on Picovskaya's Medium.

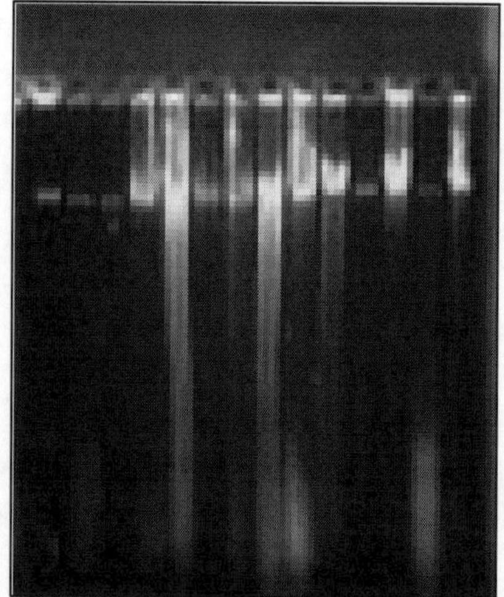

Figure 46.4: Agarose Gel Electrophoresis of the Genomic DNA of Chitinolytic Bacteria.

Genomic DNA isolation

The genomic DNA isolated was found to be of appreciable yield and quality as shown in Figure 46.4. The DNA was used as a template for the amplification of the 16S rRNA gene.

Amplification of 16S rRNA Gene

The PCR products were resolved on a 1.2 per cent agarose gel and electrophoresis was carried out at 50 V and constant current. The molecular weight of the 16S rRNA PCR product was found to be 1.5 Kb as determined using Alpha Innotech software and this is shown in Figure 46.5.

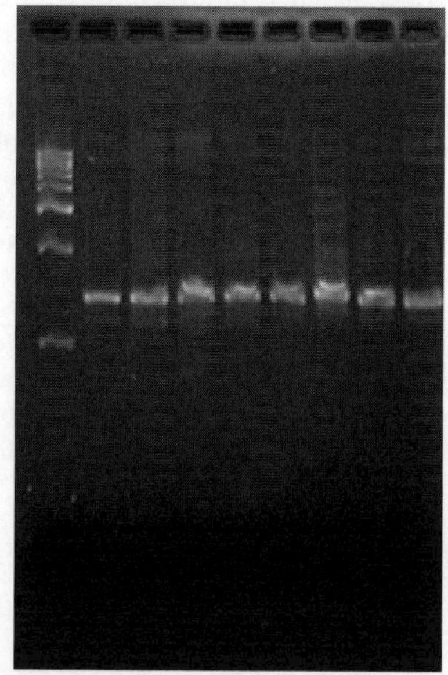

M: Molecular weight marker (1-10 Kb).
1 to 8: Amplified PCR products of various bacteria after clean up with Sigma PCR clean-up kit.

Figure 46.5: PCR Products of the 16S rRNA Gene on 1.2 per cent Agarose Gel.

Discussion

Many micro-organisms are used as biofertilizers or control agents for agriculture improvement. Chitinolytic Bacteria have been reported to act as biocontrol agents against pathogenic fungi and insect pests (Chernin *et al.*, 1997; Sampson and Gooday, 1998). It has been reported that chitinases are widely distributed in *Bt* strains and that some of the chitinase-producing strains can enhance the insecticidal activity of *Bacillus thurengensis* (*Bt*) (Liu *et al.*, 2002). Shapira *et al.*,(1989) cloned *S. marcescens* chitinase gene in *E. coli* and the chitinase preparation obtained was found to be effective in reducing disease incidence caused by *S. rolfsii* in beans and *R. solani* in cotton under greenhouse conditions. When the insect-pathogenic fungi, *Beauveria bassiana, B. brongniartii* and *Verticillium lecanii* were grown on chitin containing medium, cuticle degrading enzymes were produced. It was also found that the pre-treatment of insects with the chitinase enzyme solution was reported to be useful. In the present study, we are presently studying the chitinase activity of the Bacteria against pathogenic insects and fungi. Many materials have been evaluated, including different types of coals, bentonite, corn oil, mineral soils, peat, peat moss, vermiculite, and perlite. Peat is commonly used material for inoculant carrier. This would make the Bacterium an ideal candidate for the preparation of bioinoculum using low cost agricultural carriers.

Phosphate solubilizing microbes (PSM) have been reported to solubilize the rock phosphate through the production of organic acids, ion chelation and exchange reaction in the growth environment (Yadav and Dadarwal, 1997). As a result of this activity, PSM play an important role in supplementing P to the plants, allowing a sustainable use of phosphatic fertilizers. The effect of a combined inoculation of *Rhizobium*, a phosphate solubilizing *Bacillus megaterium* sub sp. *phospaticum* strain-PB and a biocontrol fungus *Trichoderma* spp. On growth, nutrient uptake and yield of chickpea were studied under glasshouse and field conditions. Combined inoculation of these three organisms showed increased germination, nutrient uptake, plant height, number of branches, nodulation, pea yield, and total biomass of chickpea compared to either individual inoculations or an uninoculated control (Rudresh *et. al*, 2004).

This gives the present study scope to develop a potent bioinoculum which can solubilize inorganic phosphate available in the soil to help plant growth and at the same time, act as a biocontrol agent against pathogenic bacteria, fungi and pests due to its chitinolytic activity.

References

1. Chernin LS, De la Fuente L, Sobolev V, Haran S, Vorgias CE, Oppenheim AB and Chet I (1997). Molecular cloning, structural analysis, and expression in *Escherichia coli* of a chitinase gene from *Enterobacter agglomerans. Appl. Environ. Microbiol.*, **63**: 834-839.

2. Cohen-Kupiec R, Chet I (1998). The molecular biology of chitin digestion. *Curr Opin Biotechnol* **9**: 270–277.

3. Deshpande MV (1986). Enzymatic degradation of chitin and its biological applications. *J Sci Ind Res* **45**: 273–81.

4. Gooday GW, Woodman J, Casson EA, Browne CA (1985). Effect of nikkomycin on chitin spine formation in the diatom *Thalassiosira luviatilis*, and observations on its peptide uptake. *FEMS Microbiol Lett* **28**: 335–40.

5. Gyaneshwar P, Naresh Kumar G, Parekh LJ, Poole PS (2002). Role of soil micro-organisms in improving P nutrition of plants. *Plant and Soil* **245**:83-93.

6. Hsu SC and Lockwood JL (1975). Powdered chitin agar as a selective medium for enumeratin of actinomycetes in water and soil. *Appl Microbiol.***29**: 422-426.

7. Liu M, Cai QX, Liu HZ, Zhang BH, Yan JP and Yuan ZM (2002). Chitinolytic activities in *Bacillus thuringiensis* and their synergistic effects on arvicidal activity. *J Appl Microbiol* **93**: 374-379.

8. Nawani NN, Kapadnis BP, Das AD, Rao AS and Mahajan SK (2002). Purification and characterization of a thermophilic and acidophilic chitinase from *Microbispora* sp.V2. *Appl Microbiol* **93:** 965-975.

9. Patil RS, Ghormade V, Deshpande MV (2000). Chitinolytic enzymes: an exploration. *Enzyme Microb Technol* **26**: 473-83.

10. Pikovskaya RI (1948). Mobilization of phosphorus in soil connection with the vital activity of some microbial species. *Microbiologiya*, 17.

11. Pospiech A. and Neumann B. (1995). A versatile quick-prep of genomic DNA from Gram-positive bacteria. *Trends in Genetics* **11:** 217–218.

12. Rudresh D.L., M.K. Shivaprakash, R.D. Prasad. 2004. Effect of combined application of Rhizobium, phosphate solubilizing bacterium and *Trichoderma* spp. on growth, nutrient uptake and yield of chickpea (Cicer *aritenium* L.). *Appl Soil Ecol* **28**: 139-146.

13. Sampson MN and Gooday GW (1998). Involvement of chitinase of *Bacillus thuringiensis* during pathogenesis in insects. *Microbiol* **144**: 2189-2194.

14. Shaikh SA, Deshpande MV (1993). Chitinolytic enzymes: their contribution to basic and applied research. *World J Microbiol Biotechnol* **9**: 468–75.

15. Shapira R, Ordentlich A, Chet I, Oppenheim AB (1989).Control of plant disease by chitinase expressed from cloned DNA in *Escherichia coli*. *Phytopathol* **79**: 1246–9.

16. Wang SL, Chang WT (1997). Purification and characterization of two bifunctional chitinase/lysozyme extracellulary produced by *Pseudomonas aeruginosa* K-187 in a shrimp and crab shell powder medium. *Appl. Environ. Microbiol.*, **63**: 380–386.

17. Yadav KS and KR Dadarwal (1997). Phosphate solubilization and. mobilization through soil micro-organisms., Biotechnological Approaches in Soil.

Chapter 47

Towards Describing River Water Quality Fuzzily Straightway in Linguistic Terms: Is it Achievable?– A Paradigm Shift

Kedar Rijal[1], Ashok Deshpande[2] and Vikram Ghole[3]

[1]Department of Environmental Sciences, Tribunal University, Nepal
[2]Berkeley Initiative in Soft Computing (BISC),
Special Interest Group (SIG),
Environment Management Systems (EMS), and Bioinformatics Center,
University of Pune, Pune
[3]Department of Environmental Sciences,
Vasatadada Sugar Institute, Pune

ABSTRACT

One of the most disturbing prospects facing water quality scientists is how to turn often very complex water quality data into information which is understandable and usable by non-scientists, managers, and public at large. The present practice of communicating the status of river water quality to policy makers and the end users is with Water Quality Index (WQI). In spite of the development of numerous water quality indices over the last four decades, there has been no universal acceptance.

What do we understand by the term water quality, and more particularly excellent river water quality for bathing? Different people mean differently as water quality has no precise meaning-it is fuzzy term! Why should water quality index value should be worked out first on an interval scale, and then describe its meaning in linguistic terms? Human brain thinks exactly the opposite. Is it not possible to straightway describe river water quality in linguistic terms with some degree of certainty attached to each linguistic term, and is it achievable? The answer is: yes. A paradigm shift in describing fuzzy terms (such as water quality, environmental quality, and alike) is the departure from the conventional approach. In this sequel, fuzzy logic based formalism is presented with a case study.

Keywords: *Dissolved oxygen, Water quality index, Indrayani River, Fuzzy sets, Linguistic variable, Degree of match, Fuzzy rule base system.*

Introduction

You and a friend walk outside on December morning in Pune. You announce that the weather is "mild". Your friend declares that it is "cold". Who is wrong? Or are you both right?

People recognize that language can be imprecise and that concepts such as cold, hot, or mild do not have well-defined boundaries. In 1965, Lotfi Zadeh introduced fuzzy logic, a means of pro-cessing data by extending classical set theory to handle partial membership (1). Classical set theo-ry deals with sets that are *crisp* in the sense that members are either in or out according to rules of binary logic. For example, binary logic dictates that apples in a basket are red OR not red. With fuzzy log-ic, some of the apples could be categorized as red AND not red.

In everyday life and in fields such as environmen-tal health, people deal with concepts that involve factors that defy classification into crisp sets—*safe, harmful, acceptable, unacceptable*, and so on. A classic example is a regulator carefully ex-plaining the result of a detailed quantitative risk as-sessment to a community group, only to be asked over and over again, *But are we safe?* In this case, *safe* defies crisp classification because it is a mul-tivariate state with gradations that vary among dif-ferent individuals and groups.

Fuzzy logic has become a common way of deal-ing with information in various fields, such as control theory, smart machines, and investment analysis. But fuzzy sets have also been applied to environmental science and policy. Anyone who has worked on health and environmental issues deals constantly with fuzzy concepts, such as hazardous, acceptable, or safe. Even concepts such as carcino-gen and neurotoxin define fuzzy sets whose mem-bers are selected by experts after reviewing and judging conflicting toxicological or epidemiologi-cal data. Despite the relevance of fuzzy logic and early efforts to promote its use in risk assessment (2), fuzzy logic applications in risk assessments are still rare. The paper relates to the application of fuzzy logic to river water quality classification.

Depletion of freshwater resources, rapid urbanization and industrialization coupled with look warm approach towards environment pollution control, calls for the effective water quality management programs. One of the most disturbing prospects facing water quality scientists is how to turn often very complex water quality data into information which is understandable and usable by many institutional players such as non-scientists, managers, and public at large. Water quality index is in essence, a function used to simplify large quantities of data into a more useful form in order to convey an image of overall quantity to a variety of different users. The most critical deficiency of the selected indices is the lack of dealing with uncertainty and subjectivity present in this complex environmental problem (Chang *et al.*, 2001; Mpimpas *et al.*, 2001; Silvert, 2000).

What do we understand by the term water quality, and more particularly excellent river water quality for bathing? Different people mean differently as water quality has no precise meaning-it could be labelled as - fuzzy term! Why should water quality index value to be worked out firstly on an interval scale, and then describe its meaning in linguistic terms? Human bring thinks just the opposite! Is it not possible to straightway describe river water quality in linguistic terms with some degree of certainty attached to each linguistic term, and is it achievable? The answer is: yes. A paradigm shift in describing fuzzy terms (such as water quality, environmental quality, and alike) is the departure from the conventional approach. In this sequel, fuzzy logic based formalism for straightway describing water quality in linguistic terms with some degree of certainty for the specific usage is presented, and its superiority over the conventional water quality indexing system is demonstrated in a case study.

Water Quality Classification

The efforts on the development of Water Quality Indices (WQI) started in 1990s and are considered a subset of indices used to measure environmental quality on an interval scale. Some of widely used indexing systems are discussed in brief.

Conventional Methods

In general five different types of water quality indices have been considered for the calculation and comparison of the water quality. However, the two commonly used water quality indices are considered in the present study. The first WQI is an index proposed by Horton hereafter called as the arithmetic water quality index (WQI_A), which is denoted as:

$$WQI_A = \sum_{i=1}^{n} w_i q_i \tag{1}$$

Where n is the number of variables, w_i is the relative weight of the ith parameter such that $\sum_{i=1}^{n} w_t = 1$ and q_i is the quantity rating of ith parameters. The second WQI is a multiplicative form of index proposed by Brown *et al.*, Multiplicative water quality index (MQI_M) is defined as:

$$WQI_M = \prod_{i=1}^{n} q_i^{W_i} \tag{2}$$

WQI is classified in linguistic terms as: (0-25)-Not acceptable; (26-50) - Just acceptable; (51-75) and Acceptable; (76-100) -Highly acceptable. There is no degree of certainty attached to these linguistic descriptions.

Fuzzy logic Based Method

In spite the development of numerous water quality indices over the last three decades, there has been no universal acceptance. There exists aleatory uncertainty due to randomness in water quality parametric data, and epistemic uncertainty in the perception of experts about rating these parameters (partial belief). The problem of water quality classification can, therefore, be approached using fuzzy logic based approach. In this sequel, we have followed the conceptual framework of the model presented by Deshpande, in his paper (1994) and made an attempt to compare the conventional and fuzzy logic based approach in a case study. It is important to understand some of the important terms used in the paper.

Matching between Fuzzy Values

The fuzzy number for field data (A) on parameters and the fuzzy number characterizing linguistic terms (A') are matched together to arrive a measure called a Degree of Match (DM) defined by

$$DM(A,A')=\frac{\int \mu_{A\cap A'}(x)dx}{\int \mu_A(x)dx}, x \in X \tag{3}$$

in which X denotes the universe, and $\mu_{A\cap A'}(x)$ is membership grade for $A \cap A'$. The degree of match between A and A' resent in Figure 47.1 shows the extent to the assertion is included in the Antecedent part of rule. Furthermore, if A and A' is the discrete possibility distribution the measure is defined as:

$$DM(A,A')=\frac{\sum \mu_{A\cap A'}(x)}{\sum \mu_A(x)}, x \in X \tag{4}$$

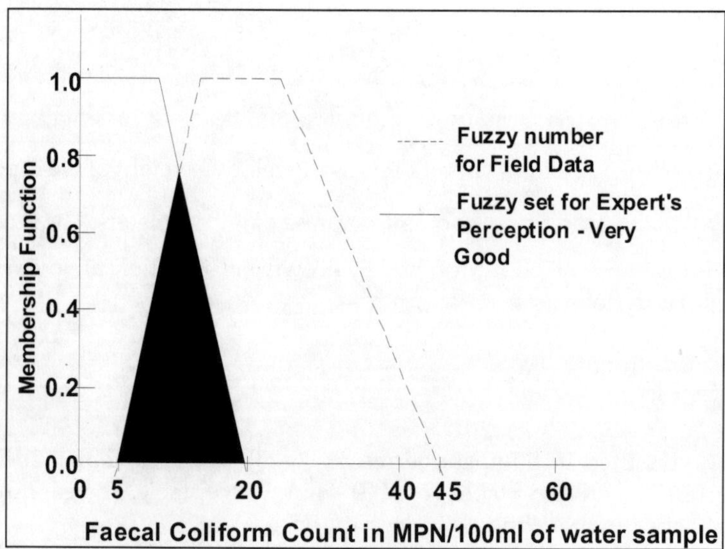

Figure 47.1: Fuzzy Matching between *A* and *A'*.

In the present study, we consider relevant water quality parameters. One such parameter in the context of bathing in river water is Most Probable Number (MPN expressed as MPN/100ml of water sample).

Fuzzy Rule Based System

A set of rule is constructed for classifying water quality as highly acceptable, just acceptable, not acceptable (rejected) in order to aggregate the set of attributes. Each rule has antecedent propositions connected together using AND operator, resulting in some consequences. The assertions related to its antecedent part are obtained from the users, which are imprecise or fuzzy. Therefore, a fuzzy rule based system can be developed for the knowledge representation or reasoning process. Here the partial matching is allowed and the analyst can estimate the extent to which the assertion satisfies the antecedent part of the rule contrary to the rule-base system which examines as to whether the antecedent part is satisfied or not (Lee, 1992). Figure 47.2 presents a hierarchical structure for water classification in a set rule.

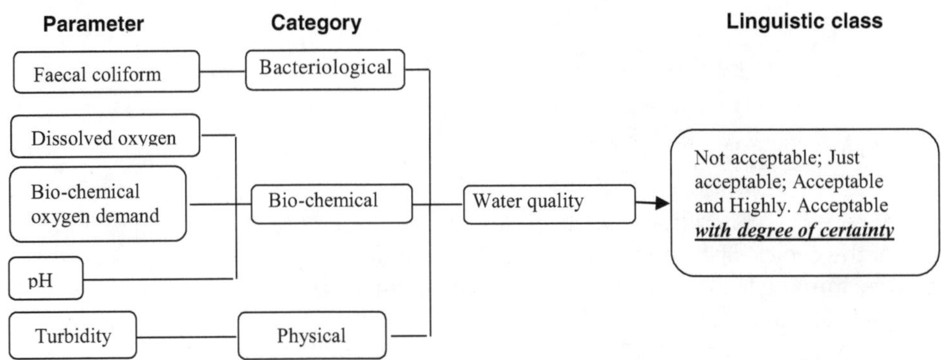

Figure 47.2: Hierarchical Structure for Water Quality Classification.

The chemical status of water is judged in the first hierarchical level of knowledge. While the second level characterizes bacteriological, chemical and physical status of water to arrive at the ultimate acceptable strategy of water quality for bathing purpose. Following are examples rules stored at two different hierarchical levels of the knowledge base:

Fuzzy Rules

If DO is *<fair>* and BOD is *<good>* and pH is *<very good>*

Then chemical status of water is *<good>*.

The rule at the nest level could be

If Bacteriological status of water is <*fair*> and Chemical status of water is *<good>* and Physical status of water is *<fair>*

Then water quality for bathing is *<just acceptable>* with certain degree of certainty.

Relevance of Bootstrap in the Present Study

All parametric mean values obtained from bootstrap technique have been used for computation of convention water quality index approach and fuzzy logic It was observed that high pollution level persists in Indrayani River in different seasons, and there was no significant variation in the concentration of the pollution indicating parameters. Collection of several samples for the analysis in all the seasons would have been a drain on the financial resources. It was, therefore, considered appropriate to use bootstrapping procedure to arrive at the conclusion on the average parametric concentration.

As an allusion to lifting oneself up by one's own bootstraps, the term *bootstrapping* means using a special process to perform a task that one would be unable to do in general. Furthermore, Bootstrap sampling procedures has adopted for re-sampling by increasing sample size up to (n=1000) with replacement procedure without any loss of generosity of mother sample. The boot estimation was done by using the formula;

$$\bar{x} = \frac{1}{n}\sum_{i=1}^{n} X_i \text{ for the bootstrap mean} \quad (5)$$

$$\overline{X}_b = \frac{1}{n}\sum_{i=1}^{n} X(i) \quad (6)$$

It could be stated that bootstrapping has increased the data precision without loss of the generosity of the mother sample. It could be argued that the relevance of this re-sampling technique will be helpful for solving environmental issues based on very limited data sets.

Case Study

Using WQI formalism, it can be stated that the WQI values of both the sampling locations (S_1 and S_2) are almost same revealing same pollution status of the river at both locations, and linguistically described as *fair* category. The computational details of WQIs at both locations in Indrayani River are presented in the Table 47.1.

Table 47.1: Water Quality Index using Bootstrap Mean Values.

Parameters	*Bootstrap Mean*		*Weight (w_i)*	*WQI at Station (S_1)*		*WQI at Station (S_2)*	
	S_1	S_2		Q_1	Q_1^{W1}	Q_2	$Q_2^{W_2}$
FC (MPH/100ml)	742	709	0.26	22	2.23	25	2.31
DO (mg/l)	4	4.29	0.21	30	2.04	32	2.07
BOD (mg/l)	15.2	14.15	0.14	18	1.50	20	1.52
pH	7.56	7.66	0.18	95	2.27	95	2.27
Turbidity (ppm)	16.16	16.54	0.11	65	1.58	63	1.58
Temperature (°C)	19.9	19.9	0.10	5	1.17	5	1.17
$\prod_{t=1}^{n} Qi^{Wi}$			**1**		**28.63**		**30.49**

Limited water samples were collected in different seasons and the bootstrap mean values were worked out (Table 47.1). Indrayani river water quality for bathing remains almost the same in rainy season and fair seasons and could be linguistically described as fair for bathing with no degree of certainty attached to the description. It is very hard to communicate such information to the public or end users with confidence.

The problem of river water quality classification was, therefore, approached using combination of degree of match (DM) and the fuzzy rule-based system (Deshpande *et al.*, 1994).In a fuzzy rule based system, the experts represent their knowledge concerning the classification of the object (water quality) in the form of rules. Each rule has a set of antecedent propositions comprising to parameters names for example: pH, Dissolved Oxygen, etc. and attributes values or linguistic description like very good, good, etc. These linguistic descriptions are invariably fuzzy or vague keeping in view the inadequate information on the health implications of each parameter on the users and the integrated effect of all the parameters on human health. In order to describe the aggregated effect of water quality a hierarchical structure for the overall water quality classification straightway in linguistic terms is developed.

Table 47.2: Degree of Match with Fuzzy Values for each Linguistic Class.

Sampling Stations	*Parameters*	*Linguistic Class*			
		Very Good	*Good*	*Fair*	*Poor*
S_1	FC	0	0	0	1
	DO	0.2	1	0.98	0.1
	BOD	0	0	0	1
	pH	1	0.28	0	0
	Turbidity	0.02	1	0.03	0
S_2	FC	0	0	0	1
	DO	0.6	1	0.8	0
	BOD	0	0	0	1
	pH	1	0.7	0.001	0
	Turbidity	0.016	1	0.33	0

Table 47.2 presents the degree of match for each parameter for both the sampling stations (ghats). The dominating parameters namely, FC and BOD exhibit *poor* linguistic class description with degree of certainty 1. But the linguistic value of parameters pH and DO is in class *very good* and *good* meaning thereby that these parameters might not be responsible for the deterioration of for Indrayani River water quality for bathing purpose.

Table 47.3 presents fuzzy description of river water quality for bathing with degree of certainty. The parameters FC and BOD contribute towards worsening water quality status for bathing in Indrayani. The water quality is *not acceptable/poor* for bathing with degree of certainty 1.00 at two ghats of the Indrayani River, near Pune, India (Table 47.3).

Table 47.3: Fuzzy Description of River Water Quality for Bathing with Degree of Certainty.

Water Quality Description	*Degree of Certainty*
Highly Acceptable/very good	0
Acceptable/good	0
Just Acceptable/fair	0
Not acceptable/poor	1

The calculated values of QWI_M are 28.63 and 30.49 at the stations S_1 and S_2 in fare season, respectively (Table 47.1) indicating water quality as fair for bathing at these two stations. The results obtained from two different approaches are contradictory. More importantly, the computational procedure for indexing system does not attach any degree of certainty to the statement water quality is fair for the specific usage-bathing". Fuzzy logic based approach, straightway describes water quality at the ghat as poor or rejected for bathing with degree of certainty 1.00.

Concluding Remarks

The fuzzy logic based describes water quality straightway in linguistic terms with degree of certainty which, in turn, would help the decision makers. This is what is expected of modelling exercises and this is how human brain works. In the real world, no one works out first a number, and then say that the water quality is *good/ bad* for bathing. Fuzzy logic based method with a case study delineated in the paper is the complete departure from indexing system, and have received severe criticism in the past. There is an urgent need to exploit the potential of fuzzy logic based approaches in many areas of science and technology in general and environmental management system in particular. The biggest room in this world is the room for improvement- so we would continue to carry out further research.

References

1. APHA (1992). Standard methods for the examination of water and wastewater 18th Edition. American Public Health Association, American Water Works Association, Water Environment Federation, p949.

2. Wiener, V., Cyrus, D.P., Vermeulen, L.A., and Wade, P. (2006). Development of a water quality index for estuarine water quality management in South Africa.WRC Report No. 1163/1/06. www.Wrc.org.za/download.

3. Chang, N., Chen, H.W, and Ning, K.S. (2001). Identification of river water quality using the fuzzy synthetic evaluation approach. *Journal of Environmental Management, 63,* 293-305.

4. Silvert, W. (2000). Fuzzy indices of environmental conditions. *Ecol Model, 130,* 111-119.

5. Deshpande, A.W., Raje, D.V., and Khanna, P. (1994). Fuzzy description of river water quality, paper for International conference–EUFIT.

6. Efron, B. 1981. *Censored Data and Bootstrap. Journal of American Statistical Association, 76*, 312-319.

7. Miller, R., and Turkey J. (1974). The Jackknife - A Review. *Biometrika, 61*, 1-15

8. Rijal, K., Deshpande, A., and Ghole, V. (2008). Bathing in polluted rivers, water borne diseases, and fuzzy measure: A case study in India. International journal of environment and Waste Management (IJEWM). In press-manuscript number -2809.

9. Ross, T. J. (2004). Fuzzy logic with engineering application. John Wiley and Sons.

Chapter 48

A Review of Technologies to Unleash Power from Wastewater Methane Emissions

P. Aswale[1], *M. Karthik*[1], *T. Nandy*[1] *and V. Mhaisalkar*[2]

[1]*CSIR-National Environmental Engineering Research Institute, Nehru Marg, Nagpur*

[2]*Visveswariya National Institute of Technology, Nagpur*

ABSTRACT

Energy consumption is key input to the progress of the mankind. Increasing and changing life style has increased the global demand for energy. It is expected to increase significantly in the near future. The world has started running out of oil and coal reserves and estimates prove that current stocks would last only a few decades. Therefore, time has come to give utmost importance to harnessing non-conventional and renewable energy sources. Unlikely source of energy production needs to be explored and exploited to reduce the impact on the environment. Wastewater offers a likely source for power generation and waste reduction and has been practiced insensitively globally. The carbon and hydrogen present in wastewater offer an opportunity to process into a fuel or energy based on existing and advanced technologies. In India about 22,900 MLD of domestic and 13,500 MLD of industrial wastewaters is being generated each year and increasing. The most favorable option to limit the greenhouse impacts due to waste management are those that higher global warming potential gases are reduced and by replacing fossil fuels in energy production. This chapter presents various opportunities and limitations present for power or fuel generation from wastewaters.

Keywords: *Wastewater, Non conventional energy, Methane, Power generation.*

Introduction

The rapid growth of science and technology and their products has led to the piling up of waste materials of diverse nature from anthropogenic activities. The wastes have to be processed and disposed scientifically without causing environmental damages. Pollution caused by indiscriminate discharge of municipal sewage from various cities and towns are the primary cause for the degradation of our water and soil resources and pose a great threat to human health. Conventional sewage management is a logistical nightmare and capital intensive for developing nations. The limited capacity of the environment to absorb these wastes without seriously damaging ecology has led to development of efficient, scientific and safer methods for the treatment, processing and disposal of wastewaters. About 22,900 MLD of domestic sewage is discharged from major cities and towns in India and is the biggest source of pollution of water bodies. In the recent year's technologies has been developed which not only reduces waste quantity but generates useful and decentralized form of energy.

As the world becomes more environmentally aware, there is a growing recognition that waste is a valuable commodity. Traditional methods of dealing with waste such as landfilling or burning and burying produce unacceptable harmful emissions. Instead, different waste streams should be regarded as differentiated resources which can be re-used or re-manufactured. Human waste can be significantly processed to generate power recovering their value to provide electricity, heat or produce transport fuels. This is an alternative way of waste treatment and more environmentally sound solution than recycling or landfilling. Waste to Energy (WtE) primarily uses combustion technology and is the most popular 'renewable' technology which realistically meets the commitments for energy generation while at the same time also providing significant quantities of electric power. WtE used optimally can be an additional power source without the need for new conventional power plants.

Currently India ranks fifth in electricity generation (above 600 billion kWh) with demand shortage pegged at 10 per cent. This shortage can be achieved by producing electricity from unlikely sources such as wastewater. Wastewater contains enough organic matter which can be processed to get a concentrated form of fuel such as biogas and syngas. Fuels could be processed through anaerobic digestion, pyrolisis to obtain gaseous fuel.

Waste-to-Energy Processes

Waste-to-energy (WtE) refers to any waste treatment that creates energy in the form of electricity or heat from a waste source. WtE is a form of energy recovery. Most

Table 48.1: Thermal and Non-Thermal Technologies for Waste-to-Energy.

WtE Technologies	*Advantages*	*Disadvantages*	*Product Form*
Thermal technologies			
Gasification	☆ Generate electricity directly ☆ Using the syngas is potentially more efficient than direct combustion. ☆ Gasification can also begin with materials that are not otherwise useful fuels, such as biomass or organic waste.	☆ The technology requires large energy inputs.	☆ Hydrogen ☆ Synthetic fuel
Pyrolysis	☆ Break down organic materials in the absence of oxygen,	☆ Gas generated can not used directly in combustion turbine generators	· Methane · Complexes hydro-carbon · Hydrogen · Carbon monoxide
Plasma arc gasification	☆ Plasma-arc torch to produce temperatures as high as 7204°C. This extreme heat breaks down wastes, forming syngas which can be converted into a variety of marketable fuels–including ethanol, natural gas (methane), and hydrogen–or it can be used to generate electricity directly	☆ The technology requires large energy inputs and remains early stage ☆ Used as nuclear and hazardous waste processing facilities.	· Syngas · Hydrogen · Crbon monoxide

Contd...

Table 48.1–*Contd...*

WtE Technologies	*Advantages*	*Disadvantages*	*Product Form*
Non-thermal technologies			
Anaerobic digestion	☆ The organic content of the sludges is significantly reduced by conversion into gaseous end-products. ☆ The obnoxious odour of the sludge is removed ☆ Significant reduction in the quality of pathogenic bacteria ☆ The biogas that is formed is a mixture of carbon dioxide (CO_2) and methane (CH_4) that can be used for digester heating or to generate power.	☆ High initial capital cost is involved ☆ The slow rate of bacterial growth requires long periods of time for start-up and limits the flexibility of the process to adjust to changing feed loads, temperatures and other environmental conditions. ☆ The process is prone to upsets if not regularly monitored and if corrective action is not taken in time	· Methane · Carbon dioxide
Fuel Cells			
PAFC	☆ Quiet, Low emissions, High efficiency, Proven reliability	☆ High costs, low energy density	Convert chemical energy from fuel (methane) directly into electricity
MCFC	☆ Quiet, Low emissions, High efficiency, Self reforming	☆ High costs, Need to demonstrate long term dependability	
SOFC	☆ Quiet, Low emissions, High efficiency, High energy density, Self reforming	☆ High costs	
PEMFC	☆ Quiet, Low emissions, High efficiency	☆ High costs, limited field test experiences	

WtE processes produce electricity directly through combustion, or produce a combustible fuel commodity, such as methane, methanol, ethanol or synthetic fuels. There are number of new and emerging technologies that are able to produce energy from waste and other fuels without direct combustion. Many of these technologies have the potential to produce more electric power from the same amount of fuel by increasing efficiencies. This is mainly due to the separation of corrosive components (ash) from the converted fuel, thereby allowing a higher combustion temperature in *e.g.* boilers, gas turbines, internal combustion engines and fuel cells. Some are able to efficiently convert the energy into liquid or gaseous fuels. The WtE can be broadly classified as thermal and non thermal technologies. Table 48.1 presents the processes available for power generation from wastewaters. It is easier to produced methane for energy recovery being the cheaper for operation. Methane is produced during anaerobic digestion and its formation at three steps: hydrolysis, volatile acid fermentation and methane formation. Each step carried out by a different group of micro-organisms. In the anaerobic process a certain amount of gas is produced per pound of volatile solids broken down or destroyed by the bacteria. This is 'gas yield'. Average values of gas generation from various wastewaters are given in the Tables 48.2 and 48.3.

Table 48.2: Potential Gas Production Swine, Dairy, Poultry, Beef and Human Waste.

Live Stock	*Gas Yield, ft^3/Pound Volatile Solids Destroyed*	*Potential Gas Production ft^3/Animal Unit per day*	*Energy Production Rate, Watt. Hour/Animal*
Swine (150 pounds)	12	4.1	30
Cattle (1200 pounds)	7.7	22.7	166
Poultry (4 pound bird)	8.6	0.21	2
Beef (1000 pounds)	15	31	227

Table 48.3: Gas Production from Industrial Wastewater Streams Suitable for Methane Production.

Effluent Streams	*COD (mg/l)*	*OLR (kg/m^{3}d^1)*	*HRT (Hours)*	*Gas Generation (l/d)*
Black liquor from pulp and paper industry	2000-3000	7	48	10
Spent wash from distillery	110000-190000	23	192	7.2
Dairy industry	1100-3500	8	8	0.41
Slaughter houses waste	3000-5000	1.8	5	240

In India methane emissions from wastewater account only 2 per cent of total methane emissions. It is still considered to be the most important source of emissions due to the phenomenal growth in population and industries. The population growth and the corresponding methane emissions from domestic wastewater in India for the period of 2000-2007 are shown in the Table 48.4.

Table 48.4: Population Growth and Corresponding Methane Emissions from Domestic Selected Industries.

Sl.No.	Year	Total Domestic CH_4 Emissions (million tones/yr)	Total Industrial CH_4 Emission (million tones/yr)
1.	2000	0.405	1.230
2.	2001	0.415	1.275
3.	2002	0.426	1.323
4.	2003	0.437	1.319
5.	2004	0.449	1.383
6.	2005	0.461	1.200
7.	2006	0.474	1.057
8.	2007	0.486	1.075

* Using Inter Governmental Panel for Climate Change 2006 Guidolines.

Methane gas is a valuable energy resource and leading anthropogenic contributor to global warming after carbon dioxide. Methane's overall contribution to global warming is significant because, over a 100-year time frame it is estimated to be 21 times more effective at trapping heat in the atmosphere than carbon dioxide. Over the last two centuries, methane's concentration in the atmosphere has more than doubled from about 700 parts per billion by volume (ppbv) to 1730 ppbv in 1997. Increase in atmospheric methane concentrations is largely due to increasing emissions from anthropogenic sources. Global methane emissions from wastewater ranged between 30-40 Teragram per annum (Tgpa), constituting to about 8-11 per cent of total global anthropogenic methane emissions. The methane so produced can be used to generate electricity and cooking, illumination, heating and as automobile fuel while the hydrogen produced can be used to make hydrogen fuel cells which can also be used to generate electricity. According to the California Energy Commission, there is approximately 6300 MW of anaerobic digestion capacity worldwide with Asia accounting for over 95 per cent. Under all circumstances methane represents about 20 per cent of the U.S. energy supply. Compared to other fossil fuels, methane produces few atmospheric pollutants and generates less carbon dioxide per unit energy released. Methane is a major source of electricity generation use of gas turbines and steam turbines. High efficiencies can be achieved through combining gas turbines with a steam turbine in combined cycle mode. Combined cycle power generation using methane is thus one of the the cleanest source of power available using fossil fuels, and this technology is widely used wherever gas can be obtained at a reasonable cost. Fuel cell technology may eventually provide cleaner options for converting methane gas into electricity, but as yet it is expensive.

Engines for Waste to Energy

Power generation using anaerobic digestion process has proved to be economically attractive in many cases. It not only offers safe waste disposal but also provides an opportunity to generate clean electric power. The wastewaters if treated properly have a tremendous potential to generate electricity. Technological advances in anaerobic digestion have led to increase a methane yields from effluents, which can be further utilized to generate electricity.

In a power plant, the steam rotates the turbine and in turn generates electricity. Various technologies are available in the market for conversion of waste into power. Some of these technologies are as follows:

Combustion Turbine

Combustion turbine is a rotary engine that extracts energy from a flow of combustion gas. It has an upstream compressor coupled to a downstream turbine, and a combustion chamber in-between. Energy is added to the gas stream in the combustor, where air is mixed with fuel and ignited. Combustion increases the temperature, velocity and volume of the gas flow. This is directed through a nozzle over the turbine's blades, spinning the turbine and powering the compressor. Energy is extracted in the form of shaft power, compressed air and thrust, in any combination, and used to power aircraft, trains, ships, generators, and even tanks

Microturbines

Microturbines are small gas turbines, self-contained sources that provide a controlled on-site power. They can operate in parallel with other units or operate alone. Natural gas is the primary fuel utilized although some renewable applications for biogas are being pursued.

Microturbines are a developed technology that holds promise of higher efficiencies and potentially lower operating cost. A factor that affects market for microturbines attractive to widespread development includes:

- ✰ The initial price of microturbines which moves down the price volume curve.
- ✰ Reliability and cost of fuel supply.
- ✰ The degree of customer attraction to a "high tech" product.
- ✰ The extent to which units demonstrate long-term operating reliability.
- ✰ The primary environmental concerns with microturbines are noise and air emissions. CO_2 and other emissions from natural gas (NG) are relatively low compared to other fossil fuels.

Internal Combustion Engines

At present, the use of internal combustion (I/C) engines outnumbers all other distributed generation technologies combined. I/C engines can be fueled by diesel, gasoline, methane, or natural gas. Historically, most I/C engines used for electric power generation use diesel fuel. Gasoline engine generator sets are usually not selected for DG applications because of costs. They are available as emergency generators where limited run time of less than 100 hours per year is expected.

Natural gas engines have emerged as a competitive prime mover for the power generation market. Sales have grown tremendously for power generation since 1990s. The use of natural gas in I/C engines is expected to increase at the expense of diesel and small turbines. This growth is due to improvements in cost, efficiency, reliability, and emissions.

Stirling Engines

A Stirling engine is a heat engine that operates by cyclic compression and expansion of air or other gas, (*working fluid*), at different temperature levels such that there is a net conversion of heat energy to mechanical work. The engine is like a steam engine in that all of the engines heat flows in and out through the engine wall. Unlike the steam engine's use of water in both its liquid and gaseous phases as the working fluid, the Stirling engine encloses a fixed quantity of permanently gaseous fluid such as air or helium. As in all heat engines, the general cycle consists of compressing cool gas, heating the gas, expanding the hot gas, and finally cooling the gas before repeating the cycle.

The Stirling engine is noted for its high efficiency, quiet operation, and the ease with which it can utilize almost any heat source. This compatibility with alternative and renewable energy sources has become increasingly significant as the price of conventional fuels rises, and also in light of concerns such as peak oil and climate change. The advantages and disadvantages of these technologies are presented in Table 48.5.

Future Research Areas

The following areas are required to be addressed for increased power generation from wastewater.

- ☆ Technical barriers for commercialization of pyrolysis and gasification are bed material handling, reliability/troubleshooting and syngas clean up etc.
- ☆ Reciprocating or IC engines have a major advantage in terms of both the economy of scale and product track record. The emissions are slightly higher than those of micro-turbines or fuel cells and hence new designs must reduce emissions to compete in the long run.
- ☆ Improve micro-turbine for use of alternative fuels.
- ☆ Design of combustion chamber to withstand high temperature and others include reliable gear mechanism to translate the stirling cycle piston motions vis-à-vis conventional engine motions.
- ☆ Fuel cell technologies that combine the benefits of PEMFC and PAFC into a single membrane and operates at temperatures between 90 - 160°C. Improving rapid start-up and shutdown.

Conclusion(s)

As the population grows environmental degradation resulting in resources depletion. For sustainable development, preservation of the biosphere, production of

Table 48.5: Advantages and Disadvantages of Engines.

Technologies/Engines	*Advantages*	*Disadvantages*	*Generation Capacity*
Combustion Turbine	☆ High efficiency and low cost particularly in large systems. ☆ Readily available over a wide range of power output. ☆ Capability of producing high-temperature steam using exhaust heat. ☆ Proven reliability and availability ☆ High power-to-weight ratio	☆ Reduced efficiencies at part load. ☆ Sensitive to ambient conditions like temperature, altitude etc. ☆ Smaller systems are costlier and not efficient as larger systems.	☆ Capacity 500 Kw–25 MW ☆ Efficiency 20- 45 per cent depending on the size.
Internal Combustion (I/C) or Reciprocating Engines	☆ Low capital cost. ☆ Higher reliability and fuel flexibility. ☆ Small footprint required for a given amount of energy. ☆ Not affected by lower gas pressures.	☆ Involves lot of atmospheric emissions. ☆ Very noisy. ☆ O and M required at frequent intervals.	☆ Capacity 5 Kw–7 MW ☆ Efficiency 25- 45 per cent
Microturbines	☆ Small number of moving parts ☆ Compact size, Light-weight and Good efficiencies in cogeneration ☆ CO_2 and nitrogen oxides (NO_x) emissions are relatively low than other I/C engines. ☆ Can utilize waste fuels ☆ Reliable and continuous fuel supply. ☆ Low operating pressures. ☆ High operating speeds. ☆ Lubrication and maintenance is less.	☆ Initiation and establishment cost is high. ☆ Low fuel to electricity efficiencies ☆ Machining operations are very tedious. ☆ Less efficient than reciprocating engines. ☆ Loss of power output and efficiency with higher ambient temperatures.	☆ Capacity 25-500 kW ☆ Efficiency 20- 30 per cent
Stirling engine	☆ Low noise and vibrationless operation ☆ Low emissions ☆ Low maintenance, and high reliability ☆ Relatively few moving parts, mechanically simple ☆ Multi-fuel capability, including solar power ☆ Long life	☆ High cost ☆ Low efficiencies	☆ Capacity 1–25 kW ☆ Efficiency 12–20 per cent

food and energy by technologies that have minimal environmental impact. Power requirement has grown tremendously since the last two decades in India. Reliance on conventional power sources are responsible for increased greenhouse gas emissions and associated global warming. Non conventional energy sources need to be tapped to reduce dependence on conventional energy sources. Wastewater offers a likely source for treatment of waste and simultaneously power generation. Efficiency of the engines involve in converting waste into energy/power is necessary with research and development of advanced engines to produce power and reduce overall impact of human growth to environment.

References

1. IPCC Guidelines for National Greenhouse Gas Inventories, http://www.ipcc.ch/index.htm
2. State wise urban population by sex in India (1996-2007). http://www.indiastat.com
3. National Communication (2004). India's Initial National Communication to the United Nations Framework Convention on Climate Change, Submitted by the Ministry of Environment and Forests, Government of India. can be accessed at http://www.natcomindia.org
4. Status of Sewage Treatment in India, Central Pollution Control Board (CPCB), 2005.
5. U.S. Methane Emission 1990-2020: Inventories, Projections and Opportunities for reduction, 1999.
6. V. Venkateswara Sarma, Future models of green technologies for sustainable development.
7. Czepiel, P.M., Crill, P.M., and Harris, R.C. (1993). Methane Emissions from Municipal Wastewater Treatment Processes. Env. Sci. Tech. 27 2472-2477.
8. DeHollander, G.R., 1998.Gaseous emissions from wastewater facilities. Water Environment Research 70(4), 580-584.
9. Doorn, M.R.J., Stratrit, R.P., Barnard, W.R., Eklund, B., 1997. Estimates of global greenhouse gas emissions from industrial and domestic wastewater treatment, EPA-600/R-97-019.US EPA.
10. Inventorisation of Methane Emission from Domestic and Key Industries Wastewater- India's Second National Communication to the United Nations Framework Convention on Climate Change (NEERI Report, 2008).
11. NEERI Report (2003). GHG Emissions from wastewater.
12. Ragsakthi S. S. Sundarvel, Urbanization and Methane Emission in India.
13. Chynoweth, D. P. "Global Significance of Biomethanogenesis" in Global Environmental Chemistry, Dunnette, D. and O' Brian, Eds., ACS Symposium series.

14. Chynoweth, D. P. and Isaacson, H. R., eds., Anaerobic Digestion of Biomass, Elsevier Applied Science Publishers Ltd., New York, 1987.
15. Using Sewerage System to Generate Electricity, Jalees Asghar
16. Distributed Generation Technologies http://en.wikipedia.org
17. Khan A. 1995. Municipal wastewater treatment and energy recovery.
18. India Development report, 1997. Available from Ministry of Environment and Forests (MoEF), Government of India at http://mospi.nic.in
19. Sumana Bhattacharya and Mitra MP (1998). Greenhouse gas emissions in India for the Base year 1990, SASCOM and Centre on global change, NPL, New Delhi, Scientific report number11:79-81

Chapter 49

Properties of Recycled Aggregate Concentrate with Micro-silica as Partial Cement Replacement

Veerendra Kumar, Rajesh Kumar, Chaudhary Ashish and Navin Kumar

Department of Civil Engineering, IIT (BHU),Varanasi

ABSTRACT

One of the greatest challenges of the present time is to evolve strategies for the utilization of large amounts of construction and demolition (C and D) wastes produced due to the development of modern society. Main reasons for C and D wastes are demolition of structures which have become obsolete either in serving the basic functions or due to structural deterioration, demolition of structures for better economic gain (through new construction), due to natural disasters like earthquake, cyclone and floods, and war-inflicted damages. Most of the construction and demolition wastes are disposed by dumping them as landfills or for reclaiming land. But with the demand of land increasing day by day, the locations, capacity and width of the land that can receive waste materials are becoming limited. Hence, reuse of demolition waste appears to be an effective solution to produce aggregates that can be reused in concrete. From economic point of view, recycling of C and D waste is good option when recycled aggregate is competitive with natural aggregate in relation to cost and quality. Thus, by adopting this, twin objective of conservation of natural resources and pollution free environment may be achieved. The experimental work has been carried out to examine the feasibility to use recycled coarse aggregate instead of fresh or virgin coarse aggregate with micro-silica as part replacement of cement, and various properties of recycled aggregate are studied and discussed in the present chapter.

Keywords: *Recycled aggregates, Micro-silica, Construction, Cement, Concrete.*

Introduction

Recycling of demolition waste was first carried out after Second World War in Germany to tackle the problem of disposing large amounts of demolition waste caused by war. Recycling and reuse of reclaimed asphalt pavement into new asphalt pavement aggregate, aggregate base and sub-base for roads and granular fill is in common place in USA. In certain cases, federal and state highways contracts for new highway construction, by using recycled materials (1). Hong Kong generates about 14 million tones of demolition waste each year and faces problem in accommodating this surplus material. Low grade concrete applications (up to M20) with 100 per cent recycled coarse aggregate are allowed for mass concreting and minor concrete structures. However, use of recycled fine aggregate is not permitted. For higher grade applications (up to M35), the current specification of Hong Kong allows a maximum of 20 per cent replacement of normal coarse aggregate by recycled aggregate and concrete can be used for general concrete applications except in water retaining structures (1). In Japan, the target for recycling ratio of demolished concrete in the year 2000 was set at 90 per cent and recycling upto 96 per cent of demolished concrete waste was done but entirely used as sub-base material for road pavement (1). In 1997, the amount of construction and demolition waste per person per year in Spain stood at 326kg and only 5 per cent of which was recycled. This represents a clear contrast with other countries such as Holland and Belgium, where due to series of political initiatives, along with the appropriate legislation and control, recycling level reached up to 90 per cent and 87 per cent respectively, for the 718 and 657 kg demolished waste generated per person per year in these two countries (2). In England, it has been reported that in 2001, 220 million tones of aggregates were used for construction, a quarter were recycled materials (3). In Taiwan (1999), after severe earthquake, an estimated 30 millions tones of construction and demolition waste was generated. A comprehensive plan for management of construction and demolition waste was initiated and under this plan a recycling plant was established. This recycling plant recycled about 80 per cent of these construction waste materials used in landfills and 30 per cent of the construction waste material used in road base (3). Estimates of demolished concrete in European countries and U.S.A each year are close to 100 million tones (4). In Indian context, according to Central Pollution Control Board, construction industry accounts for 12 to 14.7 million tones of demolition waste and these wastes are disposed by dumping in low- lying areas (1).

Thus, the experimental work has been carried out to examine the feasibility to use recycled coarse aggregate instead of fresh or virgin coarse aggregate with micro-silica as part replacement of cement. Strength and durability of concrete are parameters considered in this study. The properties of concrete at each stage get affected by the properties of its constituent materials. The materials used in experimental work are cement, fine aggregate, fresh or virgin coarse-aggregate, recycled aggregate, micro-silica and water.

Recycled Aggregate

Aggregates made from Construction and Demolition Waste

Recycled concrete aggregate can be produced from precast elements and cubes after testing, and demolished concrete buildings. In the former case the aggregate could be relatively clean, with only cement paste adhering to it while in the latter case the aggregate could be contaminated with salts, bricks and tiles, sand and dust, timber, plastics, cardboard, paper, and metals. As a natural aggregate, the quality of recycled aggregates, in terms of size distribution, absorption, abrasion etc. also needs to be assessed before using the aggregate (3).

Properties of Concrete made with Recycled Aggregate

Concrete mixes using recycled aggregates can be designed in the same way as those using natural aggregates, provided the extra absorption in the former is appropriately accounted for when determining the unit water content (3).

Treatment Process of Recycled Aggregate

Two methods were proposed to improve the quality of the recycled aggregates (5).

1. Impregnation with a solution of micro-silica that is intended to add a thin layer of silica fume particles over the surface of the recycled aggregate.
2. Ultrasonic cleaning of the recycled aggregate in order to remove the loose particles and improve the bond between the new cement paste and the recycled aggregate.

Preparation of Recycled Aggregates in Laboratory

In this experimental work, recycled aggregates are obtained from waste cubes that are tested in departmental concrete laboratory. The various steps involved are:

1. Cubes are manually broken by hammer in such a manner that coarse aggregates and mortar are separated. Mortar adhered to coarse aggregates surface can be seen.
2. The material obtained from broken cubes is then sieved to separate coarser from loose mortar. The material is sieved through sieve of opening 4 mm.
3. Abrasion machine is used to remove adhered mortar from aggregates (approx. 250 revolutions).
4. The materials obtained from abrasion machine are again sieved with sieve opening 4 mm.
5. The coarse aggregates obtained are washed and then dried for use.

Properties of Materials Used in the Experimental Work

Cement

Ordinary Portland cement of 43 Grade has been used in this experiment [(6)]. The consistency was 33 per cent with initial and final setting time of 70minutes and 280 minutes respectively with soundness of 1mm.

Fine Aggregate

Locally available Sone river sand has been used in this experimental work [(6–7)] conforming to Zone = II with specific gravity of 2.57.

Coarse Aggregate

Crushed stone of maximum size 20mm and 10mm has been used in this experimental work. The properties of aggregate are determined as per test mentioned in IS Code [(7–8, 9)]. Sieve Analysis results have been shown in Table 49.1.

Table 49.1: Sieve Analysis of Coarse Aggregates.

IS sieve no.	*Cumulative per cent of Weight Retained (20mm)*	*Cumulative per cent of Weight Retained (10 mm)*
20mm	26.6	0
10mm	88.7	46.7
4.75mm (ISS 480)	100	94.05
2.38mm (ISS 240)	100	99.65
1.18mm (ISS 120)	100	100
< 1.18mm	100	100

Recycled Coarse Aggregate

Recycled coarse aggregates are obtained from breaking waste cubes tested in Structural Engineering Laboratory [(7–8, 9)].

The comparisons of various properties of regular with recycled aggregates are given in Table 49.2 and discussed below:

1. As expected, water absorption of recycled coarse aggregate is more than that of fresh or virgin coarse aggregate because of adhered mortar.
2. Recycled coarse aggregate does not satisfy the criteria of elongation index. There are two reasons for this. Firstly, coarse aggregates initially used for concrete work are not with in the limit as per codal provision. Secondly, during the extraction process to get recycled coarse aggregate from waste concrete cubes; aggregates getting crushed. It is undesirable to have elongation index more than 15 per cent.
3. Recycled coarse aggregate satisfy the flakiness index. Reason for this is aggregates initially used are with in the limit as per codal provision. It should not exceed 25 per cent.

4. The reason of high abrasion value of recycled coarse aggregate is mortar adhered to its surface.
5. Impact and crushing value of recycled aggregate is better than that of fresh aggregate or virgin aggregate. There are two reasons for this. Firstly, coarse aggregates initially used for concrete work are within the limit as per codal provision. Secondly, during the extraction process to get recycled coarse aggregate from waste concrete cubes; there was change in shape. The shape of aggregate obtained is semi-angular or semi-round.

Table 49.2: Comparison of Properties of Regular Coarse Aggregate and Recycled Coarse Aggregate

	Regular (20 mm)	*Regular (10 mm)*	*Recycled*
Specific gravity	2.82	2.84	2.71
Water absorption	0.348 per cent	0.200 per cent	1.098 per cent
Elongation index	16.17 per cent	37 per cent	21.1 per cent
Flakiness index	6.5 per cent	23.4 per cent	17.53 per cent
Abrasion value	15.28 per cent	19.6 per cent	22.84 per cent
Crushing value	23.08 per cent	18.15 per cent	22.18 per cent
Impact value	16.62 per cent	19.58 per cent	14 per cent

Micro-Silica

Micro-silica, a product of M/s Elkem India Pvt. Limited has been used in the experimental work.

Water

Water used in the experimental work for mixing and curing was clean and free from injurious amount of oils, acids alkalis, salts, sugar, organic materials or other substance that may be deleterious to concrete. Potable water is generally considered satisfactory for mixing and curing concrete.

Experimental Work

In this experimental work M35 grade concrete were cast with fresh or virgin coarse aggregate and with recycled coarse aggregate. Moreover, in case of recycled coarse aggregate; micro-silica replaces cement by 2.5 per cent. Recycled coarse aggregate obtained were mixture of angular and semi angular shape aggregates. Thus different concrete mix design were done considering shape of recycled coarse aggregate as angular and semi- angular. The experimental work can be divided into two broad categories. In the first category, change in properties of cement partly replaced with micro-silica has been studied. In the second category, the effect of micro-silica on the strength and durability of concrete have been studied. For this work, five sets of cubes (30 samples in each set) were cast and the identification of sets of cubes cast are given in Table 49.3.

Table 49.3: Identification of Sets of Cubes

Sets of Cubes	Description
FA	Fresh or Virgin coarse aggregate
RA	Angular recycled coarse aggregate
RAF	Angular recycled coarse aggregate with 2.5 per cent micro-silica as cement replacement
RSA	Semi-angular recycled coarse aggregate
RSAF	Semi-angular recycled coarse aggregate with 2.5 per cent micro-silica as cement replacement

Properties of Cement Partially Replaced by Micro-silica

Normal Consistency

There is increase of normal consistency with replacement of cement with 2.5 per cent of micro-silica (Table 49.4). The increase in consistency is most probably due to the micro-silica, which has large surface area as it is finer than cement.

Initial and Final Setting Time

From Table 49.4 it is found that setting time decreases in sample containing micro-silica.

Soundness

It is observed that soundness of cement and cement with micro-silica found to be within permissible limit (Table 49.4).

Table 49.4: Properties of Cement Partially Replaced by Micro-Silica.

Type of Cement	Consistency	Initial Setting Time (minutes)	Final Setting Time (minutes)	Soundness (mm)
C	33 per cent	70	280	1
CSF	34.5 per cent	62	270	1

C: Cement; CSF: Cement with 2.5 per cent micro-silica as cement replacement.

Effect of Micro-Silica on Strength of Cement

The compressive strength of cement with partial replacement by micro-silica is determined as per codal provisions. It is observed that the strength of cement decreases with addition of micro-silica in cement (Table 49.5 and Figure 49.1). The strength of mortar cubes with micro-silica as 2.5 per cent cement replacement is less than mortar cubes without micro-silica.

Concrete Mix Design

The detailed concrete mix design for various sets is done as per IS Code [10]. The water cement ratio of 0.375 is maintained constant in each set. For increasing slump,

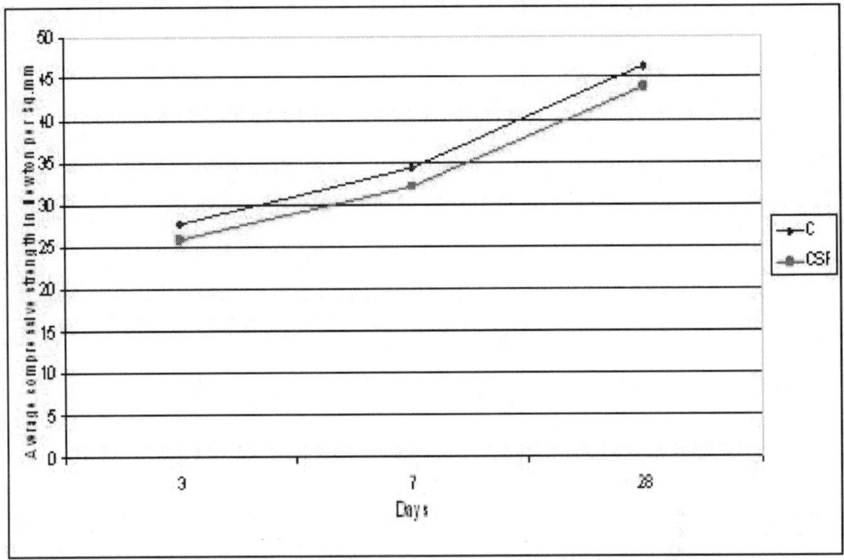

Figure 49.1: Compressive Strength of Cement.

water and cement quantities were increased to keep water cement ratio constant. Slumps obtained for concrete without micro-silica were in range of 40-60mm. Low slumps (20-40mm) were obtained in case of recycled aggregate concrete with micro-silica. Mix proportion of various mixes for per cubic meter of concrete has been given in Table 49.6

Table 49.5: Average Compressive Strength of Cement (N/mm^2).

Type of Cement	*3 Days*	*7 Days*	*28 Days*
C	27.8	34.4	46.5
CSF	25.4	32.1	43.9

Table 49.6: Mix Proportion for different Sets of Cubes (For one cubic meter concrete)

	Water (kg)	*Cement (kg)*	*Micro-silica (kg)*	*Fine Aggregate (kg)*	*Coarse Fresh/ Recycled aggregate (kg)*
FA	202.96	541.22	0	528.85	1126.45
RA	203.73	543.28	0	527.60	1079.95
RAF	203.73	529.70	13.58	527.60	1079.95
RSA	196.196	523.19	0	484.20	1163.43
RSAF	196.196	510.11	13.08	484.20	1163.43

Properties of Concrete with Fresh Aggregate and Recycled Aggregate

Compressive Strength

In this part of study, the cubes of 10cm x 10cm x 10cm were cast for all mixes. The cubes were tested as per the requirement of IS: 516-1959. The effect of age on the compressive strength has been studied on the cubes of all mix design at 7, 28, 60, 90 and 120 days. The compressive strength of different mix are given in Table 49.7.

Table 49.7: Compressive Strength of different Mix (N/mm²).

	7 Days	*28 Days*	*60 Days*	*90 Days*	*120 Days*
FA	50.8	52.9	55.3	57.5	59.1
RA	46.2	50.0	54.0	56.6	58.7
RAF	52.6	55.3	59.6	63.2	66.1
RSA	47.1	50.6	54.4	57.0	59.4
RSAF	53.3	56.0	60.1	64.0	67.0

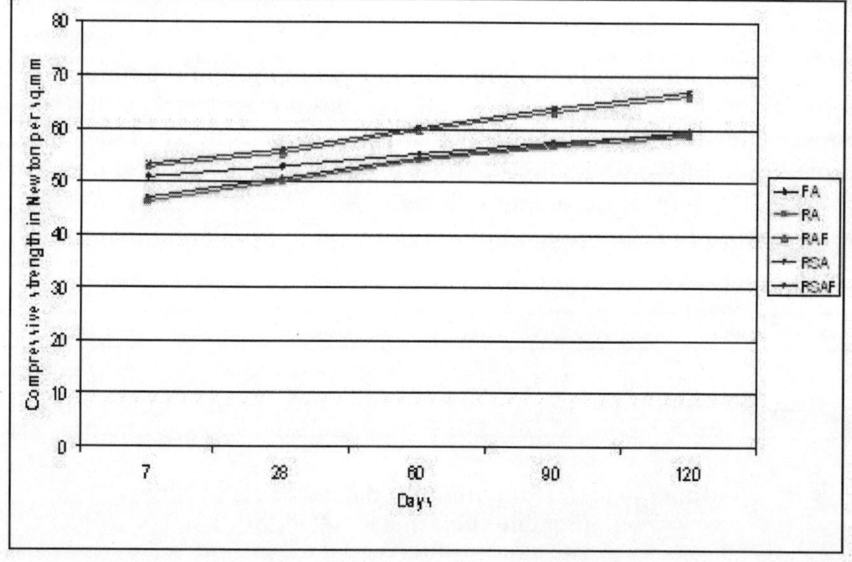

Figure 49.2: Compressive Strength of various Concrete Mixes.

Table 49.7 and Figure 49.2 shows that the strength of RAF and RSAF is greater than that of FA at all ages. However, strength of RA and RSA is less than that of FA but at later ages, strength of RA and RSA is almost equal to that of FA.

Water Absorption

Water absorption of various concrete mixes is measured as per ASTM C-642:06. The water absorption after immersion at different age of concrete mixes are shown in Table 49.8.

Table 49.8: Water Absorption after Immersion (in percentage) of Concrete Mixes.

	7 Days	*28 Days*	*60 Days*	*90 Days*	*120 Days*
FA	3.761	3.497	3.095	2.246	1.612
RA	4.019	3.781	3.419	2.547	1.905
RAF	4.088	3.856	3.501	2.634	1.980
RSA	4.035	3.793	3.425	2.561	1.918
RSAF	4.101	3.882	3.550	2.659	1.993

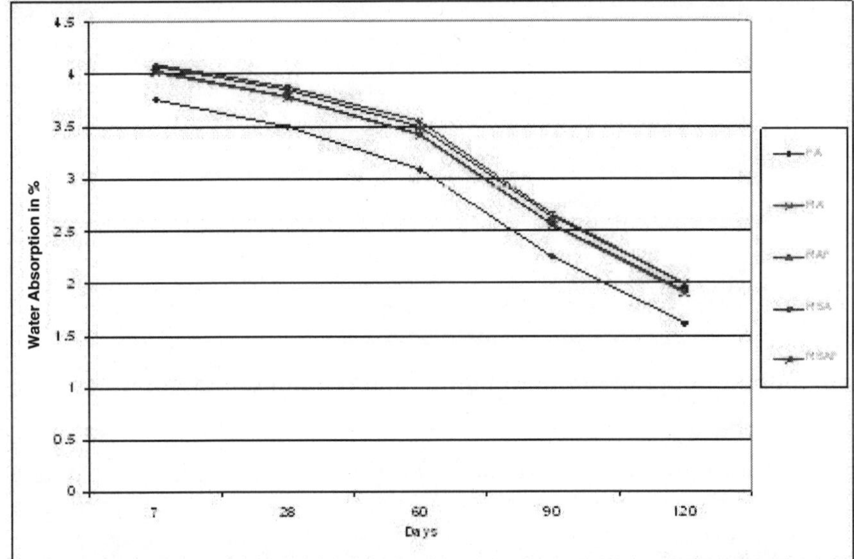

Figure 49.3: Water Absorption in various Mixes after Immersion.

As expected the water absorption of recycled aggregate concrete is more than that of fresh aggregate concrete (Table 49.8). However, concrete with micro-silica has maximum water absorption (Table 49.8 and Figure 49.3). According to Tiwari [11], water absorption in micro-silica concrete decreases with time when desired slump is obtained by increasing the amount of superplasticizer in order to keep water quantity constant.

Table 49.9: Water Absorption after Immersion and Boiling (in percentage) of Concrete Mixes.

	7 Days	*28 Days*	*60 Days*	*90 Days*	*120 Days*
FA	3.855	3.570	3.136	2.277	1.645
RA	4.081	3.827	3.440	2.570	1.930
RAF	4.137	3.898	3.534	2.658	2.010
RSA	4.090	3.833	3.442	2.580	1.941
RSAF	4.162	3.926	3.567	2.685	2.018

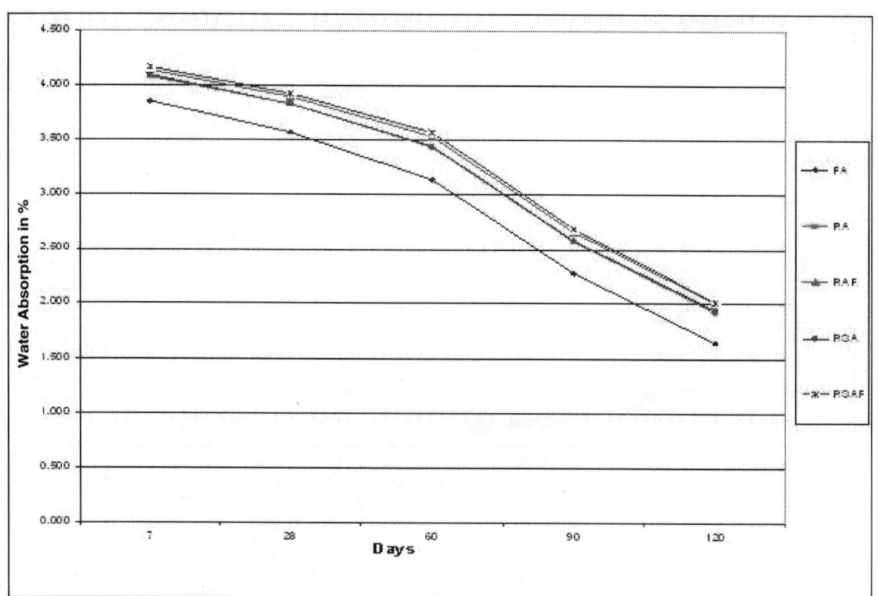

Figure 49.4: Water Absorption in various Concrete Mixes after Immersion and Boiling.

Water absorption after immersion and boiling at different age of concrete mixes is given in Table 49.9.

It is clear from Table 49.9 and Figure 49.4 that water absorption after immersion and boiling in concrete reduces with time. On the other hand, water absorption is increasing in all concrete mixes after immersion and boiling (Table 49.8 and 49.9).

Bulk Density

Bulk density of various concrete mixes is measured as per ASTM C-642:06. Bulk density dry of various concrete mixes is shown in Table 49.10; bulk density after immersion of various concrete mixes is given in Table 49.11 and bulk density after immersion and boiling of various concrete mixes in Table 49.12. The bulk density of recycled aggregate concrete is less than fresh or virgin aggregate concrete while it is increasing with time for all concrete mixes (Table 49.10).

Table 49.10: Bulk Density (Dry) in g/cc.

	7 Days	*28 Days*	*60 Days*	*90 Days*	*120 Days*
FA	2.446	2.459	2.480	2.505	2.531
RA	2.398	2.408	2.424	2.440	2.460
RAF	2.388	2.395	2.405	2.421	2.440
RSA	2.394	2.403	2.418	2.435	2.455
RSAF	2.379	2.388	2.401	2.416	2.435

Table 49.11: Bulk Density after Immersion in g/cc.

	7 Days	*28 Days*	*60 Days*	*90 Days*	*120 Days*
FA	2.538	2.547	2.560	2.570	2.582
RA	2.494	2.496	2.500	2.509	2.517
RAF	2.484	2.487	2.492	2.500	2.510
RSA	2.491	2.493	2.497	2.503	2.512
RSAF	2.479	2.483	2.489	2.495	2.504

Table 49.12: Bulk Density after Immersion and Boiling in g/cc.

	7 Days	*28 Days*	*60 Days*	*90 Days*	*120 Days*
FA	2.540	2.549	2.562	2.573	2.584
RA	2.497	2.499	2.503	2.512	2.520
RAF	2.488	2.491	2.495	2.504	2.513
RSA	2.494	2.496	2.500	2.507	2.515
RSAF	2.482	2.486	2.491	2.500	2.508

From the above results, the bulk density after immersion and boiling of recycled aggregate concrete is less than fresh or virgin aggregate concrete. However, density after immersion and boiling of recycled aggregate concrete with micro-silica is least among concrete with fresh aggregate and concrete with recycled aggregate. Bulk density after immersion and boiling of all concrete mixes is increasing with time (Figures 49.5–49.7 and Tables 49.10–49.12).

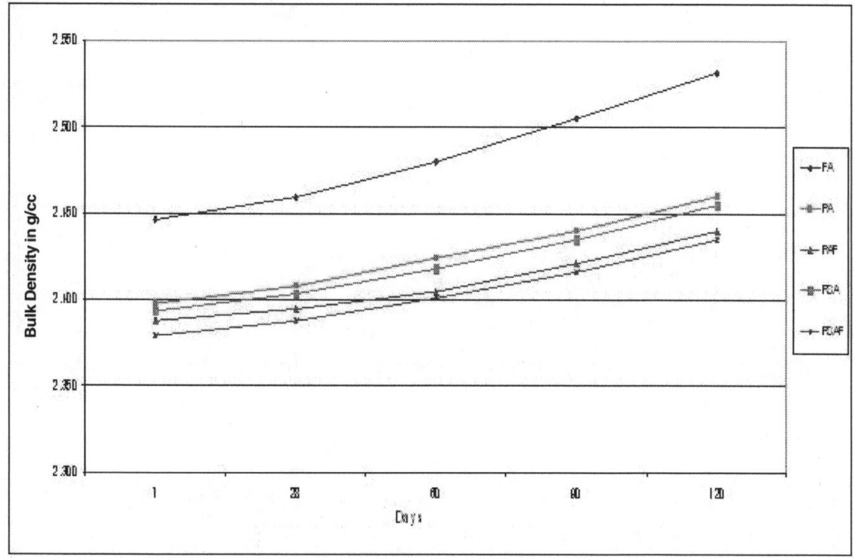

Figure 49.5: Bulk Density (Dry) of Various Concrete Mixes.

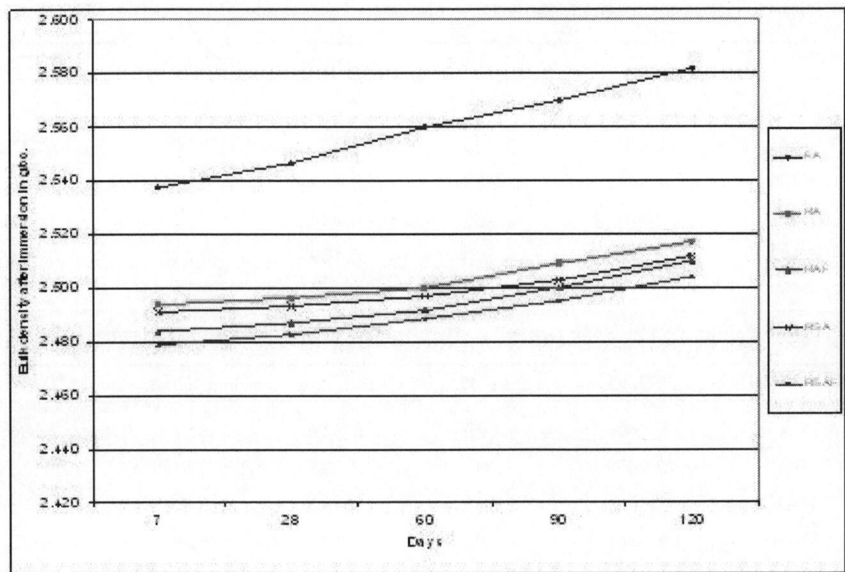

Figure 49.6: Bulk Density of Various Mixes after Immersion.

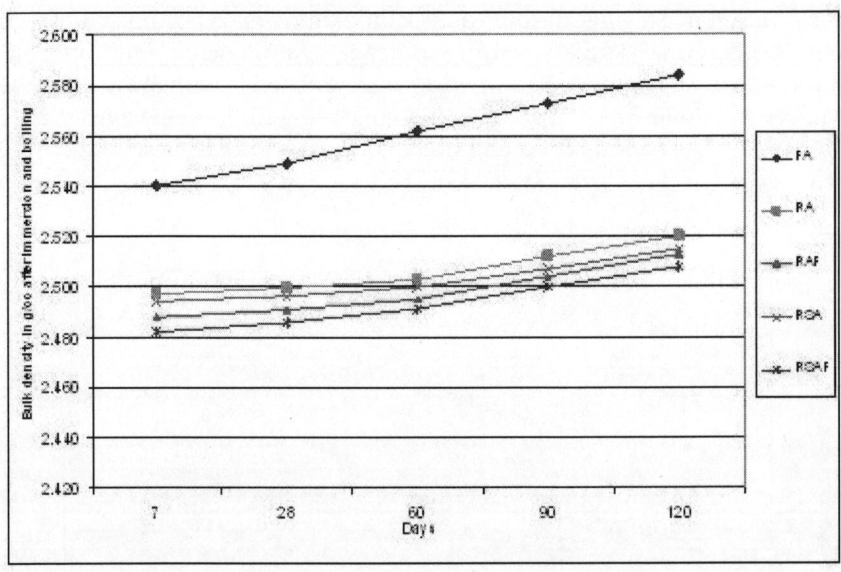

Figure 49.7: Bulk Density of Various Mixes after Immersion and Boiling.

Voids

Voids in various concrete mixes are measured as per ASTM C-642:06. Voids of various concrete mixes are given in Table 49.13.

Table 49.13: Voids in Percentage.

	7 Days	*28 Days*	*60 Days*	*90 Days*	*120 Days*
FA	3.712	3.446	3.041	2.219	1.603
RA	3.921	3.722	3.418	2.565	1.900
RAF	3.953	3.756	3.456	2.583	1.948
RSA	3.930	3.731	3.428	2.574	1.907
RSAF	3.996	3.788	3.471	2.593	1.954

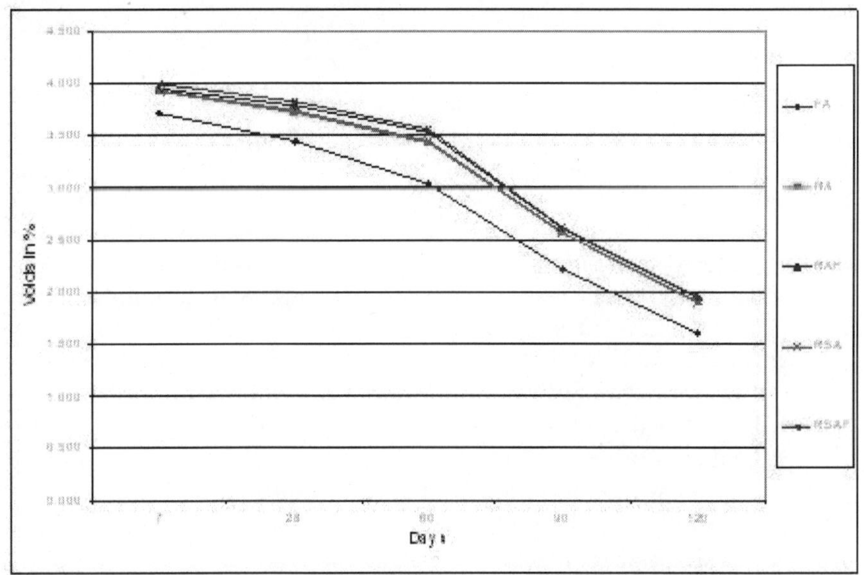

Figure 49..8: Percentage Voids in various Concrete Mixes.

It is observed that the voids in recycled aggregate concrete are more than fresh or virgin aggregate concrete. However, voids of recycled aggregate concrete with micro-silica are maximum among concrete with fresh aggregate and concrete with recycled aggregate (Table 49.13 and Figure 49.8). The higher percentage voids in recycled aggregate concrete with micro-silica is because of increased water content in order to get desired workability by keeping water cement ratio constant (when no superplasticizer is used).

Conclusion(s)

The various conclusions were drawn from the whole experimental work:

1. The amount of water required for normal consistency of cement increases with addition of micro-silica.
2. Initial and final setting time of cement decreases due to micro-silica.
3. It is found that strength of cement decreases with the partial replacement of cement by micro-silica.

4. The workability of concrete decreases with recycled aggregate and recycled aggregate with micro-silica.
5. The compressive strength of recycled aggregate concrete (when shape of aggregate considered angular and semi-angular) is less than fresh aggregate concrete. However, the strength of recycled aggregate concrete (when shape of aggregate considered angular and semi angular) with micro-silica is more than fresh aggregate concrete. Recycled aggregate concrete, where shape of aggregate considered semi-angular, has maximum compressive strength. The reason for higher compressive strength of semi-angular shaped recycled aggregate concrete (with and without micro-silica) more than angular shaped recycled aggregate concrete is lower water content for same water cement ratio.
6. Recycled aggregate concrete with micro-silica has maximum water absorption than recycled aggregate and fresh aggregate concrete without micro-silica. Recycled aggregate concrete with micro-silica, where shape of aggregate considered semi-angular, has maximum water absorption after immersion. For recycled aggregate concrete where shape of aggregate is considered semi-angular has water absorption less than recycled aggregate concrete with micro-silica where shape of aggregate is angular. It is to be noted that recycled aggregate concrete with and without micro-silica when shape of aggregate is semi-angular has lowest water content for same water cement ratio among concrete with fresh or virgin aggregates and concrete with recycled aggregate with and without micro-silica when shape of aggregate considered is angular.
7. Water absorption after immersion and boiling increases in all concrete mixes and it is maximum for recycled aggregate concrete with micro-silica where shape of aggregate considered semi-angular.
8. Bulk density of recycled aggregate concrete is less than fresh aggregate concrete. It further reduces for recycled aggregate concrete with micro-silica.
9. Similarly bulk density obtained after immersion and bulk density obtained after immersion and boiling shows similar results as of bulk density.
10. Percentage voids are least in fresh aggregate concrete and maximum for recycled aggregate with micro-silica. Recycled aggregate concrete with micro-silica when shape of aggregate is considered semi- angular has maximum percentage voids. Recycled aggregate concrete without micro-silica when shape of aggregate is considered semi- angular has percentage voids more than fresh aggregate concrete and recycled aggregate concrete without micro-silica when shape of aggregate considered angular but less than recycled aggregate concrete with micro-silica when shape of aggregate considered angular. It is to be noted that recycled aggregate concrete with and without micro-silica when shape of aggregate is semi-angular has lowest water content for same water cement ratio among concrete with fresh or virgin aggregates and concrete with recycled aggregate with and without micro-silica when shape of aggregate considered angular.

References

1. Rao, A., Jha, N.K. and Misra, S., 'A framework for use of construction and demolition waste as recycled aggregate in India.' The Indian Concrete Journal, 2006: January, pp. 12-19.

2. Fonteboa-Gonzalez, B. and Abella- Martinez, F., 'Concrete with aggregates from demolition waste and silica fume: Materials and mechanical properties.' Building and Environment, 2008:43, pp. 429-437.

3. Rao, A., Jha, N.K. and Misra, S., 'Use of aggregates from recycled construction and demolition waste in concrete' Resources, Conservation and Recycling, 2007:50, pp. 71-81.

4. Masood, A., Ahmad, T., Arif, M. and Mahadi, F., 'Waste management strategies for concrete.' Environment Engineering Policy, 2002:3, pp. 15-18.

5. Katz, Amnon., ' Treatments for the improvement of recycled aggregate' Journal of materials (ASCE), 2004:November/December, pp. 597-603.

6. "Methods of physical tests for hydraulic cement ", IS: 4031-1968, Bureau of Indian Standards, New Delhi.

7. "Methods of test for aggregate for concrete–Particle size and Shape", IS: 2386-1963 Part I, Bureau of Indian Standards, New Delhi.

8. "Methods of test for aggregate for concrete–Specific gravity, Density, Voids, Absorption and Bulking", IS: 2386-1963 Part III, Bureau of Indian Standards, New Delhi.

9. "Method of test for aggregate for concrete–Mechanical Properties", IS: 2386-1963 Part IV, Bureau of Indian Standards, New Delhi.

10. "Recommended guidelines for concrete mix design", IS: 10262-1982, Bureau of Indian Standards, New Delhi.

11. Tiwari, H.K., 'Effect of micro-silica on strength, workability and durability of concrete' M. Tech. Dissertation, Department of Civil Engineering, IT, BHU, Varanasi, 2005.

Chapter 50

Conceptual Model for e-Waste Recycle and Reuse

Anjali Srivastava

CSIR-National Environmental Engineering Research Institute, Kolkata Zonal Centre i-8, Sector C, East Kolkata Area Development Project, P.O. East Kolkata Township, Kolkata

ABSTRACT

In the present global scenario management of electronic waste (e-waste) has become a cause of concern. India presently is also generating a huge amount of e-waste (thanks to its recent growth in economic standards and newer taxing policies in favour of electronic items). The estimate of the e-waste generate avalible in literature are based on sale figures. No scientific data on quantification of the e-waste generated is presently available.

Electronic waste includes computers, entertainment electronics, mobile phones and other items that have been discarded by their original users. While there is no generally accepted definition of electronic waste, in most cases electronic waste consists of electronic products that were used for data processing, telecommunications, or entertainment in private households and businesses that are now considered obsolete, broken, or unrepairable.

Electronic waste is of concern largely due to the toxicity of some of the substances if processed improperly. The toxicity is in part due to lead, mercury, cadmium and a number of other substances. A typical computer monitor may contain more than 6 per cent lead by weight. Up to thirty-eight separate chemical elements are incorporated into electronic waste items. These metals can seep into and pollute soil and groundwater, with dangerous consequences: just 1/70th of a teaspoon of mercury can contaminate 20 acres of a lake, making the fish unfit to eat. In many states, it is now illegal to dump hazardous electronic waste into landfills. The unsustainability of discarded electronics and computer technology is another reason for the need to recycle–or perhaps more practically, reuse–electronic waste.

By virtue of our economic status, Indians in general have an unrecognised expertise in recycling and reusing various materials, e-waste is not an exeception to this. Every growing urban settlment today has significant number of people engaged in recycling electronic components. However no organized effort has been made to handle e-waste. The present paper deals with the possible methods of organizing, recycling, and reuse of e-waste.

Keywords: *Electronic, e-waste, e-products, Hazards, Waste management, Toxic.*

Introduction

E-waste is a popular, informal name for electronic products nearing the end of their "useful life." Computers, televisions, VCRs, stereos, copiers, and fax machines are common electronic products.

Electronic waste or e-waste is most rapidly growing waste problem in the world. It is a crisis not only of quantity but also a crisis born of toxic ingredients (like Hg, Pb, Cd, Cr, PVC's, BFR's) that poses both an occupational and environmental health threat. Till date, initiatives taken for handling waste problem have been woefully inadequate. Continued negligence from all quarters has made this issue into a major environmental issue today.

Recent policy changes in India have led to an additional tremendous in flux of leading multinational companies to setup electronic manufacturing facilities and research and development centers for hardware and software in the sector. The domestic market is characterized by strong economic growth and rapidly changing consumption pattern. This growth has significant socio economic impact. The increase in consumption rates of electrical and electronic products and higher obsolescence rates are leading to growing generation of e-waste, Waste Electronic and Electrical Equipment (or WEEE). The present practices of e-waste management in India suffer from a number of drawbacks like the, unhealthy conditions of informal recycling, inadequate legislation, poor awareness and reluctance on part of the corporate to address the critical issues. One of the major consequences are adverse effects on the environment and human health. Unhealthy conditions are developed during the informal recycling. More over resources are wasted when economically valuable materials are dumped. The total WEEE generation in India is approx. 1,46,000 tones per year. Three categories of WEEE account for almost 90 per cent generation:-

- ☆ Large, household appliances 42.1 per cent
- ☆ Information and Communication Technology equipment 33.9 per cent
- ☆ Consumer, Electronics 13.7 per cent

About 25,000 workers are employed at scrap-yards in Delhi alone, where 10,000 to 20,000 tons of e-waste are handled every year, with computers accounting for 25 per cent of it. Other e-waste scrap-yards exist in Meerut, Ferozabad, Chennai, Bangalore and Mumbai. About 80 per cent of the e-waste generated in the US is exported to India, China and Pakistan, and unorganised recycling and backyard scrap trading forms close to 100 per cent of total e-waste processing activity. Many of

India's corporations burn e-waste such as PC monitors, PCBs, CDs, motherboards, cables, toner cartridges, light bulbs and tube-lights in the open along with garbage, releasing large amounts of mercury and lead into the atmosphere. The following chart summaries the flow of e-waste in India (Figure 50.1).

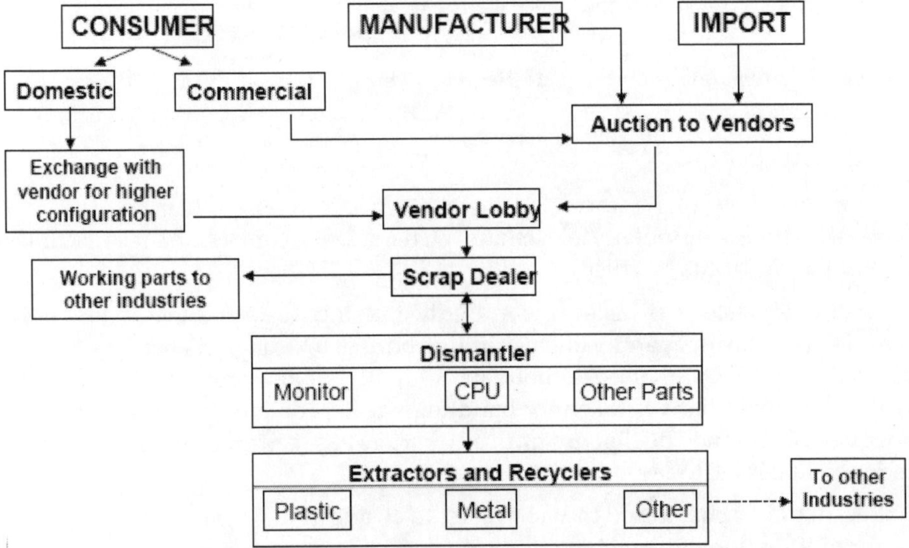

Figure 50.1: e-Waste Flow.

E-waste is also imported into India by various means. These include scraps both declared and mislabeled, donations, and second hand computers (Figure 50.2).

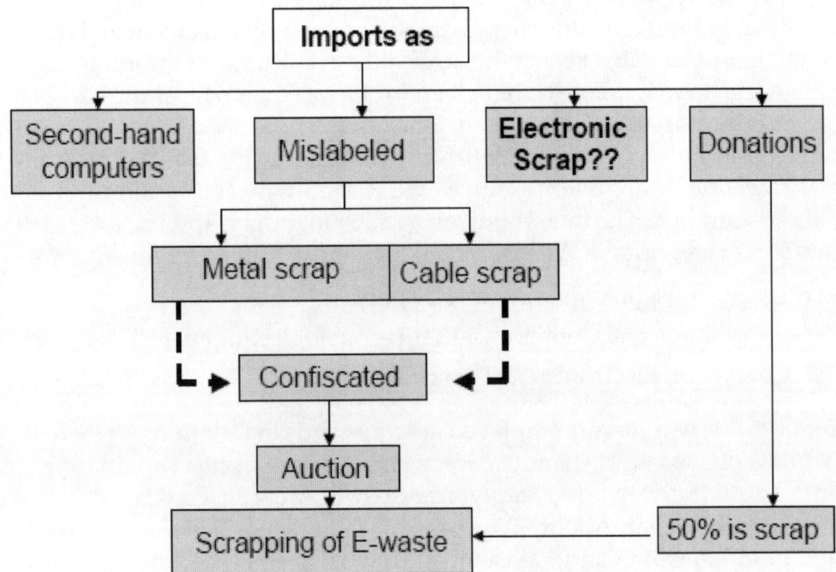

Figure 50.2: e-Waste Imported in Various Forms.

The top states in order of highest contribution to WEEE include Maharastra, Andhra Pradesh, Tamil Nadu, Uttar Pradesh, West Bengal, Delhi, Karnataka, Gujarat, Madhya Pradesh and Punjab. The city wise ranking of largest WEEE generators is Mumbai, Delhi, Bangalore, Chennai, Kolkatta, Ahmedabad, Hyderabad, Pune, Surat andNagpur. In India, recycling of e-waste is almost left to the informal sector which does not have adequate means and infrastructure to handle either the increasing quantities or scientific processes, leading to intolerable risk for human health. So the need of the hour is to frame a newer and more efficient model that will ensure effective material and financial flow, reduce environmental and occupational health problems and create new income genenerating oppurtunities.

e-Waste Constituents

- ☆ Circuit boards, which contain heavy metals like lead and cadmium
- ☆ Computer batteries contain cadmium
- ☆ Each computer and television contains on an average 4-8 pounds of lead-about 20 per cent of the weight of a monitor glass.
- ☆ Cathode ray tubes contain lead oxide
- ☆ Brominated flame retardants, used on printed circuit boards, cables and plastic casing
- ☆ Copper cables and plastic computer casings contain PVC
- ☆ Refrigerators contain CFCs
- ☆ Dry cell, lead acid, carbon-zinc, nickel-metal hydride, zinc chloride, zinc-carbon, alkaline, lithium, zinc-cadmium and button cell batteries contain heavy metals and mercury
- ☆ Mercury is also present in fluorescent and other energy-efficient lamps.
- ☆ Discarded electrical equipment or its parts, bulbs, LEDs, iron box, geysers, refrigerators, washing machines.

Prevalent Methods of e-Waste Management

E-waste management practices comprise of various means of final disposal of used electronic and electrical equipments (Figure 50.3). In the hierarchy of disposal

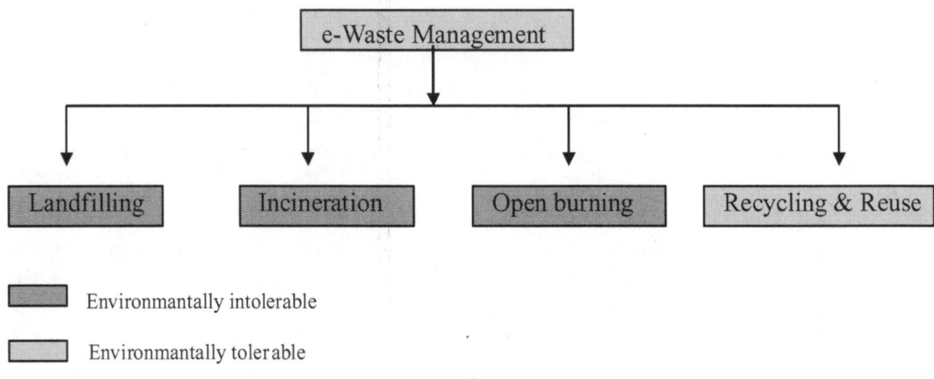

Figure 50.3: Prevalent Methods of e-Waste Management.

methods landfilling is considered most harmful and recycling and reuse, the most environmantally sound.

Landfilling

Landfilling is one of the most widely used methods of waste disposal. However, it has major disadvantages:

- ☆ Even state-of-the-art landfills, which are sealed to prevent toxins from entering the ground, are not completely leak proof in the long-term.
- ☆ Older landfill sites and uncontrolled dumps pose a much greater danger of releasing hazardous emissions.
- ☆ Mercury, Cadmium and Lead are among the most toxic leachates.
  a. Mercury, for example, will leach when certain electronic devices such as circuit breakers are destroyed.
  b. Lead has been found to leach from broken lead-containing glass, such as the cone glass of cathode ray tubes from TVs and monitors.
  c. When brominated flame retarded plastics or plastics containing cadmium are landfilled, both PBDE and cadmium may leach into soil and groundwater.
- ☆ Landfilled condersers emit hazardous PCB's.
- ☆ Besides leaching, vaporisation is also of concern in landfills. For example, volatile compounds such as mercury or a frequent modification of it, dimethylene mercury can be released.
- ☆ Landfills are also prone to uncontrolled fires, which can release toxic fumes.

Significant impacts from landfilling could be avoided by conditioning hazardous materials from e-waste separately and by landfilling only those fractions for which there are no further recycling possibilities and ensure that they are in state-of-the-art landfills that respect environmentally sound technical standards.

Incineration

Incineration is the process of destroying waste through burning. Because of the variety of substances found in e-waste, incineration is associated with a major risk of generating and dispersing contaminants and toxic substances.

- ☆ The gases released during the burning and the residue ash is often toxic. Studies of municipal solid waste incineration plants have shown that copper, which is present in printed circuit boards and cables, acts a catalyst for dioxin formation when flame-retardants are incinerated. These brominated flame-retardants when exposed to low temperature (600-800°C) can lead to the generation of extremely toxic polybrominated dioxins (PBDDs) and furans (PBDFs). PVC, which can be found in e-waste in significant amounts, is highly corrosive when burnt and also induces the formation of dioxins.

The release of toxic gases and the formation of toxic ash can be reduced by preventing co-incineration of e waste and by prior treatment of flue gases purification. The loss of valueable trace elements can be minimised by sorting, recovery and processing seperately prior to incineration.

Open Burning

Since open fires burn at relatively low temperatures, they release many more pollutants than in a controlled incineration process at an MSWI-plant. Chronic exposure to open fire emissions may lead to diseases such as emphysema and cancer. For example, burning PVC releases hydrogen chloride, which on inhalation mixes with water in the lungs to form hydrochloric acid. This can lead to corrosion of the lung tissues, and several respiratory complications. Often open fires burn with a lack of oxygen, forming carbon monoxide, which poisons the blood when inhaled. The residual particulate matter in the form of ash is prone to fly around in the vicinity and can also be dangerous when inhaled.

Recycling and Reuse

In order to minimize environmental and public health hazards due to improper e waste disposal. The only sensible method that can be applied is the recycling and reuse of e waste. Reuse involve second hand use or use after slight modification made into the original functioning equipment like upgrading the memories. These older units obviously have a limited life span and end up as waste sooner or later in these developing countries. So our last resort is recycling of e-waste. Table 50.1 presents hazardous waste produced at various stages of recycling for select e-waste.

Major Types of Recycling in Practice for Select e-Waste

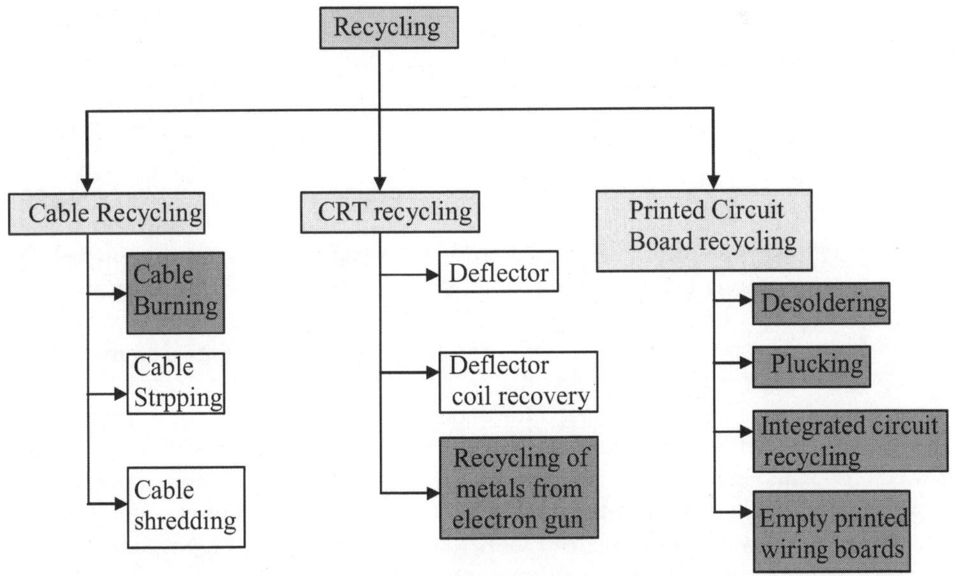

Figure 50.4: Recycling Practice for Select e-Waste.

Table 50.1: Hazardous Waste Produced at Various Stages of Recycling for Select e-Waste.

Type of Recycling	*Steps*	*Hazards*
Cable Recycling	Cable Burning	**Air:** Dioxins and furans, PACs. **Water, Soil, Air:** Hydrocarbon Ash, metals (copper or aluminium)
	Cable Stripping	**Air:** Dioxins and furans
	Cable Sredding	**Air:** Metal (copper or aluminium) and PVC dust dioxins and furans. **Water:** Wastewater
CTR Recycling	Deflector Coil and electron gun removal	**Air:** Dust containing lead, cadmium, phosphor deflector coil recovery
	Electron gun recycling	**Air:** Dust containing lead, cadmium, phosphorus and lead oxide. **Soil Water:** Lead ions
Circuit board Recycling	Desoldering	**Air:** Lead and tin fumes Brominated flame retardants Dioxins and furans
	Plucking	
	Integrated circuit recycling	**Air:** Acid fumes. **Water and soil:** Acid Heavy metals (*e.g.* copper, lead, cadmium)
	Empty printed wirring board recycling	**Air:** Dioxins and furans Brominated flame retardants Heavy metals (*e.g.* copper, lead, cadmium) Respirable suspended particulates (RSP). **Water and soil:** Heavy metals (*e.g.* copper, lead, cadmium)

Considering the present state of e-waste management in Indian and prevailing socioeconomic conditions a model is being proposed for management of e-waste in urban areas (Figures 50.5 and 50.6).

Conclusion(s)

India like many other developing countries, is facing a huge crisis with managing its growing e waste generation for e waste is not only being generated in enormous amounts within the country itself, but also due to the significant amount dumped by the developed nations. Over the last five years, the Indian IT industry has recorded a CAGR (Compounded Annual Growth Rate) of more than 42.4 per cent, which is almost double the growth rate of IT industries in many of the developed countries. In the IT action plan, the government has targeted to increase the present level of penetration, from 5 per 500 people to 1 for 50 people, by 2008. This envisages applying IT in every walk of the economic and social life of the country. About 80 per cent of the e-waste generated in the US is exported to India, China and Pakistan, and unorganised recycling and backyard scrap-trading forms close to 100 per cent of total e-waste processing activity In India e waste management is still restricted mainly to the informal sectors, which lacks effective organisation, with limited involvement of

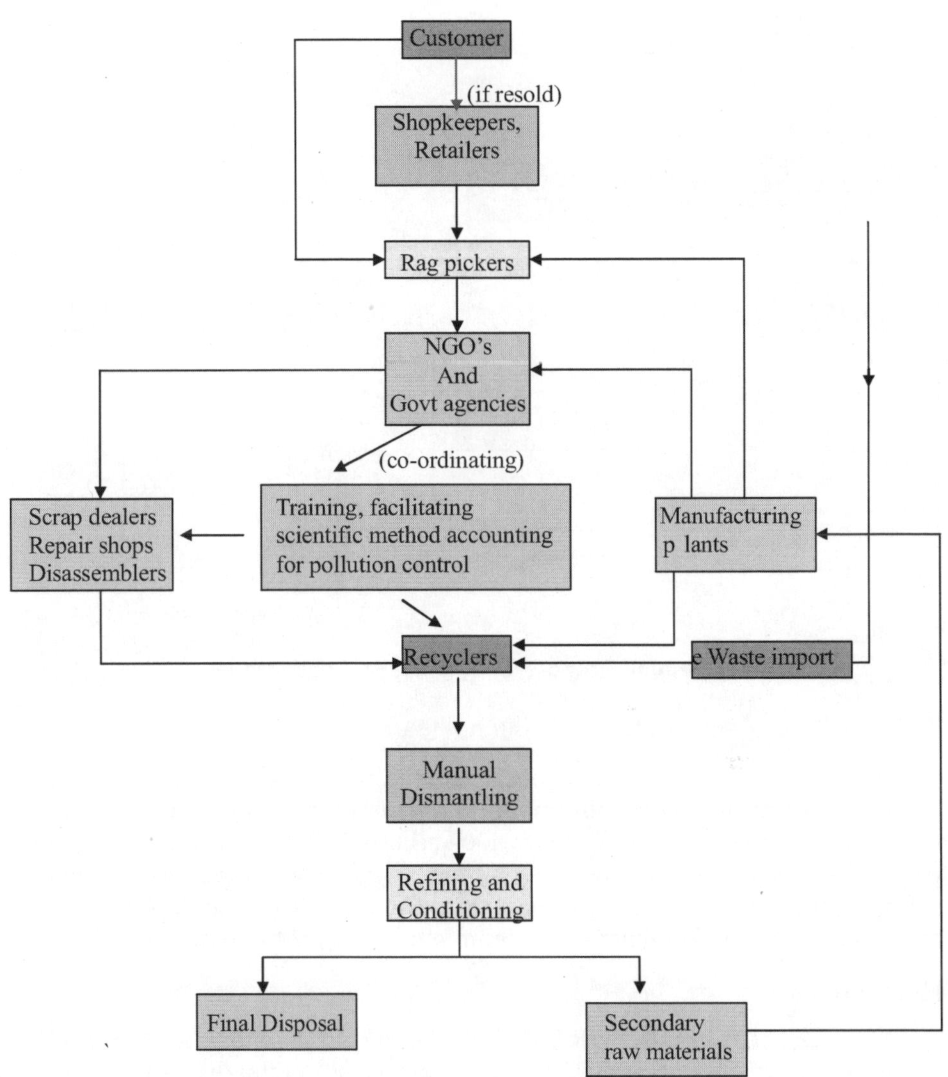

Figure 50.5: Proposed Model for Flow of e-Waste in Urban Area.

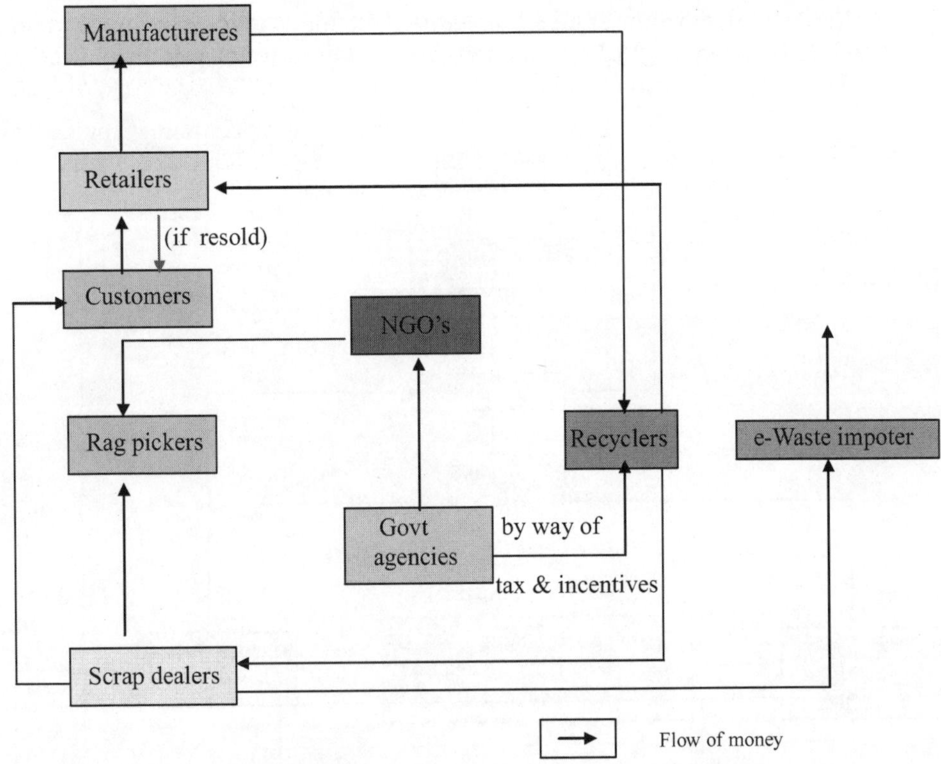

Figure 50.6: Flow of Money in the Proposed Model.

Government and non-government organizations. In the proposed model an effort has been made to impart a certain amount of organisation in the existing method of e waste management, by involving various sectors including Government, non-government, private sectors and the common people. It will help in an eco friendly management of e waste, ensure an even financial flow in every strata of the society and in addition may also generate new employment oppurtunities.

References

1. Macauley, M, Palmer, K, Shih, JS (2003). "Dealing with electronic waste: modeling the costs and environmental benefits of computer monitor disposal" J Environ Manag 68: 13-22
2. Report on Asia-Pacific Regional Expert Group Meeting on e-Wastes (22-23 June, 2000). Retrieved on February 10th, 2005 from http://www.icteap.org/events/e-waste/bgpaperppt.pdf
3. Preparatory Project Report on The Asia- Pacific Regional Scoping Workshop on Environmentally Sound Management of Electronic Wastes. Retrieved on February 10th, 2005 from www.icteap.org/events/e-waste/lijinhui.pdf

4. Richards, B. Environmental Management in Electronics Manufacturing. Retrieved on February 10th, 2005 from http://miranda.emeraldinsight.com/vl=5837266/cl=86/fm=html/nw=1/rpsv/cw/mcb/03056120/v23n4/s4/p16
5. M. N. Mundada, Sunil Kumar, A. V. Shekdar, A new challenge for waste management in India, Volume 61, Number 3/265–279. Issued on June 2004 from International Journal of Environmental Studies.
6. http://www.ewaste.com
7. http://www.ewaste.ch
8. www.expresscomputeronline.com
9. www.toxicslink.org
10. www.uneptie.org

Chapter 51

Impact of Electrooxidation on Combined Physico-chemical and Membrane Treatment Processes: Treatment of High Strength Chemical Industry Wastewater

Smita Masid, Sujata Waghmare, Nitin Gedam, Rohit Misra, Rita Dhodapkar, N.N. Rao and Tapas Nandy

CSIR-National Environmental Engineering Research Institute (Council of Scientific and Industrial Research), Nehru Marg, Nagpur

ABSTRACT

Industrial wastewaters containing high concentrations of organics, particularly of recalcitrant nature, demand unique type of treatment approaches often requiring two or more processes operated in tandem. In this chapter, the effectiveness of combined treatment processes *viz.*, coagulation/flocculation (C/F), electrooxidation (EO) and membrane (M) processes in that order for tackling organic load in a segregated chemical industry wastewater is reported. The wastewater is characterized by strong color, high concentrations of salinity, total kjeldhal nitrogen (TKN) and chemical oxygen demand (COD). These identified processes were investigated independently as well as their combinations. The C/F, adsorption, microfiltration (MF), ultrafiltration (UF), and EO were studied with the aim of reducing organic load on the final stage of reverse osmosis (RO) membrane treatment. Three different combined process; i) UF-RO (CP-I); ii) C/F-EO-UF-RO (CP-II) and iii) C/F-EO1-EO2-UFRO (CP-III) were investigated. The overall COD and TDS removal efficiencies in that order were: CP-I, 73 per cent and 82 per cent; CP-II, 84 per cent and 85 per cent, and CP-III, 93 per cent and 87 per cent. The per cent COD and TDS removals at each stage of the combined process was evaluated and the impact of EO process on the overall performance of the process schemes was discussed. The variation in permeate flux in a membrane processes was also monitored and discussed in terms of membrane damage due to impact of organics. It emerged that intermediate level EO process considerably reduces organic load on the RO membrane, while large part of TDS removal can be attributed to final RO treatment. It is also observed that sequential electrooxidation units as in CP-III lead to substantial fragmentation of organic molecules which possibly have molecular weights < 200 Daltons, the molecular weight cut-off (MWCO) of RO membranes. This was inferred from negligible increase in per cent organics removal through UF-RO combination after EO2 in CP-III. The scheme C/F-EO1-EO2-UF-RO (CP-III) is useful to reduce initial COD of wastewater by 93 per cent, and simultaneously 87 per cent of salinity is also reduced. Thus, the combined treatment processes appear to be promising for obtaining greater efficiency of organic and salinity load removals.

Keywords: *Chemical industry wastewater, Coagulation-flocculation, Electrooxidation, Microfiltration, Ultrafiltration, Reverse osmosis.*

Introduction

Chemical industry wastewater usually has strong color, odor and contains large amounts of hazardous, toxic organic compounds and high concentration of salts. Direct discharge is strictly regulated due to it's potentially sever impact on the receiving aquatic systems and may alter the ecology of receiving stream. Although in wastewater treatment biological process have often been found to be economical and simple, but this type of wastewaters are difficult to enhance the quality by direct biological or any other single stage treatment. Therefore, there is need for developing combinations of various processes targeting reduction in pollution loads. This enables one to emerge with feasible treatment schemes targeting treatment of high strength wastewater.

The combination of pretreatment *viz.* adsorption, physico-chemical oxidation, biological oxidation, striping, including membrane separation which has been increasingly used in the field of water and wastewater treatment and these are adequate for producing water suited for different kinds of applications. Recently membrane separation has been increasingly applied for wastewater treatment [1-3]. RO systems have also been successfully used in the treatments of high strength wastewater [1, 4, and 5]. An increasing trend towards combining several process operations, including chemical pretreatments with membrane systems is found in the literature [6-12]. In many instances this approach is adopted mainly to enhance membrane stability together with enhancing overall treatment efficiency.

In the present study, the C/F, adsorption, microfiltration (MF), ultrafiltration (UF), and EO were studied with the aim of reducing organic load on the final stage of reverse osmosis (RO) membrane treatment. Three different combined process; i) UF-RO (CP-I); ii) C/F-EOUF- RO (CP-II) and iii) C/F-EO1-EO2-UF-RO (CP-III) were investigated and the results were analyzed on the basis of process principles of intermediate units, particularly electrooxidation.

Materials and Method

Materials

Sodium hydroxide (NaOH), Poly-aluminum chloride (PAC), Polyelectrolyte (PE), Sulphuric acid used were of GR grade. Granular activated carbon (GAC) was obtained from Industrial Carbons Pvt. Ltd., Baroda (India). The segregated high strength chemical industries wastewater used in this study was obtained from a CETP site in Gujarat. The physico-chemical characteristics of the wastewater samples in plastic carboys stored at room temperature showed were comparable on replication of analysis on different days.

Microfiltration membranes were purchased from Whatman India Pvt. Ltd., India. Ultrafiltration and RO membranes were obtained from CSMCRI, Bhavnagar. Table 51.1 presents membrane characteristics and applied pressures to achieve filtration in each case. The RO membranes constitute cast thin film composite coating derived from polymerization of metaphenylenediamine (MPD) and mesoylchloride (TMC). The polysulfone membranes were cut into discs of suitable diameter, preserved in 3 per cent Formalin solution, washed and used for filtration.

Table 51.1: Characteristics of Membrane Studied for Filtration.

Process	*Pressure# (Bar)*	*MWCO (Daltons)*	*Pore Size (μm)*	*Materials*	*Disc Size (cm^2)*
Microfiltration*	≥ 1	1,00,000-50,000	0.2-0.45	Nylon	8.04
Ultrafiltration**	6.9	20,000-1,000	10^{-2}-10^{-2}	Polysulfone	8.04
RO**	35	≤ 200	10^{-4}-10^{-3}	TFC/RO (Polysulfone)	8.04

*: Whatmann®; **: CSMCRI, Bhavnagar.

#: Operating pressure in the present study.

MWCO: Molecular Weight Cut-off.

Experimental Procedures

Chemical Coagulation

Chemical coagulation experiments were conducted using PAC as coagulant and PE as coagulant aid in a Jar test apparatus (Phipps and Bird, USA). Dose optimization was done separately and the optimized doses of poly aluminum chloride (1ml of 10 per cent PAC / L) and anionic polyelectrolyte (RF8160, 1 ml of 0.1 per cent PE / L) was applied. The experiment involved rapid mixing of wastewater and coagulant at 100 rpm for 1 min., slow mixing at 30 rpm for 30 min followed by settling for 30 min.

Adsorption

The adsorption experiment was carried out using powder activated carbon. A 1g powder activated carbon was added to the 1 L wastewater sample and mixed thoroughly for 2h contact time. After carbon settling, the supernatant was withdrawn and filtered through 0.45 μm prior to analysis of physico-chemical parameters.

Electro-chemical Oxidation (ECO)

The electro-chemical oxidation of segregated chemical industries wastewater was carried out as per the reported method of three-dimensional carbon bed electrochemical reactor (TDR) [13]. Approximately 1.10 L effluent sample was taken into TDR during each run and electric potential was applied using Testronix 34-D DC power supply (Mumbai, India). The reactor was operated in batch mode. Common salt was added as supporting electrolyte and source of chloride ions for desired anode activity. A 1-h contact time was allowed initially for the establishment of adsorption equilibrium prior to supplying DC power. The electro-oxidation experiments were run at constant current mode and the cell voltage was 4.00 ± 0.20 V at 2 A for 1h. Several test samples were collected and variation in concentrations of COD, TOC and TKN was determined as a function of reaction time.

Membrane Filtration: Testing Set Up

Membrane filtration was done in a Sterilitech dead end module (DEM). The active surface area of the membrane used was 8.04 cm^2. The shearing to the surface of the membrane is assured by a magnetic agitator, which permits a better homogenization of the solution and also prevents clogging of the membrane. The total volume of the cell was 300 ml. Membrane filtration was carried out in DEM, using different membrane discs like microfiltration (MF), Ultrafiltration (UF) and RO as the case may be. The membrane was initially stabilized for 20-30 min for opening of the pores in the membrane surface. The circular membrane disc was fixed at the bottom of the dead end module and the segregated wastewater sample was then introduced into the filtration assembly. Filtration was carried out under pressure using compressed nitrogen gas cylinder. The applied operating pressure ranges for different membranes were given in Table 51.1. The permeate was collected from the outlet of module and the retentate in the assembly was recovered.

A certain reduction in water flux (FR) was considered to indicate fouling of membrane. This was calculated by comparing the water flux before and after the filtration as follows.

$$FR_W = \frac{WF_b - WF_a}{WF_b} \times 100$$

Where, WF_a and WF_b are the water fluxes before and after filtration.

Analysis

The surfaces of membrane, before and after use, were analyzed for the changes in chemical structure of MPD-TMC layer and PS support membrane by AutoImage FTIR (Spectrum 100, Perkin Elmer). Infrared spectra are highly specific probes of chemical structure and presence of certain molecular groups. The method of attenuated total reflection. (ATR-FTIR), in which the film is firmly attached or pressed against an ATR crystal, confines the probed region to a layer ~1 m thick, containing the whole top layer and a part of support (polysulfone).

The concentration of various pollution parameters were determined as per the procedures given in Standard Methods [14]. The measurement and adjustment of pH was done using Orion pH meter (USA). The per cent COD and TDS removal efficiencies were estimated using the formula.

$$\text{Per cent efficiency} = \frac{(C_i - C_f)}{C_i} \times 100$$

where, C_i = Initial COD or TDS concentration, C_f = Final COD or TDS concentration

Results and Discussion

Characteristics of Chemical Industries Wastewater

The physicochemical characteristics of studied chemical industries effluent are presented in Table 51.2. The concentration of soluble organic contents represented by COD was 19600 mg/L and salinity (TDS) concentration was 70804 mg/L. The univalent ions, such as chloride (14653 mg/L) and bivalent ion like sulphate (602 mg/L) are also present. In addition the concentrations of suspended solids (SS), NH_3-N, TKN and Alkalinity are also high. The high TKN to COD ratio (0.51) suggests presence of heterocyclic, nitro-, and amino-aromatic compounds. A shift in the TKN/COD ratio in the influent means a change in the type of organic compounds entering the system, which can impact the effectiveness of the process.

Table 51.2: Physico-chemical Characteristics of Chemical Industries Effluent.

Sl.No.	*Parameter*	*Concentration*	*Sl.No.*	*Parameter*	*Concentration*
1.	pH	8.1	2.	Alkalinity	3000
3.	SS	1,125	4.	TDS	70,804
5.	COD	19,600	6.	TKN	10,080
7.	NH_3-N	1,456	8.	Chloride	14,653
9.	Sulphate	602	10.	Phosphate	2.1
11.	BOD	8,500			

*All the parameters are in mg/l, except pH.

Biochemical oxygen demand (BOD) is an important parameter for determining the amount of organic pollution useful in measuring waste loadings of treatment plants and in evaluating treatment efficiency. Biochemical Oxygen Demand (BOD) is a definitive indicator of required treatment in wastewater, high influent BOD requires extensive treatment to provide the oxygen necessary to break down the water's organic contents. The chemical industry effluent was found to have high concentrations of COD and BOD with low biodegradability index (BI, BOD/COD = 0.43). Though the

BOD/COD ratio (~0.43) suggests possibility of biological treatment, the chemical industry effluent contains heterocyclic, nitro-, and amino-aromatic compounds that are toxic to micro-organisms. This renders significant difficulty in maintaining the required MLSS concentration in the activated sludge process and it also entails poorer performance. It is known that effluents with BI < 0.3 are hardly biodegradable and excellent biodegradation may take place at BI > 0.6 only [15]. In the present context, pretreatments which would remove or transform the residual organics in the chemical industry effluent as well as reduce the salts concentration were required to be developed.

Evaluation of Treatment Processes

Chemical Treatments Combined with Microfiltration – Coagulation + MF, Adsorption + MF and Lime + MF

The comparison of per cent reduction in SS and COD concentrations as a result of Microfiltration, Coagulation + MF, Adsorption + MF and Lime + MF is shown in Figure 51.1. Only 87 per cent SS was removed using MF. The Coagulation + MF, Adsorption + MF and Lime + MF treatments resulted removal of 15-18 per cent COD also, while SS removal slightly improved. Due to coagulation the COD removal efficiency was moderately high *i.e.* 17 per cent and hence coagulation followed by MF can result in reduction of concentration of organic compounds.

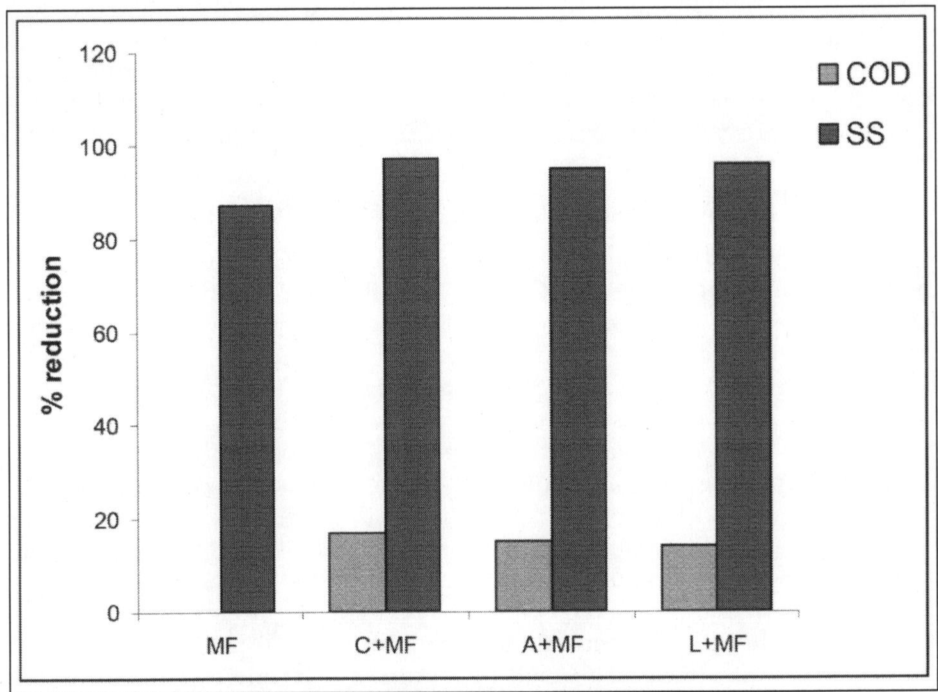

C- Coagulation, A- Adsorption, L- lime

Figure 51.1: Comparison of COD and SS Removal Efficiencies using Microfiltration (MF) and Combined Chemical Treatment + MF.

Membrane Processes

The chemical industry effluent sample was first filtered through different membranes like MF, UF and RO to test their stability to organics containing effluent. The comparison of per cent SS, COD and TDS removals based on the characteristics of permeates from each membrane filtration and changes in permeate fluxes that indicate fouling status of membranes were presented in Table 51.3. In the case of MF and UF significant removals of SS were observed. Additionally, 10 per cent COD and 6 per cent TDS removal was observed with UF membrane. However, the water flux decreased by 8-12 per cent in each case upon reuse indicating deposition of solids on the membrane takes place even under the applied agitation conditions. Since the filtration takes place perpendicular to the plane of membrane disc, the deposition of fine colloidal matter on the disc is possible and clogging of the pores can be envisaged. The recovered MF and UF discs indeed contained slimy and brown deposit. On the other hand, filtration of the chemical industry effluent sample using RO membrane resulted in 97 per cent SS, 63-65 per cent COD and 72-78 per cent TDS removals. However, the water flux increased by 5-9 per cent, implying that the organic compounds in the effluent attacked the TFC coating on the membrane. Hence, direct application of membranes to treat the chemical industries wastewater is infeasible. Hence, we tried membrane filtration through UF followed by RO.

In UF - RO combined process; the wastewater was first filtered through the UF membrane where SS and colloidal matter was removed along with 10 and 6 per cent reduction in COD and TDS, respectively. The ultra filtered sample when applied to RO membrane, 73 per cent COD and 83 per cent TDS were removed. As the effluent after UF still contained large amount of organic compounds, it attacked the TFC/RO membrane. This can be inferred from 7 per cent increase water flux upon reuse of the membrane, due to damage to the membrane.

Single Stage Electro-chemical Oxidation - C/F-EO-UF-RO (CP-II)

In this combined process, referred as C/F-EO-UF-RO (CP-II), single stage electro oxidation was introduced between C/F and UF with the aim of reducing the organic load into subsequent membrane processes. The coagulation/flocculation was carried out with optimized doses of PAC and PE. The effluent from C/F process was subjected to electrochemical oxidation in the 3D-carbon bed reactor (V = 4.3V; I = 2A; t = 1h). The effluent from EO unit was treated through UF followed by RO. The stage wise per cent COD and TDS removals are given in Table 51.4. As expected the introduction of EO unit in the scheme contributed more than 50 per cent COD removal, compared to the COD of raw effluent sample. Ultrafiltration, improved reduction in COD by 3 per cent, while the final RO treatment resulted in an overall 84 per cent COD removal, while >80 per cent TDS reduction was observed with RO process only. Thus, the observed 30 per cent COD and 80 per cent TDS removal through RO application implies that the organic compounds contributing to 30 per cent COD after electrooxidation have molecular weight >200 Daltons based on the MWCO of the RO membrane. The concentration of organics after UF in this scheme was still high at 10800 mg/L that might damage RO membrane. Indeed the RO membrane got damaged and the pure water flux across the used membrane increased by 5 per cent. Hence, more reduction in COD to protect the RO system from possible attack of organics was targeted by providing a second stage electro-chemical oxidation unit as in Scheme C/F-EO1- EO2-UF-RO (CP-III).

Table 51.3: Performance of different Membranes Towards Filtration of Raw Effluent

Membrane	*Per cent Removal Efficiency*			*Remarks*
	SS	*COD*	*TDS*	
Microfiltration (MF)	87	-	-	Membrane fouled and flux reduced by 8 per cent
Ultrafiltration (UF)	92	10	6	Membrane fouled and flux reduced by 12 per cent
Reverse Osmosis (RO)				
Cycle 1	97	66	78	Membrane damaged and flux increased by 5 per cent
Cycle 2	97	65	75	Membrane damaged and flux increased by 8 per cent
Cycle 3	97	63	72	Membrane damaged and flux increased by 9 per cent
UF-RO	98	73	83	Membrane damaged and flux increased by 7 per cent

Two Stage Electro-chemical Oxidation- C/F-EO1-EO2-UF-RO (CP-III)

In this combined process, C/F-EOI-EOII-UF-RO; the effluent from C/F process was subjected to electro-chemical oxidation in two consecutive stages, followed by UF and RO treatments, with the aim of further reducing the organic load on the RO membrane. The results are given in Table 51.4. It was observed that after second stage electrochemical oxidation, the COD of the effluent was reduced by 92 per cent, leaving behind 1830 mg/L only. This was accompanied by a modest decrease in concentration of dissolved solids (5 per cent). Through treatment of effluent from EO2 using UF only 0.5 and 2 per cent reduction in COD and TDS was observed from second stage electrochemical oxidation. However, when the UF permeate was introduced in RO system for further filtration the increase in per cent COD reduction was very low *i.e.* 1.0 and 0.5 per cent (compared to EO2 and UF stages), but drastic removal observed in TDS *i.e.* 80 per cent, as expected. The overall removal efficiency of CP-III combined processes was 93 and 87 per cent, respectively for COD and TDS. From very small increase in COD removal through RO treatment in CP-III after introducing EO2, compared to observed 30 per cent increase in CP-II scheme with single stage EO, we may conclude that the second stage EO effectively fragments the organics in to compounds having molecular weight < 200 Daltons, which pass through RO membrane. As the organic load was lowered after introduction of second stage electrochemical oxidation, the membrane was not damaged as inferred from 6 per cent decrease in the pure water flux. The flux decrease is common because some salts and organics deposit on the RO membrane and hinder filtration.

Table 51.4: Comparison of Performance of Combined Processes.

Combine Processes	Parameter		Per cent Removal Efficiency	
	COD mg/L	TDS mg/L	COD	TDS
CP- I (UF-RO)				
Raw sample	23000	75000	-	-
UF	20470	70500	11	6
RO	6210	13500	73	82
CP- II (C/F-EO-UF-RO)				
Raw sample	23000	75000	-	-
Coagulation	18170	-	21	-
ECO	11600	73200	50	2
UF	10800	71000	53	5
RO	3550	11230	84	85
CP – III (C/F-EO 1-EO 2-UF-RO)				
Raw sample	23000	75000	-	-
Coagulation	18170	-	21	-
ECO-I	11600	73200	50	2
ECO-II	1830	71500	92	5
UF	1720	70000	92.5	7
RO	1560	10230	93	87

ATR-FTIR Evidence for RO Membrane Damage

From the measurements of pure water fluxes prior to and after use of RO membranes, it can be concluded that the effluent having high concentration of organics can potentially damage the RO membrane. The RO membranes used in the present study comprise a thin film composite coating of polymer from reactions of metaphenylenediamine and trimesoylchloride (MPD-TMC polymer coating) over the polysulfone (PS) base membrane. We expected that the organics from the chemical industry effluent first attack the MPD-TMC layer. The originally glossy membrane looses its texture upon use for filtration. However, the subtle changes in the chemistry of MPD-TMC layer on RO membrane surface may be understood using FTIR technique. The FTIR spectral segments of interest for RO membrane samples before and after filtration were shown in Figure 51.2. The –C=O of -COCl band at 1763 cm-1 is absent, instead a sharp peak at 1742 cm-1 appeared which may be due to ester carbonyl. The band at 1686 cm-1 is present, which is characteristic of -C = O band of an acylamide group (amide I). All other peaks characteristic of –NH stretching (3430 cm-1), –NH bending (1509 cm-1), O=S=O symmetric and asymmetric stretches at 1175-1156 and 1329 cm-1 were also observed. The presence of O=S=O bands indicate that the MPD-TMC layer is very thin, and the polysulfone membrane at the bottom of MPD-TMC also absorbs IR radiation. These bands were present in either of the membranes *i.e.*, before and after use, however, the peaks at 1686, 1509, 3430 cm-1 showed distortion

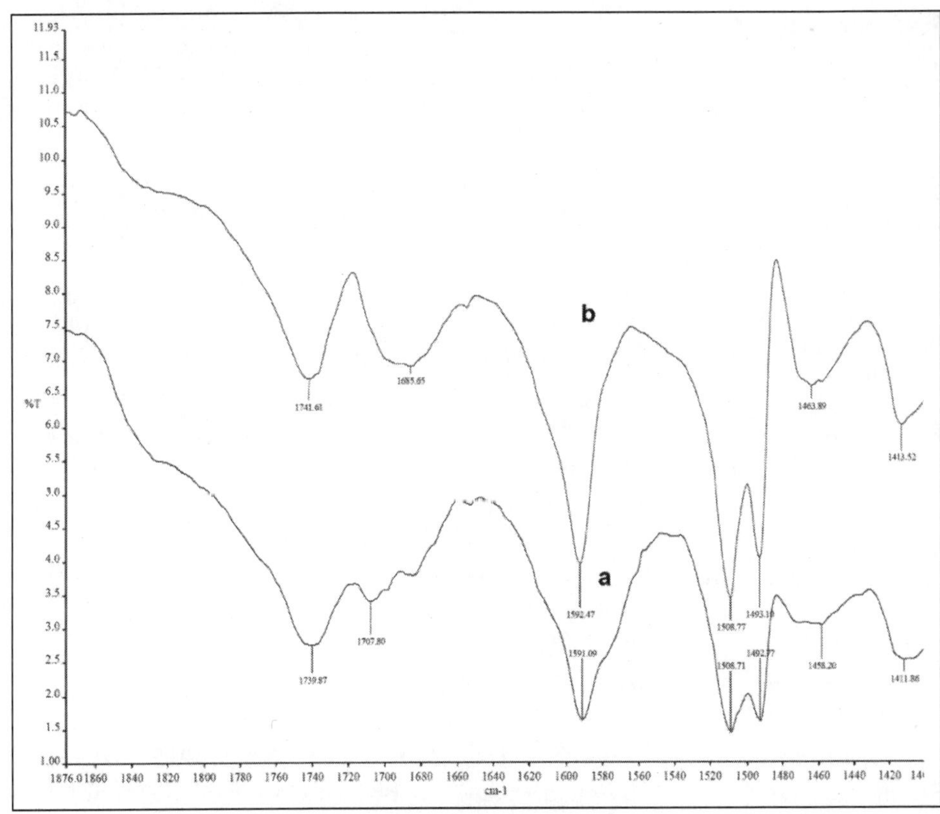

Figure 51.2: FTIR Spectra for RO Membranes: (a) Before use, (b) After use in Filtration of Chemical Industry Effluent.

in the used membrane sample. This may signify that chemicals in the effluent attack the MPD-TMC membrane.

Conclusion

The chemical industry effluent with high concentration of organics can potentially damage TFC/RO membranes by attacking the MPD-TMC thin coating because of which substantial increase in flux may be noticed. A two stage electrochemical oxidation pretreatment as in the combined process, C/F-EOI-EOII-UF-RO is useful to reduce concentration of organics and hence deleterious effects of the compounds on the RO membrane can be minimized.

Acknowledgments

The authors wish to thank Dr. Tapan Chakrabarti, Acting Director, NEERI, Nagpur for encouragement. The work is carried out with financial assistance received under SIP-16(3.5) and NWP-47.

References

1. I. Galambos, J.M. Molina, P. Jaraya, G. Vatatia and E.B. Molnara, High organic content industrial wastewater treatment by membrane filtration, Desalination, 162 (2004) 117-120.

2. G.J. Zhang, Z.Z. Liu, L.F. Song, J.Y. Hu, S.L.Ong and W.J. Ng, Post treatment of banknote printing workes wastewater ultrafiltration concentrate, Water Res., 38 (2004) 3587-3595.

3. A. Bodalo-Santoyo, J.L. Gomez-Carrasco, E. Gomez- Gomez, F. Maximo-martin and A.M. Hidalgo-Montesinos, application of reverse osmosis to reduce pollutants present in industrial wastewater, Desalination, 115(2) (2003) 101-108.

4. T.A. Peters, Purification of landfill Leachate with reverse osmosis and nanofiltration, Desalination, 119 (1998) 289-193.

5. M. Into, A.S. Jonsson and G. Lengden, Reuse of industrial wastewater following treatment with reverse osmosis, J.Membr.Sci.,242 (2004) 21-25.

6. C.Guigui, J.C.Rouch, L. Durand-Bourlier, V. Bonnelye and P. Aptel, Impact of coagulation condition on the in-line coagulation / UF process for drinking water production, Desalination, 147(2002) 95-100.

7. K. Konieczny, M. Bodzek, K. Keller and M.Rajca, Coagulation "in-line"- UF/MF hybrid system in the removal of pollutants from natural water, in: Proc. Water supply and water quality, Zakopane 2006, vol. 1, pp. 565-576 (in polish)

8. K. Konieczny, D. Sakol, J. Plonka, M.Rajca, M. Bodzek, Coagulation-Ultrafiltration system for river water treatment, Desalination, 240, (2009), 151-159.

9. D. Abdessemed, G. Nezzal, R. Ben Aim, Coagulation-Adsorption-Ultrafiltration for wastewater and reuse, Desalination, 131, (2000), 307-317.

10. J. Labanda, J. Llorens, Wool scouring waste treatment by combination of coagulation floculation process and membrane sepration technology, Chemical Engineering and Processing, 47 (2008), 1061-1068.

11. Tak-Hyun Kim, C. Park, J. Lee, Eung-Bai Shin, Sangyong Kim, Piolet scale treatment of textile wastewater by combined process (fluidized biofilm process-chemical coagulation-electrochemical oxidation), Water Research, 36 (2002), 3979-3988

12. Munir Cheryan, Ultrafiltration and Microfiltration handbook, a TECHNOMIC publication.

13. N. N. Rao, Misra Rohit, Gedam Nitin, P.N. Parameswaran, J. K. Astik, Kinetics of electroxidation of landfill Leachate in a three-dimensional carbon bed electrochemical reactor, Chemosphere 76 (2009) 1206–1212.

14. APHA, Standard methods for the examination of water and wastewater, 2005, 21st addition.

15. J. L. Morais, P. P. Zamora, Use of advanced oxidation processes to improve the biodegradability of mature landfill leachates, J. Hazard. Mater. 123 (2005) 181-186.

Author Index

Subject Index

V

W

Z

Y